THE ECONOMY
OF NATURE

THE ECONOMY OF NATURE

THIRD EDITION

A Textbook in Basic Ecology

ROBERT E. RICKLEFS
University of Pennsylvania

W. H. FREEMAN / NEW YORK

Cover photographs:
Burchell's Zebras © 1993 G. Dimijian, MD/Photo Researchers, Inc.
Inset © 1993 DeWitt Jones/The Stock Market

Library of Congress Cataloging-in-Publication Data

Ricklefs, Robert E.
    The economy of nature : a textbook in basic ecology / Robert E.
Ricklefs. — 3d ed.
        p. cm.
    Includes bibliographical references (p.   ) and index.
    ISBN 0-7167-2409-X
    1. Ecology.   I. Title.
QH541.R54 1993
574.5 — dc20   92-39616
                                                                    CIP

Printed in the United States of America

4 5 6 7 8 9 0 VB 9 9 8 7 6 5 4

CONTENTS

Preface   vii

1 Introduction   1

PART 1 *Life and the Physical Environment*   23

  2 The Physical Environment   25
  3 Adaptation to Aquatic and Terrestrial Environments   49
  4 Climate, Topography, and Soils   65
  5 The Diversity of Biological Communities   89

PART 2 *Ecosystems*   101

  6 Energy in the Ecosystem   103
  7 Pathways of Elements in the Ecosystem   123
  8 Nutrient Regeneration in Terrestrial and
    Aquatic Ecosystems   141
  9 Regulation of Ecosystem Function   162

PART 3 *Organisms*   175

  10 Homeostasis, Acclimation, and
    Developmental Response   177
  11 Behavior in Heterogeneous Environments   195
  12 Life Histories   207
  13 Sex, Family, and Society   221

PART 4 *Populations*   241

  14 Population Structure   243
  15 Population Growth and Regulation   262

v

16 Temporal and Spatial Dynamics 281

17 Population Genetics and Evolution 296

PART 5 *Species Interactions* 317

18 Relationships Among Species 319

19 Competition 341

20 Predation 361

21 Evolutionary Responses and Coevolution 385

PART 6 *Communities* 405

22 Community Structure 407

23 Community Development 427

24 Biodiversity 451

25 History and Biogeography 476

PART 7 *Ecological Applications* 493

26 Extinction and Conservation 495

27 Development and Global Ecology 513

Glossary 535

Appendix A: International System of Units 563

Appendix B: Conversion Factors 566

Index 568

PREFACE

The response to previous editions of *The Economy of Nature* has been very gratifying. I particularly appreciate the many comments and suggestions from readers and their encouragement to revise this book. Ecology has become such a broad discipline that it is difficult for one person to keep up with its many developments, and I have been helped very much in this task by discussions with students and colleagues who have generously shared their own knowledge and insights.

This third edition remains a basic textbook covering the most important principles of ecology for undergraduate courses. While discussing basic ecological patterns and processes, I have endeavored to make the writing accessible to the student who has little background in biology, to emphasize the major points with detailed examples, to provide a feeling for the natural history of the earth's varied habitats, and to illustrate the text with clear graphs and photographs that are closely integrated with the narrative. In emphasizing the relationships of plants, animals, and microbes to their environments and to one another, I have also tried to convey a sense of humankind's position in the earth's ecology and how our activities affect the global environment and influence the quality of our — and other organisms' — lives.

In the ten years since publication of the second edition, the discipline of ecology has undergone considerable change and notable advances have been made. We now have a deeper understanding of the genetic and evolutionary aspects of populations and communities, such as coevolution (Chapter 21) and the roles of evolutionary change and adaptive radiation in the development of biological communities (Chapters 23 and 24). Ecologists have explored fascinating connections between behavior and ecology, including the rules that govern decisions about foraging and habitat selection (Chapter 11), as well as mate choice and social interaction (Chapter 13). The physiological and morphological bases of organism adaptations have also been the subject of much study, resulting in a better understanding of the interrelationship between different organ systems and functions in adaptation (Chapter 12). The dynamics of ecosystem function (Chapter 9) and the historical and geographic components of the structure of present-day ecological systems (Chapter 25) are other areas of fruitful research. Taken together, these advances reflect the broadening perspective of ecology and the increasing integration and synthesis of diverse aspects of the science. The third edition has been updated in light of this new knowledge and gives prominence to new directions of inquiry, such as the ecological bases of extinction and the influence of climate change on biological communities and ecosystems (Chapters 26 and 27).

Although the book's design has been changed appreciably, many familiar and favored features of previous editions have been retained. This edition, like its predecessors, has a solid grounding in natural history. I have continued to include detailed examples of physiological adaptations, social behavior, interactions between populations, ecological succession, and other ecological phenomena, which are drawn from a variety of ecosystems and types of organisms. This edition also stresses the idea that organisms are the most fundamental and natural units of ecology. Their activities and their relationships to their environments and to one another are the foundation upon which populations, communities, and ecosystems are built. For example, feeding relationships determine the flux of energy and the cycling of nutrients through the ecosystem, underlie the regulation of population size, and influence the coexistence of species within biological communities. Therefore, this edition begins by considering the ecological relationships of individual plants, animals, and microbes. Finally, I continue to emphasize evolutionary thinking as central to the study of ecology. Ecological systems change, and they respond to change by means of evolution within the populations that make up these systems. The concepts of Darwinian fitness and natural selection provide a rationale for understanding the incredible diversity of the natural world.

This third edition is somewhat longer than its predecessor but is still suitable for a course of one semester or one quarter. A slightly different arrangement of chapters brings some topics closer together (for example, the population and evolutionary consequences of interactions between species) and provides a more logical development of the subject matter. The book is organized into seven parts, or sections. The first (Chapters 2–5) introduces the relationships of organisms to their physical environments, and the second (Chapters 6–9) explores how these relationships tie the living and nonliving realms into ecosystems. The next section (Chapters 10–13) covers the life history strategies that organisms employ to cope with variation in the physical environment and with social interactions within populations. Following this, the book addresses increasingly complex levels of organization in ecology — how population processes reflect the activities of individuals (Chapters 14–17), how populations of different species interact (Chapters 18–21), and how these interactions influence the coexistence of species in biological communities (Chapters 22–25). The concluding section on human ecology (Chapters 26 and 27) highlights the application of basic principles to environmental problems.

Of course, no single topical order is suitable for every course. The major sections of this book stand alone and can be read in different sequences. Some of the more quantitative information has been placed in boxes to avoid breaking the flow of the narrative. Each chapter now has a summary and a brief list of suggested readings for reference or additional background and thought.

Above all, I hope that *The Economy of Nature* will continue to encourage students to appreciate the natural world they live in, the increasing role of human activities in this world, and our progress toward achieving a balanced ecological position based on the rational application of knowledge and understanding.

ROBERT E. RICKLEFS
Philadelphia
January 1993

THE ECONOMY
OF NATURE

1

INTRODUCTION

The English word *ecology* is taken from the Greek *oikos*, meaning "house," our immediate environment. In 1870 the German zoologist Ernst Haeckel gave the word a broader meaning: the study of the natural environment and of the relations of organisms to one another and to their surroundings. Haeckel wrote:

> By ecology, we mean the body of knowledge concerning **the economy of nature** — *the investigation of the total relations of the animal both to its organic and to its inorganic environment; including above all, its friendly and inimical relation with those animals and plants with which it comes directly or indirectly into contact — in a word, ecology is the study of all the complex interrelationships referred to by Darwin as the conditions of the struggle for existence.*

Thus **ecology** is the science by which we study how organisms (animals, plants, and microbes) interact in and with the natural world.

The term *ecology* came into general use only in the late 1800s, when European and American scientists began to call themselves ecologists. The first societies and journals explicitly devoted to ecology appeared in the early decades of this century. Since that time, ecology has undergone immense growth and diversification, so much so that those who are devoting their professional lives to ecology now number in the tens of thousands. With the dual crises of rapid growth of the human population and accelerating deterioration of the earth's environment, ecology has assumed the utmost importance. Management of biotic resources in a way that sustains a reasonable quality of human life depends on the wise application of ecological principles not merely to solve or prevent environmental problems but also to inform our economic, political, and social thought and practice.

This text, *The Economy of Nature,* presents the basic principles of the scientific discipline of ecology. These principles have been defined through more than a century of observation, experimentation, and theoretical exploration of natural systems. They explain processes that maintain the structure and function of these systems; they tell us how each part fits into the whole, emphasizing the interrelatedness of all of nature; and they help us understand why the functioning of natural systems can break down under certain stresses. In this capacity, they offer guidelines for the preservation of biodiversity and management of the environment for sustained use.

This introductory chapter provides a general framework for the study of ecology. We shall first discuss several ways of organizing ecological knowledge and insight. Following this is a brief exposition of some of the most important principles of ecology; we shall pay particular attention to thermodynamics, population dynamics, and natural selection and evolution. Finally, a few examples will illustrate the intimate connection between ecology and some pressing environmental problems. Ecology provides a framework for interpreting the overwhelming abundance of information on the natural environment that is available to us every day. It also gives us the insight we need to envision the consequences of our interactions with natural systems. Haeckel's analogy of the economy of nature emphasizes that everything on the surface of the earth is interconnected in the same way that human enterprises are interrelated and defined by economic principles. We and our enterprises directly affect the outcome of natural processes. Thus humankind itself is an important part of the economy of nature.

Levels of ecological organization

The **organism** is the most fundamental unit of ecology. No smaller unit in biology, such as the organ, cell, or molecule, has a separate life in the environment (although, in the case of single-celled protists and bacteria, cell and organism are synonymous). The structure and function of the organism — whether it is a plant, animal, or microbe — are determined by a set of genetic instructions inherited from its parents and by influences in the environment in which the organism lives. Every organism is bounded by a membrane or other covering across which it exchanges energy and materials with its surroundings. Its success as an ecological entity depends on its having a positive balance of energy and materials to support its maintenance, growth, and reproduction. In Part 1 of this book, we will examine factors that influence exchanges between the organism and the physical environment, and we will see how organisms cope with problems involving temperature, water loss, salt balance, and other environmental challenges.

In the course of their lives, organisms transform energy and process materials in a variety of ways as they metabolize, grow, and reproduce. In doing so, they modify the conditions of the environment and the amounts of resources available for other organisms; they contribute to energy fluxes and to the cycling of elements in the natural world. Organisms and their physical and chemical environments together make up an **ecosystem.** The ecosystem approach to ecology describes organisms in terms of common "currencies," their contents of energy and chemical elements. Thus it provides a framework for studying the movement of energy and the cycling of elements within ecological systems.

We may speak of a forest ecosystem, a prairie ecosystem, and an estuarine ecosystem as distinct units because relatively little exchange of energy or substances occurs *between* these units compared to the innumerable transformations going on *within* each of them. Ultimately, however, all ecosystems are linked together in a single **biosphere** that includes all the environments and organisms at the surface of the earth. The importance of movement of materials between ecosystems within the biosphere is underscored by the global consequences of human activities. Industrial and agricultural wastes spread far from their points of origin, inflicting their detrimental consequences on all regions of the earth. Ecosystem processes are the subject of Part 2 of this book.

Many organisms of the same kind together constitute a **population.** Populations differ from organisms in that they are potentially immortal, their numbers being maintained over time by the birth of new individuals that replace those that die. Populations also have collective properties, such as geographic boundaries, densities (number of individuals per unit of area), and dynamic properties (for example, evolutionary responses to environmental change and periodic cycles of numbers in some cases) that are not exhibited by individual organisms. The population approach to ecology (Parts 3 and 4) is concerned with the numbers of individuals and their changes through time.

Many populations of different kinds living in the same place constitute a **community.** The populations within a community interact in various ways, as explained in Part 5. Many consist of predators that eat other kinds of organisms. Almost all are themselves prey. Some—such as bees and the plants whose flowers they pollinate, and many symbiotic associations involving microbes living together with plants and animals—enter into cooperative arrangements, called mutualisms, in which both parties benefit from the interaction. All these interactions influence rates of change in the numbers of individuals in populations—that is, in the **dynamics** of populations.

Unlike organisms, communities have no rigidly defined boundaries; no skin separates a community from what surrounds it. The total interconnectedness of ecological systems means that interactions between populations spread across the globe as individuals and materials move between habitats and regions. Water flowing from a headwater to an estuary connects the terrestrial and aquatic communities of the watershed to those of the marine realm. The migrations of gray whales link the communities of the Bering Sea and the Gulf of California. Thus the community is an abstraction representing a level of organization rather than a unit of structure in ecology (Part 6). After discussing ecosystems, we shall trace the progression of ecological thinking from the organism to the population and community levels of organization.

Communities and ecosystems are related but quite distinct concepts that derive from different fields of study within ecology. We may speak of the forest ecosystem or the community of animals and plants that inhabit the forest. But the two terms apply to different types of processes. The study of ecosystems addresses the movement of energy and materials within the environment, as they are affected by the activities of organisms and by physical and chemical transformations in the soil, atmosphere, and water. The study of populations and communities is more concerned with the development of ecological structures and with the regulation of ecological processes by means of population growth and the interaction of populations with the environment and each other. Thus the population/community and ecosystem approaches represent different ways of looking at the natural world.

The kinds of organisms

From the standpoint of ecosystem function, different kinds of organisms play very different roles. The major types of organisms are plants, animals, fungi, protists, and bacteria. Characteristic differences in structure and function between these groups have important implications for their ecological relationships. Plants capture the energy of sunlight and use it to synthesize carbon dioxide and water into sugars and other organic molecules. These carbon-containing compounds and the energy stored in their chemical bonds accumulate as plants grow and reproduce. The organic carbon produced by green plants also provides food, either directly or indirectly, for the rest of the community. Some animals consume plants; others consume animals that have eaten plants. Many fungi and bacteria feed on the dead remains of plants and animals; others feed on living individuals, in which case we refer to them as pathogens—disease organisms. Protists are a varied group. Some of them, the algae, undertake photosynthesis; others, such as the protozoa, are single-celled analogs of the more complex animals.

Animals and plants differ in many important ways besides their source of energy (Figure 1.1). Because plants must expose large surfaces to light to capture energy efficiently, their structure emphasizes thinness in leaves and stiff supportive structures in stems. Mobility is out of the question. Plants also require vast quantities of water to replace the amount lost from their leaves; not surprisingly, they are firmly rooted in the ground in constant touch with supplies of water and nutrients. Those that aren't, such as orchids and other tropical "air plants" (epiphytes), can survive only in humid environments bathed in cloud mists much of the year.

The large surfaces that animals require to exchange energy and materials with the environment are often enclosed within the body. A set of human lungs has a surface area of about 100 square meters (m²), which is close to the

FIGURE 1.1 A giraffe browsing on a tree in eastern Africa emphasizes the fundamental differences between plants, which assimilate the energy of light and convert carbon dioxide to organic carbon compounds, and animals, which derive their energy from the production of plants.

size of a tennis court. The gut also presents a large surface across which nutrients are assimilated into the body. Internalizing exchange surfaces has enabled animals to achieve bulk and to develop the skeletal and muscular systems that make mobility possible.

Although plants and animals are similar in having coordinated, complex body plans with diverse, interacting organs, they differ in their mechanisms of growth and reproduction. Animals grow predominantly by the multiplication of cells in many tissues and organs throughout their bodies. During growth, each part of the animal maintains a precisely regulated size relative to the rest of the body. In contrast, most plants grow in a modular way with many independent growth centers (meristems) at the tips of branches and roots (Figure 1.2). This modular organization enables plants to withstand loss of tissue to grazers and continue to grow and reproduce. Each meristem produces a precisely regulated structure, but the form of the whole plant may vary depending on which of the shoot tips escape being eaten and other dangers and on which grow most rapidly.

Any one growing shoot can reproduce the entire structure of the plant. Thus a branch cut from a tree may take root when placed in soil and become a separate tree, even though it is genetically identical to the parent from which it was cut. This ability allows most plants to reproduce asexually — to engage in what biologists often refer to as vegetative reproduction, or cloning (Figure 1.3). Most animals reproduce by means of specialized sexual cells (eggs and

FIGURE 1.2 Each of the growing tips of this Joshua tree can produce reproductive structures and, potentially, an entire tree. The death of one or more of the branches does not directly affect the others. Joshua trees are native to the Mohave Desert of southern California. Photograph by J. Boucher, courtesy of the U.S. National Park Service.

FIGURE 1.3 The aspen trees in this photograph are part of a single clone. Each individual stem ("tree") has grown from a common root system that developed from a single seedling. Photograph by K. D. Swan, courtesy of the U.S. Forest Service.

sperm) produced by the sexual organs, or gonads. Despite this general difference between plants and animals, most plants also employ sexual reproduction in their life cycles, and many animals have the capacity to clone themselves, either by modifying the sexual process to eliminate fertilization or by mechanisms resembling vegetative growth. The so-called clonal animals — hydroids and corals are familiar examples — emphasize the difficulty of drawing sharp lines between animal function and plant function (Figure 1.4).

The fungi are distinctive not only because of their structure but also because of their unique ecological roles in breaking down dead plant material

FIGURE 1.4 Colony of a marine bryozoan. Each individual zooid was produced by asexual reproduction at the edge of the colony, which is about 15 mm in diameter. Courtesy of D. Harvell.

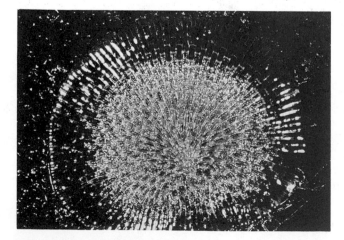

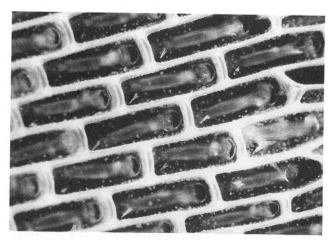

(Figure 1.5). Fungi digest their foods externally, secreting acids and enzymes into their immediate surroundings. They alone can decompose some of the most resistant chemicals in wood and other plant structures, so they account for the recycling of many of the nutrients contained in dead plant matter.

Bacteria are also biochemical specialists. Their range of metabolic capabilities enables them to accomplish biochemical transformations, such as assimilating molecular nitrogen (the common form found in the atmosphere) and using inorganic chemicals (for example, hydrogen sulfide) as sources of energy. Plants, animals, fungi, and most protists cannot accomplish these feats. Furthermore, many bacteria can live under the anaerobic (lacking free oxygen) conditions often present in mucky soils and sediments, where their metabolic activities regenerate nutrients and make them available for plants. We will have much more to say later about the special place of microorganisms in the functioning of the ecosystem. We must always keep in mind, however, that the unique qualities of different types of organisms influence their ecological relationships, the organization of their populations, and even their evolutionary responses to changes in the environment.

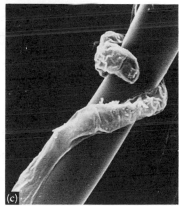

FIGURE 1.5 Many fungi, such as the *Amanita* mushroom fungus (a) and the shelf fungus (b), obtain nutrients from dead organic material. Others are pathogenic and attack living tissues of plants and animals, or even other fungi. (c) The hypha of one fungus attacks that of a larger plant pathogen. The hypha of the larger fungus has a diameter of about 5 μm. Courtesy of the U.S. Department of Agriculture.

Habitats

Habitats are the places, or physical settings, in which organisms live. Ecologists characterize habitats by their conspicuous physical features, often including the predominant form of plant life or, sometimes, animal life (Figure 1.6). Thus we speak of forest habitats, desert habitats, and coral reef habitats. How we define a habitat depends on the point of reference; it is strictly a matter of convenience. The habitat of an earthworm is the soil, whereas that of the bear treading on the soil is the forest.

Ecologists have devoted much effort to classifying habitats. For example, one may distinguish terrestrial and aquatic habitats; among aquatic habitats, freshwater and marine; among marine habitats, ocean and estuary; among ocean habitats, benthic (on or within the ocean bottom) or pelagic (free-swimming or floating). Such classifications ultimately break down, because habitat types overlap broadly and absolute distinctions between them do not exist. The idea of habitat nonetheless emphasizes the variety of conditions to which organisms are exposed at the earth's surface. Inhabitants of abyssal ocean depths and tropical rain-forest canopies experience vastly different conditions of light, pressure, temperature, oxygen concentration, moisture, and salts, not to mention food resources and enemies. Thus, although the same principles of ecological interaction and evolution apply to both habitats, the expressions of these processes differ absolutely. The variety of habitats holds the key to much of the diversity of living organisms. No one organism can live under all the conditions of the earth's different habitats; each must specialize. Thus different habitats have different types of plants, animals, fungi, protists, and bacteria.

Scale in time and space

The natural world varies in time and space. We perceive variations in our environment as the alternation of day and night and the seasonal progression of temperature and precipitation. Superimposed on these cycles are irregular and unpredictable variations in the conditions of our environment. Winter weather is generally cold and wet, but the weather at any particular time cannot be predicted much in advance; it varies perceptibly over intervals of a few hours or days as cold fronts and other atmospheric phenomena pass through. Some irregularities in conditions, such as the alternation of series of especially wet and dry years, occur over longer periods. Events of great ecological consequence, such as fires and tornadoes, strike a particular place only at very long intervals.

Each type of variation in the environment has a characteristic dimension or **scale**. Variation between night and day has a dimension of 24 hours; seasonal variation has a dimension of 365 days. Waves pound a rocky shore at intervals of seconds; winter storms bringing rain may follow one another at intervals of days or weeks; hurricanes may strike a particular coast at intervals of decades. In general, the more extreme the condition, the lower its frequency.

FIGURE 1.6 Four distinctly different habitats: (a) semiarid scrub-grassland in southern Texas, (b) red cedar forest in northern Idaho, (c) open acacia woodland in Kenya, and (d) coral reef in Panama. The dominant plant form in each of the terrestrial habitats is determined by climatic factors, particularly temperature and precipitation. Note the similarity of growth forms of the acacia trees and corals. Like the acacia, the coral, which is a colonial animal, requires light because its tissues contain symbiotic photosynthetic algae. Photographs courtesy of the U.S. Forest Service, U.S. National Park Service, and J. W. Porter.

Both the severity and the frequency of events are relative measures, depending on the organism that experiences them. Fire may touch a tree many times within its life span but may skip dozens of generations of an insect population. As we shall see in the section on populations, how organisms and populations respond to change in their environments depends on *temporal variation.*

The environment also differs from place to place. Variations in climate, topography, and type of rock underlying the soil cause large-scale heterogeneity (across meters to hundreds of kilometers). At smaller scales, much heterogeneity is generated by the structures of plants, by the activities of animals, and even by the particle structures of soil. A particular dimension of *spatial variation* may be important to one animal and not to another. The difference between the top and underside of a leaf is consequential to the aphid but not to the moose, which happily eats the whole leaf, aphid and all.

Spatial heterogeneity affects animals that move because it determines how frequently an individual may encounter new environments. That is, spatial variation is perceived as temporal variation by an animal traveling through the environment. For a plant, the dimension of spatial variation determines the variety of conditions that its roots encounter in the soil and the variety of habitats in which its offspring germinate, which depends on how far its pollen and seeds travel.

The temporal and spatial dimensions of ecologically important phenomena are generally correlated; the duration of a phenomenon usually increases with its size. For example, tornadoes last only a few minutes and affect small areas compared with the devastation inflicted by a hurricane over a period of days or weeks (Figure 1.7). In the oceans, at one extreme, small eddy currents may last only a few days; whereas at the other extreme, ocean gyres (circulating currents encompassing entire ocean basins) are stable over millennia.

Figure 1.8 depicts a general relationship between time and distance for some familiar phenomena. Compared to marine and (especially) to atmo-

FIGURE 1.7 A thunderstorm over the desert near Tucson, Arizona, may dump several inches of rain over a small area in less than an hour.

spheric phenomena, patterns of temporal variation in landforms have very long time dimensions for a given spatial dimension. The reason is simple: spatial patterns in the terrestrial realm are determined by underlying topography and geology, which change very slowly over time by such processes as mountain building, lava flows, erosion, and even continental drift. In contrast, spatial heterogeneity in the open oceans results from physical processes in water, which is obviously more fluid (changing) than the land. As you might expect, atmospheric processes have very short periods at a given spatial dimension.

A principle related to the space–time correlation states that the frequency of a phenomenon is generally inversely related to its spatial dimension or local severity. Thus tornadoes and hurricanes occur at longer intervals, on average, than do winter storms. The frequency of forest fires or brush fires is inversely related to the area they burn. Such disturbances create patches of habitat in various stages of ecological development, or succession, thereby contributing to the spatial heterogeneity of the environment.

Some general principles of ecology

All ecological systems are governed by a small set of general principles. Among the most important of these are

1. Ecological systems function according to the laws of thermodynamics.

2. The physical environment exerts a controlling influence on the productivity of ecological systems.

3. The structure and dynamics of ecological communities are regulated by population processes.

4. Over generations, organisms respond to change in the environment by evolution within populations.

FIGURE 1.8 Temporal and spatial scales of variation in (a) atmospheric and (b) marine systems. Notice the general correlation between temporal and spatial dimensions. The large spatial dimensions of droughts and El Niño events reflects their tight link to changes in ocean circulation patterns. The shorter time scales of population fluctuations compared to physical variation in the marine realm indicates the role of biological causes of population change. After J. H. Steele, *J. Theoret. Biol.* 153:425–436 (1991).

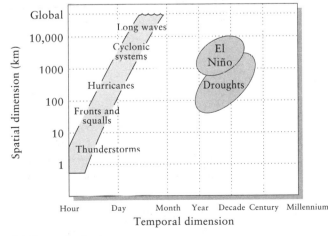

(a) Atmospheric phenomena

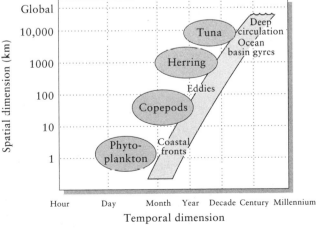

(b) Marine phenomena

The laws of thermodynamics, which apply to energy in nature, govern physical and chemical transformations in biological systems, such as the diffusion of oxygen across body surfaces and the rates of biochemical reactions. Within the broad limits imposed by physical constraints, however, life can pursue many options.

That organisms reproduce and that they die are basic facts of life. Young individuals replace old ones in populations. When births exceed deaths, populations increase; when fewer are born than die, populations decline. Thus each population of individuals has underlying properties that control its growth and that are sensitive to physical conditions, food resources, and enemies in the environment. These relationships link the dynamics of any one population to those of other species, creating a single dynamic system: the biological community. When many species interact, their dynamics can become quite complex even though they are governed by a small set of simple principles. The principles of population biology also govern the relative abundances of species — which are rare and which are common — and even how many species live together in the same place: the biodiversity of a habitat. On a more practical side, these principles must guide our efforts to understand the growth and eventual control of the human population; to determine management practices for exploited populations, such as cattle, fish, and trees; and to prevent the decline and extinction of other populations.

Natural selection and evolution

The history of life on earth has shown that the attributes of individuals change over time by the process of **evolution.** Evolution has two consequences that are crucial to ecology. First, biological systems continuously change through modification, over generations, of the structure and function of organisms within each population. Thus, although the principles of thermodynamics and population dynamics do not change, their expression in each particular ecological system never ceases to evolve. Second, the structure and function of organisms evolve in response to the features of their environments, which include both the physical conditions that prevail and the various other kinds of organisms with which each population interacts. For example, animals that are vulnerable to predators are often colored in such a way that they blend in with their backgrounds and escape notice (Figure 1.9). Plants that inhabit hot, dry climates may have thick, waxy cuticles that reduce the loss of water by evaporation across their leaf surfaces. These features of structure and function that suit an organism to the conditions of its environment are called **adaptations.**

The close correspondence between organism and environment is no accident. It derives from a fundamental and unique principle of biological systems: **natural selection.** Only those individuals that are well suited to the environment survive and produce offspring. The inherited traits that they pass on to their progeny are preserved. Unsuccessful individuals do not survive or they produce fewer offspring, and their less suitable traits therefore disappear from the population as a whole. Charles Darwin was the first to recognize that this process allowed populations to respond, over periods of many generations, to changes in their environments.

Natural selection expresses three properties of life and its relationship to the environment: (1) **genetic variation** among individuals within populations, (2) the **inheritance** of traits from parents by offspring, and (3) the influence of the environment on survival and reproduction, which defines the evolutionary **fitness** of the individual. The particular design that we see in the close match of organisms to their environments is not inherent in the process of natural selection. Rather, the environment itself is the template for design in that interactions between organisms and their environments result in differences in survival and reproductive success among individuals with different traits. Whether a rabbit runs slowly or rapidly is inconsequential; only the influence of its swiftness on the number of its descendants — its fitness — matters. Only in the presence of predators, which act as agents of selection, does speed have consequence for survival.

Inheritance and genetic variation within populations are facts of genetics obvious to anyone who has noticed the variability of physique, facial appearance, eye color, and hair among humans — and the tendency of related individuals to share many of these traits. Plant and animal breeders have known for centuries that by carefully selecting breeding lines for a desired trait, they could alter the appearance of a population toward a desired end: longer wool, increased egg and milk production, sweeter fruit. Selection was practiced on domesticated plants and animals long before biologists appreciated its role in shaping the structure and function of organisms in natural systems.

The first evidence that environmental change could foster new traits in a population came from early attempts to control insect pests. In southern California, cyanide gas was used in the early part of this century to control populations of scale insects on citrus crops. By 1914, however, populations in some groves had become tolerant of the gas, and fumigation was no longer effective. Laboratory experiments demonstrated not only that tolerance was inherited but also that some individuals naturally resisted cyanide poisoning in areas where the fumigant had never been used. This resistance was caused by a **mutation** — a random change in the DNA molecule that determines genetic inheritance — that appeared rarely and sporadically in individual scale insects. In the absence of cyanide, the mutation for cyanide resistance has no value to

FIGURE 1.9 The green coloration of the mantid (a) and tree frog (b) blend in with the backgrounds of these tropical animals and reduce their risk of being seen by hunting predators. Note the leaflike projections from the thorax of the mantid and the speckling on the frog, which resembles the fungi and other blemishes on natural leaves. Photographs by C. C. Hansen, courtesy of the Smithsonian Tropical Research Institute.

the organism and occurs randomly and infrequently at low levels in the population. But where the gas treatment was used to control scale insects, individuals carrying the resistant trait survived while all others died. In addition to performing their intended functions, cyanide and more recently developed pesticides and antibiotics have often inadvertently conferred a marked selective advantage on resistant individuals, yielding populations of pests that are all the more difficult to control.

Artificial selection, the responses of populations to changes in the environment brought about by human activities, and examples of natural selection in natural settings have convinced biologists of the pervasiveness of natural selection and evolutionary change. For the most part, the properties of organisms—including their morphology, physiology, and behavior—are adapted to the particular environment of each organism. Selection arises from the interaction of organisms with physical and chemical factors, from social interactions with relatives and unrelated individuals of the same species, and from relationships with other species that may be food resources or enemies. Evolution by natural selection is described in Part 4, "Populations," and its applications to understanding physiological adaptations of organisms, social behavior, and interactions between species are covered in Part 3, "Organisms" and Part 5, "Species Interactions."

Generation of ecological diversity

The fundamental principles of thermodynamics, population dynamics, and evolution apply equally to all types of organisms and systems. Even so, the most distinctive quality of biological systems is their **diversity.** The earth is inhabited by millions of different kinds, or species, of organisms. Presumably, all these descended from a much smaller number (perhaps only one) at some remote time in the past. The process by which species proliferate, **speciation,** involves the isolation of subpopulations from a single population and their independent evolutionary change. Eventually, sufficient differences evolve to prevent the isolated individuals from interbreeding successfully should the subpopulations renew their contact. This sequence, repeated over and over again, has produced a mind-boggling number of living things.

Two factors cause isolated subpopulations to diverge ecologically. First, different habitats or other environmental factors steer evolutionary change along different paths. Thus a part of biological diversity results from the variety of environments at the surface of the earth. This process is self-accelerating, because biological structures create environmental heterogeneity and establish opportunities for further evolutionary diversification. Second, biological interactions within habitats promote diversification. Competitors and various enemies (predators, herbivores, and disease organisms) constantly exert selective pressure and thus bring about evolutionary change in other species. Hence a small number of general processes operating in varied ecological settings have given rise to all the wonderful variety of life on earth. Ecological issues surrounding the concept of diversity are discussed in Part 7, "Communities."

The study of ecology

Ecologists are concerned with how environments and organisms interact to create the diversity of habitats and biological communities that occupy different regions of the earth. To understand natural diversity, ecologists must learn both the common principles of function that all ecological systems share and the manner in which these systems respond to variations in the environment.

Like other scientists, ecologists employ many methods to learn about nature. Most of these methods reflect four aspects of scientific investigation: (1) observation and description, (2) development of hypotheses or explanations, (3) testing of these hypotheses, and (4) application of general knowledge to solve specific problems. Most research programs begin with a set of facts about nature that invite explanation. Usually these facts are related to a pattern in nature. For example, measuring plant growth over several years, during which amounts of rainfall also are observed, might reveal a correlation between precipitation and plant production. To cite another example, exploration during the last century established that the number of animal and plant species in tropical latitudes greatly exceeds that in temperate regions. This fact of nature, which relates biodiversity to latitude, grew out of comparisons of the accumulated observations of many scientists until it became confirmed as a general pattern.

Hypotheses are ideas about how a system works. If correct, a hypothesis enables us to understand how an observed pattern is produced. Suppose we observe that male frogs sing on warm nights after it rains. This is a pattern. If a reasonable amount of observation yields no exceptions to the pattern, it may be regarded as a generalization that enables us to predict the behavior of frogs from knowledge of the weather. Having established the existence of a pattern, we may wish to understand it better. For example, we may wish to know *how* a frog responds to temperature and rainfall; we may also wish to know *why* a frog responds the way it does. The "how" part of this particular phenomenon involves details of sensory perception, interplay with hormone systems, and neuromotor effectors — in other words, it involves more physiology than ecology. The "why" part deals with the costs and benefits of the behavior to the individual; it is more ecological and evolutionary in nature. If we also suspect that male frogs attract mates by singing, we may entertain the idea that males sing after rains because that is when females look for mates. If frogs chorused at other times, they might attract few mates (low benefit) but still expose themselves to predation or other risk (high cost) in the attempt.

If we are to convince ourselves of the truth of a hypothesis, we must put it to the test. Only rarely can a particular idea be proved beyond a doubt, but our confidence increases the more we explore the implications of a hypothesis and find it to be consistent with the facts. If our hypothesis about frog singing were true, we would expect to observe larger numbers of receptive females on nights following rains than on nights following fair weather. This is a **prediction,** which is a statement that follows logically from a hypothesis. If subsequent observations of the activity of females confirm this prediction, then the hypothesis is strengthened; if not, the hypothesis is weakened or perhaps may be rejected altogether.

The strongest tests of hypotheses are often **experiments,** in which one or a small number of variables are manipulated independently of others. Thus we may design experiments to test predictions of a hypothesis specifically and unambiguously. In the frog example, the number of females hopping about is not a direct test of male mating success, which is the point of the hypothesis. We would prefer to determine whether mating success is lower when a male sings after fair weather than it is when a male sings after rains, but males don't sing after fair weather. Perhaps, by some suitable manipulation, we could trick a male into singing on the "wrong" night. Another experiment that comes to mind would be to record the songs of males on a tape, play them through speakers on different nights, and tally the numbers of females that are attracted to the calls.

Tests of predictions generate new information that often initiates additional rounds of hypothesis formation and testing. For example, if we find that female frogs are more active after rainy weather, we have discovered a new pattern that invites explanation. In this way scientific discovery builds upon itself, generating a rich understanding of the workings of natural systems.

Although the general processes of acquiring scientific knowledge and understanding appear to be straightforward, many pitfalls exist. For example, the existence of correlation between variables does not imply causal relationship; the mechanism of causation must be determined independently by suitable investigation. Different hypotheses may explain a particular observation equally well, and one must make predictions that distinguish among the alternatives. In addition, many predictions cannot be tested by experimental methods. Thus we may be unable to determine the consequences of variation in particular factors for the structure or functioning of a system. These limitations become particularly critical with patterns that have evolved over long periods and with systems that are too large for practical manipulation.

For instance, the observation that biodiversity decreases at higher latitudes has suggested as many explanations as there are physical attributes of the environment that vary with latitude. As one travels north from the equator, average temperature and precipitation decrease, light intensity and biological production decrease, and seasonality increases. Each of these factors may interact with biological systems in ways that could affect the number of species that can coexist in a locality. Isolating the effects of each of these factors has proved difficult because each tends to vary in parallel with the others.

Faced with these difficulties, ecologists have resorted to several alternative approaches to hypothesis testing. One of these is the **microcosm** experiment, an attempt to replicate the essential features of a system in a simplified laboratory setting. Thus an aquarium with five animal species may behave like the more complex natural system in a pond or even like ecological systems more generally; if so, experimental manipulations of the microcosm may yield results that can be generalized to the larger system. Another approach is to construct **mathematical models** of a complex system. Instead of using natural systems or microcosms, the investigator represents a system as a set of mathematical equations. These equations portray our understanding of how a system works in the sense that they describe the relationship of each of the system's components to other components and to outside influences.

A mathematical model is a hypothesis; it provides an explanation of the observed structure and functioning of a system. Models can be tested by comparing the predictions they yield with what is actually observed. Most models make predictions about attributes of the system that have not been measured or about the response of the system to perturbation. Whether these predictions are consistent with observation determines whether the hypothesis on which they are based is confirmed or rejected. We shall explore several examples of the contribution of microcosms and models to the development of ecological knowledge and understanding, particularly in the sections on ecosystems (Part 2), species interactions (Part 5), and communities (Part 6).

Human ecology

To say the least, humans are a prominent part of the biosphere. The human population exceeds five billion individuals; its development of technology has resulted in its consuming energy and resources far in excess of the needs dictated by biological metabolism. This flagrant consumption of resources and the attendant production of wastes has caused two related problems of global dimension. The first is the impact of human activity on natural systems, including the disruption of ecological processes and the extermination of species. The second is the steady deterioration of the human environment as we push the limits of sustainable development. Understanding ecological principles is a necessary step in dealing with these problems, as the following examples illustrate.

Several years ago, the Nile perch was introduced into Lake Victoria in East Africa. This was done with the well-intentioned purpose of providing additional food for the peoples living in the area and additional income from export of the surplus catch. Because simple ecological principles were ignored, however, the result was virtual destruction of the lake's entire fishery. Until the introduction of the Nile perch, Lake Victoria supported endemic fishing for a variety of species, mostly of the family Cichlidae, which themselves feed primarily on detritus and plants. Nile perch eat other fish: the smaller cichlids in this instance. Because energy is lost with each step in the food chain, predatory fish cannot be harvested at so high a rate as herbivorous prey species. Furthermore, the perch was alien to Lake Victoria, and the local cichlids had not evolved behavior to escape predation. Inevitably, the perch annihilated the cichlid populations, destroying the native fishery and all but eliminating its own food supply. Consequently, the perch's voracious habits among defenseless prey brought about its own demise as an exploitable fish. To be sure, the native fishery was already precariously overexploited as a result of the burgeoning local human population and the recent application of advanced fishing technology. However, the appropriate solution to these problems would have been better management of the cichlids, not introduction of an efficient predator upon them.

Other consequences of the introduction of the Nile perch turn the story into a tragicomedy of errors. The flesh of the perch is not favored by the peoples living near the shores of Lake Victoria, who prefer the familiar texture and flavor of the native fishes. Moreover, the oily meat of the perch must be preserved by smoking rather than sun-drying, so local forests are being cut rapidly for firewood. Because the larger perch must be fished with larger and more elaborate nets, local subsistence fishermen have not been able to compete with more prosperous outsiders who are equipped for commercial fishing. The lesson is painful yet simple: humans are an integral part of the ecology of the Lake Victoria area. Traditional local fishing had been sustained for thousands of years, until population pressure and the opportunity to develop an export fishery prompted an ecologically unsound decision that spelled economic and social disaster.

Half a world away, efforts to save the California sea otter illustrate the intricate intermingling of ecology and other human concerns. The sea otter was once widely distributed around the North Pacific Rim from Japan to Baja California; in the 1700s and 1800s, intense hunting for fine otter pelts reduced the population to near extinction. Predictably, the fur industry collapsed as it overexploited its economic base. Subsequent protection enabled the California sea otter population to multiply to more than 1500 individuals by the 1980s. The sea otter's success, alas, has irked some California fishers, who claim that the otters—which do not need commercial fishing licenses—drastically reduce stocks of valuable abalone, sea urchins, and spiny lobsters. Matters deteriorated to the marine equivalent of a range war between the fishing industry and conservationists, with the otter caught in the line of fire, often fatally.

Ironically, the otters benefit a different commercial marine enterprise, the harvesting of kelps, which are large seaweeds used in making fertilizer. Kelps grow in shallow waters in stands called kelp forests, which provide refuge and feeding grounds for larval fish (Figure 1.10). Kelps are also grazed by sea urchins, which, when abundant, can denude an area of its plant life. The sea otter is a principal predator of sea urchins. When the expanding otter population has spread into new areas, urchin populations have been controlled, allowing kelp forests to regrow. Human involvement in the ecosystem of the sea otter requires that management of otter populations and various commercial enterprises be balanced in an economically and socially acceptable manner. This will be impossible without a better understanding of the complex ecological role of the otter in the system.

Although the plight of endangered species may arouse us emotionally, ecologists are increasingly realizing that the only effective means of preserving and using natural resources is through the conservation of entire ecological systems and of broad-scale ecological processes in a sustainable fashion. Individual species, including the ones that humans rely on for food and other products, themselves depend on the maintenance of a healthy ecological support system. We have already seen how predators such as the Nile perch and sea otter may assume key roles in the functioning of natural systems, for better or worse depending on the circumstances. On land, pollinators and seed dispersers are necessary to the long-term maintenance of plant formations. Therefore, it is ecologically naive to try to manage forests without paying attention to maintaining the populations of insects, birds, and mammals on which the persistence of the forest depends. Natural systems are diverse and highly evolved in the sense that each species is adapted to its particular role within the system; in turn, the system's integrity depends to some degree on

FIGURE 1.10 The kelp forest provides feeding grounds and refuge for many species of fish and invertebrates. The integrity of the habitat depends on the presence of sea otters, which eat sea urchins, which in turn eat the young kelp plants. When sea urchins are abundant, they may prevent the regeneration of the kelp forest. Photograph by E. Hanauer, courtesy of P. Dayton.

each player acting out its role. Reduction in biodiversity entails the risk of upsetting the balance of a system.

The natural balance of nature includes continuous change. Most natural systems sustain cycles of disturbance (fire, hurricanes, disease) and rejuvenation. Natural restoration depends on there being patches of intact habitat from which species may recolonize disturbed areas. But as habitats become more and more fragmented, as forests in some regions have been by the spread of agriculture and urban development, disturbances can so thoroughly disrupt an area as to leave it little chance of complete recovery, even with substantial human help. Disturbance cycles, whatever their spatial and temporal dimensions, are as much a part of the natural landscape as are key species, and they must be taken equally into account in any effort to maintain an ecosystem.

Many advances in technology appear to improve the quality of human life, but they also make humans efficient consumers who increasingly strain the earth's ecological support systems and ultimately *reduce* the quality of life. Viewed on any scale, the deterioration of the earth's environment is alarming. In partially enclosed systems, such as the estuaries of major rivers, we may readily identify the causes of much of the change. San Francisco Bay, for example, has gradually but profoundly changed under the pressures of population and agriculture, to the point where it would be unrecognizable to nineteenth-century inhabitants of the region. The diking of wetlands to create farmland, filling to create land for urban construction, pollution from sewage and from agricultural and industrial waste, and the introduction of exotic species have transformed a healthy, productive, and economically valuable system into an ecological disaster, whose greatly diminished value lies entirely in recreation (which is itself diminishing in appeal) and in commercial shipping. The bay's once generous biotic resources are practically gone. True, the present human population of the San Francisco Bay area may be incompatible with a pristine estuary. However, the worst deterioration came about through state subsidy of unprofitable agricultural ventures, a dearth of ecologically and economically sensible city planning, careless introduction of pest species, use of the bay as a cheap sewer, and uncontrolled harvesting of fish and shellfish. Ecological disaster need not have happened.

The multiplication of technology and population has brought ecological crisis to far larger regions of the earth than San Francisco Bay. One of the most critical is the Sahel region of Africa. The Sahel is a vast band of semiarid habitat lying at the southern edge of the Saharan Desert and extending from Gambia in the west to Ethiopia in the east. Population pressure, and with it intensified grazing and collection of firewood, have caused extensive **deforestation** of the entire region. As humans and domesticated animals clear vegetation over such vast areas, climate is affected. Rainfall decreases because less water vapor is recirculated to the atmosphere and because the heating and cooling patterns of the land surface are changed. Even when plentiful rains result in good crop yields, the abundance may be snatched away by plagues of grasshoppers and locusts, whose populations grow rapidly under the moister conditions. The Sahel is at present truly devastated; the quality of human life there, not to mention the state of natural ecosystems, is very poor.

In wetter parts of the world, tropical rain forests are being cleared at an alarming rate for forest products and largely unsuitable agriculture, bringing profit to a few in the short term but predictable disaster in the long term. The destruction of rain forests dismantles an intricate, self-sustaining system that cannot easily be recreated. Species, many of them unknown to science and some of them with potential commercial value, are being driven to extinction. Burning of cut vegetation adds greatly to the already worrisome level of carbon dioxide in the atmosphere. Rampant erosion strips the unprotected land of valuable topsoil and chokes rivers with silt. In this instance, the ecological principles that should govern rational use of the rain forest are generally understood. The destruction of tropical rain forests, as well as the deterioration of other habitats and resources, derives from the short-term need to provide for a poor and growing population and from the failure to add long-term costs to the immediate price of exploitation. We are borrowing heavily on our future without the prospect of ever being able to pay back the debt.

The destruction of rain forests is scarcely alone in its far-reaching effect. Automobile emissions reduce tree and crop production over large areas. Sulfurous gases from burning coal have acidified the rain and snow over large areas of eastern North America and Europe, killing forests and degrading freshwater ecosystems. Chlorofluorocarbons (CFCs) used to pressurize spray cans and as refrigerants have reduced the ozone concentration in the upper atmosphere, increasing the amount of dangerous ultraviolet radiation that reaches the earth's surface.

Industrial growth and deforestation have caused carbon dioxide, methane, and CFCs in the atmosphere to increase to such levels that they are changing the earth's climate. In the so-called **greenhouse effect,** long-wave radiation from the earth's surface is absorbed and retained by the atmosphere. As a result, the temperature of the earth is increasing and is expected to rise by 2 to 6 degrees Celsius (°C) during the next century. Such change will cause a rise in sea level as the polar ice caps melt; it will disrupt agriculture and shift ecological habitats across the landscape as temperature and moisture patterns change; and it may cause widespread extinction of forms of life that are unable to adjust to the temperature change or to its indirect effects.

Our capacity to wreak havoc on ourselves and the world in which we live is virtually unlimited. The most appalling evidence of this ability is our means of waging global nuclear war, wherein destruction of ecological systems and extinction of species, possibly including ourselves, would be virtually without parallel in the history of the earth. Nuclear war remains unthinkable even if we ignore the ecological consequences, and its likelihood has been greatly reduced in recent years. But tragically, our gradual degradation of a livable planet may in the end be just as punishing. To understand ecology will not by itself solve our environmental problems in all their political, economic, and social dimensions. However, as we contemplate the need for global management of natural systems, we are doomed to fail if we do not understand their structure and functioning, an understanding that depends on knowing the principles of ecology.

≋ SUMMARY

1. Ecology is the study of the natural environment and of the relationships of organisms to each other and to their surroundings.

2. Organism, population, community, and ecosystem represent levels of organization of ecological structure and functioning. They form a hierarchy of progressively more complex entities.

3. Different kinds of organisms play different roles in the functioning of ecosystems. Plants fix the energy of sunlight; animals consume biological forms of energy; bacteria and fungi are largely responsible for the breakdown of biological materials and the regeneration of nutrients.

4. A plant's or animal's habitat is the place in which that plant or animal lives. The concept emphasizes the structure of the environment as it is experienced by each type of organism.

5. Ecological processes and structures have characteristic dimensions of time and space, which ecologists refer to as their scale.

6. Ecological systems function within the constraints of thermodynamic laws governing energy transformations. Other attributes of communities and ecosystems are regulated by the interactions among their constituent populations.

7. All ecological systems are subject to evolutionary change, which results from the differential survival and reproduction, within populations, of individuals that exhibit different genetically

determined traits. Because natural selection maximizes individual reproductive success, ecological systems evolve to function at levels close to the limits imposed by physical and thermodynamic considerations.

8. Diversity of ecological systems is generated by the proliferation of species in a heterogeneous environment. Spatial variation in conditions promotes differences between the species that live in different habitats. Interactions among populations within habitats also promote local diversification of species.

9. Ecologists study natural systems with a variety of techniques. The most important of these are observation, the development of hypotheses to explain observations, and the testing of hypotheses by attempting to confirm the predictions they generate. Experiments are an important tool in testing hypotheses. When natural systems lend themselves readily to experimentation, ecologists may work with microcosms or mathematical models of systems.

10. Humans are a major element of the biosphere, and our activities have created an environmental crisis of global proportions. Solving our acute environmental problems will require the intelligent application of general principles of ecology within the framework of political, economic, and social considerations.

🐦 SUGGESTED READINGS

Barel, C. D. N., et al. 1985. Destruction of fisheries in Africa's lakes. *Nature* 315:19–20. (Introduction of the Nile perch into Lake Victoria.)

Bartholomew, G. A. 1986. The role of natural history in contemporary biology. *BioScience* 36:324–329.

Booth, W. 1988. Reintroducing a political animal. *Science* 241:156–158. (The ecological role of sea otters in kelp communities.)

Franklin, J. F., C. S. Bledsoe, and J. T. Callahan. 1990. Contributions of the long-term ecological research program. *BioScience* 40:509–523.

Harley, J. L. 1972. Fungi in ecosystems. *Journal of Animal Ecology* 41:1–16.

Margulis, L., D. Chase, and R. Guerrero. 1986. Microbial communities. *BioScience* 36:160–170.

Margulis, L., and K. V. Schwartz. 1987. *Five Kingdoms—An Illustrated Guide to the Phyla of Life on Earth* (2nd ed.). W. H. Freeman, New York.

Nichols, F. H., J. E. Cloern, S. N. Luoma, and D. H. Peterson. 1986. The modification of an estuary. *Science* 231:567–573. (San Francisco Bay.)

Sai, F. T. 1984. The population factor in Africa's development dilemma. *Science* 226:801–805.

Sinclair, A. R. E., and J. M. Frywell. 1985. The Sahel of Africa: ecology of a disaster. *Canadian Journal of Zoology* 63:987–994.

Urban, D. L., R. V. O'Neill, and H. H. Shugart, Jr. 1987. Landscape ecology. *BioScience* 37:119–127.

Life and The Physical Environment

THE PHYSICAL
ENVIRONMENT

This chapter considers some of the most basic features of organisms and the environments they inhabit. We often speak of the living and the nonliving as opposites: biological versus physical and chemical, organic versus inorganic, biotic versus abiotic, animate versus inanimate. But although we can easily distinguish these two great realms of the natural world, they do not exist in isolation from each other. Life depends on the physical world. Living beings also affect the physical world: soils, the atmosphere, lakes and oceans, and many sedimentary rocks owe their properties in part to the activities of plants and animals.

Although they are distinct from physical systems, life forms nonetheless function within limits set by physical laws. Like internal combustion engines, they transform energy to perform work. The automobile engine burns gasoline chemically; it transmits power from the cylinder to the tires mechanically. When the organism metabolizes carbohydrates and moves its appendages, it follows related chemical and mechanical principles.

Biological systems, then, operate on the same principles as physical systems, but one important difference sets them apart. In physical systems, energy transformations always follow the path of least resistance and so act to even out variations in energy level throughout the system. In biological systems, the organism transforms energy in such a way that it remains *out of equilibrium* with the forces of gravity, heat flow, diffusion, and chemical reaction. The physical conditions inside the organism's body (temperature is a good example) often differ dramatically from those of its surroundings. When organisms move, they overcome gravity and work against the resistance of the physical world. Thus the physical world both provides the context for life and constrains its expression.

In a sense, the organism's use of energy is its secret of life. Consider a simple physical example. A boulder rolling down a steep slope releases energy during its descent, but it performs no useful work. The source of the energy — in this case, gravity — is external, and as soon as the boulder comes to rest in the valley below, it regains an equilibrium with the forces in its physical environment. In distinct contrast, a bird in flight constantly expends energy to maintain itself aloft against the pull of gravity. The source of the bird's energy — the food it has assimilated — comes from within, and the bird uses that energy to perform useful work: pursuit of prey, escape from predators, or migration.

The ability to act against external physical forces distinguishes the living from the nonliving. A bird in flight supremely expresses this quality, but plants just as surely perform work to counter physical forces when they absorb soil minerals into their roots and synthesize the highly complex carbohydrates and proteins that make up their structure.

Life depends totally on the physical world. Not only do organisms ultimately receive their energy from sunlight, but they also must tolerate the extremes of temperature, moisture, salinity, and other physical factors that occur in their surroundings. The heat and dryness of deserts exclude most species, just as the bitter cold of polar regions discourages all but the most hardy. Nor must we look to extreme conditions for evidence; the form and function of all plants and animals have evolved partly in response to conditions prevailing in the physical world. In this chapter, we shall explore those attributes of the physical environment that are most consequential for life. Because life processes take place in an aqueous environment, and because water makes up the largest part of all organisms, water seems a logical place to start.

Properties of water

Water is abundant at the earth's surface, and within the temperature range usually encountered, it is liquid. Many thermal and solvent properties of water certainly favor life. For example, one must add or remove a large amount of heat energy to change the temperature of water, and water conducts heat rapidly. Because of these two properties, the temperatures of organisms and aquatic environments tend to remain relatively constant and homogeneous. Water also resists change of state between solid (ice), liquid, and gaseous (water vapor) phases. Over 500 times as much energy must be added to evaporate a quantity of water as to raise its temperature by 1°C! Freezing requires the removal of 80 times as much heat as that needed to lower the temperature of the same quantity of water by 1°C (Box 2.1). Another curious thermal prop-

Box 2.1 *Thermal properties of water*

Specific heat is the quantity of heat energy required to raise the temperature of 1 g of water 1°C: 1 calorie (cal) or 4.2 joules (J).

Heat of melting is the quantity of heat energy that must be added to ice to melt 1 g of water at 0°C: 80 cal or 335 J.

Heat of vaporization is the quantity of heat energy that must be added to evaporate 1 g of water: 597 cal or 2498 J at 0°C, 536 cal or 2243 J at 100°C.

Thermal conductivity is the flux of heat through a 1 cm² cross section at a gradient of 1°C cm⁻¹ (units are J cm⁻¹ s⁻¹ °C⁻¹): 0.0055 at 0°C, 0.0060 at 20°C, 0.0063 at 40°C, and 0.022 for ice at 0°C.

Density is the mass per unit volume:

$$
\begin{aligned}
\text{Water at } 30°C &= 0.99565 \text{ g cm}^{-1} \\
20°C &= 0.99821 \\
10°C &= 0.99970 \\
4°C &= 0.99997 \text{ (maximum density)} \\
0°C &= 0.99984 \\
\text{Ice at } 0°C &= 0.917
\end{aligned}
$$

erty of water is that whereas most substances become more dense when they are cooled, water becomes less dense as it cools below 4°C. Water also expands upon freezing. Consequently ice floats (Figure 2.1), which not only makes ice skating possible but also prevents the bottoms of lakes and oceans from freezing and enables aquatic plants and animals to find refuge there in winter.

FIGURE 2.1 Water becomes less dense as it freezes, and ice therefore floats. But because the density of ice is 0.92 g cm⁻³ (not very far from that of water, at 1 g cm⁻³), more than 90 percent of the bulk of this antarctic iceberg lies below the surface.

Water has an immense capacity to dissolve inorganic compounds, making them accessible to living systems and providing a medium within which they can react to form new compounds. The formidable solvent properties of water derive from the strong attraction of water molecules for other compounds. Molecules consist of electrically charged atoms or groups of atoms called **ions.** Common table salt, sodium chloride (NaCl), contains a positively charged sodium atom (Na^+) and a negatively charged chlorine atom (Cl^-). When in contact with salt, water molecules attract the charged sodium and chlorine atoms so powerfully, compared with the bonds that hold the molecule together, that the salt molecule readily dissociates into its component atoms — another way of saying that the salt dissolves.

Water and soil

Most terrestrial plants obtain the water they need from the soil. The amount of water that soil holds and its availability to plants vary with the physical structure of the soil. Soil consists of grains of clay, silt, and sand and particles of organic material. Grains of clay, produced by the weathering of minerals in certain types of bedrock, are the smallest; grains of sand, derived from quartz crystals that remain after minerals more susceptible to weathering dissolve out of rock, are often the largest; silt particles are intermediate in size. Collectively, these particles make up the soil skeleton. As the name implies, the **soil skeleton** is a stable component that influences the physical structure of the soil and its water-holding ability but does not play a major role in its chemical transformations.

Water is sticky. The capacity of water molecules to cling to one another and to surfaces they touch accounts for the familiar phenomena of surface tension and the rising of water against gravity in capillary tubes. Water clings tightly to surfaces of the soil skeleton. Because the total surface area of particles in a given volume of soil increases as their size decreases, clay soils and silty soils hold more water than coarse sands, through which water drains quickly.

The amount of water in the soil only partly determines its availability. Plant roots easily take up water that clings loosely to soil particles by surface tension, but water near the surface of sand and silt particles binds tightly to the soil skeleton by stronger forces. These forces are called the **water potential** of the soil. Soil scientists describe soil water potential and the strength with which the cells of root hairs can absorb water from the soil in terms of equivalents of atmospheric pressure (Box 2.2). Capillary attraction holds water in the soil with a force equivalent to a pressure of about 0.1 atmosphere (atm), or 10^4 pascals (Pa). Water that is attracted to soil particles with a force of less than 0.1 atm (water in the middle of interstices between large soil particles, generally more than 0.005 mm from their surfaces) drains out of the soil under the pull of gravity and joins the groundwater in the crevices of the bedrock below. The amount of water held against gravity by forces of attraction greater than 0.1 atm is called the **field capacity** of the soil.

A force equivalent to 0.1 atm can raise a column of water nearly 1 m. We know that plant roots exert a much greater pull on water in the soil, because

Box 2.2   *Units of pressure*

Pressure is a force (weight) per unit area. In the English system of measurement, the customary unit of pressure is the pound per square inch (lb in.$^{-2}$). The pressure of the atmosphere at sea level — that is, the total weight of atmosphere above a square inch of surface — is 14.7 lb in.$^{-2}$. This value is often expressed in terms of the height of a column of mercury (Hg) having the same weight (29.9 in. or 760 mm). A pressure of 1 mm Hg is sometimes referred to as 1 torr. In the International System of Units (SI) — see Appendix A — the unit of pressure is the pascal (Pa), which is the weight of 1 newton (N) per square meter.

Expressed in these various units, sea-level atmospheric pressure is equal to

1	atmosphere of pressure (atm)
14.7	pounds per square inch (lb in.$^{-2}$)
29.9	inches of mercury (in. Hg)
760	millimeters of mercury (mm Hg)
760	torr
101,325	pascals (Pa)
101.3	kilopascals (kPa)
0.101	megapascals (MPa)
1.01	bar (b; 1 b = 100,000 Pa)

water rises in the tallest trees to leaves more than 100 m above the ground. In fact, most plants can exert a maximum pull of about 15 atm on soil water. After plants under drought stress have taken up all the water in the soil that is held by forces weaker than 15 atm, they can no longer obtain water and they wilt, even though water remains in the soil. Thus ecologists refer to a water potential of 15 atm as the **wilting coefficient,** or **wilting point,** of the soil.

As soil water decreases, the remainder is held by increasingly stronger forces, on average, because a greater proportion of the water lies close to the surfaces of soil particles. Figure 2.2 shows the relationship between water content (in percent of the soil consisting of water) and water potential for a typical soil with a more or less even distribution of soil particle sizes from clay (up to 0.002 millimeter, mm) through silt (0.002–0.05 mm) to sand (0.05–2.0 mm). Such soils are called loams. When saturated, a loam holds about 45 grams (g) of water per 100 g of dry soil (thus it is 45 percent water). The field capacity is about 32 percent, the wilting coefficient about 7 percent. The difference between the field capacity and the wilting coefficient, about 25 percent in this case, measures the water available to plants. Of course, plants obtain water most readily when the soil moisture is close to the field capacity.

In soils with predominantly smaller particles, the soil skeleton has a relatively large surface area; such soils hold more water at both the wilting coefficient and the field capacity, and a correspondingly larger proportion of soil water is held by forces greater than 15 atm. Soils with predominantly larger

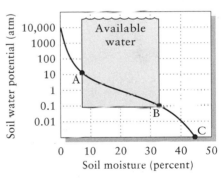

FIGURE 2.2 Relationship between water content of a loam and the average force of attraction of the water to soil particles (water potential). The difference between the soil water content at field capacity (B, 0.1 atm) and the wilting coefficient (A, 15 atm) is the water available to plants. Point C is the saturation capacity of the soil. After N. C. Brady, *Nature and Properties of Soils* (8th ed.). Macmillan, New York (1974).

skeletal particles have less surface area and larger interstices between particles (Figure 2.3). More of the soil water is held loosely and is thus available to plants, but such soils have lower field capacities. Plants can obtain the most water from soils that exhibit a variety of particle sizes between sand and clay.

Temperature

Life processes, as we know them, occur only within the range of temperatures at which water is liquid: 0 to 100°C at the earth's surface. Relatively few organisms can survive body temperatures above 45°C. The photosynthetic cyanobacteria tolerate temperatures as high as 75°C, and some bacteria occur in hot springs close to the boiling point of water (Figure 2.4). The properties that make existence possible at high temperatures are not well understood. Compared to those of most others, the proteins of thermophilic ("heat-loving") bacteria have subtly different proportions of amino acids; as a result, the structure of these proteins remains stable at temperatures up to 95°C.

Although temperatures on the earth rarely exceed 50°C, except in hot springs and at the soil surface in hot deserts, temperatures below the freezing point of water occur commonly over large portions of the earth's surface. When living cells freeze, the crystal structure of ice disrupts most life processes

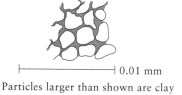

├────────────────┤ 0.01 mm

Particles larger than shown are clay

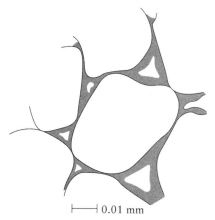

├──────┤ 0.01 mm

Particles larger than shown are sand

▨ Water held by capillary attraction

FIGURE 2.3 These drawings represent the relative sizes of the smallest and the largest soil particles that constitute what we call silt. Clay consists of particles smaller than the smallest of these. Particles that exceed the larger make up sand. Water held by capillary attraction (at least 0.1 atm water potential), indicated by shading, occupies a larger fraction of a clay soil than of a silty soil, and a sandy soil holds less water yet.

FIGURE 2.4 A hot spring in Yellowstone National Park. Even though the temperature of the water approaches the boiling point, some thermophilic bacteria thrive in this environment.

and may damage delicate cell structures, leading rapidly to death. Many species successfully cope with freezing temperatures either by maintaining their body temperatures above the freezing point of water or by activating mechanisms that enable them to resist freezing or to tolerate its effects. The freezing point of water may be depressed by dissolved substances that interfere with the formation of ice. For example, the freezing point of seawater, which contains about 3.5 percent dissolved salts, is −1.9°C.

The blood and body tissues of most vertebrates contain less than half the salt content of seawater and thus may freeze at a higher temperature than the freezing point of the ocean. Saltier blood would enable vertebrates to live in polar seas; a 10 percent solution of sodium chloride, for example, reduces the freezing point of water by 10.5°C. But the sensitivity of protein structure and function to high salt concentrations makes this an impractical solution. Instead, the freezing points of the body fluids of many marine organisms are reduced by large quantities (up to 30 percent in some terrestrial invertebrates) of glycerol and glycoproteins, which act like antifreeze. Their presence in the blood and tissues allows antarctic fish, for example, to remain active in seawater that is colder than the normal freezing point of the blood of fish inhabiting temperate or tropical seas (Figure 2.5); a 10 percent glycerol solution lowers the freezing point by about 2.3°C.

Supercooling provides a second solution to the problem of freezing. Under certain circumstances, fluids can cool below the freezing point without ice

FIGURE 2.5 In the antarctic fish *Trematomus*, blood and tissues are prevented from freezing by the accumulation of high concentrations of glycoproteins, which lower the freezing point of body fluids to below the minimum temperature of seawater (−1.8°C) and prevent ice crystal formation. Photograph courtesy of P. Dayton.

crystals developing. Ice generally forms around some object, called a seed, which can be a small ice crystal or other particle. In the absence of seeds, pure water may cool to more than 20°C below its melting point. Supercooling has been recorded to −8°C in reptiles and to −18°C in invertebrates. Glycoproteins in the blood of these cold-adapted animals impede ice formation by coating developing crystals (seeds).

Finally, some organisms can tolerate the freezing of most or all of the water in their bodies. Such organisms restrict ice formation to the spaces between cells and avoid it within them; as a result, ice does not destroy cell structure. However, because salts are excluded from ice and concentrate in the liquid water within cells, such freezing-tolerant organisms must cope with extremely high salt levels in their tissues during the winter.

Temperature has several opposing effects on life processes. First, heat increases the kinetic energy of molecules and thereby accelerates chemical reactions; the rate of biological processes commonly increases between two and four times for each 10°C rise in temperature throughout the physiological range (Figure 2.6). Second, enzymes and other proteins become less stable at high temperatures and may not function properly or retain their structure. Third, the level of heat energy in the cell influences the conformations of proteins, which balance the natural kinetic motions induced by heat and the forces of chemical attraction between different parts of the molecule. Similarly, the physical properties of fats, which reside in cell membranes and which many animals accumulate as a reserve of food energy, also depend on temperature. When cold, fats become stiff (picture the fat on a piece of meat taken from the refrigerator); when warm, they become fluid.

FIGURE 2.6 Examples of temperature dependence in organism function. (a) Rate of oxygen consumption in the Colorado potato beetle. At lower temperatures the rise in oxygen consumption is exponential, increasing by a factor of about 2.5 for each 10°C. After K. Marzusch, *Zeits. vergl. Physiol.* 34:75–92 (1952). (b) Rate of development (the inverse of the larval development period) in the butterfly *Pieris rapae*. After N. Gilbert, *J. Anim. Ecol.* 53:589–597 (1984).

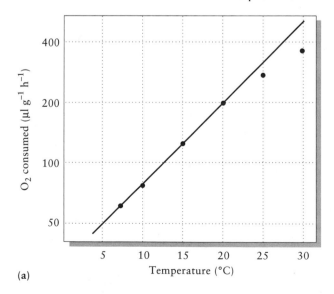

(a)

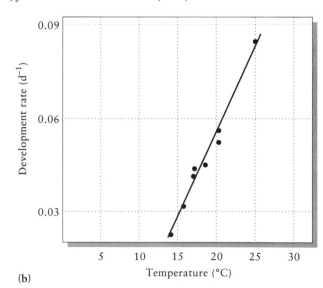

(b)

Enzymes function well only when they assume the proper shape. Too hot, and the molecule may open its structure and tend to unfold; too cold, and it may close up, preventing substrates from binding properly to it. In short, the structures of enzymes and other molecules enable them to function best within the normal range of body temperature of the organism.

Carbon, oxygen, and biological energy transformations

Plants and animals consist of many elements joined together into organic molecules that constitute the individual organism. Such organic compounds also provide the energy, in the form of chemical bonds between atoms and molecules, needed to maintain the organism. These energy-containing bonds arise from chemical changes among the atoms of various elements. In biological systems, one of the most prevalent of these transformations is the chemical **reduction** of carbon, which is accomplished when electrons are added to the carbon atom. **Oxidation** reactions, on the other hand, strip electrons from atoms; carbon dioxide (CO_2) is an oxidized form of carbon. During **photosynthesis,** plants chemically reduce the carbon atom in carbon dioxide. This altered atom forms new compounds, such as the carbohydrate glucose ($C_6H_{12}O_6$), that have elevated energy levels. The added energy that is thus *stored* comes from light. To release this stored energy for other purposes, both plants and animals undo the results of photosynthesis by oxidizing carbon back to carbon dioxide. This transformation *releases* energy, a portion of which organisms harness for their own purposes; the rest escapes as heat.

Photosynthesis and respiration involve the complementary reduction and oxidation of carbon and oxygen. Oxygen's common oxidized state is molecular oxygen (O_2), which occurs as a gas in the atmosphere and dissolved in water. In a reduced state, oxygen readily forms water molecules (H_2O). Thus, as carbon is reduced during photosynthesis, oxygen is oxidized from its form in water to its molecular form. During **respiration,** inhaled or absorbed oxygen is reduced to the form contained in the compound we call water as carbon is oxidized to the form it exhibits in carbon dioxide. Why does the coupling of an oxidation reaction to a reduction reaction result in a net release of energy? The reduction of oxygen is thermodynamically more favorable (it requires less energy input) than the reduction of carbon. Therefore, the oxidation of carbon releases more energy than the reduction of oxygen requires.

Plants reduce more carbon than they oxidize (otherwise they would not grow), so they require an external source of carbon. The only practical source of inorganic carbon, carbon dioxide, has an extremely low concentration in the atmosphere (about 0.03 percent). As a result, the concentration of CO_2 in the atmosphere, which drives the movement of CO_2 into plant cells, is much, much less than the gradient of pressure that sucks water vapor from the plant into the surrounding atmosphere. This makes water conservation a special problem for plants, especially in arid environments, and it accounts for the fact that plants transpire (give off into the air) 500 g of water, more or less, for every gram of carbon assimilated (Figure 2.7).

Water vapor
pressure
up to
0.1 atm

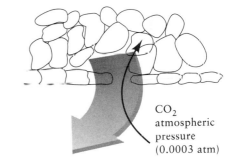

CO_2
atmospheric
pressure
(0.0003 atm)

Figure 2.7 Schematic diagram of the different pressures that carbon dioxide and water exert in the tissue of a leaf and the surrounding air. Because the plant uses carbon dioxide in photosynthesis, the concentration of that gas remains at a low level in the leaf. The vapor pressure of water in the atmosphere may be as low as 0 atm (dry air) and at the surface of leaf tissues as high as 0.1 atm (at 45°C).

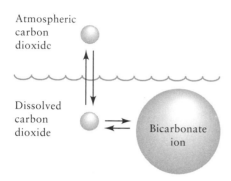

Atmospheric carbon dioxide

Dissolved carbon dioxide

Bicarbonate ion

FIGURE 2.8 Schematic diagram suggesting the relative amounts of carbon dioxide that are available in the atmosphere, that are dissolved in water, and that occur in the form of bicarbonate ion in water at equilibrium. The fact that bicarbonate ion forms when carbon dioxide dissolves in water and itself readily dissolves in water greatly increases the reservoir of carbon dioxide available to aquatic plants.

Carbon availability is less of a problem for aquatic plants than for terrestrial plants. The solubility of carbon dioxide in fresh water and under ideal conditions is 0.0003 cubic centimeter per cubic centimeter (cm^3 cm^{-3}), or 0.03 percent by volume—about the same as its concentration in the atmosphere. When carbon dioxide dissolves, however, some of the molecules react with water to form carbonic acid (H_2CO_3) and associated compounds, which provide a reservoir of inorganic carbon. Depending on the acidity of the water, carbonic acid molecules dissociate into bicarbonate ions (HCO_3^-) and carbonate ions (CO_3^{2-}). Within the range of acidity of most natural waters (pH values between 6 and 9), bicarbonate ion is the more common form, and it dissolves readily in water (69 g of $NaHCO_3$ per liter of water, for example). As a result, seawater normally contains concentrations of bicarbonate ion equivalent to 0.03 to 0.06 cm^3 of carbon dioxide gas per cubic centimeter of water (3–6 percent), over 100 times the concentration of the dissolved gas (Figure 2.8). The bicarbonate ion readily enters the cells of aquatic plants, where it releases carbon dioxide at the site of photosynthesis.

Just as getting enough carbon dioxide is a problem for plants in terrestrial environments, oxygen availability limits animals in aquatic habitats. Compared to its concentration of 0.21 cm^3 cm^{-3} in the atmosphere, the solubility of oxygen reaches a maximum (at 0°C in fresh water) of 0.01 cm^3 cm^{-3}—only 14 parts per million (ppm) by weight. Furthermore, below the limit of light penetration in deep bodies of water and in waterlogged sediments and soils, no aquatic plants exist to produce oxygen photosynthetically; therefore, animal and microbial respiration may severely deplete dissolved oxygen. Deeper layers of water in lakes and the mucky sediments of marshes frequently become totally deprived of oxygen (**anaerobic** or **anoxic**). Similar conditions in waterlogged soils of swamps pose problems for terrestrial plants, whose roots need oxygen for respiration just as animals do. Under such conditions, plants may have specialized vascular tissues, called **aerenchyma,** that conduct air directly from the atmosphere to the roots. The roots of cypress trees and many mangroves grow vertical extensions that project above the anoxic soil and take up oxygen directly from the atmosphere (Figure 2.9).

Inorganic nutrients

Organisms assimilate a wide variety of chemical elements. After hydrogen, carbon, and oxygen, the elements required in greatest quantity are nitrogen, phosphorus, sulfur, potassium, calcium, magnesium, and iron (Table 2.1). Certain groups of organisms need other elements in abundance. For example, diatoms construct their glassy shells of silicon compounds, tunicates accumulate vanadium in high concentration, and nitrogen-fixing bacteria require molybdenum. Plants need such elements as boron and selenium in minute quantities, but the physiological functions of these **micronutrients** are not well understood.

Plants acquire mineral nutrients—other than oxygen, carbon, and some nitrogen—from water that is in contact with their roots. They obtain nitrogen in the form of ammonia ion (NH_4^+) or nitrate ion (NO_3^-), phosphorus in

FIGURE 2.9 The knees of these bald cypress trees in a freshwater swamp in South Carolina conduct air from the atmosphere to roots growing in waterlogged, anaerobic sediments. Other marsh and swamp plants have air-conducting tissues (aerenchyma) in their stems. Courtesy of the U.S. Forest Service.

the form of phosphate ion (PO_4^{3-}), calcium and potassium in the form of their elemental ions (Ca^{2+} and K^+), and so on. The availability of these elements varies with their chemical form in the soil and with temperature, acidity, and the presence of other ions. Phosphorus, in particular, often limits plant production; even when it is abundant, most of the compounds it forms in the soil do not dissolve easily.

TABLE 2.1	*Major nutrients required by organisms, and some of their primary functions*
Element	**Function**
Nitrogen (N)	Structural component of proteins and nucleic acids
Phosphorus (P)	Structural component of nucleic acids, phospholipids, and bone
Sulfur (S)	Structural component of many proteins
Potassium (K)	Major solute in animal cells
Calcium (Ca)	Structural component of bone and of material between woody plant cells; regulator of cell permeability
Magnesium (Mg)	Structural component of chlorophyll; involved in the function of many enzymes
Iron (Fe)	Structural component of hemoglobin and many enzymes
Sodium (Na)	Major solute in extracellular fluids of animals

All natural waters contain some dissolved substances. Although nearly pure, rainwater acquires some dissolved minerals from dust particles and droplets of ocean spray in the atmosphere. Most lakes and rivers contain 0.01 to 0.02 percent dissolved minerals and roughly 0.025 to 0.05 percent of the average salt concentration of the oceans (3.4 percent by weight), in which salts and other minerals have accumulated over the millennia.

The minerals dissolved in fresh water and salt water differ in composition as well as quantity. Seawater abounds in sodium and chlorine and has significant amounts of magnesium and sulfate. Fresh water contains a greater variety of ions, but calcium usually makes up most of the **cations** (ions carrying a positive charge) and carbonate and sulfate most of the **anions** (those that carry a negative charge). The composition of fresh water differs from that of salt water because of the different rates of solution and solubilities of substances. Few compounds reach their maximum solubilities in fresh water; their concentrations reflect the compositions and rates of solution of materials in the rock and soil that the water contacts. Limestone consists primarily of calcium carbonate, which dissolves quickly; thus water in limestone areas contains abundant calcium ions, making it "hard." Granite is composed of such minerals as quartz and feldspar, which do not contain calcium and which dissolve slowly; water flowing through granitic areas contains few dissolved substances and accordingly is "soft." Nitrogen (mostly nitrate and dissolved organic nitrogen compounds) enters bodies of fresh water in relative abundance in the runoff from surrounding terrestrial systems. In contrast, most of the phosphorus in streams, ponds, and lakes readily complexes with iron and precipitates out of the system. As a result, phosphorus, rather than nitrogen, usually limits plant production in freshwater systems.

The oceans function like large stills, concentrating minerals as nutrient-laden water arrives via streams and rivers and as pure water evaporates from the surface. Here the concentrations of some minerals, particularly calcium carbonate ($CaCO_3$), reach limits set by their maximum solubilities. Calcium carbonate dissolves only to the extent of 0.000014 g per gram (about 1 cm³) of water. Its concentration in the oceans reached this level eons ago, and excess calcium ions entering the oceans each year from streams and rivers precipitate to form limestone sediments. At the other extreme, the solubility of sodium chloride (0.36 g per gram of water) far exceeds its concentration in seawater; most of the sodium chloride washing into ocean basins remains dissolved. The iron content of seawater tends to be low, so relatively little phosphorus sediments out of the water column of the oceans in the form of insoluble complexes with iron. In contrast to the situation in fresh water, nitrogen generally limits marine productivity to a greater extent than does phosphorus, because it readily leaves the system by sedimentation in the form of organic detritus.

Osmotic potential

Organisms obtain nutrients from the soil, water, or their food. Often these nutrients are much more concentrated in the tissues than in the surroundings, so organisms must assimilate them against the prevailing gradient. But organisms also must exclude from their bodies many substances that are abundant

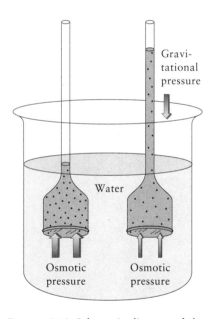

FIGURE 2.10 Schematic diagram of the water (osmotic) potential developed by solutes enclosed within a membrane permeable to water but not to the solutes (a semipermeable membrane). At left, because the solutes are at high concentration, water tends to move across the membrane into the inverted funnel. Within the funnel the increasing volume pushes fluid up the stem. Eventually, the osmotic pressure of the fluid, which decreases as the solutes become more diluted, is balanced by the gravitational pressure exerted by the fluid in the stem.

in the environment and are metabolically useless or even toxic at high concentration. When a surface permits the influx of desirable substances, it can keep out others only by **selective permeability** (if wanted and unwanted substances differ greatly in size or electrical charge) or by **active transport** of unwanted substances out across the surface.

Left to their own devices, ions diffuse across the surfaces of organisms from the side where they are more abundant to the side where they are less abundant, tending to equalize their concentrations between the organism and its surroundings. Water also moves across permeable membranes (the process is called **osmosis**) toward regions of higher ion concentrations (that is, lower water concentrations), tending to equalize the concentrations of dissolved substances on both sides of the membrane. This tendency of a solution to attract water is known as its **osmotic potential,** and it can be expressed as a pressure (Figure 2.10). A molar concentration of a substance in solution ideally exerts an osmotic pressure of 21 atm. Seawater exerts a high osmotic pressure (about 12 atm), and fresh water exerts practically none. The body fluids of vertebrate animals, exerting an osmotic pressure 30–40 percent that of seawater, occupy an intermediate position. Gradients of osmotic potential pose different problems for freshwater fish, which tend to gain water and lose solutes, and saltwater fish, which tend to gain solutes and lose water (Figure 2.11). Most organisms solve such problems by pumping ions in one direction or the other across various body surfaces (skin, kidney tubules, and gills), expending considerable energy in the process.

Certain environments pose special osmotic problems. Aquatic environments with salt concentrations greater than that of seawater occur in some landlocked basins, particularly in arid zones where evaporation is great. The Great Salt Lake (20 percent salt) in Utah and the Dead Sea (23 percent salt), lying between Israel and Jordan, are well-known examples. The osmotic potential of such environments, being well over 100 atm, would suck the water from most animals and plants, but a few aquatic creatures, such as brine shrimp (*Artemia*), can survive in salt water concentrated to the point of crystallization (300 g per liter, or 30 percent). Brine shrimp excrete salt at a prodigious rate to keep the salts in their body fluids less concentrated than those in their surroundings.

Light

Light is the primary source of energy for the ecosystem. Green plants absorb light and assimilate its energy by photosynthesis. But plants cannot use all light striking the earth's surface. Rainbows and prisms show that light consists of a spectrum of wavelengths that we perceive as different colors. Wavelengths of light are generally expressed in micrometers, μm (one-millionth of a meter, 10^{-6} m), or nanometers, nm (one-billionth of a meter, 10^{-9} m). The visible spectrum extends between wavelengths of about 400 nm (violet) and 700 nm (red). Light of shorter wavelength than 400 nm makes up the ultraviolet part of the spectrum; we refer to light of longer wavelength than 700 nm as infrared. The energy content of light varies with wavelength and hence with color;

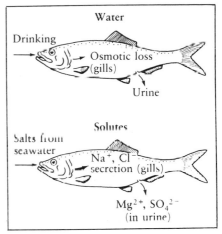

(a) Marine

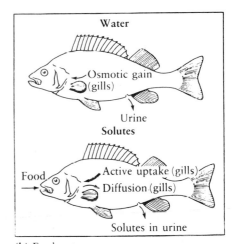

(b) Freshwater

FIGURE 2.11 Pathways of exchange of water and solutes by (a) marine fish, whose body fluids are **hypo-osmotic** (contain less concentrated salt than the surrounding water), and (b) freshwater fish, whose body fluids are **hyperosmotic**. The gills and kidneys actively exclude or retain solutes to maintain salt balance. Marine fish must drink to acquire water. After K. Schmidt-Nielson, *Animal Physiology. Adaptation and Environment.* Cambridge Univ. Press, Cambridge (1975).

shorter wavelength blue light has a higher energy level than longer wavelength red light.

Light that reaches the upper part of the earth's atmosphere from the sun extends far beyond the visible range: through the ultraviolet region toward the short-wavelength, high-energy X rays at one end of the spectrum, and through the infrared region to such extremely long-wavelength, low-energy radiation as radio waves at the other end of the spectrum. Because of its high energy level, ultraviolet light can damage exposed cells and tissues. As light passes through the atmosphere, however, most of its ultraviolet components are absorbed, primarily by a molecular form of oxygen known as **ozone** (O_3) that occurs in the upper atmosphere. The atmosphere thus shields life at the earth's surface from the most damaging wavelengths of light (Figure 2.12). Certain pollutants in the atmosphere, particularly the chlorofluorocarbons (CFCs) used as refrigerants and as propellants in aerosol cans, chemically remove ozone from the upper atmosphere. This degradation has progressed so far over some parts of the earth as to greatly increase the levels of dangerous ultraviolet radiation that reach the surface.

Vision and the photochemical conversion of light energy to chemical energy by plants occur primarily within that portion of the solar spectrum at the earth's surface that contains the greatest amount of energy. Absorption of radiant energy depends on the nature of the absorbing substance. Water only weakly absorbs light whose characteristic wavelengths fall in the visible region

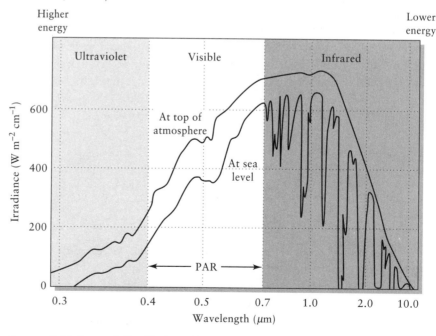

FIGURE 2.12 Spectral distribution of direct sunlight at the top of the atmosphere and at sea level. The difference between the two, which is proportionately greatest in the ultraviolet region of the spectrum, represents light absorbed by the atmosphere. After D. M. Gates, *Biophysical Ecology*. Springer-Verlag, New York (1980).

of the spectrum of energies; as a result, a glass of water appears colorless. Dyes and pigments strongly absorb some wavelengths in the visible region, while reflecting or transmitting light of definite colors that become identifying characteristics. Plant leaves contain several kinds of pigments, particularly **chlorophylls** (green) and **carotenoids** (yellow), that absorb light and harness its energy. Carotenoids, which give carrots their orange color, absorb primarily blue and green (Figure 2.13) and reflect light in the yellow and orange regions of the spectrum. Chlorophyll absorbs red and violet light while reflecting green and blue.

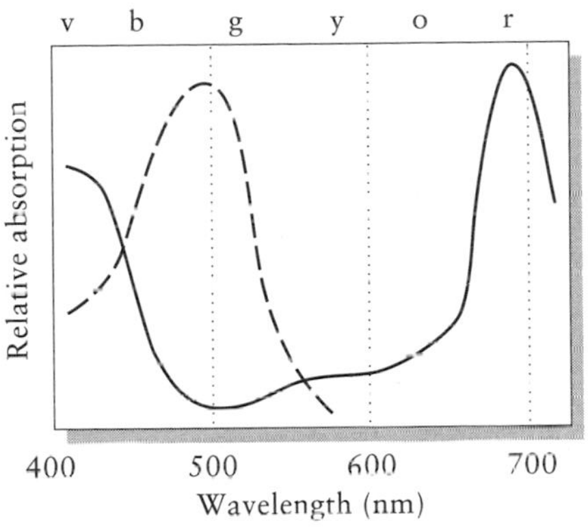

Light intensity

Ecologists measure the intensity of light as the energy content within the **photosynthetically active region** of the spectrum (**PAR**) — that between wavelengths of 400 and 700 nm. Intensity may be expressed in a variety of units, including the langley, watt, and einstein, all named after major figures in physics (Table 2.2). The intensity of light reaching the outer limit of the atmosphere — the so-called **solar constant** — is approximately 2 cal cm² min⁻¹ (2880 ly d⁻¹ or 1395 W m⁻²). In reality, far less light energy reaches any area on the surface of the earth. This is because of nighttime periods without light, the low incidence of light early and late in the day and at high latitude, and cloud cover. A temperate habitat on a clear day in summer might typically receive 350 W m⁻² of PAR, or about 700 ly d⁻¹ (that is, about a quarter of the solar constant).

At low light levels, usually less than one-quarter that of full sunlight, the rate of photosynthesis varies in direct proportion to light intensity. Brighter light saturates the photosynthetic pigments, however, and as intensity increases thereafter, the rate of photosynthesis increases more slowly or levels off. The response of photosynthesis to light intensity has two reference points (Figure 2.14). The first, called the **compensation point,** is the level of light intensity at which photosynthetic assimilation of energy just balances respira-

FIGURE 2.13 Absorption of light of different wavelengths by two groups of plant pigments — chlorophylls and carotenoids — that capture light energy used in photosynthesis. The colors of the spectrum are violet (v), blue (b), green (g), yellow (y), orange (o), and red (r). After R. Emerson and C. M. Lewis, *J. Gen. Physiol.* 25:579–595 (1942).

TABLE 2.2	*Intensity of photosynthetically active sunlight at the surface of the earth on a clear day*	
Measurement	Units	Typical value
langley (ly)	1 cal cm⁻²	700 ly d⁻¹
watt (W)	1 J s⁻¹	1000 W m⁻²
einstein (E)	6 × 10²³ photons	2000 μE m⁻² s⁻¹
		200 nE cm⁻² s⁻¹

Source: M. G. Barbour, J. H. Burk, and W. D. Pitts, *Terrestrial Plant Ecology.* Benjamin Cummings, Menlo Park, Calif. (1980).

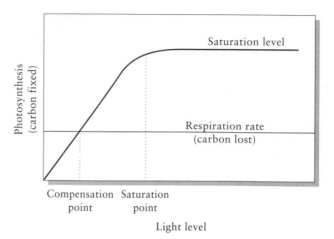

FIGURE 2.14 Relationship between carbon dioxide assimilation (photosynthesis) and light intensity, illustrating the compensation point and the saturation point of the photosynthetic process. After M. G. Barbour, J. H. Burk, and W. D. Pitts, *Terrestrial Plant Ecology.* Benjamin/Cummings, Menlo Park, Calif. (1980).

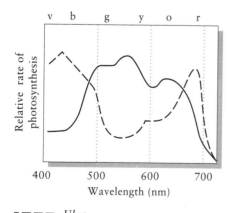

FIGURE 2.15 Relative rates of photosynthesis by the green alga *Ulva* and the red alga *Porphyra* as a function of the color of light. After F. T. Haxo and L. R. Blinks, *J. Gen. Physiol.* 33:389–422 (1950).

tion. Above the compensation point, the energy balance of the plant is positive; below it, that energy balance is negative. The second reference point is the **saturation point,** above which the rate of photosynthesis no longer responds to increasing light intensity. Among terrestrial plants, the compensation points of species that normally grow in full sunlight (a maximum of about 500 W m^{-2}) occur between 1 and 2 W m^{-2}. The saturation points of such species usually are reached between 30 and 40 W m^{-2}—less than a tenth of the energy level of bright, direct sunlight. As one might expect, the compensation and saturation points of plants that typically grow in shade occur at lower light intensities.

Water absorbs or scatters enough light to limit the depth of the sunlit zone of the sea. The transparency of a glass of water is deceptive. In pure seawater, the energy content of light in the visible part of the spectrum diminishes to 50 percent of its surface value within 10 m depth, and it drops to less than 7 percent within 100 m depth. Furthermore, water absorbs longer wavelengths more strongly than shorter ones; virtually all infrared radiation disappears within the topmost meter of water. Short wavelengths of light (violet and blue) tend to scatter when they strike water molecules and thus also do not penetrate deeply. As a consequence of the absorption and scattering of light by water, green light predominates with increasing depth. The photosynthetic pigments of aquatic plants parallel this spectral shift. Plants near the surface of the oceans, such as the green alga *Ulva* (sea lettuce), have pigments resembling those of terrestrial plants and best absorb blue and red light. The deep-water red alga *Porphyra* has additional pigments that enable it to use green light more effectively in photosynthesis (Figure 2.15).

Absorption of light limits the depth at which plants exist in the oceans to a fairly narrow zone close to the surface, in which photosynthesis can exceed plant respiration. This area is called the **euphotic zone.** The lower limit of the euphotic zone, where photosynthesis just balances respiration, is the compensation point. It may be defined in terms of either depth or light level. When algae in the phytoplankton sink below the compensation point or are carried below it by currents, and do not soon return to the surface on upwelling currents, they die.

In some exceptionally clear ocean and lake waters, the compensation point may lie 100 m below the surface, but this is a rare condition. In productive waters with dense phytoplankton or in water turbid with suspended silt particles, the euphotic zone may be as shallow as 1 m.

The thermal environment

Much of the solar radiation absorbed by objects is converted to heat. The earth warms each day and cools by night. As the days lengthen and the sun rises higher in the sky toward summer, the environment becomes warmer; more heat is added each day than is lost. The warmth of the sun heating the atmosphere drives the winds. Light absorbed by water is the major source of heat for evaporation. Each object and each organism continually exchanges heat with its surroundings. When the temperature of the environment exceeds that of the organism, the organism gains heat and becomes warmer. When the environment is cooler, the organism loses heat and cools. The heat budget of an organism includes several avenues of heat gain and several of heat loss (Figure 2.16). Temperature reaches an equilibrium when gains equal losses. When gains exceed losses, energy is stored or accumulated in the body and temperature rises. When losses exceed gains, temperature falls.

The most important avenues of heat transfer between an organism and its environment are discussed in the following paragraphs.

Radiation is the absorption or emission of electromagnetic radiation. Sources of radiation in the environment include the sun, the sky (scattered

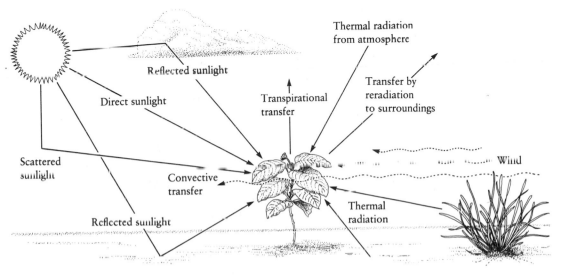

FIGURE 2.16 Pathways of exchange of heat between a plant and its environment by radiation, convection, and transpiration. After D. M. Gates, *Biophysical Ecology.* Springer-Verlag, New York (1980).

light), and the landscape. At night, objects that have warmed up in the sunlight radiate their stored heat to colder parts of the environment and, eventually, to space. Although we cannot see this infrared radiation, the bodies of organisms, especially warm-blooded birds and mammals, often are the "brightest" objects in the night. Because radiation increases with the fourth power of thermodynamic temperature (which is measured in kelvins — K — that is, °C + 273), we radiate tremendous quantities of energy to the clear, black night sky. We can also receive radiation from atmospheric water vapor and from vegetation, which balances some of our nighttime radiation loss.

Conduction is the transfer of the kinetic energy of heat between substances in contact. Thus a vacuum conducts no heat. Because of its greater density, water conducts heat more than 20 times faster than air. The rate of conductance between two objects, or between the inside and the outside of an organism, depends on the **insulation** of the surface (its resistance to heat transfer), its surface area, and the temperature gradient. An organism may either gain or lose heat by conduction, depending on its temperature relative to that of the environment.

Convection is the movement of liquids and gases of different temperatures, particularly over surfaces across which heat is transferred by conduction. Air conducts heat poorly. In still air, a **boundary layer** of air forms over a surface. A warm body tends to warm this boundary layer to its own temperature, effectively insulating itself against heat loss. A current of air flowing past a surface tends to disrupt the boundary layer and to increase the rate of heat exchange by conduction. This convection of heat away from the body surface is the basis of the "wind chill factor" we hear about on the evening weather report. That is to say, moving air cools like colder still air. For example, a wind blowing 32 kilometers per hour (km h^{-1}) at an air temperature of -7°C has the cooling power of still air at -23°C.

The total convective heat loss from a terrestrial organism exposed to wind varies in proportion to the one-third power of wind velocity, the two-thirds power of the organism's diameter, and the difference between the air and surface temperatures. Large animals lose heat more rapidly than small ones, but not in proportion to body mass, which varies in proportion to the cube of diameter. Thus the body temperatures of large organisms decrease less rapidly than those of small ones.

Evaporation requires heat. The evaporation of 1 g of water from the body surface removes 2.43 kilojoules (kJ) of heat at 30°C. As plants and animals exchange gases with the environment, some water evaporates from respiratory surfaces. The rate of evaporative heat loss depends on the amount of water exposed on the surface of the organism, the relative temperatures of the surface and the air, and the vapor pressure of the atmosphere. Like heat, moisture can be trapped in the boundary layer of air that forms around bodies, so convection increases evaporative as well as conductive heat loss. Warm air holds more water than cold air (Figure 2.17); hence it has greater potential for evaporating water than does cold air. In hot climates, water evaporating from the skin and respiratory surfaces cools many animals. For warm-blooded animals in cold climates, evaporation can become an unavoidable problem as

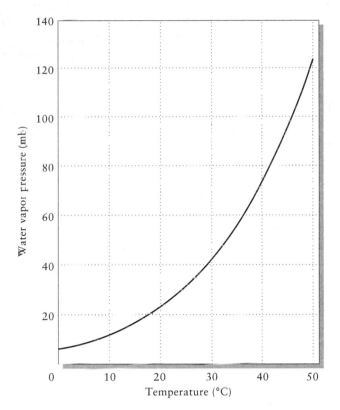

FIGURE 2.17 The relationship between temperature and water vapor pressure (in millibars, mb) of the atmosphere at saturation. Data from R. J. List, *Smithsonian Meteorological Tables* (6th ed.). Smithsonian Institution, Washington, D.C. (1966).

cold air containing little water warms in contact with the body surface. We see evidence of such water loss on winter days when water evaporated from the warm surfaces of the lungs condenses as our breath mingles with the cold atmosphere.

The buoyancy and viscosity of water and air

Because water is dense (800 times denser than air), it provides considerable support for organisms — which, after all, are themselves mostly water. But animals and plants also contain bone, proteins, dissolved salts, and other materials denser than salt water or fresh water. These materials would cause organisms to sink were it not for a variety of mechanisms that reduce their density or retard their rate of sinking. Many fish have a swim bladder, a small gas-filled structure whose size can be adjusted to make the density of the body equal to that of the surrounding water. Some large kelps, a type of seaweed found in shallow waters, have analogous gas-filled organs. The kelps are attached to the bottom by holdfasts, and gas-filled bulbs float their leaves to the sunlit surface waters.

Most fats and oils have densities between 0.90 and 0.93 g cm^{-2} (90 to 93 percent of the density of pure water). Many microscopic, unicellular plants (phytoplankton), which float in great numbers in the surface waters of lakes and oceans, contain droplets of oil that compensate for the natural tendency of cells to sink. The buoyancy of fish and other large marine organisms is also enhanced by accumulated lipids. Reduced skeleton, less musculature, and perhaps even the reduced salt concentration of body fluids further lighten the bodies of aquatic organisms. It has been argued that aquatic vertebrates maintain low osmotic concentrations in their blood and body fluids (about one-third to one-half that of seawater) because such low concentrations reduce density. Unlike bony fish, sharks and rays lack a swim bladder; the density of their bodies is reduced by their not depositing mineral salts in most of their skeleton. Calcium carbonate and calcium phosphate, the principal mineral components of bone, have densities close to three times that of water; the density of cartilage is closer to that of water.

The high viscosity of water helps to buoy up some organisms that would otherwise sink more rapidly, but it hampers the movement of others. Tiny marine animals often have long, filamentous appendages that retard sinking (Figure 2.18), just as a parachute slows the fall of a body through air. The "wings" of maple seeds, the spider's silk thread, and the tufts on dandelion and milkweed seeds serve a similar function and increase the dispersal range of terrestrial species. But to reduce the drag encountered in moving through a medium as dense and viscous as water, fast-moving animals must assume streamlined shapes. Mackerel and other swift fish of the open ocean closely approach the hydrodynamicist's body of ideal proportions. Of course, air offers far less resistance to movement, having less than $1/50$ the viscosity of water. But the atmosphere provides little buoyancy. To provide lift against the pull of gravity, birds and other flying organisms expend prodigious amounts of energy.

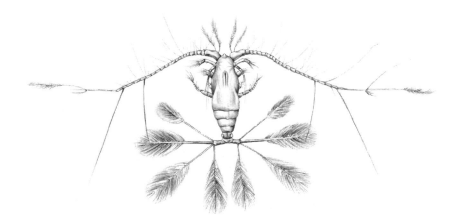

FIGURE 2.18 Filamentous and feathery projections from the body of the tropical marine planktonic crustacean *Calocalanus pavo*. Overall length is about 1.2 mm. After R. S. Wimpenny, *The Plankton of the Sea*. Faber and Faber, London (1966).

Allometry and the consequences of body size

Allometry is the study of the change in proportion of various parts of an organism as a consequence of growth. Size changes everything in ecology. In aquatic habitats, drag on the body depends on size and speed. Whereas a whale can coast on its momentum, a copepod stops dead in the water as soon as its power stroke ends; its momentum isn't sufficient to break through the viscosity of water! On land, the mechanical properties of moving appendages depend on their length; thus body size greatly affects locomotion. All these considerations influence the manner in which plants and animals adapt to the conditions of the environment.

Big organisms relate differently to their environments than do small ones because the scale of various of their physical processes is out of proportion to their size. In ecology, rates of processes and dimensions of objects are often related as follows:

$$Y = aX^b$$

where Y is being compared to X and a and b are constants pertaining to the relationship. Biologists refer to this equation as expressing an allometric relationship. Y might be the heart rate of an organism and X its body mass. The constant b is called the **allometric constant.** When b is 1, Y is directly proportional to X, and their relationship in this special case is referred to as **isometric.** When b exceeds 1, Y increases proportionately more rapidly than X, so the ratio of Y to X increases with larger X. When b is less than 1, the ratio of Y to X decreases with larger X; when b is less than 0, the absolute value of Y decreases with larger X. Figure 2.19 offers an example: the relationship between heart rate and body mass for mammals ranging over many orders of magnitude in size. The equation for the line that best fits the points has a slope, or allometric constant, of less than 0 ($b = -0.23$); this is in accordance with the slower heart rates of larger species. In contrast, the resting metabolic rates (RMR) of mammals increase with larger size, but less rapidly than their mass itself; the allometric constant of the relationship between RMR and body mass is 0.73 (Figure 2.20).

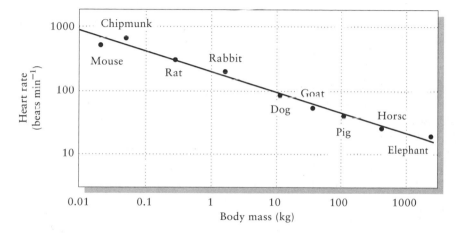

FIGURE 2.19 Allometric relationship between heart rate and body mass in a variety of mammals ranging in size from mouse to elephant. Data from P. L. Altman and D. S. Dittmer (eds.), *Biology Data Book.* Fed. Amer. Soc. Exp. Biol., Washington, D.C. (1964).

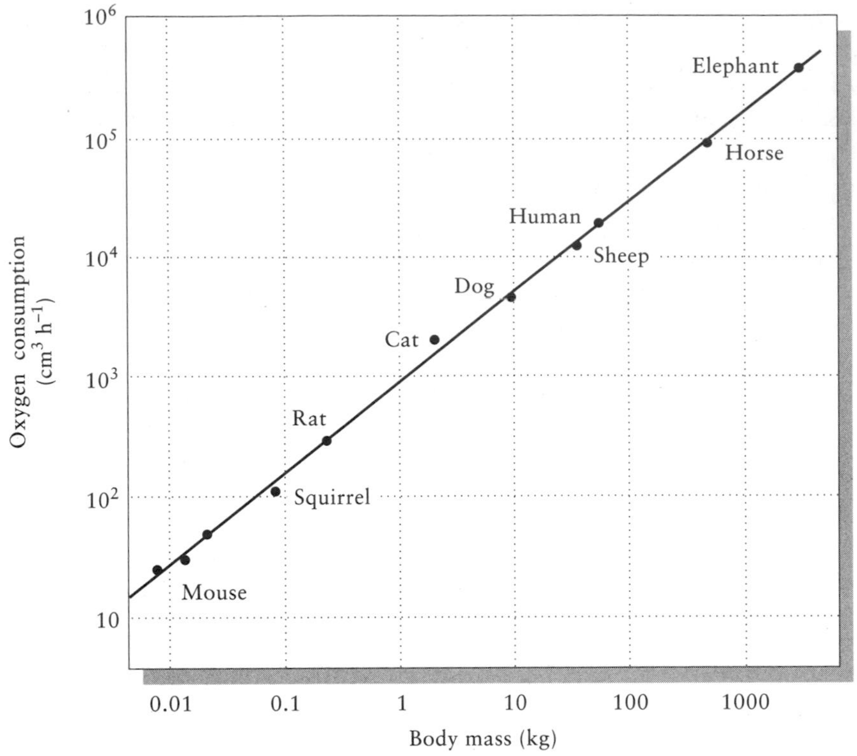

FIGURE 2.20 Allometric relationship between resting metabolic rate and body mass in several mammals. Data from P. L. Altman and D. S. Dittmer (eds.), *Biology Data Book*. Fed. Amer. Soc. Exp. Biol., Washington, D.C. (1964).

The allometry of certain relationships arises from simple physical or geometrical considerations. For example, the volume of a sphere increases in proportion to the cube of its diameter, but its surface area increases in proportion only to the square of the diameter. Hence the relationship between surface (S) and volume (V) has an allometric constant of $2/3$, or

$$S = aV^{2/3}$$

Therefore, larger organisms have relatively smaller surfaces compared to the bulk of their bodies. A sphere of radius r has a surface area of $4\pi r^2$ and a volume of $4\pi r^3/3$; the ratio of surface area to volume is $3/r$. Thus a sphere that has a diameter of 1 mm, which is comparable in size to a small water flea, has 1000 times as much surface per unit of volume as a bear-sized sphere 1 m in diameter. This makes the uptake of oxygen relatively easy for the smaller organism, but a high rate of heat loss would make it impossible for that small organism to maintain a high body temperature. In response to the need to obtain enough oxygen and to distribute it, larger species evolved vast, elaborate interior surfaces (lungs) and circulatory systems. But avoiding sudden changes in body temperature and conserving water pose less severe problems. In general, because large animals have great bulk per unit of surface compared with smaller animals, they are less vulnerable to changes in the environment.

Organisms are physical systems, and they must obey physical laws and operate within limits set by the physical environment. By expending energy, however, plants and animals can maintain themselves out of equilibrium with their physical surroundings. As we have seen, each type of habitat creates unique problems for life, and each type of organism devises, over evolutionary time, its own unique solutions. We shall consider some of these solutions in the next chapter.

🦋 SUMMARY

1. Water is the basic medium of life. It is liquid within the range of temperatures encountered over most of the earth, and it has an immense capacity to dissolve inorganic compounds. These properties, and its abundance at the earth's surface, make water an ideal medium for living systems.

2. Because water clings tightly to soil surfaces, its availability depends in part on the physical structure of the soil. Soils made up of small particles—clays, for example—hold water more strongly than do sandy soils. Most plants cannot remove water from the soil when the water potential exceeds 15 atm.

3. Most kinds of organisms cannot survive temperatures greater than 45°C, but thermophilic bacteria grow in hot springs at up to 95°C. They apparently tolerate such temperatures via increased intramolecular forces that hold proteins together.

4. Organisms in cold environments withstand freezing temperatures by metabolically maintaining elevated body temperatures; by lowering the freezing point of their body fluids with salts, glycerol, or glycoproteins; by supercooling their body fluids; or by tolerating freezing.

5. Biological energy transformations are based largely on the chemistry of carbon and oxygen. Energy is assimilated during photosynthesis as carbon is reduced and, in a coupled reaction, oxygen is oxidized from its form in water to molecular oxygen. The energy stored in carbohydrates is released by the oxidation of carbon to carbon dioxide (respiration).

6. Carbon dioxide is scarce in the atmosphere (0.03 percent) but is more abundantly distributed in aquatic systems, where it forms soluble carbonates. Oxygen, abundant in the atmosphere, is relatively scarce in water, where its solubility and rate of diffusion are low, and it may be depleted by bacterial respiration of organic matter (producing anoxic conditions) in deep, stagnant layers.

7. Organisms assimilate many elements necessary to life processes and biological structure. The availability of these elements relative to the amounts of them that are needed varies tremendously among habitats. The scarcity (relative to need) of nitrogen and phosphorus often limits plant growth.

8. In aquatic habitats, differences in the concentrations of dissolved salts establish osmotic gradients between the organism and its surroundings. Hyperosmotic organisms, which have greater salt

concentrations, tend to lose salt and gain water; hypo-osmotic organisms gain salt and lose water. Organisms achieve osmotic balance by altering the osmolarity of body fluids and actively pumping salts across membranes.

9. All the energy for life ultimately comes from sunlight. Solar radiation varies over a spectrum of wavelengths. Plants extract energy primarily in the high-intensity, short-wavelength portion of the spectrum, which roughly coincides with visible light.

10. Light is attenuated by water. The depth of the euphotic zone, at the bottom of which (at the compensation point) photosynthesis balances respiration, varies from 100 m in clear waters to a few tens of centimeters in turbid or polluted water.

11. The thermal environment of organisms, especially in terrestrial habitats, is determined by radiation, conduction, convection, and evaporation.

12. Water is denser than air and provides more buoyancy, but it is also more viscous and therefore impedes movement.

13. Many factors bearing on the ecology of organisms scale disproportionately with respect to overall body size. Such relationships are described by the power (allometric constant) of the equation relating a measurement of some structure or function to body size. Surface area scales to the 0.67 power of body size when shape is held constant; hence larger organisms have proportionately smaller surface-to-volume ratios. This and other allometric relationships have important consequences for organisms of different sizes.

☙ SUGGESTED READINGS

Brock, T. D. 1985. Life at high temperatures. *Science* 230:132–138.

Gates, D. M. 1965. Energy, plants, and ecology. *Ecology* 46:1–13.

Gates, D. M. 1971. *Man and His Environment: Climate.* Harper & Row, New York.

Hochachka, P. W., and G. N. Somero. 1984. *Biochemical Adaptation.* Princeton Univ. Press, Princeton.

Hutchinson, G. E. 1957. *A Treatise on Limnology, Vol. I: Geography, Physics, and Chemistry.* Wiley, New York.

Knoll, A. H. 1991. End of the Proterozoic eon. *Scientific American* 265:64–73. (The role of primitive organisms in modifying the early environment of the earth.)

Lovelock, J. E. 1988. *Ages of Gaia.* W. W. Norton, New York. (Describes the interdependence of organisms and their environments, and treats the entire biosphere as a superorganism; controversial.)

Peters, R. H. 1983. *The Ecological Implications of Body Size.* Cambridge Univ. Press, Cambridge.

Vogel, S. 1981. *Life in Moving Fluids. The Physical Biology of Flow.* Princeton Univ. Press, Princeton.

ADAPTATION TO AQUATIC AND TERRESTRIAL ENVIRONMENTS

The physical environment includes many factors that are important to the well-being of the organism. In this chapter, we shall explore a variety of mechanisms that animal and plant species have evolved in response to the need to adjust to several prominent environmental factors. In particular, plant and animal species have evolved ways to control the movement (flux) of heat and various substances across their surfaces. These include seeking appropriate external environments, adjusting the internal environment in response to surrounding conditions, and modifying the quality and area of the body surface. Such modifications, which better suit the organism to its particular environment, are called adaptations.

A common theme, then, revolves around manipulation of the variables that affect flux — surface area, conductance, and concentration gradient (Figure 3.1) — but this theme manifests itself differently in adaptations to each type of environmental factor. Furthermore, the relationship of an organism to any one factor depends on its relationship to all others. This point may be illustrated by the coupled problems of regulating salt balance and water balance, with which we shall begin our discussion.

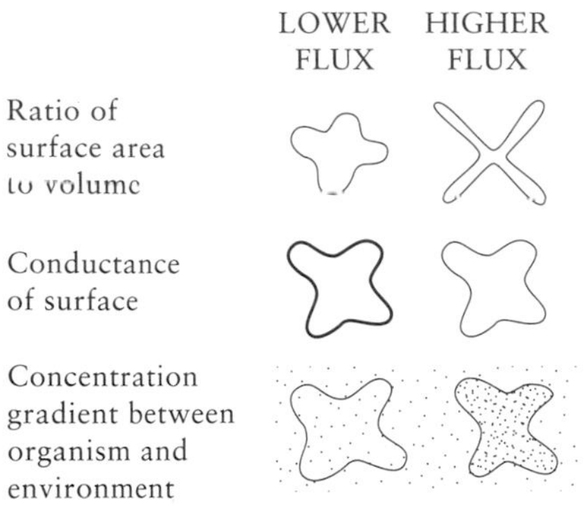

LOWER HIGHER
FLUX FLUX

Ratio of
surface area
to volume

Conductance
of surface

Concentration
gradient between
organism and
environment

FIGURE 3.1 The influence of surface area, conductance, and gradient on the flux of materials or energy between the organism and the environment.

○ Salt   • Urea

Salt excreted
by gills and kidneys

FIGURE 3.2 Schematic diagram illustrating the role of urea in balancing the osmotic potential of the blood of sharks and rays with that of the surrounding seawater.

Salt balance and water balance

Left to their own devices, ions diffuse across cell membranes from regions of high concentration to regions of low concentration, thereby tending to equalize these concentrations. Water also moves across membranes (osmosis) toward regions of high ion concentrations, tending to dilute dissolved substances. Maintaining an ionic imbalance between the organism and the surrounding environment (osmoregulation) against the physical forces of diffusion and osmosis requires energy and often is accomplished by organs specialized for salt retention or excretion.

Ion retention is critical to terrestrial and freshwater organisms. Freshwater fish, for example, continuously gain water by osmosis through the mouth and gills, which are the most permeable tissues exposed to the surroundings. (The skin is relatively impermeable.) To counter this influx, individuals eliminate water in the urine. But, of course, if fish did not also selectively retain dissolved ions, they would soon become lifeless bags of water. The kidneys of freshwater fish retain salts by actively removing ions from the urine and feeding them back into the bloodstream. In addition, the gills can selectively absorb ions from the surrounding water and secrete them into the bloodstream. Terrestrial animals acquire mineral ions in the water they drink and the food they eat, although sodium deficiencies in some areas force animals to obtain salt directly from such mineral sources as salt licks; plants absorb ions dissolved in soil water.

Some marine sharks and rays exhibit a rather elegant solution to the problem of osmotic balance. Sharks retain urea $[CO(NH_2)_2]$—a common nitrogenous waste product of metabolism in vertebrates—in the bloodstream instead of excreting it in the urine (Figure 3.2). The urea raises the ionic concentration of the blood to the level of seawater without there being any increase in the concentration of sodium chloride. Although sharks and rays must regulate the diffusion of specific ions into and out of their bodies, high levels of urea in the blood effectively cancel the tendency of water to leave by osmosis. And inasmuch as sharks do not have to drink water to replace osmotic losses of water, they also do not ingest large quantities of salt. The fact that the few freshwater species of sharks and rays do not accumulate urea in their blood emphasizes the osmoregulatory role of urea in marine species.

The small copepod *Tigriopus* lives in pools high in the splash zone along rocky coasts. The pools receive fresh seawater infrequently, and as the water evaporates, the salt concentration rises to high levels. *Tigriopus* solves its water loss problem by increasing the osmotic potential of its body fluids. It accomplishes this by synthesizing certain amino acids abundantly. These small molecules increase the osmotic potential of the body to match that of the habitat, without the deleterious physiological consequences of high levels of salt.

Water balance and salt balance are intimately related in terrestrial as well as aquatic organisms. As we have seen, plants transpire hundreds of grams of water for every gram of dry matter produced. Plants inevitably take up salts along with the water that passes into their roots. In saline environments, plants actively pump excess salts back into the soil across the root surfaces, which therefore function as the plant's "kidneys," or secrete them from the surfaces of their leaves (Figure 3.3). Most animals obtain more salts in their food than

FIGURE 3.3 (a) The roots of mangrove vegetation immersed in salt water. Some species excrete excess salt from their roots. (b) Specialized glands in the leaves of the button mangrove *Conocarpus erecta* excrete salt, which precipitates on their outer surfaces.

they need. Where water abounds, such animals can drink large quantities of water to flush out salts that would otherwise accumulate in the body. Where water is scarce, however, animals must produce a concentrated urine to conserve water. And so, as one would expect, desert animals have champion kidneys. For example, whereas humans can concentrate salt ions in their urine to about 4 times the level in blood plasma, the kangaroo rat's kidneys produce urine with a salt concentration as high as 14 times that of its blood, and the Australian hopping mouse, a desert-adapted species, 25 times that of its blood.

Nitrogen excretion

Most carnivores, regardless of whether they eat crustacea, fish, insects, or mammals, consume excess nitrogen in their diets. This nitrogen, ingested in the form of proteins and nucleic acids, must be eliminated from the body when these compounds are metabolized (Figure 3.4). Most aquatic organisms produce the simple metabolic by-product ammonia (NH_3). Although ammonia is mildly poisonous to tissues, aquatic organisms eliminate it rapidly in a copious, dilute urine, or directly across the body surface, before it reaches a dangerous concentration. Terrestrial animals, on the other hand, cannot afford to use large quantities of water to excrete nitrogen. Instead, they produce protein metabolites that are less toxic than ammonia and therefore can accumulate in the blood and urine without danger. In mammals, this waste product is urea [$CO(NH_2)_2$], the same substance that sharks and rays produce and retain to achieve osmotic balance in marine environments. Because urea dissolves in water, excreting it requires some urinary water loss — how much depends on the concentrating power of the kidneys. Birds and reptiles have carried adaptation to terrestrial life one step further: they excrete nitrogen in the form of uric acid ($C_5H_4N_4O_3$), which crystallizes out of solution and then may be excreted as a paste in great concentration in the urine.

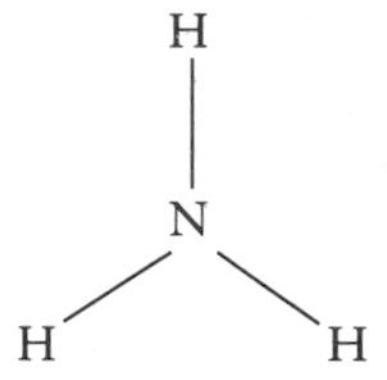

Temperature and water conservation

When air temperature approaches or exceeds the maximum permissible body temperature, individuals can dissipate heat only by evaporating water from their skin and respiratory surfaces. In deserts, the scarcity of water makes evaporative heat loss a costly mechanism; organisms often must reduce activity, seek cool microclimates, or undertake seasonal migrations to cooler regions (Figure 3.5). Many desert plants orient their leaves in such a way as to

Ammonia

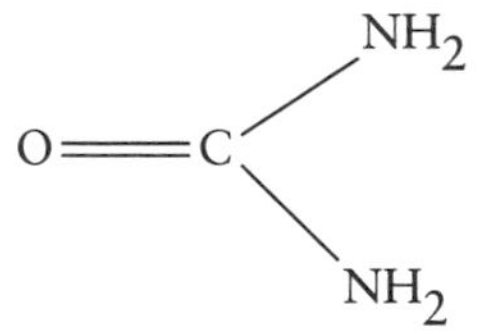

Urea

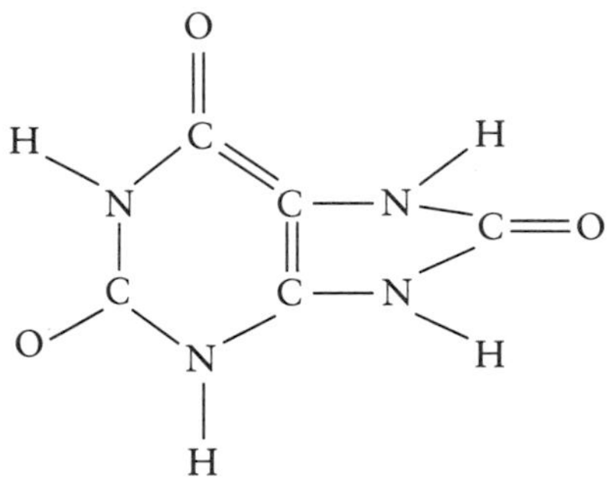

Uric acid

FIGURE 3.4 Chemical structures of common nitrogenous waste products.

FIGURE 3.5 A jackrabbit seeking refuge from the hot sun of southern Arizona in the shade of a mesquite tree. The large ears and long legs of desert-inhabiting jackrabbits effectively radiate heat when the environment is cooler than the body. Courtesy of the U.S. Fish and Wildlife Service.

FIGURE 3.6 The behavior and physiological features of the kangaroo rat adapt it exquisitely to desert environments. Courtesy of the U.S. Fish and Wildlife Service.

avoid the direct rays of the sun; others shed their leaves and become inactive during periods of combined heat and water stress.

Among mammals, the kangaroo rat is well suited to life in a nearly waterless environment (Figure 3.6). Its large intestine resorbs water from waste material so efficiently as to produce virtually dry feces. Kangaroo rats also recover much of the water that evaporates from the lungs via condensation in enlarged nasal passages. When the kangaroo rat inhales dry air, moisture in its nasal passages evaporates, cooling the nose and saturating the inhaled air with water. When moist air is exhaled from the lungs, much of its water condenses on the cool nasal surfaces. By alternating condensation with evaporation during breathing, the kangaroo rat minimizes its respiratory water loss.

Kangaroo rats avoid the desert's greatest heat by restricting their feeding to the cooler nighttime periods and by spending the daylight hours below ground in relatively cool, humid burrows. In sharp contrast, ground squirrels remain active during the day. They conserve water by restricting evaporative cooling. As a consequence, their body temperatures rise when they forage above ground, exposed to the hot sun. But before their body temperatures become dangerously high, they return to their cool burrows, where they dissipate heat by conduction and radiation rather than by evaporation. Thus, by shuttling back and forth between their burrows and the surface, ground squirrels extend their activity into the heat of the day and pay a relatively small price in water loss. Like the ground squirrel, the camel's body temperature also rises in the heat of the day — by as much as 6°C. Large body size gives the camel a distinct advantage, however. Because of its low surface-to-volume ratio, the camel heats so slowly that it can remain in the sun most of the day. Excess heat is dumped at night by conduction and radiation to the cooler surroundings.

Plant–water relationships

Water is held in the soil surrounding roots by a variety of forces, whose strengths may be expressed in terms of **water potentials,** or pressures. When water sticks tightly to the surfaces of soil particles, it is referred to as **hygroscopic water** and is held with a pressure of at least −50 bars (the bar, abbreviated b, is approximately equal to atmospheric pressure at sea level). By convention, water potentials, which are the converse of osmotic potentials, are given negative values. The larger the negative value, the greater the attraction for water. Water within the spaces between soil particles is held by capillary attraction of water molecules to one another, which generates forces as low as about −0.15 b. Water in large pores in the soil and at great distance from the surfaces of soil particles, held with pressures of less than −0.15 b, usually drains through the soil under the pull of gravity. Plants first tap the soil water that is held with the least potential; it is easiest to extract. But as soils dry out, the remaining water is held with greater and greater water potential, eventually causing water stress. As a general rule, plants cannot obtain water held with pressures more negative than −15 b, the so-called permanent wilting point of the soil.

Plants obtain water from the soil by osmosis, so the ability of the roots to take up water depends on their osmotic potential. Osmotic potential, in turn, is a function of the concentration of dissolved molecules (including sugars) and ions in the cell sap. A 1 molar (M) solution, for example, has an osmotic potential of about 22 b (that is, a water potential of −22 b). By manipulating the osmotic potential of root cells, plants can alter their ability to remove water from soil. Plants growing in dry areas and salty areas have been shown to increase the water potential of their roots to as much as −60 b by increasing the concentrations of amino acids, carbohydrates, or organic acids in cells. They pay a high metabolic price, however, to maintain such concentrations of dissolved substances.

Plants conduct water to their leaves through xylem elements, the empty remains of cells connected end-to-end, which function pretty much as pipes to conduct water over long distances (Figure 3.7). For water to move from the root to the leaves, the water potential in the xylem must be more negative than that in the surrounding epidermal and cortex cells of the roots into which water enters from the soil. Similarly, the water potential of the leaves must exceed that of the roots enough to draw water upward against the osmotic potential of the roots, the pull of gravity, and the resistance of the xylem. Leaves generate water potential by transpiring water to the atmosphere. Dry air has a water potential of as much as −1000 b, which is more than enough to do the job, given reasonably low resistance to water flow in the stem.

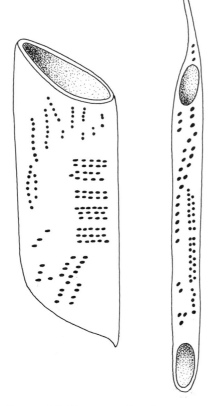

FIGURE 3.7 Two examples of narrow vessel members in the xylem of oak trees conduct water through the stem. Individual vessel elements are the walls of wood cells from which the living matter has disappeared. The cells are arranged vertically in the wood, and water flows from cell to cell through openings in the ends. After K. Esau, *Anatomy of Seed Plants.* Wiley, New York (1960).

C₄ and CAM photosynthesis

Plants require water primarily to replace losses that accompany gas exchange with the atmosphere. Because the atmosphere contains little carbon dioxide (0.03 percent), the concentration gradient for water loss exceeds that for carbon dioxide assimilation by several orders of magnitude. And because the vapor pressure of water increases with temperature, heat magnifies the problem of water loss.

Heat- and drought-adapted plants have anatomical and physiological modifications that reduce transpiration across plant surfaces, reduce heat loads, and enable plants to tolerate high temperatures. When plants absorb sunlight, they heat up. Plants can minimize overheating by increasing their surface area for heat dissipation and protecting their surfaces from direct sunlight with dense hairs and spines (Figure 3.8). Spines and hairs also produce a still boundary layer of air that traps moisture and reduces evaporation. Plants may further reduce transpiration by covering their surfaces with a thick, waxy cuticle that is impervious to water and by recessing the stomates in deep pits, often themselves filled with hairs (Figure 3.9).

In some plants, the biochemical mechanism of carbon assimilation also displays modifications that conserve water. Most species in environments with adequate water assimilate the carbon in carbon dioxide into an organic molecule in a single biochemical step in a pathway known as the Calvin cycle (Figure 3.10). We can represent this step as

$$CO_2 + RuBP \rightarrow 2\ PGA$$

where RuBP (ribulose bisphosphate) is a 5-carbon compound and PGA (phosphoglycerate) is a 3-carbon compound. (Because the immediate product of carbon assimilation is a 3-carbon compound, biologists call this mechanism **C₃ photosynthesis**.) Several biochemical steps beyond the production of PGA,

FIGURE 3.8 Cross section (a) and surface view (b) of the pubescent leaf of the desert perennial herb *Enceliopsis argophylla*. Courtesy of J. R. Ehleringer. From J. R. Ehleringer, in E. Rodriguez, P. Healy, and I. Mehta (eds.), *Biology and Chemistry of Plant Trichomes*. Plenum Press, New York (1984), pp. 113–132.

FIGURE 3.9 In oleander, a drought-resistant plant, stomates lie deep within hair-filled pits on the leaf's undersurface (magnified about 500 times). The hairs reduce water loss by slowing air movement and trapping water. Courtesy of M. V. Parthasarathy.

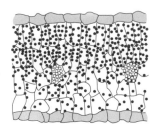

(a)

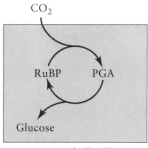

Mesophyll cell

(b)

FIGURE 3.10 (a) An idealized cross section of a leaf of a C_3 plant, illustrating the general dispersion of chloroplasts throughout the mesophyll. (b) The major steps of the Calvin cycle of carbon dioxide assimilation. Courtesy of J. R. Ehleringer.

the Calvin cycle regenerates 1 molecule of RuBP while making 1 carbon atom available for the synthesis of glucose. Inasmuch as glucose contains 6 carbons, each molecule of glucose produced requires 6 turns of the Calvin cycle.

The enzyme responsible for assimilation of carbon — RuBP carboxylase — has a very low affinity for carbon dioxide. As a result, at the concentration of CO_2 in the atmosphere and its resulting concentration in the plant mesophyll cell, plants assimilate carbon inefficiently. To achieve high rates of assimilation, plants must pack their cells with large amounts of RuBP carboxylase. Furthermore, this enzyme facilitates the oxidation of RuBP in the presence of high oxygen and low carbon dioxide concentrations, especially at elevated leaf temperatures. That is, oxidation of RuBP partially undoes what RuBP carboxylase accomplishes when it reduces carbon, thus making photosynthesis inefficient. In fact, carbon assimilation tends to inhibit itself unless CO_2 achieves a high concentration in the cell. Plants can accomplish this only when their stomatal resistance remains low — which, of course, leads to high water loss.

Many plants in hot climates exhibit a modification of C_3 photosynthesis that involves (1) an additional step in the assimilation of carbon dioxide and (2) the spatial separation of the initial assimilation step and Calvin cycle pathways within the leaf. Biologists call this modification **C_4 photosynthesis** because the assimilation of CO_2 initially results in a 4-carbon compound:

$$CO_2 + PEP \rightarrow OAA$$

where PEP (phosphoenol pyruvate) contains 3 carbons and OAA (oxaloacetic acid) contains 4. The assimilatory reaction is catalyzed by the enzyme PEP carboxylase, which, unlike RuBP carboxylase, has a high affinity for CO_2. Assimilation occurs in the mesophyll cells of the leaf, but in most C_4 plants, photosynthesis (including the Calvin cycle) takes place in specialized cells surrounding the leaf veins (Figure 3.11). These are called bundle sheath cells. Oxaloacetic acid diffuses into the bundle sheath cells, where it breaks down to produce CO_2 and pyruvate, a 3-carbon compound. The pyruvate moves back into the mesophyll cells, where enzymes convert it to PEP to complete the carbon assimilation cycle.

The CO_2 released by metabolism of OAA in the bundle sheath cells enters the Calvin cycle, as it does in C_3 plants. C_4 photosynthesis confers an advantage because CO_2 can be concentrated within the bundle sheath cells to a level, determined by the concentration of OAA, that far exceeds its equilibrium established by diffusion from the atmosphere. At this higher concentration, the Calvin cycle operates more efficiently. Because the enzyme PEP carboxylase has a high affinity for CO_2, it can bind CO_2 at a lower concentration in the cell, thereby allowing the plant to increase stomatal resistance and reduce water loss.

Certain succulent plants in desert environments use the same biochemical pathways as C_4 plants but segregate CO_2 assimilation and the Calvin cycle between day and night. The discovery of this arrangement in plants of the family Crassulaceae (the stonecrop family; sedum is one example) and the initial assimilation and storage of carbon dioxide in certain 4-carbon organic acids (malic acid and OAA), led to the name **crassulacean acid metabolism,** or **CAM.**

CAM plants open their stomates for gas exchange during the cool desert night, at which time transpiration of water is minimal. CAM plants initially assimilate CO_2 in the form of **4-carbon** OAA and malic acid, which the leaf tissues store in high concentration in vacuoles within the cells (Figure 3.12). During the day, the stomates of the leaves close, and the stored organic acids are gradually recycled to release CO_2 to the Calvin cycle. The assimilation of CO_2 and the regeneration of PEP are regulated by enzymes that have different temperature optima. CAM photosynthesis results in extremely high water use efficiencies and enables some types of plants to exist in habitats too hot and dry for other, more conventional species.

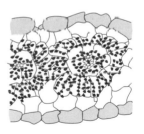

(a)

The procurement of oxygen by animals

Most animals release the chemical energy contained in organic compounds primarily by oxidative metabolism. The biochemical pathways involved — the converse of photosynthesis in many respects — collectively constitute respiration. Because oxygen plays such an important role in releasing energy, its lack of availability in the environment can restrict metabolic activity, particularly in aquatic habitats where low solubility and slow diffusion limit oxygen concentrations. Even for terrestrial organisms, which breathe an atmosphere containing abundant oxygen, the delivery of oxygen to tissues requires its solution and transport through the aqueous medium of the body.

Diffusion can satisfy the oxygen needs of tiny aquatic organisms, but the centers of organisms larger than about 2 mm in diameter are too far from the external environment for diffusion to ensure a rapid supply of oxygen. In insects, conduit systems (tracheae) carry air directly to the tissues. Other animals have blood circulatory systems to distribute oxygen from the respiratory surfaces to the body. To increase oxygen-carrying capacity, the blood of most animals contains complex protein molecules, such as **hemoglobin,** to which oxygen molecules readily attach for transport. When oxygen binds to hemoglobin — 4 molecules of oxygen for each molecule of hemoglobin — oxygen leaves the solution, making room for the diffusion of more oxygen into the blood across the blood–gas boundary. At low oxygen concentrations, the binding process reverses and releases oxygen to the tissues. Though plasma itself carries only limited oxygen in solution (about 1 percent by vol-

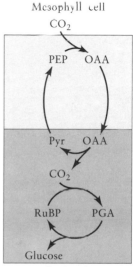

(b)

FIGURE 3.11 (a) An idealized cross section of a leaf of a C_4 plant, illustrating the concentration of chloroplasts in cells of the bundle sheath (Kranz anatomy). (b) The major steps of carbon dioxide assimilation, including the transport of carbon from the mesophyll to the bundle sheath. Courtesy of J. R. Ehleringer.

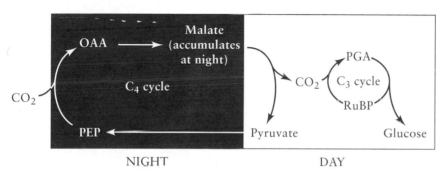

FIGURE 3.12 Outline of the photosynthetic pathway in CAM plants. After J. B. Harbourne, *Introduction to Ecological Biochemistry.* Academic Press, New York (1982).

ume, or 14 ppm by weight), whole blood transports up to 50 times more oxygen bound to oxygen-carrying molecules.

Active organisms require an abundant supply of oxygen for cell respiration. Hemoglobin and other oxygen-carrying pigments, such as hemocyanin, act as go-betweens for the uptake of oxygen from the surrounding environment and its release to the cells. The functions of uptake and release conflict, however, because the hemoglobin molecule that readily binds oxygen in the lungs or gills also holds oxygen tenaciously when it must unbind to supply active tissues.

The compromise between the oxygen-uptake and oxygen-release functions of hemoglobin, or other blood pigments, expresses itself in the **oxygen dissociation curve** (Figure 3.13). The dissociation curve portrays the amount of oxygen bound to hemoglobin, expressed as a percentage of the total possible (saturation), in relation to the concentration of oxygen in the blood plasma. Oxygen binds to hemoglobin reversibly, the percentages of bound and unbound oxygen reaching an equilibrium. As oxygen diffuses into the blood plasma in the gills or lungs, the equilibrium shifts to the right, and additional oxygen binds to hemoglobin. As active tissues deplete the blood of oxygen, the equilibrium shifts toward the left, and oxygen detaches from the oxyhemoglobin complex. In general, the proportion of hemoglobin that carries bound oxygen increases with greater concentration of dissolved oxygen in the blood plasma until the hemoglobin becomes saturated.

Tissues receive oxygen at a rate proportional to the difference between the partial pressures of the oxygen in the tissues and in the blood: the larger the difference, the higher the rate of diffusion. When dissolved oxygen leaves the bloodstream to enter the tissues, the partial pressure of oxygen in the plasma decreases. As the difference between oxygen concentrations in the blood and in the tissues decreases, diffusion slows. But as the oxygen concen-

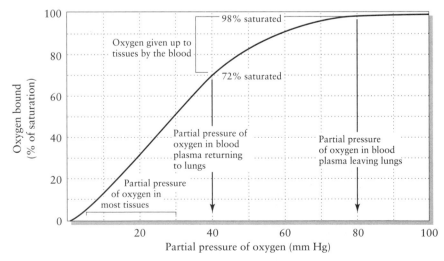

FIGURE 3.13 The oxygen dissociation curve for human hemoglobin. The blood releases about 25 to 30 percent of the oxygen bound to hemoglobin to the tissues, which have oxygen tensions of 5 to 30 mm Hg.

tration in the blood plasma decreases, hemoglobin releases some bound oxygen, thereby tending to restore the oxygen concentration. Thus oxygen drawn from the blood plasma into the tissues is partially replenished by that bound to hemoglobin.

Small changes in the amino acid sequence of the hemoglobin molecule may modify its oxygen-binding capacity with respect to both the availability of oxygen in the environment and its requirement by the organism. Because small animals have high oxygen demands, their oxygen dissociation curves are generally shifted to the right so that more oxygen is released. The dissociation curve of a mammalian embryo is to the left of that of its mother, because its oxygen supply derives from the mother's blood, which has a lower partial pressure of oxygen than the air she breathes.

The total amount of hemoglobin in the bloodstream is also subject to adjustment. For example, the blood of the sedentary goosefish has a total oxygen capacity of 5 percent by volume; that of the more active mackerel is 16 percent. This difference reflects the hemoglobin concentration in the blood and parallels the sizes of the gills in the two species, the ratio of surface area of the gills to body weight in the mackerel being 50 times that of the goosefish.

Adaptations for procuring oxygen illustrate a set of solutions to the problems that organisms confront at the interface between themselves and their environments. The design of the hemoglobin molecule, which influences simultaneously its oxygen-binding and its oxygen-releasing properties, further emphasizes the fact that adaptation often requires compromise.

Countercurrent circulation

A recurrent mechanism in aquatic organisms that enhances the uptake of dissolved oxygen from water is **countercurrent circulation,** an arrangement of the structure of gills whereby water and blood flow in opposite directions (Figure 3.14). In a countercurrent system, as blood picks up oxygen from the water

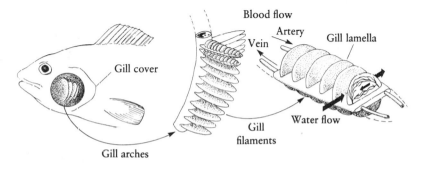

FIGURE 3.14 A fish's gill consists of several gill arches, each of which carries two rows of filaments. The filaments bear thin lamellae (leaflike structures) oriented in the direction of the flow of water through the gill. Within the lamellae, blood flows in the opposite direction to the movement of water past the surface. From D. J. Randall, *Amer. Zool.* 8:179–189 (1968).

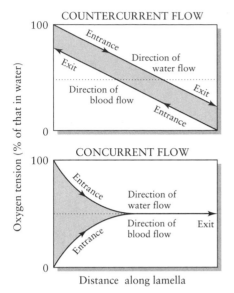

FIGURE 3.15 Changes in oxygen tension of blood and water in countercurrent and concurrent systems. The former maintains a constant gradient for oxygen to diffuse across; as a result, the oxygen tension of the blood leaving the system approaches that of the incoming water. After K. Schmidt-Nielsen, *Animal Physiology*. Cambridge Univ. Press, New York (1975).

flowing past, it comes into contact with water that has progressively greater oxygen concentration, because the water has flowed past a progressively shorter distance of the gill lamella (Figure 3.15). With this arrangement, the oxygen concentration of the blood plasma can approach the concentration in the surrounding water. If blood and water were to flow together through the gill (concurrent circulation), an equilibrium oxygen concentration would eventually be established with equal, intermediate levels in the blood and water flowing past the gill. The countercurrent system keeps the blood and water out of equilibrium and maintains a constant gradient across which oxygen can flow.

The countercurrent principle appears frequently in adaptations that increase the flux of heat or materials across surfaces. Among terrestrial organisms, birds have a unique lung structure that, unlike the lungs of mammals, results in a one-way flow of air opposite to the flow of blood. This adaptation allows birds to achieve, with lungs whose weight and volume are small enough for flight, the high rates of oxygen delivery required by their active lives.

Uptake of soil nutrients

Plants acquire mineral nutrients — nitrogen, phosphorus, potassium, calcium, and others — from dissolved forms of these elements in soil water. In the case of abundant elements that diffuse rapidly in the soil solution, such as calcium and magnesium, uptake is limited primarily by the absorptive capacity of the root. Plants compensate for reduced levels of a nutrient in the soil by increasing the extent of the root system — that is, the absorptive surface of the roots — and by active uptake. In laboratory experiments, barley and beet roots passively took up phosphorus by diffusion when its concentration in the water surrounding the root exceeded a certain critical level, 0.2 to 0.5 millimolar (mM). Under these conditions, phosphorus had a lower concentration in root tissues than in the soil solution. At soil concentrations below 0.2 to 0.5 mM, roots actively transported phosphorus across their surfaces and concentrated the element within the root cortex. This active absorption requires the expenditure of energy as the root tissue moves ions against a concentration gradient.

Plants may also respond to the decreased availability of soil nutrients by increasing root growth at the expense of shoot growth (Figure 3.16). This brings the nutrient requirements of the plant into line with nutrient availability by reducing the nutrient demand created by the leaves, by increasing the absorptive surface area of the root system, and by growing roots into new areas of soil from which the plant has not already removed scarce minerals.

Crop plants and wild species growing on fertile soils have a great capacity to absorb nutrients across their root surfaces and to vary their growth rates in response to variations in soil nutrient levels. Species adapted to nutrient-poor soils are much more conservative. They cope with low nutrient availability by allocating a large fraction of their biomass to roots; by establishing symbiotic relationships with fungi, which enhance mineral absorption; and by growing slowly and retaining leaves for long periods, thereby reducing nutrient demand. Such species typically cannot respond to artificially increased nutrient levels by increasing their growth rates. Instead, their roots absorb more nutrients than the plant requires and store them for subsequent use when the soil nutrient availability declines.

Optimum temperature

Unlike consumable resources, such as carbon dioxide, oxygen, and soil nutrients, temperature influences organisms thermodynamically through its effects on the rates at which physical and biochemical processes proceed and on the structures of such biologically important molecules as proteins and lipids. As a rule, within the temperature range between the freezing point of water (0°C) and the upper limit for most life forms (40 to 50°C), higher temperature quickens the pace of life by increasing the **kinetic energy** of the organism and its surroundings. Temperature affects physical processes, such as diffusion and evaporation, as well as biochemical reactions. For example, increased temperature speeds the diffusion of gases through the shell of a bird's egg and that of mineral ions through the soil solution. Temperature also influences the shapes of enzymes and other proteins, because the kinetic energy imparted to such molecules can break the weak forces holding the structure together. Each enzyme functions best over a narrow range of temperature determined by the amino acid sequences of protein chains and the way they affect the bonds that hold proteins in certain configurations. These sequences are, of course, subject to evolutionary modification.

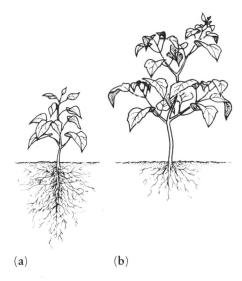

(a)   (b)

FIGURE 3.16 Different strategies of allocation between root tissue and shoot tissue during the development of a plant. Allocation often favors roots when supplies of soil nutrients or water are limited (a); in the absence of these restrictions, larger shoots allow more rapid growth (b).

Temperature adaptation is strikingly revealed by comparing organisms taken from environments with different characteristic temperatures. When we examine the performances of such organisms over a range of temperatures, each typically exhibits its maximum activity at a temperature corresponding to that of its normal environment. Many fish in the freezing oceans surrounding Antarctica swim just as actively as, and consume oxygen at a rate comparable to, fish living among tropical coral reefs (Figure 3.17). Put a tropical fish in cold water, however, and it becomes sluggish and soon dies; conversely, antarctic fish cannot tolerate temperatures warmer than 5 to 10°C.

How can fish from cold environments swim as actively as fish from the tropics? Metabolism consists of a series of biochemical transformations, most of which are catalyzed by enzymes. Because a given transformation occurs more rapidly at high temperature than at low temperature, the compensation observed in cold-adapted organisms must involve either a quantitative increase in the amount of substrate (substance catalyzed) or in the amount of enzyme that catalyzes each step, or a qualitative change in the enzyme itself. Laboratory studies of the function of isolated enzymes have shown, in many cases, that a particular enzyme obtained from a variety of organisms exhibits different catalytic properties when tested over ranges of temperature, pH, salt concentration, and substrate abundance. (Catalytic ability is often expressed as a constant abbreviated K_m). For example, the temperature dependence of the catalytic function of the lactate dehydrogenase enzymes (LDH) of three species of barracudas (*Sphyraena*) results in similar levels of activity in the dif-

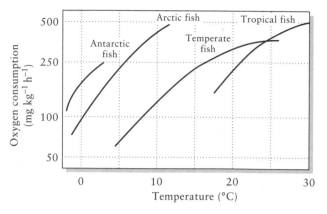

FIGURE 3.17 Temperature compensation in the rates of oxygen consumption of fish from different thermal regimes. Although metabolism in each increases with temperature, the curves are offset such that fish from different thermal environments have similar metabolic rates at the temperatures that prevail in their native habitats. After P. W. Hochachka and G. N. Somero, *Biochemical Adaptation*. Princeton Univ. Press, Princeton (1984).

ferent species within the different ranges of temperatures normally encountered (Figure 3.18).

This picture of metabolic compensation is greatly simplified, of course. Adapting to changes in the environment requires a complete adjustment of metabolic pathways, which may involve changes in enzyme structure or concentration or the use of alternative metabolic pathways. The example of temperature adaptation and all the other examples considered in this chapter emphasize the unity of organism structure and the interrelatedness of all facets of structure and function.

☙ *SUMMARY*

Adaptations to the physical environment involve the modification of the variables that regulate flux across an organism's surface and the adjustment of metabolic pathways with respect to conditions prevailing in the internal environment.

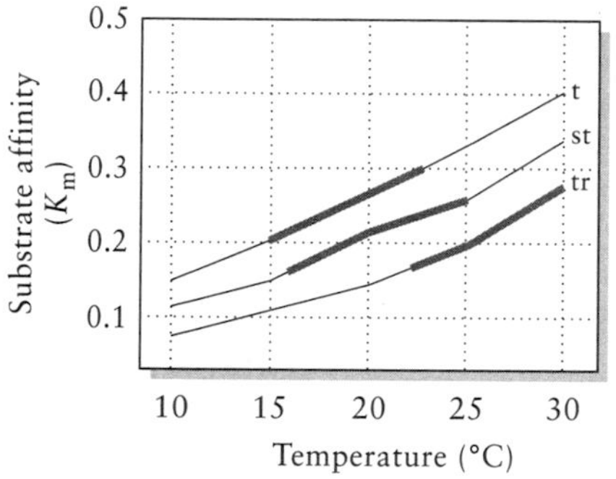

FIGURE 3.18 Relationship between substrate affinity (which is proportional to activity) and temperature for the LDH enzymes of three species of barracudas (*Sphyraena*) from temperate (t), subtropical (st), and tropical (tr) waters. Because of temperature compensation, the substrate affinities of the three species are nearly identical within the normal temperature ranges of their environments. After P. W. Hochachka and G. N. Somero, *Biochemical Adaptation*. Princeton Univ. Press, Princeton (1984).

1. To maintain salt balance and water balance, marine organisms whose internal environment is hypo-osmotic actively exclude salts; freshwater organisms, which are hyperosmotic, retain salts while excreting the water that continuously diffuses into the body; terrestrial organisms minimize water loss in part by concentrating salts and nitrogenous waste products in their urine.

2. Nitrogenous waste products of protein metabolism are excreted as ammonia by most aquatic organisms, as urea by mammals, and as uric acid by birds and reptiles. Because uric acid crystallizes out of solution, birds and reptiles may excrete it at high concentrations and thereby gain considerable economy of water use.

3. Water stress increases with temperature. In dry environments, animals seek cool microclimates, and plants increase the stomatal resistance of their leaves. Such responses uniformly reduce productivity in return for enhanced survival.

4. Plants draw water from the soil by osmotic potential in their roots, and from the roots to the leaves by water potential generated by the evaporation of water from leaves. Resistance to water conduction from the soil through the root and xylem to the leaves limits the supply of water to the leaves.

5. During photosynthesis, plants assimilate carbon through a reaction (the C_3 pathway) catalyzed by the enzyme RuBP carboxylase. This enzyme has a low affinity for carbon dioxide and brings about oxidation at high temperatures, resulting in low efficiency. Plants adapted to high temperature interpose a more efficient (C_4) carbon assimilation step, which is spatially separated from the C_3 reactions (Calvin cycle) in the leaf. In desert environments, some plants separate carbon assimilation and the Calvin cycle reactions into nighttime and daytime phases (CAM photosynthesis).

6. Oxygen diffuses through tissues too slowly to meet the metabolic needs of large animals. This problem is overcome either by conducting air directly to the tissues via a multibranched tracheal system (as in insects) or by transporting oxygen dissolved in circulating fluids. The low solubility of oxygen in water is compensated for by oxygen-binding proteins, such as hemoglobin, that take oxygen out of solution and transport it at high concentrations.

7. The uptake of oxygen by aquatic organisms is greatly facilitated by countercurrent circulation of blood through the gills in a direction opposite to that of water flowing over the gill surfaces. In this way, countercurrent circulation maintains high gradients of oxygen concentration, and the blood can achieve nearly the oxygen concentration of the surrounding water.

8. Plants obtain dissolved nutrients from soil water by passive diffusion when nutrients are abundant and by active uptake when they are scarce. When the concentrations of soil nutrients are low, plants may increase their total root surface area at the expense of shoot growth. Plants also may store nutrients when they are abundant.

9. Although most physical and biological processes are accelerated at higher temperatures, the temperature optima of metabolic processes may be adjusted to match the characteristic temperature of the environment by alteration in the structure and quantity of key enzymes.

Overall, adaptation to the physical environment depends on reaching compromises between opposing functions to both ensure the individual's survival and maximize its productivity in a particular environment.

❧ **SUGGESTED READINGS**

Chapin, F. S., III. 1991. Integrated responses of plants to stress. *BioScience* 41:29–36.

Ehleringer, J. R., R. F. Sage, L. B. Flanagan, and R. W. Pearcy. 1991. Climate change and the evolution of C$_4$ photosynthesis. *Trends in Ecology and Evolution* 6:95–99.

Feldman, L. J. 1988. The habits of roots. *BioScience* 38:612–618.

Hochachka, P. W., and G. N. Somero. 1984. *Biochemical Adaptation.* Princeton Univ. Press, Princeton.

Karov, A. 1991. Chemical cryoprotection of metazoan cells. *BioScience* 41:155–160.

Kramer, P. J. 1983. *Water Relations of Plants.* Academic Press, New York.

Lee, R. E., Jr. 1989. Insect cold-hardiness: to freeze or not to freeze. *BioScience* 39:308–313.

Louw, G. 1982. *Ecology of Desert Organisms.* Longman, New York.

Schmidt-Nielsen, K. 1990. *Animal Physiology. Adaptations and Environment* (4th ed.). Cambridge Univ. Press, London and New York.

Schulze, E.-D., R. H. Robichaux, J. Grace, P. W. Rundel, and J. R. Ehleringer. 1987. Plant water balance. *BioScience* 37:30–37.

Vogel, S. 1988. *Life's Devices.* Princeton Univ. Press, Princeton.

CLIMATE, TOPOGRAPHY, AND SOILS

The physical environment varies widely over the surface of the earth. Conditions of temperature, light, substrate, moisture, salinity, soil nutrients, and other factors have shaped the distribution and adaptations of plants, animals, and microbes. Climate zones vary widely, and even within them, such geological factors as topography and composition of bedrock further differentiate the environment on finer spatial dimensions. This chapter describes some of the important patterns of variation in the physical environment that underlie diversity in the biological components of the ecosystem.

The surface of the earth, its waters, and the atmosphere above it behave like a giant heat-transforming machine. Climate patterns originate as the earth absorbs the energy in sunlight and as winds and ocean currents redistribute this energy over the globe as heat. As the surface varies from bare rock to forested soil, open ocean, and frozen lake, its ability to absorb sunlight varies as well, thus creating differential heating and cooling. The heat energy absorbed by the earth eventually radiates back into space, after undergoing further transformations that perform the work of evaporating water and contributing to the

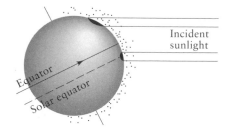

(a)

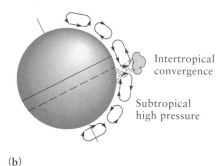

(b)

FIGURE 4.1 Influence of solar radiation on latitudinal patterns in climate. (a) The warming effect of the sun is greatest at the solar equator. (b) Warm, moist air rises in the tropics, which results in abundant rainfall. Cool, dry air descends to the surface in subtropical latitudes. The intertropical convergence refers to the latitudinal belt at the solar equator within which surface winds converge from the north and south.

circulation of the atmosphere and oceans. All these factors have created a great variety of physical conditions that, in turn, have fostered the diversification of ecosystems.

Global patterns in temperature and rainfall

The earth's climate tends to be cold and dry toward the poles and hot and wet toward the equator. Although there are many exceptions to this generalization, climate does exhibit broadly defined patterns. Global variation in climate depends on the position of the sun relative to the surface of the earth (Figure 4.1a). The sun warms the atmosphere, oceans, and land most when it lies directly overhead. A beam of sunlight spreads over a greater area when the sun approaches the horizon, and it also travels a longer path through the atmosphere, where much of its energy either is reflected or is absorbed by the atmosphere and reradiated into space as heat. The sun's highest position each day varies from directly overhead in the tropics to near the horizon in polar regions; thus the warming effect of the sun increases from the poles to the equator.

Warming air expands, becomes less dense, and tends to rise. As air heats up, its ability to hold water vapor increases and evaporation quickens; as we have seen, the rate of evaporation from a wet surface nearly doubles with each 10°C rise in temperature.

Because warm, tropical air can hold much more water than the cooler temperate and arctic air, annual precipitation is greatest in tropical regions (Figure 4.2). The tropics are wet not because there is more water in tropical latitudes than elsewhere but because water cycles more rapidly through the tropical atmosphere. The heating effect of the sun causes water to evaporate; the energy input, not the quantity of water, primarily determines latitudinal patterns in rainfall. The placement of continental land masses exerts a secondary effect. At a given latitude, rain falls more plentifully in the Southern Hemisphere because oceans and lakes cover a greater proportion of its surface (81 percent compared with 61 percent of the Northern Hemisphere). Water evaporates more readily from exposed surfaces of water than from soil and vegetation.

Energy from the sun drives the winds, distributing water vapor through the atmosphere. Indeed, wind patterns strongly influence precipitation. The mass of warm air that rises in the tropics eventually spreads to the north and south in the upper layers of the atmosphere. It is replaced from below by surface-level air from subtropical latitudes. The tropical air mass that rises under the warming sun cools as it radiates heat back into space. By the time this air has extended to about 30° north and south of the **solar equator** (the parallel of latitude that lies directly under the sun), the cooled air mass becomes denser and begins to sink back to the surface (Figure 4.1b). Condensation has already removed much of its water as precipitation in the tropics, and its capacity to evaporate and hold water increases further as it sinks and warms. As the air mass strikes the ground in subtropical latitudes and spreads to the north and south, it draws moisture from the land, creating zones of arid climate centered at latitudes of about 30° north and south of the equator (Figure 4.3). The great deserts — the Arabian, Sahara, Kalahari, and Namib of Africa; the Atacama of South America; the Mohave, Sonoran, and Chihuahuan of North America; and the Australian — all fall within these belts.

Major land masses create some exceptions to this pattern. Mountains force air upward, causing it to cool and lose its moisture as precipitation on the windward side of the range. As the air descends the leeward slopes and travels across the lowlands beyond, it picks up moisture and creates arid environments called **rain shadows** (Figure 4.4). The Great Basin deserts of the western United States and the Gobi Desert of Asia lie in the rain shadows of extensive mountain ranges.

The interior of a continent usually experiences less precipitation than its coasts, simply because the interior lies farther from the major site of water evaporation, the surface of the ocean. Furthermore, coastal (**maritime**) climates vary less than interior (**continental**) climates because the great heat-storage capacity and vertical mixing of water reduce temperature fluctuations. For example, the hottest and coldest mean monthly temperatures near the Pacific coast of the United States at Portland, Oregon, differ by only 16°C. Farther inland, this range increases to 18°C at Spokane, Washington; 26°C at Helena, Montana; and 33°C at Bismarck, North Dakota.

Ocean currents play a major role in moving heat over the surface of the earth. In large ocean basins, cold water circulates toward the tropics along the western coasts of the continents, and warm water circulates toward temperate latitudes along the eastern coasts of the continents (Figure 4.5). The cold Humboldt Current moving north from the Antarctic Ocean along the coasts of Chile and Peru creates cool, dry environments along the west coast of South America right to the equator. Conversely, the warm Gulf Stream emanating from the Gulf of Mexico carries a mild climate far to the north into western Europe and the British Isles.

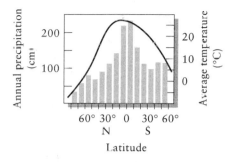

FIGURE 4.2 Average annual precipitation (vertical bars) and temperature (line) for 10° latitudinal belts within continental land masses. The illustration presents averages for many localities and so obscures the great variation within each latitudinal belt.

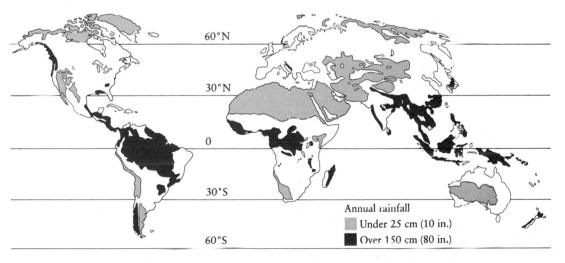

FIGURE 4.3 Distribution of the major deserts (regions with less than 25 cm annual precipitation) and wet areas (regions with more than 150 cm annual precipitation).

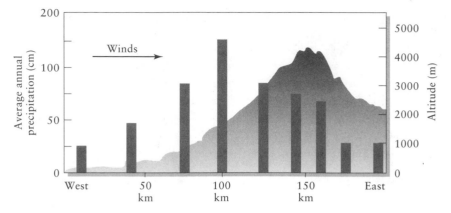

FIGURE 4.4 Influence of the Sierra Nevada mountain range on local precipitation and in causing a rain shadow to the east. Weather comes predominantly from the west across the central valley of California. As moisture-laden air is deflected upward by the mountains, it cools and its moisture condenses, resulting in heavy precipitation on the western slope of the mountains. As the air rushes down the eastern slope, it warms and begins to pick up moisture, creating arid conditions in the Great Basin. After E. R. Pianka, *Evolutionary Ecology* (4th ed.). Harper & Row, New York (1988).

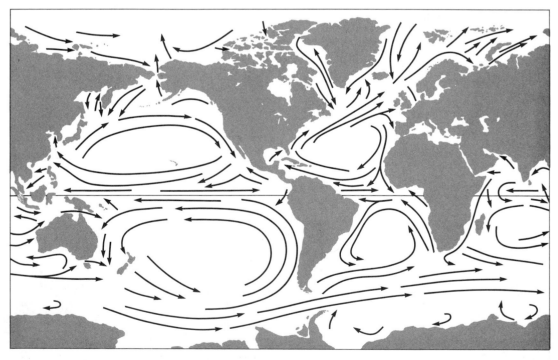

FIGURE 4.5 The major ocean currents. Water movement generally proceeds clockwise in the Northern Hemisphere and counterclockwise in the Southern Hemisphere. After A. C. Duxbury, *The Earth and Its Oceans*. Addison-Wesley, Reading, Mass. (1971).

Seasonal variation in climate

Patterns of change in climate influence plants and animals as much as do long-term averages of temperature and precipitation. Periodic cycles in climate follow astronomical cycles: the rotation of the earth upon its axis causes daily periodicity; the revolution of the moon around the earth creates lunar cycles in the amplitude of the tides; and the revolution of the earth about the sun brings seasonal change.

The equator is tilted slightly with respect to the path the earth follows in its orbit around the sun. As a result, the Northern Hemisphere receives more solar energy than the Southern Hemisphere during the northern summer, less during the northern winter. The seasonal range in temperature increases with distance from the equator (Figure 4.6). At high latitudes in the Northern Hemisphere, mean monthly temperatures vary by an average of 30°C, with extremes of more than 50°C annually, the mean temperatures of the warmest and coldest months in the tropics differ by as little as 2 or 3°C.

Latitudinal patterns in the seasonality of rainfall result in part from the seasonal northward and southward movement of belts of wet and dry climate. Seasonality of rainfall is most pronounced in broad latitudinal belts lying about 20° north and south of the equator. As the seasons change, these regions alternately come under the influence of the solar equator, bringing heavy rains, and subtropical high-pressure belts, bringing clear skies (Figure 4.7).

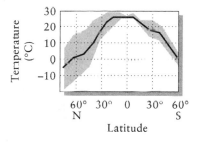

FIGURE 4.6 Annual range of mean monthly temperatures (shaded area) as a function of latitude. The mean annual temperature is indicated by the line.

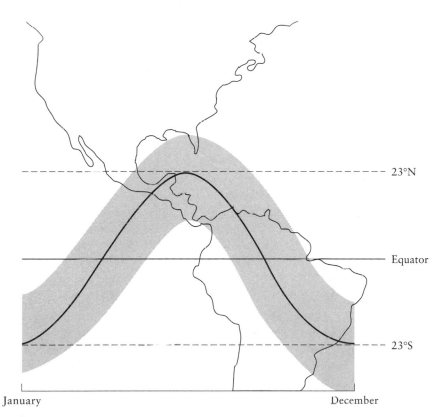

FIGURE 4.7 The seasonal latitudinal course of the intertropical convergence (see Figure 4.1b) results in two seasons of heavy precipitation on the equator and a single wet season alternating with a pronounced dry season at the edges of the tropics.

Panama, at 10°N, lies within the wet tropics, but even there the seasonal movement of the solar equator profoundly influences the climate. The major tropical belt of high rainfall remains south of Panama during most of the northern winter, but it lies directly overhead during the northern summer. Hence the winter is dry and windy, the summer humid and rainy. Panama's climate is wetter on the northern (Caribbean) side of the isthmus—the direction of prevailing winds—than on the southern (Pacific) side; mountains intercept moisture coming from the Caribbean side of the isthmus and produce a rain shadow. The Pacific lowlands are so dry during the winter months that most trees lose their leaves (Figure 4.8). Tinder-dry forests and bare branches contrast sharply with the wet, lush, more typically tropical forest that abounds during the wet season.

Farther to the north, at 30°N in central Mexico, rainfall comes only during the summer when the solar equator reaches its northward limit (Figure 4.9c). During the rest of the year this region falls within the dry, subtropical high-pressure belt. Summer rainfall extends north into the Sonoran Desert of southern Arizona and New Mexico (Figure 4.9b). This area also receives moisture during the winter from the Pacific Ocean, carried by the southwesterly winds emanating from the subtropical high-pressure belt farther south. Southern California lies beyond the summer-rainfall belt and has a winter-rainfall, summer-drought climate often referred to as a **Mediterranean climate** (Figure 4.9a).

The sun warms the seas just as it does the continents and the atmosphere, but the ocean's great mass of water acts like a heat sink to dampen daily and seasonal fluctuations in temperature. Where ocean temperature does change seasonally, it reflects seasonal movements of water masses of different temperature more often than it does local heating and cooling. During the Panamanian dry season, roughly January to April, steady winds blowing in a southwesterly direction create strong **upwelling** currents in the Pacific Ocean along the southern and western coasts of Central America. During the upwelling period, winds blow warm surface water away from the coast, where cooler water moves upward from deeper regions to replace it. As a result, water temperature varies annually three times as much on the Pacific coast of Panama as on the Caribbean coast.

Seasonal cycles in temperate lakes

Small temperate-zone lakes respond quickly to the changing seasons (Figure 4.10). In winter a typical lake has an inverted **temperature profile;** that is, the coldest water (0°C) lies at the surface, just beneath the ice. (Because the density of water increases between the freezing point and 4°C, the warmer water within this range sinks, and temperature increases to as much as 4°C toward the bottom of a lake.) In early spring the sun warms the lake surface gradually. But until the surface temperature exceeds 4°C, the sun-warmed surface water tends to sink into the cooler layers immediately below. This **vertical mixing** distributes heat throughout the water column, resulting in a uniform temperature profile. Without thermal layering to impede mixing, winds cause deep vertical movement of water in early spring (**spring overturn**), bringing nu-

FIGURE 4.8 Many trees on the Pacific slope of Panama shed their leaves during the dry season, which lasts from January through April. Photograph by M. A. Guerra, courtesy of the Smithsonian Tropical Research Institute.

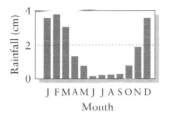

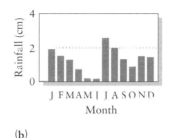

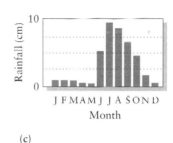

(a)   (b)   (c)

FIGURE 4.9 Seasonal occurrence of rainfall at three localities in western North America. (a) The winter rain and summer drought of the Pacific coast (Mediterranean climate type). (b) The combined climate pattern of the Sonoran Desert. (c) The summer rainy season of the Chihuahuan Desert.

trients to the surface from regions of nutrient release in bottom sediments and bringing oxygen from the surface to the depths.

Later in spring and early summer, as the sun rises higher each day and the air above the lake warms, surface layers of water gain heat faster than deeper layers, creating a zone of rapid temperature change at intermediate depth, called the **thermocline,** across which water does not mix. Now, at temperatures above 4°C, the warmer surface water literally floats on the cooler water below, a condition known as **stratification.** The depth of the thermocline varies with local winds and with the depth and turbidity of the lake. It may occur anywhere between 5 and 20 m below the surface; lakes less than 5 m deep usually lack stratification.

The thermocline demarcates an upper layer of warm water (the **epilimnion**) and a deep layer of cold water (the **hypolimnion**). Most of the primary production of the lake occurs in the epilimnion, where sunlight is intense. Oxygen produced by photosynthesis supplements oxygen entering the lake at its surface, keeping the epilimnion well aerated and thus suitable for animal life, but plants often deplete dissolved mineral nutrients and, in doing so, curtail their own production. The thermocline isolates the hypolimnion from the surface of the lake; animals and bacteria that remain below the euphotic zone of photosynthesis deplete the water of oxygen, creating **anaerobic** conditions. Thus, during late summer, temperate lakes may become biological deserts, lacking nutrients to support plant growth in surface waters and lacking oxygen to support animal life in the depths.

During the autumn, surface layers of the lake cool more rapidly than deeper layers and, becoming heavier than the underlying water, begin to sink. This vertical mixing (**fall overturn**) persists into late fall, until the temperature at the lake surface drops below 4°C and winter stratification ensues. Fall overturn causes greater vertical mixing of water than spring overturn, because temperature differences in the lake during summer stratification exceed those during winter stratification. Fall overturn speeds the movement of oxygen to deep waters and pushes nutrients to the surface. Where the hypolimnion becomes fairly warm in midsummer, deep vertical mixing may take place in late summer

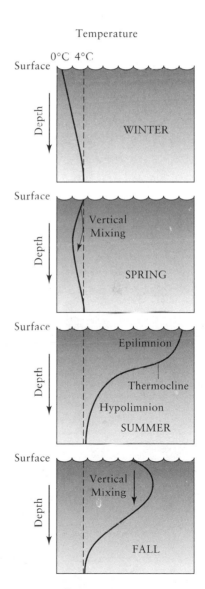

FIGURE 4.10 Seasonal changes in the temperature profile of a temperate lake.

when temperatures remain favorable for plant growth. Infusion of nutrients into surface waters at this time often causes an explosion in the population of phytoplankton — the **fall (autumn) bloom**. In deep, cold lakes, vertical mixing does not penetrate to all depths until late fall or early winter, when water temperatures are too cold to support plant growth.

Irregular fluctuations in climate

Everyone knows that weather is difficult to forecast far in advance. We often remark that a certain year was particularly dry or cold compared to others. Most aspects of climate seem unpredictable. Rainfall varies most where it is sparsest: in deserts and during the driest season in wetter localities. Year-to-year variation in temperature on a particular date is greatest where temperature fluctuates most during the year. The most extreme conditions occur infrequently, but these may affect organisms disproportionately.

Some events — earthquakes, tornadoes, volcanic eruptions, and hurricanes — create conditions that exceed the capacities of organisms to respond and can devastate the biological community. The rich Peruvian fishing industry, as well as some of the world's largest seabird colonies (Figure 4.11),

FIGURE 4.11 Nesting colony of Peruvian boobies on an island off the coast of Peru. The dense population depends on the anchovy stocks in the rich Humboldt Current. Photograph by R. C. Murphy, courtesy of the Department of Library Services, American Museum of Natural History.

thrives on the abundant fish in the rich waters of the Humboldt (Peru) Current, a mass of cold water that flows up the western coast of South America and finally veers offshore at Ecuador, north of which warm, tropical inshore waters prevail. Each year a warm countercurrent known as El Niño ("little boy" in Spanish, a name referring to the Christ child and chosen because this countercurrent appears around Christmas time) moves down the coast of Peru, sometimes strongly enough to force the cold Humboldt Current offshore, taking with it the food supply of millions of birds.

During "normal" years between El Niño events, a steady wind blows across the equatorial central Pacific Ocean from an area of high atmospheric pressure centered over Tahiti to an area of low pressure centered over Darwin, Australia. An El Niño event appears to be triggered by a reversal of these pressure areas (the so-called Southern Oscillation) and of the winds that flow between them. As a result, the westward-flowing equatorial currents stop or even reverse; upwelling off the coast of South America weakens or ceases; and warm water — the El Niño current — piles up along the coast of South America. Historical records of atmospheric pressure at Tahiti and Darwin, and of sea-surface temperature on the Peruvian coast, reveal pronounced ENSO (El Niño – Southern Oscillation) events at irregular intervals of 2 to 10 years (Figure 4.12).

The climatic and oceanographic effects of an ENSO event extend over much of the world, affecting ecosystems in such distant areas as India, South Africa, Brazil, and western Canada. A major ENSO event in 1982–1983 disrupted fisheries and destroyed kelp beds in California, caused reproductive failure of seabirds in the central Pacific Ocean, and resulted in widespread mortality of coral in Panama.

Topographical and geological causes of local variation

Variations in topography and geology can create variations in the environment within regions of uniform climate. In hilly areas, the slope of the land and its exposure to the sun influence the temperature and moisture content of the

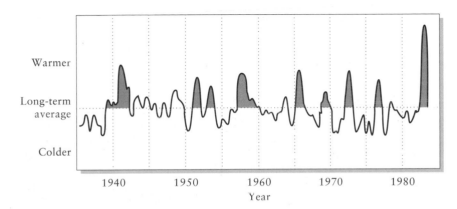

FIGURE 4.12 Major occurrences of the El Niño–Southern Oscillation (shaded peaks) are marked by large positive anomalies in sea-surface temperature in South American coastal waters. From E. M. Rasmussen, *Amer. Sci.* 73:168–177 (1985).

soil. Soils on steep slopes drain well, often causing moisture stress for plants on the hillside at the same time that water saturates the soils of nearby lowlands. In arid regions, stream bottomlands and seasonally dry riverbeds may support well-developed riparian forests, which accentuate the contrasting bleakness of the surrounding desert. In the Northern Hemisphere, south-facing slopes directly face the sun, whose warmth and drying power limit vegetation to shrubby, drought-resistant (**xeric**) forms. The adjacent north-facing slopes remain relatively cool and wet and harbor moisture-requiring (**mesic**) vegetation (Figure 4.13).

Air temperature decreases with altitude by about 6°C for each 1000-m increase in elevation. This decrease in temperature, which is caused by the expansion of air with the lower atmospheric pressure at higher altitude, is referred to as **adiabatic cooling**. Climb high enough, even in the tropics, and you will encounter freezing temperatures and perpetual snow. Where the temperature at sea level averages 30°C, freezing temperatures are reached at about 5000 m, the approximate altitude of the snow line on tropical mountains.

In north-temperate latitudes, a 6°C drop in temperature corresponds to the temperature change encountered over an 800-km increase in latitude. In many respects, the climate and vegetation of high altitudes resemble those of sea-level localities at higher latitudes. But despite their similarities, **alpine** environments usually vary less from season to season than their low-elevation

FIGURE 4.13 The influence of exposure on the vegetation of a series of mountain ridges near Aspen, Colorado. The cool and moist north-facing (left-facing) slopes permit the development of spruce forest. Shrubby, drought-resistant vegetation grows on the south-facing slopes.

counterparts at higher latitudes. Temperatures in tropical montane environments remain nearly constant and frost-free over the year, which makes it possible for many tropical plants and animals to live in the cool environments found there.

In the mountains of the southwestern United States, changes in plant communities with elevation result in more or less distinct belts of vegetation, which the nineteenth-century naturalist C. H. Merriam referred to as **life zones.** Merriam's scheme of classification included five broad zones, which he named, from low to high elevation (or from south to north), the Lower Sonoran, Upper Sonoran, Transition, Canadian (or Hudsonian), and Alpine (or Arctic – Alpine).

At low elevations in the North American Southwest, one encounters a cactus and desert-shrub association characteristic of the Sonoran Desert of northern Mexico and southern Arizona (Figure 4.14). In the woodlands along stream beds, plants and animals have a distinctly tropical flavor. Many hummingbirds and flycatchers, ring-tailed cats, jaguars, and peccaries make their only temperate-zone appearances in this area. In the Alpine zone, 2600 m higher, one finds a landscape resembling the tundra of northern Canada and Alaska. Thus, by climbing 2600 m, one experiences changes in climate and vegetation that would require a journey to the north of 2000 km or more at sea level.

Local variation in the bedrock underlying a region promotes the differentiation of soil types and enhances biotic heterogeneity. In the northern Appalachian Mountains and in mountains near the Pacific coast of the United States, outcrops of **serpentine** (a kind of igneous rock) weather to form soils having so much magnesium that plants characteristic of surrounding soil types cannot grow. Serpentine **barrens,** as they are called, support little more than a sparse covering of grasses and herbs (Figure 4.15), many of which are distinct **endemics** (species found nowhere else) that have evolved a high tolerance of magnesium. Depending on the composition of the bedrock and on the rate of weathering, granite, shale, and sandstone also can produce barrens. The extensive pine barrens of southern New Jersey, where mature trees attain no more than waist height in some areas, occur on a large outcrop of sand, which produces a dry, acid, infertile soil.

Physical characteristics of the soil and of the underlying rock also influence drainage and the soil's ability to hold moisture. For example, the extensive pine forests of the coastal plain of the southeastern United States grow on sandy soils that drain too well for most broad-leaved trees. Climate, too, plays an important role in the weathering of rock and the formation of soils. In temperate and arctic regions of the Northern Hemisphere, glaciation during the last 100,000 years has influenced soil characteristics over vast areas, scraping some areas to bare bedrock while burying others under tens of meters of wind-blown glacial dust, or **loess.**

The landscape concept

Variations in topography and soils within a region create a heterogeneous mosaic of habitat "patches" that make up the local **landscape.** The concept of landscape in ecology emphasizes environmental variation perceived over di-

Hudsonian Zone
Elevation 2500 m

Alpine Zone
Elevation 3500 m

Upper Sonoran Zone
Elevation 1500 m

Transition Zone
Elevation 2000 m

Lower Sonoran Zone
Elevation 900 m

Upper Sonoran Zone
Elevation 1200 m

FIGURE 4.14 Vegetation at different elevations in the mountains of southeastern Arizona. The Lower Sonoran Zone supports mostly saguaro cactus, small desert trees such as paloverde and mesquite, numerous annual and perennial herbs, and small succulent cacti. Agave, ocotillo, and grasses are conspicuous elements of the Upper Sonoran Zone, and oaks appear toward its upper edge. Large trees predominate at higher elevations: ponderosa pine in the Transition Zone, spruce and fir in the Hudsonian Zone. These gradually give way to bushes, willows, herbs, and lichens in the Alpine Zone above the tree line. Courtesy of the U.S. Soil Conservation Service, the U.S. Forest Service, W. J. Smith, and R. H. Whittaker. From R. H. Whittaker and W. A. Niering, *Ecology* 46:429–452 (1965).

FIGURE 4.15 A small serpentine barren in eastern Pennsylvania. The soils surrounding the barren support a forest of oak, hickory, and beech.

mensions of tens of meters to kilometers and stresses the importance of the movement of individuals and materials between habitat patches to the maintenance of ecological processes over the entire landscape. Thus populations are subdivided into local populations residing in patches of suitable habitat. Their persistence depends not only on processes within each patch but also on the movement of individuals between patches. As a consequence, the organization of the landscape strongly influences population dynamics, a principle to which we shall return later.

The exchange of materials between habitats makes the landscape a functional level of organization; **landscape ecology** is dedicated to its study. The importance of the habitat mosaic becomes clear when we consider the movement of individuals between habitat types (for example, breeding in one habitat and feeding in another) and the movement of materials between habitat types (for example, leaves falling into streams).

Human activities alter landscape patterns as well as individual habitats. Cutting trees not only reduces the area of forest but also creates a mosaic of forest and field patches. The sizes and proximity of these patches influence the movements of individuals between suitable areas and the maintenance of their populations within the region. If patches become small enough, a population may become extinct despite the presence of suitable habitat within the landscape. We shall consider landscape topics later. At this point, it is enough to understand the impact of habitat heterogeneity on local ecosystems and populations. In this instance, the landscape is equal to more than the sum of its habitat parts.

Integrated descriptions of climate

Climate factors such as temperature, humidity, precipitation, wind, and solar radiation interact in their effect on life. For example, seasonal rainfall promotes plant growth more strongly during warm months than during cold months, wind and solar radiation interact with temperature to determine thermal stress, and temperature and humidity together influence water balance.

Ecologists have devised a variety of techniques to describe combinations of climate factors. One such approach is the **climograph,** which portrays seasonal changes in temperature and precipitation simultaneously. Figure 4.16 shows, for example, that the seasons at Colon, on the Caribbean coast of Panama, bring marked variation in rainfall but little change in temperature, just the reverse of the situation in New York City. Only during July and August in New York and April in Colon do the climates of the two localities resemble each other. Continuing to examine this climograph, we find that as we move east to west across North America from Cincinnati, Ohio, to Winnemucca, Nevada, climate becomes more arid but temperatures remain within the same range. The change in vegetation from deciduous, broad-leaved forest in Ohio to short-grass prairie in Wyoming and desert shrubland in Nevada thus depends on the water relations of different plant forms rather than on temperature tolerance.

Although San Diego's weather in January resembles that of Lincoln, Nebraska, in April, the overall climates differ as much as their vegetation. San Diego's Mediterranean climate, with hot, dry summers and cool, moist winters, favors slow-growing, drought-resistant shrubs (chaparral), whereas Lincoln's wet summers and cold, dry winters favor the development of tall-grass prairie.

The climograph provides a useful basis for comparing localities, but it fails to combine the effects of temperature and precipitation in a biologically meaningful way, and it does not reveal cumulative effects of weather. For example, during dry seasons, both direct evaporation and **transpiration** (the evaporation of water from leaves) remove water from the soil. When rainfall

FIGURE 4.16 Climographs of representative localities in North and Central America. Each point represents the mean temperature and precipitation for one month. Lines connecting the months at each locality trace seasonal change in climate.

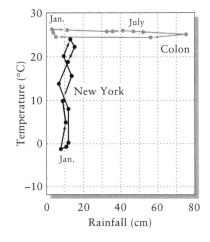

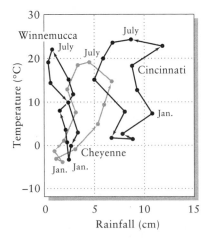

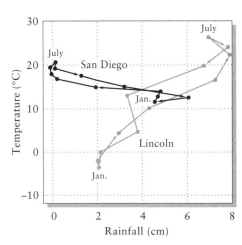

does not balance evaporation and transpiration losses, the water deficit in the soil steadily increases, perhaps for months at a time; soil water reflects last month's conditions as well as more recent ones.

We can estimate the seasonal availability of water in the soil by comparing the rate at which plants and evaporation draw water from the soil with the rate at which precipitation restores it. The method described here was developed by the geographer C. W. Thornthwaite, who used seasonal temperatures and inputs of water to the soil to estimate the sum of evaporation and transpiration, or **evapotranspiration.** Agronomists and ecologists have demonstrated that total plant growth within a habitat is directly proportional to the level of evapotranspiration. Evaporation and transpiration increase with temperature by a factor of nearly 2 for each 10°C rise in temperature, other things being equal, although the character of the soil and vegetation cover also influence water loss. Frequently, the availability of water in the soil limits evapotranspiration. Under such conditions, **potential evapotranspiration,** the amount of water that would be drawn from soil in the absence of any moisture limit, can be estimated from temperature and precipitation. The realized evapotranspiration, limited by precipitation input, is known as the **actual evapotranspiration (AE).** Because potential evapotranspiration increases with temperature, temperature and water stress go hand in hand. Thus boreal regions receiving 25 to 50 cm of precipitation each year have more favorable "water budgets" for plant production than tropical regions with similar levels of precipitation.

When precipitation input to the soil (rainfall minus surface runoff) exceeds potential evapotranspiration at all seasons, the soil remains saturated with water throughout the year, as at Brevard, North Carolina (Figure 4.17). At Bar Harbor, Maine, precipitation falls below potential evapotranspiration during the warm summer months, so the soil lacks available water during late summer and early fall. Canton, Mississippi, receives about the same amount of rain as Bar Harbor, but the hotter climate increases the potential evaporation of water, resulting in serious water deficits during the summer. Manhattan, Kansas, receives much less rain than Canton, but because most rain falls during the summer period of maximum potential evapotranspiration, soils do not develop serious water deficits. In the dry climate of Grand Junction, Colorado, soils lack water most of the year and become saturated only briefly, after heavy rains. The habitat produces correspondingly little vegetation.

Soil development

Climate affects plants and animals indirectly through its influence on the development of **soil,** which provides the substrate within which plant roots grow and many animals burrow. The characteristics of soil determine its ability to hold water and make available the minerals required for plant growth. Thus its variation provides a key to understanding the distribution of plant species and the productivity of biological communities.

Soil defies simple definition, but we may describe it as the layer of chemically and biologically altered material that overlies rock or other unaltered materials at the surface of the earth. It includes minerals derived from the parent

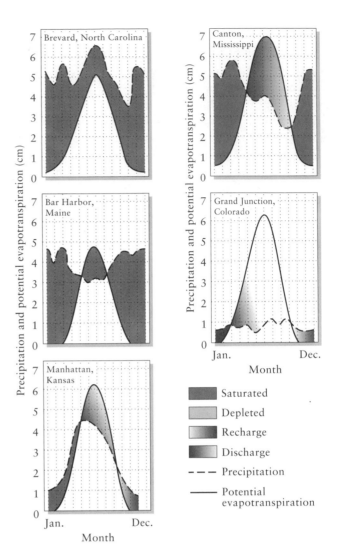

FIGURE 4.17 The relationship of precipitation and potential evapotranspiration to changes in the availability of soil moisture. When evapotranspiration exceeds precipitation, water is withdrawn from the soil until the deficit exceeds 10 cm, the average amount of moisture that soils can hold. After C. W. Thornthwaite, *Geog. Rev.* 38:55–94 (1948).

rock; altered minerals formed anew within the zone of alteration; organic material contributed by plants; air and water within the pores of the soil; living roots of plants; microorganisms; and the larger worms and arthropods that make the soil their home. Five factors largely determine the characteristics of soils: climate, **parent material** (underlying rock), vegetation, local topography, and to some extent age.

Soils exist in a dynamic state, changing as they develop on newly exposed rock material. And even after soils achieve stable properties, they remain in a constant state of flux. Groundwater removes some material; other material enters the soil from vegetation, in precipitation, as dust from above, and from the rock below. Where little rain falls, the parent material decomposes slowly and plant production adds little organic detritus to the soil. Thus arid regions typically have shallow soils, with bedrock lying close to the surface (Figure 4.18). Soils may not form at all where altered bedrock and detritus erode as rapidly as they form. Soil development also stops short on **alluvial** deposits, where fresh layers of silt deposited each year by flood waters bury older material. At the other extreme, soil formation proceeds rapidly in parts of the

FIGURE 4.18 Profile of a poorly developed soil in Logan County, Kansas, illustrating shallow soil depth and absence of soil zonation. Courtesy of the U.S. Soil Conservation Service.

humid tropics, where chemical alteration of parent material may extend to depths of 100 m. Most soils of temperate zones are intermediate in depth, extending to a rough average of about 1 m.

Soil horizons

Where a recent roadcut or excavation exposes soil in cross section, one often notices distinct layers called **horizons** (Figure 4.19). A generalized, and somewhat simplified, **soil profile** has four major divisions: O, A, B, and C horizons; the A horizon has two subdivisions (A_1 and A_2). Arrayed in descending order from the surface of the soil, the horizons and their predominant characteristics are as follows:

O   Primarily dead organic litter. Most soil organisms inhabit this layer.

A_1 A layer rich in **humus,** consisting of partly decomposed organic material mixed with mineral soil.

A_2 A region of extensive **leaching** of minerals from the soil. Because minerals are dissolved by water (mobilized) in this layer, plant roots are concentrated here.

B   A region of little organic material whose chemical composition resembles that of the underlying rock. Clay minerals and oxides of aluminum and iron leached out of the overlying A_2 horizon are sometimes deposited here.

Soil horizons

A

B

C

Parent material

Soil depth (ft)

0

1

2

3

4

(a) (b)

FIGURE 4.19 Soil profiles from the central United States, illustrating distinct layers, or horizons. (a) This profile, from eastern Colorado, is weathered to a depth of about 2 ft, where the subsoil contacts the original parent material, which consists of loosely aggregated, calcium-rich, wind-deposited sediments (loess). The A_1 and A_2 horizons are not clearly distinguished except that the latter is somewhat lighter in color. The B horizon contains a dark band of redeposited organic materials that were leached from the uppermost layers of the soil. The C horizon is light-colored and has been leached of much of its calcium. Some of the calcium has been redeposited at the base of the C horizon and at greater depths in the parent material. (b) This profile is of a typical prairie soil from Nebraska. Rainfall is sufficient to leach readily soluble ions completely from the soil. Hence there are no B layers of redeposition, as in the drier Colorado soil, and the profile is more homogeneous. The A horizon is weakly subdivided into a darker upper layer and a lighter lower layer. The weathered soil lies on a parent material composed of loess, the wind-blown remnants of glacial activity.

C   Primarily weakly altered material, similar to the parent rock. Calcium and magnesium carbonates accumulate in this layer, especially in dry regions, sometimes forming hard, impenetrable layers.

Soil horizons demonstrate the decreasing influence of climate and biotic factors with increasing depth. Critical to soil formation is the movement of mineral elements upward and downward through the soil profile. But before considering these processes in detail, we shall examine the initial altering of the bedrock and how it influences soil characteristics.

Weathering

Weathering, the physical and chemical alteration of rock material near the earth's surface, occurs where surface water penetrates. The repeated freezing and thawing of water in crevices physically breaks rock into smaller pieces and exposes greater surface area to chemical action. Initial chemical alteration of the rock occurs when water dissolves some of the more soluble minerals, especially sodium chloride (NaCl) and calcium sulfate ($CaSO_4$). Other materials — particularly the oxides of titanium, aluminum, iron, and silicon — dissolve less readily.

The weathering of granite exemplifies some basic processes of soil formation. The minerals making up the grainy texture of granite — feldspar, mica, and quartz — consist of various combinations of oxides of aluminum, iron, silicon, magnesium, calcium, and potassium, along with other, less abundant compounds. The key to weathering is the displacement of certain elements in these minerals — notably calcium, magnesium, sodium, and potassium — by hydrogen ions, followed by the reorganization of the remaining oxides of aluminum, iron, and silicon into new minerals.

Feldspar, which consists of aluminosilicates of potassium, weathers rapidly because of the displacement of potassium ions (K^+) by hydrogen ions (H^+) to form new insoluble materials, particularly **clay** particles. Mica grains consist of aluminosilicates of potassium, magnesium, and iron. As in feldspar, the potassium and magnesium are displaced readily during weathering, and the remaining iron, aluminum, and silicon form various kinds of clay particles. Quartz, a type of silica (SiO_2), is relatively insoluble and therefore remains more or less unaltered in the soil as grains of sand. Changes in chemical composition as granite weathers from rock to soil in different climates show that weathering is most severe under tropical conditions of high temperature and rainfall (Figure 4.20).

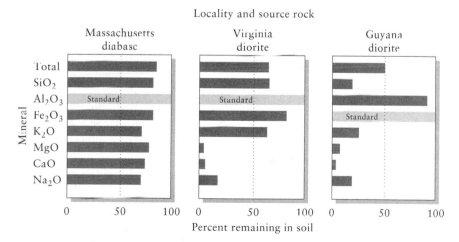

FIGURE 4.20 Differential removal of minerals from granitic rocks as a result of weathering in Massachusetts, Virginia, and Guyana. Values are compared to either aluminum or iron oxides (these standards = 100 percent), which are assumed to be the most stable components of the mineral soil. After E. W. Russell, *Soil Conditions and Plant Growth* (9th ed.). Wiley, New York (1961).

An important factor in the initial weathering of parent material, regardless of the chemical nature of the rock, is the presence of hydrogen ions in the water that percolates to the bedrock. These ions derive from two sources. All precipitation contains dissolved carbon dioxide, which, as we have seen, forms carbonic acid. Some of the carbonic acid dissociates to hydrogen ions (H^+) and bicarbonate ions (HCO_3^-). In regions not affected by pollution-caused acidification, concentrations of hydrogen ions in rainwater produce a pH of about 5. This acidity is supplemented by hydrogen ions generated by the oxidation of organic material in the soil. For example, the metabolism of carbohydrate produces carbon dioxide, and dissociation of the resulting carbonic acid generates additional hydrogen ions. In the Hubbard Brook Forest of New Hampshire, these internal processes account for about 30 percent of the hydrogen ions needed for weathering of bedrock; the remainder comes from precipitation. In the tropics, however, internal sources of hydrogen ions produced by biological oxidation of organic substrates in the soil assume greater importance and may lead to more rapid weathering.

Clay, humus, and the cation exchange capacity of soil

Plants obtain mineral nutrients from the soil in the form of dissolved ions, whose solubility derives from their electrostatic attraction to water molecules. Because ions are dissolved in water, those not immediately taken up by plants and fungi may wash out of the soil if they do not adhere strongly to stable soil particles. Clay and humus particles, separately or associated in complexes, are large enough to form a stable component of the soil. These particles and complexes, known as **micelles,** have negative electrical charges at their surfaces that hold the smaller, more mobile ions in the soil (Figure 4.21). The number of sites on soil particles available for binding positively charged ions (cations) is referred to as the **cation exchange capacity.**

The bonds between soil particles and such ions as potassium (K^+) and calcium (Ca^{2+}) are relatively weak, so they constantly break and re-form. When a potassium ion dissociates from a micelle, its place may be taken by any other positive ion close by. Some cling more strongly to micelles than others: in order of decreasing tenacity, hydrogen, calcium, magnesium, potassium, and sodium. Hydrogen ions thus tend to displace calcium and all other cations in the soil. If cations were not added to or removed from the soil, the relative proportions of the various cations associated with clay–humus particles would achieve a steady state. But carbonic acid in rainwater and organic acids produced by the decomposition of organic detritus continuously add hydrogen ions to the upper layers of the soil; these readily displace other cations, which are then washed out of the soil and into the groundwater. The influx of hydrogen ions in water percolating through the soil largely accounts for the mobility of ions in the soil and for the differentiation of layers in the soil profile.

Negative ions that are important to plant nutrition, such as nitrate, phosphate, and sulfate, can be adsorbed onto clay particles by means of ion "bridges." These bridges form under acid conditions by the association of an

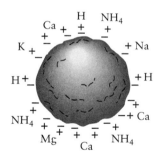

FIGURE 4.21 Schematic representation of a clay or humus particle (micelle) with hydrogen ions and mineral ions attracted by negative charges at its surface. After S. R. Eyre, *Vegetation and Soils* (2nd ed.). Aldine, Chicago (1968).

additional hydrogen ion with a functional group such as the hydroxyl group (OH); for example, $-OH + H^+$ produces the positively charged $-OH_2^+$, which in turn makes possible the binding of anions (for example, $-OH_2^+ + NO_3^-$).

Because of the cation-binding properties of clay and humus particles, the potential long-term fertility of soil—its capacity for storing nutrients—depends in large part on its clay content. Furthermore, because hydrogen ions displace others from soil particles and influence the electrical charge on the surfaces of these particles, the retention of ions in the soil and the immediate availability of ions to plants depend to a large degree on acidity.

Podsolization

The qualities of soils depend on the underlying parent rock and on the climate (Table 4.1). Under mild, temperate conditions of temperature and rainfall, sand grains and clay particles resist weathering and form stable components of the soil skeleton. In acid soils, however, clay particles break down in the A horizon of the soil profile, and their soluble ions are transported downward and deposited in lower horizons. This process, known as **podsolization,** reduces the ion exchange capacity, and therefore the fertility, of the upper layers of the soil.

Acid soils (**spodosols**) occur primarily in cold regions where coniferous trees dominate the forests. The slow decomposition of plant litter produces organic acids. In addition, rainfall usually exceeds evaporation in regions of podsolization. Under these moist conditions, because water continuously moves downward through the soil profile, little clay-forming material is transported upward from the weathered bedrock below.

In North America, podsolization advances farthest under spruce and fir forests in New England and the Great Lakes region and across a wide belt of southern and western Canada. A typical profile of a podsolized soil (Figure 4.22) reveals striking bands corresponding to regions of leaching and redeposition. The topmost layers of the profile (O and A_1) are dark and rich in organic matter. These are underlain by a light-colored horizon (A_2), which has been leached of most of its clay content. As a result, A_2 consists mainly of sandy skeletal material that holds neither water nor nutrients well. One usually finds a dark band of deposition immediately under the A_2 horizon. This is the uppermost layer of the B horizon, where iron and aluminum oxides are redeposited. Other, more mobile minerals may accumulate to some extent in lower parts of the B horizon, which then grades almost imperceptibly into a C horizon and the parent material.

Laterization

The warm, wet climate of many tropical regions weathers the soil to great depths, and the deeper the ultimate source of nutrients in the unaltered bedrock, the poorer the surface layers. Rich soils do develop in many tropical re-

TABLE 4.1	*The major soil groups*

Name	Character	Distribution in the United States
Alfisols	Moist, moderately weathered mineral soils	Ohio Valley, Great Lakes region, Rocky Mountains, central California
Aridosols	Dry mineral soils with little leaching and accumulations of calcium carbonate	Great Basin, southwestern deserts
Entisols	Recent mineral soils lacking development of soil horizons	On hard rock in the Rocky Mountains, on sands in the Southwest
Histosols	Organic soils of peat bogs, mucks	Northern Minnesota, Mississippi Delta, Florida Everglades
Inceptisols	Young, weakly weathered soils	New York, Pennsylvania, West Virginia, ullivial soils of Mississippi Valley, northwestern states, Alaska
Mollisols	Well-developed soils high in organic matter and calcium; very productive	Most of the prairie soils of the Great Plains
Oxisols	Deeply weathered, lateritic soils of moist tropics	Tropical South America and Africa; not present in the United States
Spodosols	Acid, podsolized soils of cool, moist climates with a shallow leached horizon and a deeper layer of deposition	Forested regions of New England; northern Michigan and Wisconsin
Ultisols	Highly weathered soils of warm, moist climates with abundant iron oxides	Most of the southeastern United States, the Pacific Northwest
Vertisols	High content of swelling-type clays developing deep cracks in dry seasons	Southern Texas

gions, particularly in mountainous areas where erosion continuously removes nutrient-depleted surface layers of the soil. But the soils of other areas, especially those in low-lying regions (the Amazon Basin, for example) and those that develop on parent material deficient in quartz (SiO_2) but rich in iron and magnesium (basalt, for example), contain little clay and therefore do not hold nutrients well. Clay fails to form in any abundance because of a lack of silicon. Instead, oxides of iron and aluminum predominate in the soil horizon. This type of weathering is known as **laterization;** oxides give lateritic soils (**latisols**) their typical red color.

FIGURE 4.22 Profile of a podsolized soil in Plymouth County, Massachusetts. The light-colored A_2 horizon and the dark-colored B_1 horizon immediately below it form distinct bands. Compare the general absence of roots in the A_2 horizon with their presence in the lower B_1 horizon. Courtesy of the U.S. Soil Conservation Service.

🖛 SUMMARY

Each type of plant and animal flourishes within a narrow range of environmental conditions. Climate therefore determines both the distributions of organisms and the characteristic plants within each region.

1. Global patterns of temperature and rainfall result from the local absorption of solar radiation and from the redistribution of heat energy by winds and ocean currents. Prominent features of terrestrial climates include a band of warm, moist climate over the equator and bands of dry climate at about 30° north and south latitude.

2. Seasonality is caused by the annual progression of the sun's path northward and southward and by the latitudinal movement of associated belts of temperature, wind, and precipitation. At high latitudes, the seasons primarily reflect annual cycles of temperature. Within the tropics, seasonality of precipitation influences the activity of animals and plants.

3. Seasonal warming and cooling profoundly change the characteristics of lakes in the temperate zones. During summer, such lakes are stratified, with a warm surface layer (epilimnion) separated from a cold bottom layer (hypolimnion) by a sharp thermocline. In spring and autumn, the profile of temperature with depth becomes more uniform, allowing vertical mixing.

4. Irregular and unpredictable variations in climate, such as El Niño – Southern Oscillation events, may cause major disruptions of biological communities on a global scale.

5. Topography and geology superimpose local variation in environmental conditions on more general climate patterns. Mountains intercept rainfall, creating arid rain shadows in their lees. Conditions at higher altitudes resemble conditions at higher latitudes. Soil characteristics reflect the quality of the underlying bedrock and sometimes foster specialized floras, such as those of serpentine barrens.

6. Ecologists refer to mosaics of habitat patches as landscapes. The movement of individuals and materials between patches influences populations and ecosystems within the landscape as a whole.

7. Descriptions of climate, such as the climograph and Thornthwaite's analysis of seasonal water availability, integrate the effects of both temperature and precipitation.

8. Nutrient regeneration in terrestrial systems takes place in the soil. The characteristics of soil reflect the influences of the bedrock below and the climate and vegetation above. Weathering of bedrock results in the breakdown of some native minerals (feldspar and mica) and their re-formation into clay particles, which mix with organic detritus entering the soil from the surface. These vertically graded processes usually result in distinct soil horizons.

9. The clay and humus content of the soil determines its ability to retain nutrients required by plants. Clay and humus particles (micelles) have negative charges on their surfaces that attract cations (Ca^{2+}, K^+, NH_4^+) directly and anions (PO_4^{3-}, NO^{3-}) indirectly under acid conditions. Hydrogen ions tend to displace other cations on micelles and thereby reduce soil fertility.

10. In acid, temperate-zone (podsolized) soils and deeply weathered (laterized) tropical soils, clay particles break down and the fertility of the soil is much reduced.

🖛 SUGGESTED READINGS

Barber, R. T., and F. P. Chavez. 1983. Biological consequences of El Niño. *Science* 222:1203–1210.

Barry, R. G., and R. J. Chorley. 1976. *Atmosphere, Weather, and Climate* (3rd ed.). Methuen, London.

Brady, N. C. 1974. *Nature and Property of Soils* (8th ed.). Macmillan, New York.

Bunting, B. T. 1967. *The Geography of Soil* (rev. ed.). Aldine, Chicago.

Jenny, H. 1980. *The Soil Resource. Origin and Behavior.* Springer-Verlag, New York.

Philander, G. 1989. El Niño and La Niña. *American Scientist* 77:451–459.

Rasmussen, E. M. 1985. El Niño and variations in climate. *American Scientist* 73:168–177.

Shelford, V. E. 1963. *The Ecology of North America.* Univ. of Illinois Press, Urbana.

Waring, R. H., and J. Major. 1964. Some vegetation of the California coastal region in relation to gradients of moisture, nutrients, light, and temperature. *Ecological Monographs* 34:167–215.

THE DIVERSITY OF BIOLOGICAL COMMUNITIES

Climate, topography, and soil, and the equivalent influences in aquatic environments, determine the character of plant and animal life over the surface of the earth. Every biological system exhibits similar exchanges of energy and materials (and other interactions) with the physical world. Yet the expression of these processes differs according to the specific environmental conditions at each particular place. For example, loss of water to the environment is of less consequence for plants and animals in tropical montane forests than for those in subtropical deserts; salt balance poses different problems for organisms in fresh and salt water. This variety of relationships to the environment generates the biological diversity of the natural world.

This chapter emphasizes two general principles. First, because of evolutionary adaptation, the forms of plants and animals vary in concert with the conditions of the environment. As a result, biological communities in different climate zones take on different appearances. Second, the distributions of all types of plants, animals, and microbes are confined to a narrow range of climate zones to which the organisms are particularly well suited. Thus, traveling from place to place, we leave some species behind and encounter new ones that have slightly different attributes of form and function.

Climate and plant distribution

The range of the sugar maple, a common forest tree in the northeastern United States and southern Canada, is limited by cold winter temperatures to the north, by hot summer temperatures to the south, and by summer drought to the west (Figure 5.1). Attempts to grow sugar maples outside their normal range have shown that they cannot tolerate average monthly temperatures above about 24°C or below about −18°C. The western limit of the sugar maple, determined by dryness, coincides with the western limit of forest in general. Because temperature and rainfall interact to determine the availability of moisture, sugar maples tolerate lower annual precipitation at the northern edge of their range (about 50 cm) than at the southern edge (about 100 cm). To the east, the range of the sugar maple stops abruptly at the Atlantic Ocean.

Differences in the distributions of the sugar maple and other tree-sized species of maples — black, red, and silver — suggest differences in ecological tolerances (Figure 5.2). Where their geographic ranges overlap, maples exhibit distinct preferences for local environmental conditions created by differences in soil and topography. Black maple frequently occurs together with the closely related sugar maple, but it prefers drier, better drained soils higher in calcium content (and therefore less acidic). Silver maple occurs widely in the eastern United States but prefers the moist, well-drained soils of the Ohio and Mississippi river basins. Red maple grows best either under wet, swampy conditions or in dry, poorly developed soils.

Topography, soils, and local distribution of plants

The distributions of plants clearly reveal the effects of different factors, which vary over different scales of distance. Climate, topography, soil chemistry, and soil texture exert progressively finer influences on geographic distribution. Elevation, slope, exposure, and underlying bedrock — factors that modify the

FIGURE 5.1 The range of the sugar maple in eastern North America. After H. A. Fowells, *Silvics of Forest Trees of the United States.* U.S. Department of Agriculture, Washington, D.C. (1965).

FIGURE 5.2 The ranges of black, red, and silver maples in eastern North America. The range of the sugar maple is outlined on each map to show the area of overlap. After H. A. Fowells, *Silvics of Forest Trees of the United States.* U.S. Department of Agriculture, Washington, D.C. (1965).

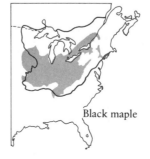

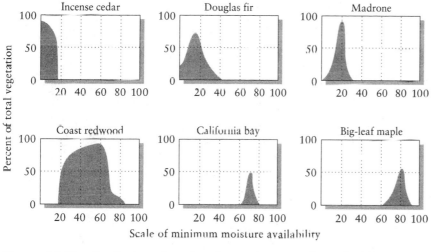

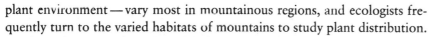

Scale of minimum moisture availability

FIGURE 5.3 The distribution of species of trees along a gradient of minimum available soil moisture in the northern coastal region of California. After R. H. Waring and J. Major, *Ecol. Monogr.* 34:167–215 (1964).

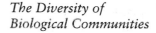
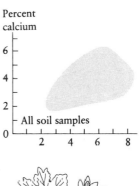
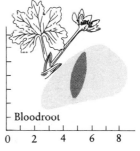

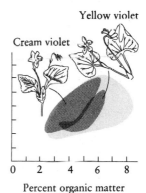

Percent organic matter

FIGURE 5.4 The occurrence of four forest-floor plants with respect to the calcium and organic matter contents of the soil in woodlands of eastern Indiana. After E. W. Beals and J. B. Cope, *Ecology* 45:777–792 (1964).

plant environment — vary most in mountainous regions, and ecologists frequently turn to the varied habitats of mountains to study plant distribution.

Along the coast of northern California, mountains create conditions that support a variety of plant communities ranging from dry coastal chaparral to tall forests of Douglas fir and redwood. When localities are ranked on scales of available moisture, the distribution of each species of plant among the localities exhibits a distinct optimum (Figure 5.3). The coast redwood dominates the central portion of the moisture gradient and frequently forms pure stands. Cedar, Douglas fir, and two broad-leaved evergreen species with small thick leaves — manzanita and madrone — occur at the drier end of the moisture gradient. Three deciduous species — alder, big-leaf maple, and black cottonwood — occupy the wetter end.

Change in one environmental condition usually brings about changes in others. Increasing soil moisture alters the availability of nutrients. Variations in the amount and source of organic matter in the soil create parallel gradients of acidity, soil moisture, and available nitrogen. Such factors often interact in complex ways to determine the distributions of plants. Figure 5.4 relates the distributions of some forest-floor shrubs, seedlings, and herbs in woodlands of eastern Indiana to levels of organic matter and calcium in the soil. These soils contain between 2 and 8 percent organic matter and between 2 and 6 percent exchangeable calcium. Within the range of soil conditions in these woodlands, each species shows different preferences. Black cherry seedlings occur only within a narrow range of calcium but tolerate considerable variation in the percentage of organic matter. Bloodroot is narrowly restricted by the percentage of organic matter in the soil but is insensitive to variation in calcium. The distributions of yellow violets and cream violets extend more broadly over varying levels of organic matter and calcium in the soil, but the two species do not overlap. Cream violets prefer soils higher in calcium and

Pinus bolanderi   *Cupressus pygmaea*   *Cupressus sargentii*

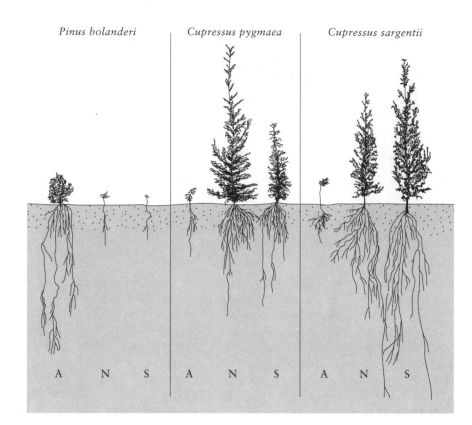

FIGURE 5.5 Seedling growth of lodgepole pine, pygmy cypress, and Sargent cypress in acid (A), "normal" (N), and serpentine (S) soils. After C. McMillan, *Ecol. Monogr.* 26:177–212 (1956).

lower in organic matter than do yellow violets; where one occurs, the other usually does not.

In the coastal ranges of northern California, several species of pines and cypresses are restricted to serpentine soils, whereas others occur only on extremely acid soils. When grown on soils from different localities, seedlings of these endemics often do best when planted in soil from the native habitat. Thus lodgepole pine grows only in acid soils, and Sargent cypress, a serpentine endemic, grows somewhat better on serpentine soil than on "normal" soil and not at all on acid soil (Figure 5.5). But not all endemics perform best on their home soil. When given the chance in an experimental garden, pygmy cypress, normally restricted to acid soils, grows much better on "normal" and serpentine soils. Where does it derive its tolerance for serpentine soils? What factors exclude mature pygmy cypress from soils on which its seedlings grow vigorously? Clearly, factors other than **edaphic** conditions (those inherent in the soil) influence the distributions of these species. Such factors include competitors and pathogens.

Environment, form, and function

The adaptations of an organism—as they are exhibited in its form, physiology, and behavior—cannot easily be separated from the environment in which it lives. Insect larvae from stagnant aquatic environments in ditches and sloughs can survive longer without oxygen than related species from well-aerated streams and rivers; species of marine snails that occur high in the intertidal zone, where they are frequently exposed to air, tolerate desiccation better than do species from lower levels.

Compare the leaves of deciduous forest trees with those of desert species. The former are typically broad and thin, providing a large surface area for light absorption and also, unavoidably, for water loss. Desert trees have small, finely divided leaves — or sometimes none at all (Figure 5.6). Leaves heat up in the desert sun. Structures lose heat by convection most rapidly at their edges, where wind currents disrupt the insulating boundary layers of still air. The more edges, the cooler the leaf and the lower the water loss; small size means that a large proportion of each leaf is close to its edge. Even on a single plant, leaves exposed to full sun may be differently shaped to dissipate heat and conserve water better than shade leaves (Figure 5.7).

The vertical distributions of species of algae within the intertidal range closely parallel their rates of photosynthesis in water and when exposed to air. Along the coast of central California, the tide varies over a vertical range of approximately 2.5 m. Height within this range determines the proportion of time that algae are exposed to air on an annual basis. Exposure is about 10 percent at 0 m above designated sea level, 40 percent at 1 m, and 90 percent at 2 m. Algae from the lower part of the tide range (*Prionitus* and *Ulva*) reduce their photosynthetic rate while exposed; algae from the middle and upper parts of the tidal range elevate their rates of photosynthesis in air (Table 5.1). The

Figure 5.6 Leaves of some desert plants from Arizona. (a) Mesquite (*Prosopis*) leaves are subdivided into numerous small leaflets, which facilitate the dissipation of heat when exposed to sunlight. (b) The paloverde (*Cercidium*) carries this adaptation even further; its leaves are tiny, and the thick stems, which contain chlorophyll, are responsible for much of the plant's photosynthesis. (Hence the name *paloverde,* which is Spanish for "green stick"). Cacti rely entirely on their stems for photosynthesis; their leaves are modified into thorns for protection. (c) Unlike most desert plants, limberbush (*Jatropha*) has broad, succulent leaves, which it produces for only a few weeks during the summer rainy season in the Sonoran Desert. Photographs are about one-quarter of actual size.

TABLE 5.1	Characteristics of intertidal species of algae from various height zones				
	SPECIES				
Characteristic	*Prionitis*	*Ulva*	*Iridaea*	*Porphyra*	*Fucus*
Mean height above sea level (ft)	−1.0	+0.5	+1.0	+3.0	+3.0
Time exposed to air (percent)	5	15	25	40	40
Leaf area per unit weight (dm^2 g^{-1} dry weight)*	0.4	3.3	1.3	1.9	0.5
Photosynthesis rate (mg CO_2 g^{-1} h^{-1})					
In air	1.1	12.2	5.2	17.7	5.8
In water	1.2	16.7	1.7	6.3	0.9
Air-to-water ratio	0.9	0.7	2.9	2.8	6.6
Photosynthesis drops to one-half submerged rate at					
Water loss (percent)	10	24	32	47	60
Hours of exposure	0.4	0.6	1.3	3.6	5.3
Rate of water loss (percent per hour)	25	40	25	13	11

*dm = decimeter; 1 dm^2 = 100 cm^2.

Source: W. S. Johnson et al., *Ecology* 55:450–453 (1974).

Sun leaf

Shade leaf

FIGURE 5.7 Silhouettes of a sun leaf and a shade leaf of white oak. The sun leaves have more edge per unit of surface area and therefore dissipate heat more rapidly. After S. Vogel, *J. Exp. Bot.* 21:91–101 (1970).

higher of the two mass-specific rates of photosynthesis (those obtained while exposed or submerged) varies directly with the surface area of the alga per unit of dry mass. How rapidly photosynthesis proceeds while the alga is exposed to air relative to the rate while it is submerged depends on its tolerance of desiccation.

When algae from the lower portion of the intertidal zone are exposed to air, photosynthesis decreases. After 24 to 36 minutes, plants lose 10 to 24 percent of the water in their tissues (at a rate of 25 to 40 percent per hour). When algae from the middle and upper intertidal zones are exposed to air, photosynthesis first increases. Only after 1 to 5 hours and a 32 to 60 percent loss of tissue water (at a rate of 11 to 25 percent per hour) does photosynthesis drop to one-half of its value while the organism was submerged. Thus the physiological attributes of these algae—low rates of water loss and relative insensitivity to desiccation—enhance their chances of surviving the long periods of exposure to air in the upper parts of the tidal range.

The water relations of coastal sage and chaparral plants in southern California illustrate the divergent forms and lifestyles of these plants. The chaparral habitat generally occurs at higher elevations than that of the coastal sage and thus has cooler and moister conditions. Both vegetation types are exposed to prolonged summer drought, but soils have greater water deficits in the sage habitat. Plants of the coastal habitat typically have shallow roots and small, delicate, deciduous leaves (Figure 5.8). Chaparral species have deep roots that often extend through tiny cracks and fissures far into the bedrock; their thick leaves have a waxy outer covering (cuticle) that reduces water loss. Most coastal sage species shed their delicate leaves during the summer drought period; the tougher leaves of chaparral plants persist.

Leaf morphology influences photosynthetic rate in tandem with its influence on transpiration (Table 5.2). The thin leaves of coastal sage species lose water rapidly, but they also assimilate carbon rapidly when water is available to replace that lost to transpiration. To demonstrate this relationship, scientists clipped leaves from plants and placed them in a chamber within which they could monitor transpiration and photosynthesis. Both functions declined as the leaves dried out and their stomates closed to prevent further water loss. Rates of both photosynthesis and transpiration of such coastal species as the black sage were high initially but shut down quickly because of rapid water loss (Figure 5.9). Photosynthetic rates of such chaparral species as the toyon (rose family) were only one-fourth to one-third those of coastal sage species at most, but the leaves resisted desiccation better and continued to be active under drying conditions for longer periods.

When chaparral and coastal sage species grow together near the overlapping edges of each other's ranges, they exploit different parts of the environment: deep, perennial sources of water versus shallow, ephemeral sources. In spite of these differences and the corresponding adaptations of leaf morphology and drought response, these species are equally productive at intermediate levels of water availability. In drier habitats, however, the prolonged seasonal

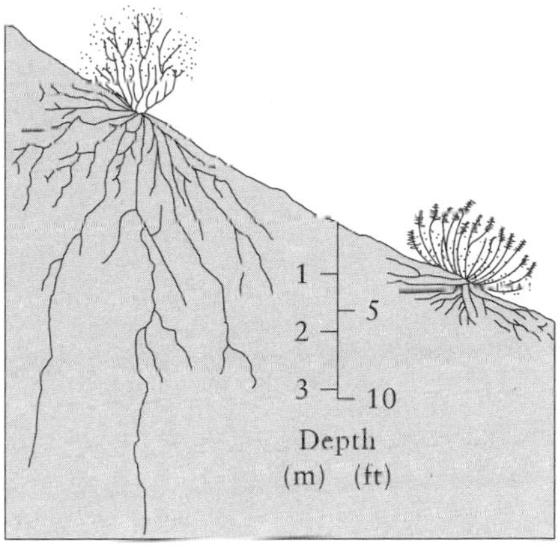

FIGURE 5.8 Profiles of the root systems of chamise (*Adenostoma fasciculatum*), a chaparral species (*left*), and black sage (*Salvia mellifera*), a member of the coastal sage community (*right*). After H. Hellmers et al., *Ecology* 36:667–678 (1955).

TABLE 5.2	Characteristics of chaparral and coastal sage vegetation in southern California	
	VEGETATION TYPE	
Characteristic	Chaparral	Coastal sage
Roots	Deep	Shallow
Leaves	Evergreen	Summer deciduous
Average leaf duration (months)	12	6
Average leaf size (cm^2)	12.6	4.5
Leaf weight (g dry weight dm^{-2})	1.8	1.0
Maximum transpiration (g H$_2$O dm^{-2} h^{-1})	0.34	0.94
Maximum photosynthetic rate (mg C dm^{-2} h^{-1})	3.9	8.3
Relative annual CO$_2$ fixation	49.8	46.8

Source: A. T. Harrison, E. Small, and H. A. Mooney, *Ecology* 52:869–875 (1971); H. A. Mooney and E. L. Dunn, *Amer. Nat.* 104:447–453 (1970).

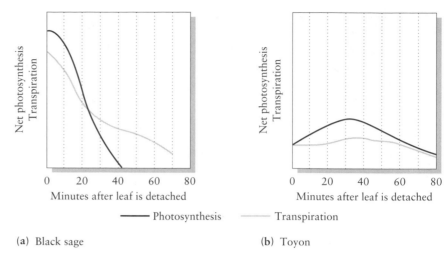

(a) Black sage (b) Toyon

FIGURE 5.9 Time courses for photosynthesis and transpiration under standard drying conditions. (a) A chaparral species (toyon, *Heteromeles arbutifolia*). (b) A coastal sage species (black sage, *Salvia mellifera*). Note that transpiration continues well after photosynthesis has been shut off; hence leaf dormancy is an ineffective long-term solution to drought. After A. T. Harrison, E. Small, and H. A. Mooney, *Ecology* 52:869–875 (1971).

absence of deep water tips the balance in favor of the deciduous coastal sage vegetation. And the increasing availability of deep water at higher elevations favors the evergreen chaparral vegetation.

In general, the adaptations of plants and animals make them well suited to the environments in which they live. Equally important, each organism's structural and functional characteristics restrict the range of environmental conditions under which it can exist. Thus one usually finds a close correlation between adaptation and environment.

Climate and plant form

Natural history, and later ecology, grew out of classification systems by which animals and plants are given names on the basis of their similarities. European botanists had described most local plant species by the end of the last century; they then began to develop systems of classification for entire communities of plants. They based most of these schemes on structure: height of vegetation, leaf or needle structure, deciduousness, and dominant plant form. Because these traits enable plants to thrive (or at least survive) in the physical environment in which they live, vegetation zones and climate correspond closely (Figure 5.10).

The earliest classifications of vegetation described the most important plants of each major association. These classifications included both a complete analysis of the species composition of communities (**floristic analysis**) and a description of plant forms, regardless of species. Floristic analysis proved useful in Europe, where botanists knew all the species and where minor differ-

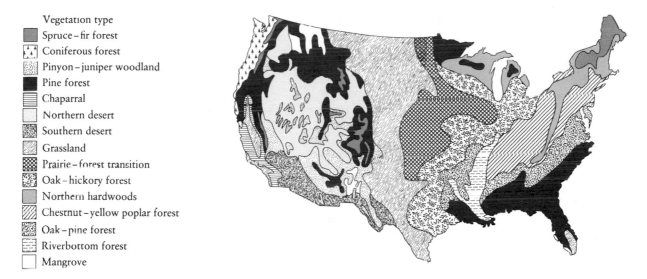

Vegetation type
- ▨ Spruce–fir forest
- ⋰ Coniferous forest
- ░ Pinyon–juniper woodland
- ■ Pine forest
- ▤ Chaparral
- ☐ Northern desert
- ▦ Southern desert
- ▨ Grassland
- ▦ Prairie–forest transition
- ▨ Oak–hickory forest
- ▨ Northern hardwoods
- ▨ Chestnut–yellow poplar forest
- ▨ Oak–pine forest
- ▤ Riverbottom forest
- ☐ Mangrove

FIGURE 5.10 Vegetation map of the United States based primarily on forest types. After H. A. Fowells, *Silvics of Forest Trees of the United States.* U.S. Department of Agriculture, Washington, D.C. (1965).

ences between communities involved the replacement of some species by others with slightly different ecological requirements. But floristic analysis proved unworkable on a global scale because geographic barriers restrict the distributions of individual species, rendering floristic comparisons ecologically meaningless. Forests in Europe and the United States, shrublands in California and Australia, and grasslands in Africa and South America, while structurally similar, have few species in common.

Worldwide classification of vegetation depends on an analysis of the form and function of plants rather than on their scientific names. Botanists have derived various sets of symbols to describe such characteristics as plant size; life form; leaf shape, size, and texture; and percent of ground coverage. In 1903 the Danish botanist Christen Raunkiaer proposed to classify plants according to the position of their buds (regenerating parts) and found that the occurrence of his major categories corresponded closely to climatic conditions. He distinguished five principal life forms, four of which are depicted in Figure 5.11.

Phanerophytes (from the Greek *phaneros,* which means "visible") carry their buds on the tips of branches, exposed to extremes of climate. Most trees and large shrubs are phanerophytes. As one might expect, this plant form predominates in moist, warm environments where buds require little protection. **Chamaephytes** (from the Greek *chamai,* "on the ground, dwarf") comprise small shrubs and herbs that grow close to the ground (prostrate life form). Proximity to the soil protects the bud, and in winter, snow cover provides additional protection from extreme cold. Chamaephytes occur most frequently

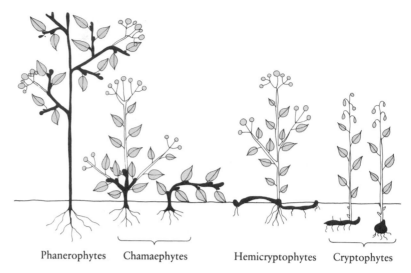

Phanerophytes   Chamaephytes   Hemicryptophytes   Cryptophytes

FIGURE 5.11 Diagrammatic representation of Raunkiaer's life forms. Shaded parts of the plant die back during unfavorable seasons, while the solid black portions persist and give rise to the following year's growth. Proceeding from left to right, the buds are progressively better protected. Therophytes, whose persistent parts are seeds, are not illustrated. After C. Raunkiaer, *Plant Life Forms*. Clarendon Press, Oxford (1937).

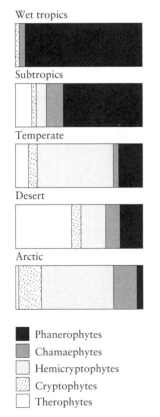

Wet tropics

Subtropics

Temperate

Desert

Arctic

■ Phanerophytes
▨ Chamaephytes
▫ Hemicryptophytes
⁙ Cryptophytes
□ Therophytes

FIGURE 5.12 The proportions of plant life forms, classified according to Raunkiaer, in various climatic regions.

in cool, dry climates. **Hemicryptophytes** (from the Greek *kryptos,* "hidden" and the Latin *hemi,* "half") persist through the extreme environmental conditions of the winter months by dying back to ground level, where the regenerating bud is protected by soil and withered leaves. This growth form is characteristic of cold, moist zones. **Cryptophytes** are further protected from freezing and desiccation because their buds are completely buried beneath the soil. The bulbs of irises and daffodils are the regenerating buds of cryptophyte plants. Like hemicryptophytes, cryptophytes are found in cold, moist climates. **Therophytes** (from the Greek *theros,* "summer") die during the unfavorable season of the year and do not have persistent buds. Therophytes are regenerated solely by seeds, which resist extreme cold and drought. The therophyte form includes most annual plants and occurs most abundantly in deserts and grasslands. Figure 5.12 shows how Raunkiaer's life forms are closely associated with different climates.

It is possible to relate plant form directly to climate by examining the distribution of vegetation types with respect to climatic variables. When we plot a sample of terrestrial localities on a graph according to their mean annual temperature and rainfall, most points fall within a triangular area whose three corners represent warm–moist, warm–dry, and cool–dry climates (Figure 5.13). Cold regions with high rainfall are rare; water does not evaporate rapidly at low temperature, and the atmosphere in cold regions holds little water vapor. Let us now superimpose, on a graph of temperature and rainfall, habitat classifications reflecting the dominant plant form (Figure 5.14).

Within the tropical and subtropical realms, with mean temperatures between 20 and 30°C, vegetation types range from true rain forest, which is wet throughout the year, to desert. Intermediate climates support seasonal forests, in which some or all trees lose their leaves during the dry season, and short, dry forests or scrublands with many thorntrees.

Plant communities in temperate areas follow the pattern of tropical communities, with the same vegetation types distinguishable in both. In colder climates, however, precipitation varies so little from one locality to another that vegetation types are poorly differentiated on the basis of climate. Where mean annual temperatures are below −5 °C, we may lump all plant associations into one type: tundra.

Toward the drier end of the rainfall spectrum within each temperature range, fire plays a distinct role in shaping the plant communities. For example, in the African savannas and midwestern North American prairies, frequent fires kill the seedlings of trees and prevent the establishment of forests, for which favorable conditions otherwise exist. Burning favors perennial grasses with extensive root systems that can survive fire. After an area has burned over, grass roots send up fresh shoots and quickly revegetate the surface. In the absence of frequent fires, tree seedlings can become established and eventually shade out prairie vegetation.

As in all classifications, exceptions appear frequently. Boundaries between vegetation types are fuzzy at best. Moreover, all plant forms do not respond to climate in the same way. For example, some species of Australian eucalyptus

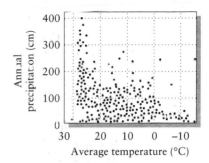

FIGURE 5.13 Average annual temperature and precipitation for a sample of localities more or less evenly distributed over the land area of the earth. Most of the points fall within a triangular region that includes the full range of climates, excluding those of high mountains.

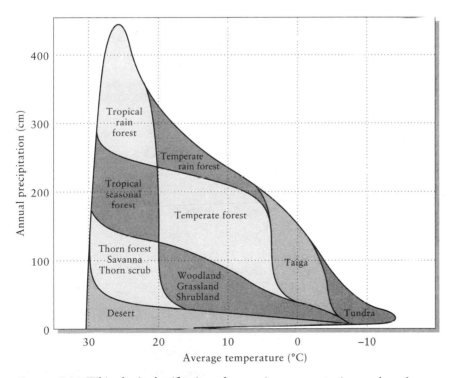

FIGURE 5.14 Whittaker's classification of vegetation types superimposed on the range of terrestrial climates. In climates intermediate between those of forested and desert regions, fire, soil, and climate seasonality determine whether woodland, grassland, or shrubland develops. From R. H. Whittaker, *Communities and Ecosystems* (2nd ed.). Macmillan, New York (1975).

trees form forests under climatic conditions that support only shrubland or grassland on other continents. Finally, plant communities reflect factors other than temperature and rainfall. Topography, soils, fire, seasonal variations in climate, and herbivory all leave their mark, further emphasizing the adaptation of life forms to the diversity of environments on the earth.

✒SUMMARY

1. The geographic distributions of plants are determined primarily by climate, whereas local distributions within regions vary according to topography and soils.

2. Climate profoundly affects the evolution of plants and animals. As a consequence, each climatic region has characteristic forms of vegetation that differ in general habit, leaf morphology, and phenology.

3. Although early schemes of classifying plant associations were based on species (floristics), those based on plant form have proved to be more broadly applicable. Raunkiaer classified plants according to the degree of protection afforded their buds and showed that the proportions of these plant types within a local flora vary systematically with climate.

4. Recognizing that plant form is directly related to climate through adaptation, we can match major types of vegetation to local temperature and precipitation. This relationship emphasizes the interaction between temperature and water availability and further acknowledges the modifying effects of soil, climate seasonality, and fire.

✒ SUGGESTED READINGS

Eyre, S. R. 1968. *Vegetation and Soils: A World Picture* (2nd ed.). Aldine, Chicago.

Forman, R. T. T. (ed.). 1979. *Pine Barrens: Ecosystem and Landscape.* Academic Press, New York.

Jaeger, E. C. 1957. *The North American Deserts.* Stanford Univ. Press, Stanford, Calif.

Levinton, J. S. 1982. *Marine Ecology.* Prentice-Hall, Englewood Cliffs, N.J.

McMillan, C. 1956. Edaphic restriction of *Cupressus* and *Pinus* in the coast ranges of central California. *Ecological Monographs* 26:177–212.

Teal, J., and M. Teal. 1969. *Life and Death of a Salt Marsh.* Little, Brown, Boston.

Terborgh, J. 1992. *Diversity and the Tropical Rain Forest.* Scientific American Library, New York.

Weaver, J. E. 1956. *Grasslands of the Great Plains.* Johnsen, Lincoln, Nebr.

Whitmore, T. C. 1990. *An Introduction to Tropical Rain Forests.* Oxford Univ. Press, New York.

Whittaker, R. H., and W. A. Niering. 1965. Vegetation of the Santa Catalina Mountains, Arizona: a gradient analysis of the south slope. *Ecology* 46:429–452.

2

Ecosystems

ENERGY IN THE ECOSYSTEM

As the diverse information gathered by naturalists during the last century coalesced into a unified picture of nature, several new concepts emerged that led the study of ecology in new directions. One of these was the realization that feeding relationships link organisms into a single functional entity, the **biological community.** Foremost among the proponents of this new ecological viewpoint during the 1920s was the English ecologist Charles Elton. Of course, every organism must feed in some manner to gain nourishment, and each may be fed upon by some other. But that these feeding relationships define an ecological unit was a novel idea early in the twentieth century.

A second concept, developed forcefully a decade later by the English plant ecologist A. G. Tansley, took Elton's idea an important step further by regarding animals and plants in associations, *together with the physical factors of their surroundings,* as a fundamental ecological system. Tansley called this the **ecosystem.** He envisioned the biological and physical parts of nature together, unified by the dependence of animals and plants on their physical surroundings and by their contribution to the maintenance of the physical world.

Lotka's thermodynamic view of the ecosystem

Working independently of the ecologists of his day, Alfred J. Lotka, a chemist by training, developed ecosystem concepts from considerations of energetics. Lotka was the first to treat populations and communities as **thermodynamic** systems. In principle, he said, each system can be represented by a set of equations that govern transformations of mass among its components. Such transformations include the assimilation of carbon dioxide into organic carbon compounds by green plants and the consumption of plants by herbivores and of animals by carnivores.

Lotka believed that the size of a system and the rate of transformations within it were determined in accordance with certain thermodynamic principles. Just as heavy machines and fast machines require more fuel to operate than their lighter and slower counterparts, and efficient machines require less fuel than inefficient ones, the energy transformations of ecosystems grow in direct proportion to their size (roughly the total masses of their constituent organisms), productivity (rate of transformations), and inefficiency. Not all the energy of sunlight enters biological pathways of transformations. In fact, most of it drives the circulation of winds and ocean currents and the evaporation of water, which make up a large, physical thermodynamic system. But part of the energy of sunlight that plants do assimilate by photosynthesis ultimately fuels all biological processes and therefore establishes the overall rate of transformations within the ecosystem.

The idea of the ecosystem as an energy-transforming system was brought to the attention of many ecologists for the first time in a paper published in 1942 by Raymond Lindeman, a young aquatic ecologist from the University of Minnesota. Lindeman's framework for understanding ecological succession on the basis of sound thermodynamic principles made a deep impression. He adopted Tansley's notion of the ecosystem as the fundamental unit in ecology and Elton's concept of the food web, including inorganic nutrients at the base, as the most useful expression of ecosystem structure.

The food chain has many links—primary producer, herbivore, carnivore—which Lindeman referred to as **trophic levels.** (The Greek root of the word *trophic* means "food.") Furthermore, Lindeman visualized a **pyramid of energy** within the ecosystem. He argued that less energy reaches each successively higher trophic level because of the work performed and because of the inefficiency of biological energy transformations on the next lower trophic level. Thus, of the light impinging on a lake, plants assimilate only a portion, the primary production of the system. Herbivores assimilate less of this energy than do plants, because the plants used some of their own production to maintain themselves before the herbivores consumed them. The ratio of production on one trophic level to that on the level below it constitutes the **ecological efficiency** of that link in the food chain.

By the 1950s, the ecosystem concept had fully pervaded ecological thinking and spawned a new branch of ecology in which the cycling of matter and the associated flux of energy through the ecosystem provided a basis for characterizing that system's structure and function. Energy, which provides a common idiom for ecological description, and the masses of elements (such as carbon) made possible the direct comparison of plants, animals, microbes, and abiotic sources of energy and elements in the ecosystem. Measurements of energy assimilation and energetic efficiencies became the tools for exploring this new thermodynamic concept of the ecosystem.

With a clear conceptual framework for the ecosystem and a "currency" of energy to describe its structure, ecologists began to measure energy flow and the cycling of nutrients in the ecosystem. One of the strongest proponents of this approach has been Eugene P. Odum of the University of Georgia, whose text *Fundamentals of Ecology,* first published in 1953, influenced a generation of ecologists. Odum depicted ecosystems as energy flow diagrams (Figure 6.1a). For any one trophic level, such a diagram consisted of a box representing the biomass (or its energy equivalent) at any given time and pathways through the box representing the flow of energy. These diagrams simplify nature but nonetheless convey the important principle that the energy passes from one link in the consumer food chain to the next, diminished by respiration and the shunting of unused foodstuffs to detritus-based food chains. Feeding relationships linked energy flow diagrams into a food web, as shown in Figure 6.1b.

Unlike energy, which ultimately comes from sunlight and leaves the ecosystem as heat, nutrients are regenerated and retained within the system. Odum elaborated energy flow diagrams to include this cycling of elements (Figure 6.2). In the development of ecosystem studies, the cycling of elements has assumed equal standing with the flow of energy. One reason for this prominence is that the amounts of elements and their movement between components can provide a convenient index to the flow of energy, which is difficult to measure directly. Carbon, in particular, bears a close relationship to energy content

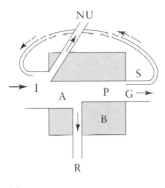

(a)

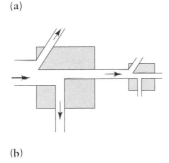

(b)

FIGURE 6.1 (a) E. P. Odum's "universal" model of ecological energy flow, which can be applied to any organism: I = ingestion; A = assimilation; P = production; NU = not used; R = respiration; G = growth; E = excreta; S = storage, as in the form of fat, for future use; and B = biomass.
(b) Representation of a food chain by Odum's energy flow models. The net production of one trophic level becomes the ingested energy of the next higher level. After E. P. Odum, *Am. Zool.* 8:11–18 (1968).

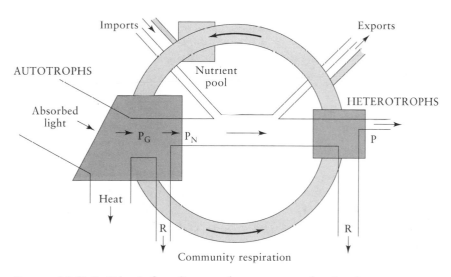

FIGURE 6.2 E. P. Odum's flow diagram of an ecosystem showing the one-way flow of energy and the recycling of materials. P_G = gross production, P_N = net production, P = heterotrophic production, and R = respiration.

because of its intimate association with the assimilation of energy via photosynthesis.

A second reason for the prominence of nutrient cycling is the fact that the levels of certain nutrients regulate primary production. In deserts, plant growth in most areas reflects the amount of water rather than the amount of sunlight or minerals in the soil. By contrast, the open oceans are deserts by virtue of their scarce nutrients, particularly nitrogen. Understanding how elements cycle between components of the ecosystem seems crucial to understanding the regulation of ecosystem structure and function.

Primary production

Plants capture light energy and transform it into the energy of chemical bonds in carbohydrates. Glucose and other organic compounds (starch and oils, for example) may be transported throughout the plant or stored conveniently for later release of their energy by respiration. Photosynthesis chemically unites two common inorganic compounds, carbon dioxide (CO_2) and water (H_2O), to form glucose ($G_6H_{12}O_6$), with the release of oxygen (O_2). The overall chemical balance of the photosynthetic reaction is

$$6CO_2 + 6H_2O \rightarrow C_6H_{12}O_6 + 6O_2$$

Photosynthesis transforms carbon from an oxidized (low-energy) state in CO_2 to a reduced (high-energy) state in carbohydrate. Because work is performed on carbon atoms, photosynthesis requires energy. This is provided by visible light. For each gram of carbon assimilated, the plant gains 39 kJ of energy. But because of inefficiencies in the many biochemical steps of photosynthesis, no more than a third (and usually much less) of the light energy absorbed by photosynthetic pigments eventually appears in carbohydrate molecules.

Photosynthesis supplies the carbohydrate building blocks and energy that the plant needs to synthesize tissues and grow. Rearranged and joined together, glucose molecules become fats, oils, and cellulose. Combined with nitrogen, phosphorus, sulfur, and magnesium, simple carbohydrates derived ultimately from glucose produce an array of proteins, nucleic acids, and pigments. Plants cannot grow unless they have all these basic building materials. Remember that chlorophyll contains an atom of magnesium; even though all other necessary elements might be present in abundance, a plant lacking magnesium cannot produce chlorophyll and thus cannot grow.

Plants require energy to build and maintain tissues. Because they use much of the energy they assimilate via photosynthesis to supply these needs, their tissues always contain substantially less energy than the total assimilated. Accordingly, ecologists distinguish two measures of assimilated energy: **gross production,** the total energy assimilated by photosynthesis, and **net production,** the accumulation of energy in plant **biomass** (including plant growth and reproduction). Because plants occupy the first position in the food chain, ecologists refer to these measures as gross or net **primary production.** The difference between gross and net production is the energy of **respiration,** the amount used for maintenance and biosynthesis (Figure 6.3).

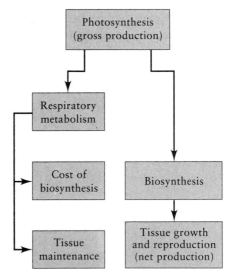

FIGURE 6.3 A diagram of the allocation of energy by plants. The metabolic costs of biosynthesis and tissue maintenance represent energy lost by way of respiration and thus unavailable to higher trophic levels.

Plant production involves fluxes of carbon dioxide, oxygen, minerals, and water on the one hand, and the accumulation of biomass on the other. In principle, the rates of any of these flows could provide an index to the overall rate of primary production. In practice, what measure is appropriate depends on habitat and growth form. Furthermore, ecologists use different measures to estimate gross and net production, or the production of an entire system and that of a small part of a single plant.

Net production can be expressed conveniently as grams of carbon assimilated, dry weight of plant tissues, or their energy equivalents. Ecologists use such indices interchangeably because they have found a high degree of correlation among them. The energy content of an organic compound depends primarily on its carbon content (assimilated at a cost of 39 kJ g^{-1}) and on the energy added or subtracted during various transformations.

In terrestrial ecosystems, ecologists usually estimate plant production by the annual increase in plant biomass. In areas of seasonal production, annual growth may be determined by cutting, drying, and weighing plants at the end of the growing season. Root growth is often ignored because roots are difficult to remove from most soils; thus harvesting measures the **annual aboveground net productivity (AANP)**, the most common basis for comparing terrestrial communities.

Because the atmosphere contains little carbon dioxide (0.03 percent), uptake by plants can measurably reduce its concentration in enclosed chambers within short periods. This concentration provides a direct measurement of photosynthetic rate under some circumstances. The most convenient application of the method is to enclose samples of vegetation (whole herbaceous plants or branches of trees) in clear chambers (light must penetrate for photosynthesis) and measure the change in concentration of CO_2 in air passed through the chamber. CO_2 uptake per gram of dry weight or per square centimeter of surface area of leaves in the chamber is then extrapolated to the entire tree or forest. Carbon dioxide flux during the light period of the day includes both assimilation (uptake) and respiration (output) and thus measures net production. Respiration can be estimated separately by measuring carbon dioxide production during the night, when photosynthesis shuts down. Gross production during the day is then estimated by adding the respiration rate—determined at night—to the daytime rate of net production.

The radioactive isotope carbon-14 (^{14}C) provides a useful variation on the gas exchange method of measuring productivity. When we add a known amount of ^{14}C-carbon dioxide to an airtight enclosure, plants assimilate the radioactive carbon atoms roughly in proportion to their occurrence in the air inside the chamber. Thus we can calculate the rate of carbon fixation by dividing the amount of ^{14}C in the plant by the proportion of ^{14}C in the chamber at the beginning of the experiment. For example, if a plant assimilates 10 mg of ^{14}C in an hour, and the proportion of isotopic carbon in the chamber is 0.05, we can conclude that the plant assimilates carbon at a rate of about 200 mg h^{-1} (10 divided by 0.05).

In aquatic systems, harvesting provides a convenient method for estimating the primary production of large plants, such as kelps, but this technique is not practical for phytoplankton. The low natural concentration of oxygen dissolved in water does, however, enable us to measure small changes in oxygen concentration in most aquatic systems.

To estimate production, samples of water containing phytoplankton are suspended in pairs of sealed bottles at desired depths beneath the surface of a lake or the ocean; one of the pair (a "light bottle") is clear and allows sunlight to enter; the other (a "dark bottle") is opaque. In the light bottles, photosynthesis and respiration occur together, and part of the oxygen produced by the first process is consumed by the second. In the dark bottles, respiration consumes oxygen without its being replenished by photosynthesis. Thus to estimate gross production, we add the change in oxygen concentration in the dark bottle (respiration alone) to that in the light bottle (photosynthesis and respiration).

In unproductive waters, such as those of deep lakes and the open ocean, the uptake of ^{14}C by plants is used to estimate carbon assimilation. The principle of the ^{14}C method is the same in aquatic systems as in terrestrial ones, except that the isotope is usually provided in the form of hydrogen carbonate ion (HCO_3^-).

Light, temperature, and photosynthesis

The productivity of terrestrial habitats generally is not restricted by the availability of light during the growing season, because light levels usually exceed the saturation points of most plants. Thus production most often is limited by water or nutrients. Like most other physiological processes, photosynthesis proceeds most rapidly within a narrow range of temperature; the optimum varies with the prevailing temperature of the environment to which plants acclimate — from about 16°C in many temperate species to as high as 38°C in tropical species — and it may depend on light intensity (Figure 6.4). Net production depends on the rate of respiration as well as on that of photosynthesis, and respiration generally increases with increasing leaf temperature.

Photosynthetic efficiency is the percentage of the energy in incident radiation that is converted to net primary production during the growing season, and it provides a useful index to rates of primary production under natural conditions. Where water and nutrients do not severely limit plant production, photosynthetic efficiency varies between 1 and 2 percent. What happens to the remaining 98 to 99 percent of the energy? Leaves and other surfaces reflect anywhere from one-quarter to three-quarters of it. Molecules other than photosynthetic pigments absorb most of the remainder, which is converted to heat and either radiated or conducted across the leaf surface or dissipated by the evaporation of water from the leaf (transpiration).

Water and production in terrestrial habitats

The tiny openings (stomates) through which leaves exchange carbon dioxide and oxygen with the atmosphere also allow the passage of water vapor (tran-

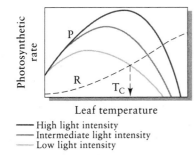

FIGURE 6.4 Photosynthetic rate as a function of leaf temperature and light intensity in the alpine heath *Loiseleuria* in Austria. Respiration (R, dashed line) increases with temperature. Comparing respiration to photosynthesis (P) shows that the temperature above which R exceeds P (the compensation point, T_C) increases with light intensity. After W. Larcher et al., in T. Rosswall and O. W. Heal (eds.), Structure and function in tundra ecosystems, *Ecol. Bull.* 20:125–139 (1975).

spiration). As the moisture content of soil decreases, it is increasingly difficult for plants to obtain water. As soil moisture approaches the wilting point, leaves close their stomates to reduce water loss. This prevents uptake of CO_2, and photosynthesis slows to a standstill. Consequently, the rate of photosynthesis depends on a plant's ability to tolerate water loss, the availability of soil moisture, and the influence of air temperature and solar radiation on the rate of transpiration.

Agronomists rate the drought resistance of plants in terms of **transpiration efficiency**, which is the ratio of net production to transpiration. In most plants, transpiration efficiencies are less than 2 grams of production per kilogram of water transpired, but they may be as high as $4 \, g \, kg^{-1}$ in drought-tolerant crops. Because transpiration efficiency varies little among a wide variety of plants, production is directly related to water availability in the environment. For example, annual production of perennial grasses in southern Arizona varies in direct proportion to summer rainfall during the year of production (Figure 6.5).

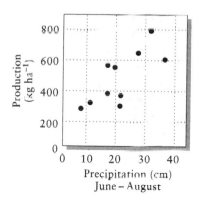

FIGURE 6.5 Relationship between production and summer precipitation for perennial grassland in southern Arizona. After D. R. Cable, *Ecology* 56:981–986 (1975).

Nutrients and plant production

Fertilizers stimulate plant growth in most habitats. For example, when nitrogen and phosphorus fertilizers were applied singly and in combination to chaparral habitat in southern California, most species responded to the application of nitrogen but not to that of phosphorus (Figure 6.6). This result suggests that the availability of nitrogen limited production. In contrast, production of the California lilac (*Ceanothus greggii*), which harbors nitrogen-fixing bacteria in root nodules, responded to the application of phosphorus but not to that of nitrogen. The productivity of annual plants (forbs and grasses) in the same habitat increased when nitrogen was applied but was depressed somewhat by the application of phosphorus alone. When the same amounts of nitrogen *and* phosphorus were added to a single plot, however, production soared. The plants could take advantage of increased phosphorus only in the presence of high levels of nitrogen.

Plants generally suffer nutrient limitation most strongly in aquatic habitats, particularly in the open ocean where the scarcity of dissolved minerals reduces production far below terrestrial levels. Even in shallow coastal waters, where vertical mixing, upwelling currents, and runoff from the land maintain nutrients at high concentrations, the addition of fertilizer (as often occurs inadvertently through water pollution) may greatly enhance aquatic production. In a study conducted along the southern coast of Long Island, New York, phytoplankton abundance closely paralleled the level of inorganic phosphorus, the latter being a general indicator of pollution. But the addition of nutrients to standard cultures of the marine alga *Nannochloris* in water samples taken from different areas demonstrated that nitrogen rather than phosphorus limited primary production in both polluted and relatively clean waters (Figure 6.7). Thus phosphorus occurred in sufficient abundance to promote algal growth, even in the unpolluted sections of coastline.

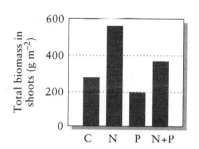

FIGURE 6.6 Response of the chaparral shrub *Adenostema* to nitrogen (N) and phosphorus (P) fertilization (C = control). After G. S. McMaster, W. M. Jow, and J. Kummerow, *J. Ecology* 70:745–756 (1982).

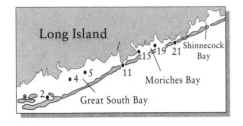

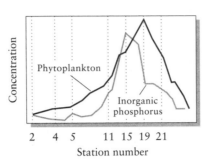

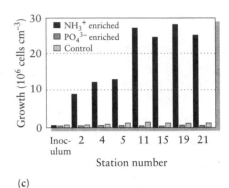

(a) (b) (c)

FIGURE 6.7 (a) Map of water-collection stations in Great South Bay and Moriches Bay, Long Island. (b) Phytoplankton and inorganic phosphorus concentrations in water collected in the summer of 1962 from stations shown in (a). (c) Growth of standard cultures of the marine alga *Nannochloris* in water, taken from each of the stations, to which ammonium or phosphate had been added. From J. H. Ryther and W. M. Dunstan, *Science* 171:1008–1013 (1971).

Global patterns in primary production

The favorable combination of intense sunlight, warm temperature, and abundant rainfall in the humid tropics results in the highest terrestrial productivity on earth. In temperate and arctic ecosystems, low winter temperatures and long winter nights curtail production. Within a given latitude belt, where light and temperature do not vary appreciably from one locality to the next, net production is directly related to annual precipitation. Above a certain threshold of water availability, net production increases by 0.4 gram dry mass per kilogram of water in hot deserts and by 1.1 g kg^{-1} in short-grass prairies and cold (Great Basin) deserts. Thus a given amount of water is more productive in cold climates than in hot climates. The productivity of forest ecosystems, where water use efficiencies vary between about 0.9 and 1.8 g kg^{-1}, is relatively insensitive to variation in water availability. Thus water generally does not limit forest production. However, forests do not develop except under moist conditions, usually in excess of at least 50 cm of precipitation annually in the cooler regions of the United States and perhaps 100 cm in the warmer eastern and southeastern areas.

Global patterns of net primary production are summarized in Table 6.1. These values come from many studies employing a wide variety of techniques, but their lack in strict comparability does not override the general patterns they reveal. Production of terrestrial vegetation is greatest in the wet tropics and least in tundra and desert habitats. Swamp and marsh ecosystems, which

				Leaf	Biomass
	Net primary production ($g\ m^{-2}\ yr^{-1}$)	Biomass ($kg\ m^{-2}$)	Chlorophyll ($g\ m^{-2}$)	surface area ($m^2\ m^{-2}$)	accumulation ratio (yr)
Habitat					
Terrestrial					
Tropical forest	1800	42	2.8	7	23
Temperate forest	1250	32	2.6	8	26
Boreal forest	800	20	3.0	12	25
Shrubland	600	6	1.6	4	10
Savanna	700	4	1.5	4	6
Temperate grassland	500	1.5	1.3	4	3
Tundra and alpine	140	0.6	0.5	2	4
Desert scrub	70	0.7	0.5	1	10
Cultivated land	650	1	1.5	4	1.5
Swamp and marsh	2500	15	3.0	7	6
Aquatic					
Open ocean	125	0.003	0.03	—	0.02
Continental shelf	360	0.01	0.2	—	0.03
Algal beds and reefs	2000	2	2.0	—	1.00
Estuaries	1800	1	1.0	—	0.56
Lakes and streams	500	0.02	0.2	—	0.04

TABLE 6.1 *Average net primary production and related dimensions of the earth's major ecosystems*

Source: R. H. Whittaker and G. E. Likens, *Human Ecol.* 1:357–369 (1973).

occupy the interface between terrestrial and aquatic habitats, can produce as much biomass annually as tropical forests.

The open ocean is a virtual desert, where scarcity of mineral nutrients limits productivity to a tenth that of temperate forests, or even less. Upwelling zones (where nutrients reach the surface from deeper waters) and continental-shelf areas (where bottom sediments in shallow water rapidly exchange nutrients with surface waters) support greater production. In estuaries, coral reefs, and coastal algal beds, production approaches the levels observed in adjacent terrestrial habitats. Primary production in freshwater habitats compares favorably with that in marine habitats, achieving the highest levels in rivers, shallow lakes, and ponds and the lowest in clear streams and deep lakes. In general, phosphorus limits production in freshwater systems, nitrogen in marine systems.

Food chain energetics

Plants manufacture their own food from raw inorganic materials. Hence ecologists refer to plants as **autotrophs** (literally, "self-nourishers"). Animals and most microorganisms, which obtain their energy and most of their nutrients by eating plants or animals or the dead remains of either, are called **heterotrophs** (literally, "nourished from others"). The dual roles of living forms as food producers and food consumers give the ecosystem a trophic structure, determined by feeding relationships, through which energy flows and nutrients cycle. The food chain from grass to caterpillar to sparrow to snake to hawk traces one particular path that energy follows through the trophic structure. With each link in the food chain, biochemical transformations dissipate much energy before organisms feeding at the next higher trophic level can consume it. All the grass in Africa piled together would dwarf a mound of all the grasshoppers, gazelles, zebras, wildebeests, and other animals that eat grass. That mound of herbivores, in turn, would overwhelm the pitiful heap of all the lions, hyenas, and other carnivores that feed on them.

As Lindeman pointed out in 1942, the amount of energy reaching each trophic level depends on both the net primary production itself and the efficiencies with which animals convert food energy to biomass energy at each trophic level. Of the light energy assimilated by photosynthesis, plants use 15 to 70 percent for maintenance, thereby making that portion unavailable to consumers. Herbivores and carnivores are more active than plants and expend correspondingly more of their assimilated energy on maintenance. As a result, the productivity of each trophic level is typically only 5 to 20 percent that of the level below it (Figure 6.8). Ecologists refer to the percentage transfer of energy from one trophic level to the next as either the **ecological efficiency** or the **food chain efficiency.**

Once consumed, food energy follows a variety of paths through the organism (Figure 6.9). Regardless of the source of food, what the organism digests

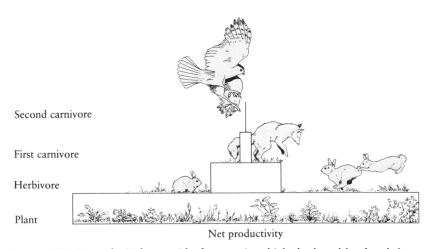

Second carnivore

First carnivore

Herbivore

Plant

Net productivity

FIGURE 6.8 An ecological pyramid of energy in which the breadth of each bar represents the net productivity of each trophic level in the ecosystem. For this particular system, ecological efficiencies are 20, 15, and 10 percent between trophic levels, but these values vary widely in different communities.

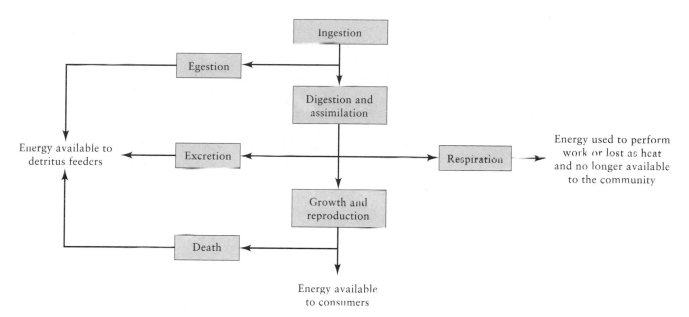

FIGURE 6.9 Allocation of energy within a link of the food chain.

and absorbs constitutes its **assimilated energy,** which supports maintenance, builds tissues, or may be excreted in the form of unusable metabolic by-products. The energy used to fulfill metabolic needs, most of which escapes the organism as heat, makes up the **respired energy.** Animals excrete a smaller fraction of assimilated energy in nitrogen-containing wastes (primarily ammonia, urea, or uric acid) that are produced when the diet contains an excess of nitrogen. Assimilated energy retained by the individual organism becomes available for the synthesis of new biomass (production) through growth and reproduction, which animals feeding at the next higher trophic level may then consume.

Many components of food resist digestion and assimilation: hair, feathers, insect exoskeletons, cartilage, and bone in animal foods, and cellulose and lignin in plant foods (Figure 6.10). These substances may be **egested** by defecation or regurgitation. Such egested materials may have been relatively unaltered chemically during their passage through the gut, but nearly all disintegrated into smaller fragments when subjected to chewing and to contractions of the stomach and intestines. This mechanical breakdown makes them more readily usable by detritus feeders.

Assimilation efficiency

Ecological efficiency is the product of the efficiencies with which organisms exploit their food resources and convert them into biomass. Because most biological production is consumed by one organism or another, the exploita-

FIGURE 6.10 Photograph of elephant dung showing the poorly digested fibrous plant material.

tion efficiency of an entire trophic level approaches 100 percent overall. Therefore, ecological efficiency depends on two factors: the proportion of consumed energy assimilated (the **assimilation efficiency**) and the proportion of assimilated energy incorporated in growth, storage, and reproduction (the **net production efficiency**). Box 6.1 summarizes several energetic efficiencies.

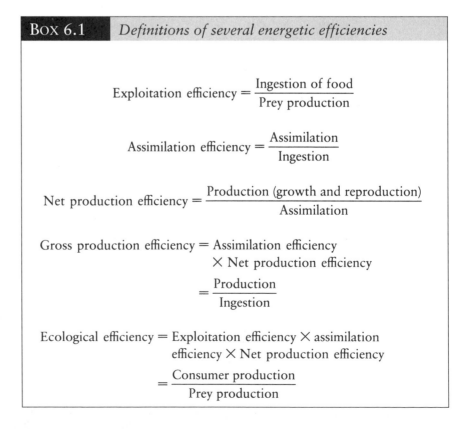

BOX 6.1	*Definitions of several energetic efficiencies*

$$\text{Exploitation efficiency} = \frac{\text{Ingestion of food}}{\text{Prey production}}$$

$$\text{Assimilation efficiency} = \frac{\text{Assimilation}}{\text{Ingestion}}$$

$$\text{Net production efficiency} = \frac{\text{Production (growth and reproduction)}}{\text{Assimilation}}$$

$$\text{Gross production efficiency} = \text{Assimilation efficiency} \times \text{Net production efficiency}$$
$$= \frac{\text{Production}}{\text{Ingestion}}$$

$$\text{Ecological efficiency} = \text{Exploitation efficiency} \times \text{assimilation efficiency} \times \text{Net production efficiency}$$
$$= \frac{\text{Consumer production}}{\text{Prey production}}$$

For plants, net production efficiency is the ratio of net to gross production. This index varies between 30 and 85 percent, depending on habitat and growth form. Rapidly growing plants in temperate zones — whether trees, old-field herbs, crop species, or aquatic plants — have uniformly high net production efficiencies (75 to 85 percent). Similar types of vegetation in the tropics exhibit lower net production efficiencies, perhaps 40 to 60 percent. Thus, as we should expect because of the higher temperature, respiration increases relative to photosynthesis in tropical latitudes.

The energy value of plants depends on the amount of cellulose, lignin, and other indigestible materials present. Herbivores assimilate as much as 80 percent of the energy in seeds and 60 to 70 percent of that in young vegetation. Most grazers and browsers (elephants, cattle, grasshoppers) extract 30 to 40 percent of the energy in their food for themselves. Millipedes, which eat decaying wood composed mostly of cellulose and lignin (and the microorganisms that occur in decaying wood), assimilate only 15 percent of the energy in their diet.

The effect of food quality on assimilation can be seen in the efficiencies with which a single animal retains energy from different portions of its diet. In studies with mountain hares (*Lepus timidus*), the efficiency of energy assimilation on a diet of small willow twigs came to 39 percent, of which the hares lost an additional 5 percent through urinary nitrogen. Lower assimilation efficiencies characterized diets of larger twigs (31 percent), presumably because of their thick, less digestible bark, and of birch twigs (23 to 35 percent), on which hares could not maintain a constant weight. By measuring the content of fiber (cellulose and lignin) in the food and feces, investigators determined that the digestibility of fiber varied between only 15 and 25 percent. Assimilation of various elements in the diet, estimated by taking similar input – output measurements, was found to vary between 9 percent for phosphorus and 81 percent for magnesium on willow twigs 3 to 5 mm in diameter.

Food of animal origin is more easily digested than that of plant origin; assimilation efficiencies of predatory species vary from 60 to 90 percent. Vertebrate prey are digested more efficiently than insect prey, because the indigestible exoskeletons of insects constitute a larger proportion of the body than the hair, feathers, and scales of vertebrates. Assimilation efficiencies of insectivores vary between 70 and 80 percent, whereas those of most carnivores are about 90 percent.

Net production efficiency

Maintenance, movement, and heat production require energy that animals otherwise could use for growth and reproduction. Active, warm-blooded animals exhibit low net production efficiencies: birds less than 1 percent and small mammals with high reproductive rates up to 6 percent. More sedentary, cold-blooded animals, particularly aquatic species, channel as much as 75 percent of their assimilated energy into growth and reproduction. The overall efficiency of biomass production within a trophic level (**gross production efficiency**) is the product of assimilation efficiency and net production efficiency. Gross production efficiencies of warm-blooded terrestrial animals rarely ex-

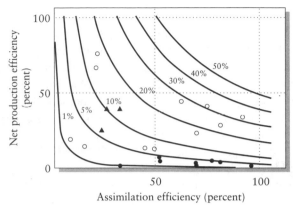

Figure legend:
○ Cold-blooded aquatic animals   ● Warm-blooded terrestrial animals
▲ Cold-blooded terrestrial animals

FIGURE 6.11 Relationships between assimilation efficiency and net production efficiency for a variety of animals. Gross production efficiencies are indicated by the curved lines on the graph. From R. E. Ricklefs, *Ecology* (2nd ed.). Chiron Press, New York (1979).

ceed 5 percent, and those of some birds and large mammals fall below 1 percent. For insects, these efficiencies lie within the range of 5 to 15 percent, and for some aquatic animals they exceed 30 percent (Figure 6.11).

Detritus feeding

Terrestrial plants, especially woody species, allocate much of their production to structures that are difficult to ingest, let alone digest. As a result, most terrestrial plant production is consumed as **detritus** — dead remains and undigestible excreta of herbivores — by organisms specialized to attack wood and leaf litter. This partitioning between herbivory and detritus feeding establishes two food chains in terrestrial communities. The first originates as relatively large animals feed on leafy vegetation, fruits, and seeds; the second originates as relatively small animals and microorganisms consume detritus in the litter and soil layer. These separate food chains sometimes mingle considerably at higher trophic levels, but the energy of detritus tends to move into the food chain much more slowly than the energy consumed by herbivores.

The relative importance of herbivore-based and detritivore-based food chains also varies greatly among communities. Herbivores predominate in plankton communities, detritus feeders in terrestrial communities. The proportion of net production that enters herbivore–predator strands of the food web depends on the relative allocation of plant tissue between structural and supportive functions on the one hand, and growth and photosynthetic functions on the other. Herbivores consume 1.5 to 2.5 percent of the net production in temperate deciduous forests, 12 percent of that in old-field habitats, and 60 to 99 percent of that in plankton communities.

Rate of energy flow through the community

Food chain efficiencies indicate the amount of energy that eventually reaches each trophic level of the community. The rate of transfer of energy or, inversely, its **residence time** in each trophic level provides a second index to the energy dynamics of the ecosystem. For a given rate of production, the residence time (or **transit time**) of energy in the community and the storage of energy in living biomass and detritus are directly related: the longer the residence time, the greater the accumulation of energy.

The average residence time in a particular link in the food chain equals the energy stored divided by the rate of flux:

$$\text{Residence time (yr)} = \frac{\text{Energy stored in biomass (kJ m}^{-2}\text{)}}{\text{Net productivity (kJ m}^{-2}\text{ yr}^{-1}\text{)}}$$

(We may calculate the residence time defined by this equation in terms of mass rather than energy, in which case it expresses the **biomass accumulation ratio**.) Wet tropical forests produce dry matter at an average rate of 1800 g m^{-2} yr^{-1} and have an average living biomass of 42,000 g m^{-2}. Inserting these values into the equation for average residence time, we obtain 23 years (42,000/1800). Average residence times of energy in plants in representative ecosystems vary from more than 20 years in forested terrestrial environments to less than 20 days in aquatic plankton-based communities (Table 6.1). These figures are only averages. Leaf eaters and root feeders consume much of the energy assimilated by forest trees during the year of production, some of it within days of assimilation by the plant.

The figures in Table 6.1 also underestimate the average residence time of energy in plant biomass, because they do not include the accumulation of dead organic matter in the litter. The residence time of energy in accumulated litter can be determined by an equation analogous to that for the biomass accumulation ratio:

$$\text{Residence time (yr)} = \frac{\text{Litter accumulation (g m}^{-2}\text{)}}{\text{Rate of litter fall (g m}^{-2}\text{ yr}^{-1}\text{)}}$$

In forests, this value varies from 3 months in the wet tropics to 1 to 2 years in dry and montane tropical habitats, 4 to 16 years in the southeastern United States, and more than 100 years in temperate mountains and boreal regions. Warm temperatures and the abundance of moisture in lowland tropical regions create optimal conditions for rapid decomposition.

Ecosystem energetics

Flux of energy and its efficiency of transfer summarize certain aspects of the structure of an ecosystem: the number of trophic levels, the relative importance of detritus feeding and herbivory, the steady-state values for biomass and accumulated detritus, and the turnover rates of organic matter in the community. The importance of these measures was argued by Lindeman, who constructed the first energy budget for an entire biological community — that of Cedar Bog Lake in Minnesota. The proliferation of energy flow studies during the 1950s and 1960s clearly reflected energy's value as a universal currency, a common denominator to which all populations and their acts of consumption could be reduced.

The overall energy budget of the ecosystem reflects the balance between income and expenditure, just as it does in a bank account. The ecosystem gains energy through the photosynthetic assimilation of light by green plants and the transport of organic matter into the system from outside. Sources of organic matter are referred to as **allochthonous** inputs (from the Greek *chthonos*, "of the earth," and *allos*, "other"); local production is referred to as **autochthonous.**

In Root Spring near Concord, Massachusetts, the energy flux of herbivores is 0.31 W m^{-2}, but the net productivity of aquatic plants is only 0.09 W m^{-2}, the balance being transported into the spring by leaf fall from nearby trees. In general, autochthonous production predominates in large rivers, lakes, and most marine ecosystems; allochthonous imports make up the largest part in small streams and springs under the closed canopies of forests; and life in caves and abyssal depths of the oceans, to which no light penetrates, subsists entirely on energy transported in from outside.

Lindeman constructed the Cedar Bog Lake energy budget (Table 6.2) from measurements of the harvestable net production on each of three trophic levels — plants (primary producers), herbivores (primary consumers), and all

TABLE 6.2	*An energy flow model for Cedar Lake Bog, Minnesota*		
	ENERGY (kcal m^{-2} yr^{-1})		
Energy production or removal	Primary producers	Primary consumers	Secondary consumers
Harvestable production*	704	70	13
Respiration	234	44	18
Removal by consumers			
Assimilated	148	31	0
Unassimilated	28	3	0
Gross production (totals)	1114	148	31

*Does not include net production removed by consumers. Actual net production, including removal by consumers, was 879 kcal m^{-2} yr^{-1} for primary producers, 104 kcal m^{-2} yr^{-1} for primary consumers, and 13 kcal m^{-2} yr^{-1} for secondary consumers.

Source: L. Lindeman, The trophic–dynamic aspect of ecology. *Ecology* 23:399–418 (1942).

carnivores lumped together (secondary consumers) — and from laboratory determinations of respiration and assimilation efficiencies. Lindeman's energy budget is somewhat startling in that organisms at the next higher trophic level consumed only 20 percent of the net primary production and 33 percent of the net secondary production. Most of the biomass not consumed sedimented out of the system, causing the lake to fill with layers of organic detritus.

Even with sedimentation, the Cedar Bog Lake ecosystem achieved a 12 percent overall ecological efficiency of energy transfer between trophic levels. After comparing similar analyses for five aquatic communities, ecologist D. G. Kozlovski concluded that (1) assimilation efficiency increases at higher trophic levels; (2) net production efficiency decreases at higher levels; (3) gross production efficiency also decreases; and (4) ecological efficiency (assimilation or gross production at level n divided by that at level $n - 1$) remains constant between trophic levels at about 10 percent.

Food chain length

Kozlovski's 10 percent generalization is not a fixed law of ecological thermodynamics. In Silver Springs, Florida, investigators measured ecological efficiencies of 17 percent between the producer and herbivore level but only 5 percent between herbivores and secondary consumers. Ecological efficiencies are usually lower in terrestrial than in aquatic habitats. A useful rule of thumb is that the top carnivores in terrestrial communities can feed no higher than the third trophic level, on average, whereas aquatic carnivores may feed as high as the fourth or fifth level. Of course, a tiny fraction of the total energy may travel through a dozen links before it is dissipated by respiration. But such high trophic levels certainly do not contain enough energy to fully support a single predator population (Box 6.2).

✨ SUMMARY

The ecosystem is the whole complex of organisms and the physical environments they inhabit. It is also a giant thermodynamic machine that continuously dissipates energy in the form of heat. This energy initially enters the biological realm of the ecosystem via photosynthesis and plant production, which provides energy for animals and nonphotosynthetic microorganisms.

1. Charles Elton described communities in terms of feeding relationships and emphasized the pyramid of numbers as a dominant organizing principle in community structure.

2. In 1935, A. G. Tansley coined the term *ecosystem* to include organisms and all the abiotic factors in the habitat.

3. A. J. Lotka, in 1925, provided a thermodynamic perspective on ecosystem function, showing that the movement and transformations of mass and energy conform to thermodynamic laws.

BOX 6.2	*Estimating the average number of trophic levels in various ecosystems from primary production, consumer energy flux, and ecological efficiencies*

We can estimate the average length of food chains in a community from net primary production, average ecological efficiency, and average energy flux of a top predator population. The energy available [$E(n)$] to a predator at a given trophic level n (plants being level 1) is equal to the product of the net primary production (NNP) and the intervening ecological efficiencies (Eff). Thus

$$E(n) = \text{NNP Eff}^{n-1}$$

where Eff is the geometric mean of the efficiencies of transfer between each level. This expression can be rearranged to give

$$n = 1 + \frac{\log[E(n)] - \log(\text{NPP})}{\log(\text{Eff})}$$

Using this equation and some rough estimates for the values on its right-hand side, we can calculate the average number of trophic levels to be about 7 for marine plankton-based systems, 5 for inshore aquatic communities, 4 for grasslands, and 3 for wet tropical forests, as shown in the following table.

Community	Net primary production (kcal m^{-2} yr^{-1})	Predator ingestion (kcal m^{-2} yr^{-1})	Ecological efficiency (percent)	Number of trophic levels
Open ocean	500	0.1	25	7.1
Coastal marine	8000	10.0	20	5.1
Temperate grassland	2000	1.0	10	4.3
Tropical forest	8000	10.0	5	3.2

Values are approximations based on many studies.

These estimates should be taken with a grain of salt, to be sure, but they do indicate the general size of the pyramid of energy built upon a base of primary production within an ecosystem.

4. Raymond Lindeman, in 1942, popularized the idea of the ecosystem as an energy-transforming system, providing a formal notation for the energy flux in trophic levels and for ecological efficiency.

5. The study of ecosystem energetics dominated ecology during the 1950s and 1960s, due largely to the influence of Eugene P. Odum, who championed energy as a common currency for describing ecosystem structure and function.

6. Gross primary production is the total energy fixed by photosynthesis. Net primary production is the accumulation of energy in plant biomass; hence it is the difference between gross production and plant respiration.

7. Primary production is measured by one or some combination of a variety of methods: harvest, gas exchange (carbon dioxide in terrestrial habitats, oxygen in aquatic habitats), and assimilation of radioactive carbon (^{14}C).

8. The rate of primary production in most terrestrial habitats is not limited by light, which generally exceeds the saturation points of most species. The efficiency of photosynthesis (gross production divided by total incident light energy) during daylight periods in the growing season is 1 to 2 percent in most habitats.

9. Because plants lose water in direct proportion to the amount of carbon dioxide assimilated, plant production in dry environments is limited by, and varies in direct proportion to, the availability of water. Transpiration efficiency, the ratio of production (in grams of dry mass) to water transpired (in kilograms), typically ranges between 1 and 2; it occasionally reaches 4 in drought-adapted species.

10. Production in both terrestrial and aquatic environments can be enhanced by the application of various nutrients, especially nitrogen and phosphorus, which indicates that nutrient availability limits production. This effect is greatest in the open ocean and deep lakes, where nutrient inputs from sediments and runoff are least.

11. The movement of energy and materials through the food chain can be characterized by the assimilation efficiency (the ratio of assimilation to digestion) and the net production efficiency (the ratio of production to assimilation). Unassimilated material enters detritus-based food chains.

12. Assimilation efficiency depends on the quality of the diet, particularly the amounts of digestion-resistant structural material (cellulose, lignin, chitin, keratin), and it varies from about 15 to 90 percent. Net production efficiency is lowest in animals whose costs of maintenance and activity are greatest, especially warm-blooded vertebrates, for which values of 1 to 5 percent contrast with the values of 15 to 45 percent that are typical of invertebrates.

13. Gross production efficiency (the ratio of production to ingestion) varies between about 5 and 20 percent in most studies.

14. The average residence time of biomass or energy in a single link of the food chain is the ratio of biomass or the energy stored in it to rate of net production. Average residence times for primary production vary from 20 years in some forests to 20 days or less in aquatic, plankton-based communities.

15. Considerations of energy flux and ecological efficiencies suggest that the highest trophic level at which a consumer population can be maintained ranges from the third level in terrestrial food chains to the seventh level in plankton-based communities of the open ocean.

❧ SUGGESTED READINGS

Clarkson, D. T., and J. B. Hanson. 1980. The mineral nutrition of higher plants. *Annual Review of Plant Physiology* 31:239–298.

Cook, R. E. 1977. Raymond Lindeman and the trophic–dynamic concept in ecology. *Science* 198:22–26.

Fenchel, T. 1988. Marine plankton food chains. *Annual Review of Ecology and Systematics* 19:19–38.

Griffin, D. H. 1981. *Fungal Physiology.* Wiley, New York.

Howarth, R. W. 1988. Nutrient limitation of net primary production in marine ecosystems. *Annual Review of Ecology and Systematics* 19:89–110.

Laws, R. M. 1985. The ecology of the Southern Ocean. *American Scientist* 73:26–40.

Lindeman, R. 1942. The trophic–dynamic aspect of ecology. *Ecology* 23:399–418.

Odum, E. P. 1968. Energy flow in ecosystems: a historical review. *American Zoologist* 8:11–18.

Tilman, G. D. 1984. Plant dominance along an experimental nutrient gradient. *Ecology* 65:1445–1453.

Webb, W., S. Szarek, W. Lauenroth, R. Kinerson, and M. Smith. 1978. Primary productivity and water use in native forest, grassland, and desert ecosystems. *Ecology* 59:1239–1247.

Whittaker, R. H., and G. E. Likens. 1973. Primary production: the biosphere and man. *Human Ecology* 1:357–369.

Wiegert, R. G. 1988. The past, present, and future of ecological energetics. In L. R. Pomeroy and J. J. Albert (eds.), *Concepts of Ecosystem Ecology. A Comparative View.* Springer-Verlag, New York, pp. 29–55.

7

PATHWAYS OF ELEMENTS IN THE ECOSYSTEM

Nutrients, unlike energy, remain within the ecosystem, where they continuously cycle between organisms and the physical environment. Most of these nutrients originate in rocks of the earth's crust or in the earth's atmosphere, but within the ecosystem they are reused over and over by plants and animals before being lost in sediments, streams and groundwater, or gases. Though all the energy assimilated by green plants is "new" energy received from outside the ecosystem, most of the nutrients assimilated by plants have been used before. Ammonia absorbed from the soil by roots might have leached out of decaying leaves on the forest floor that same day. The carbon dioxide assimilated by every green plant originated by animal, plant, or microbial respiration. On average, a molecule of carbon dioxide resides in the atmosphere for about 8 years before being reassimilated, although plants may recycle carbon dioxide almost instantaneously (for example, within the closed canopy of a forest) before it escapes to the general atmospheric circulation.

Each element follows a unique route, determined by its particular biochemical transformations, in its cycle through the ecosystem. Living systems transform elements and their compounds to furnish nutrients for building structure and to mobilize the energy required by all life processes. Over long periods, processes that transform elements from one form to another must balance those processes that restore the first form.

Elemental cycles sometimes do become unbalanced, and elements accumulate in or are removed from the system. For example, during periods of coal and peat formation, dead organic materials accumulate in the sediments of lakes, marshes, and shallow seas where anaerobic conditions slow their decomposition by microorganisms. In other cases, under intensive cultivation or following the removal of natural vegetation, erosion can wash away nutrient-laden layers of soil that took centuries to develop. By and large, however, most systems exist in a steady state in which the import of elements approximately balances their export. Furthermore, the ecosystem usually gains and loses nutrients slowly compared to the rate at which they cycle within the system.

Energy transformation and element cycling

Transformations that produce organic forms of a particular element are referred to as **assimilatory** processes. Photosynthesis, by which inorganic carbon (carbon dioxide) is changed to the organic carbon of carbohydrates, is one example of an assimilatory transformation of an element. Others involving nitrogen and sulfur, for example, assume equal importance in the ecosystem. In the overall cycling of carbon, photosynthesis is balanced by respiration, a complementary **dissimilatory** process that involves the oxidation of organic carbon with the accompanying release of energy and the return of carbon to its available inorganic form.

Not all transformations of elements in the ecosystem are biological, nor do all involve the net assimilation or release of useful quantities of energy. Many chemical reactions take place in the air, soil, and water. Some of these, such as the weathering of bedrock, release certain elements (potassium, phosphorus, and silicon, for example) to the ecosystem. Other physical and chemical processes, such as the sedimentation of calcium carbonate in the oceans, remove elements from circulation and incorporate them into rocks in the earth's crust, where they can remain locked up for eons.

Most energy transformations are associated with the biochemical oxidation and reduction of carbon, oxygen, nitrogen, phosphorus, and sulfur. In each case, an energy-releasing transformation (**oxidation**) pairs with an energy-requiring transformation (**reduction**), and energy shifts from the reactants in the first to the products in the second (Figure 7.1). Usually, oxidation releases more energy than reduction requires, and the balance escapes as heat —hence the thermodynamic inefficiency of life processes. A typical coupling between transformations might involve the oxidation of carbon in carbohydrate (perhaps glucose), with the release of energy, and the reduction of nitrate-nitrogen to amino-nitrogen (the building blocks of proteins), which requires energy. These particular transformations may involve many intermediate steps of the type shown in Figure 7.1.

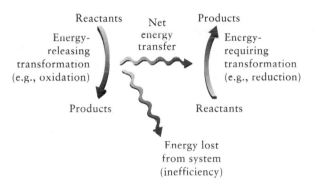

FIGURE 7.1 The coupling of energy-releasing and energy-requiring transformations is the basis of energy flow in the ecosystem.

Plants accomplish the initial input of energy into the ecosystem by an assimilatory transformation—the reduction of carbon—in which light, rather than a coupled dissimilatory process, serves as the source of energy (Figure 7.2). A portion of that energy escapes biological systems with each subsequent transformation. Some of these transformations involve the assimilation of other elements required for growth and reproduction; most are biochemical transformations required to maintain the cell environment and to effect movement. The cycling of elements between living and physical parts of the ecosystem is connected to energy flow by the coupling of the dissimilatory part of one cycle to the assimilatory part of another.

Compartment models of the ecosystem

Each form of an element can be thought of as occupying a separate **compartment** in the ecosystem, like a room in a house. In this analogy, biochemical transformations are fluxes of the elements among the compartments—that is, their movement between rooms (Figure 7.3). Carbon occurs as carbon dioxide both in the atmosphere and dissolved in water, as carbonate and bicarbonate

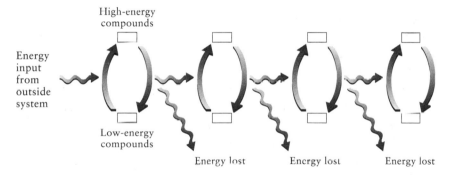

FIGURE 7.2 As energy flows through the ecosystem, elements alternate between assimilatory and dissimilatory transformations, thus going through cycles.

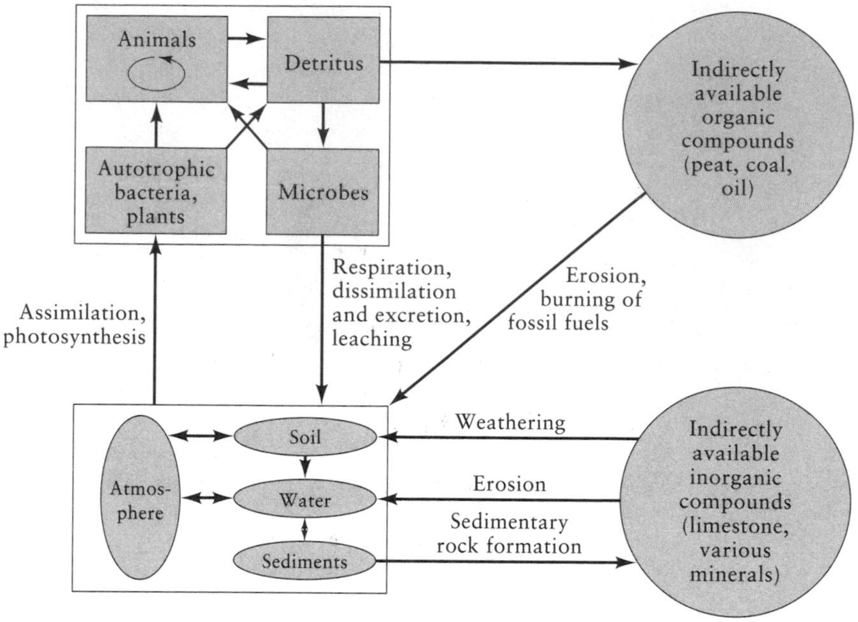

FIGURE 7.3 A generalized compartment model of the ecosystem. A more specific model can be drawn for each element that cycles in the ecosystem; Figure 7.5 is a good example.

ions dissolved in water, as calcium carbonate (limestone) in sediments, and in all organic molecules. Photosynthesis moves carbon from the carbon dioxide compartment to that containing organic forms of carbon (assimilation); respiration brings it back (dissimilation). The compartment of organic carbon has many subcompartments: animals, plants, microorganisms, and detritus. Herbivory, predation, and detritus feeding move carbon among these subcompartments.

Some transformations involve changes in energy state. Photosynthesis adds energy to carbon, which we may think of as lifting the element to the second floor of a house. In descending the respiration "staircase," carbon releases this stored chemical energy, which the individual can then use for other purposes.

Elements cycle rapidly among some compartments of the ecosystem and much more slowly among others. The movement of an element between living organisms and the inorganic forms of the element that organisms both produce and use occurs over periods ranging from a few minutes to the life span of the organism and its subsequent existence as organic detritus. Both organic and inorganic forms of elements may leave the rapid circulation within the ecosystem for compartments not readily accessible to transforming agents. For example, coal, oil, and peat contain vast quantities of carbon that has been removed from circulation in the ecosystem. Sedimentation removes inorganic (oxidized) carbon from ecosystem circulation primarily by sedimentation as calcium carbonate, which forms thick layers of limestone over large areas presently or formerly covered by seas. These forms of carbon are returned to the rapid cycling compartments only by the slow geological processes of uplift and erosion. Of course, humans have greatly accelerated one component of this flux by burning fossil fuels.

The water cycle

Although water is chemically involved in photosynthesis, most water flux through the ecosystem happens by the physical processes of evaporation, transpiration, and precipitation (Figure 7.4). Light energy absorbed by water performs the work of evaporation. The condensation of atmospheric water vapor to form clouds releases the potential energy in water vapor as heat. Thus evaporation and condensation resemble photosynthesis and respiration thermodynamically, and they provide an instructive model for the cycling of nutrients within the ecosystem more generally.

More than 90 percent of all water is locked away in rocks in the earth's core and in sedimentary deposits near the surface. Water from this compartment enters the **hydrological cycle** of the biosphere through such geological processes as volcanic outpourings of steam. Most of the water at the earth's surface originated in this way, although the great reservoirs in the interior contribute little to contemporary water flux in the ecosystem on a day-to-day basis.

Over land surfaces, precipitation (23 percent of the global total) exceeds evaporation and transpiration (16 percent). Evaporation of water from the oceans, however, exceeds replenishment from precipitation. Much of the water vapor transported by winds from the oceans to the land condenses and falls as precipitation over mountainous regions and where rapid heating of the land surface creates vertical air currents. The net flow of atmospheric water vapor from ocean to land balances runoff from the land into ocean basins.

We may calculate the energy that drives the global hydrological cycle by multiplying the total weight of water evaporated (378×10^{18} g yr^{-1}) by the energy required to evaporate 1 g of water (2.24 kJ). The product, 8.5×10^{20} kJ yr^{-1}, represents about one-fifth of the total energy income from the sun's radiation striking the earth. The remaining energy either is reflected or is absorbed and reradiated as heat. Of the total radiation striking temperate forests, for ex-

Atmospheric water vapor (13)

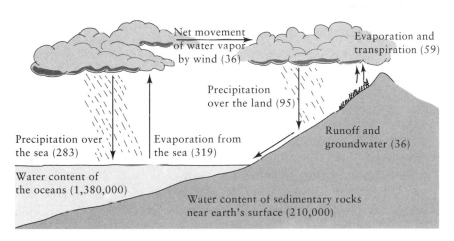

Net movement of water vapor by wind (36)

Evaporation and transpiration (59)

Precipitation over the land (95)

Precipitation over the sea (283)

Evaporation from the sea (319)

Runoff and groundwater (36)

Water content of the oceans (1,380,000)

Water content of sedimentary rocks near earth's surface (210,000)

FIGURE 7.4 The water cycle, with its major components expressed on a global scale. All pools and transfer values (shown in parentheses) are expressed as billion billion (10^{18}) grams and billion billion grams per year. After R. G. Barry and R. J. Chorley, *Atmosphere, Weather, and Climate.* Holt, New York (1970); G. E. Hutchinson, *A Treatise on Limnology,* Vol. 1: *Geography, Physics, and Chemistry.* Wiley, New York (1957).

ample, plants absorb about 40 percent, of which they dissipate two-thirds through evapotranspiration. Only a few percent, at most, is assimilated by photosynthesis.

Evaporation, not precipitation, determines the flux of water through the ecosystem. The absorption of radiant energy by liquid water couples an energy source to the hydrological cycle. Evaporation and precipitation are closely linked because the atmosphere has a limited capacity to hold water vapor; any increase in the evaporation of water into the atmosphere creates an excess of vapor and causes an equal increase in precipitation.

Water vapor in the atmosphere at any one time (the compartment size) corresponds to an average of 2.5 cm (1 in.) of water spread evenly over the surface of the earth. An average of 65 cm (26 in.) of rain or snow falls each year (the water flux), which is 26 times the average amount of water vapor. Thus the steady-state content of water in the atmosphere—the atmospheric pool—replaces itself 26 times each year on average. (Conversely, water has an average transit time of $1/26$ of a year, or 2 weeks.) The soils, rivers, lakes, and oceans contain more than 100,000 times the amount of water in the atmosphere. Fluxes through both pools are the same, however, because evaporation balances precipitation. Thus the average transit time of water in its liquid form at the earth's surface (about 3650 years) is about 100,000 times longer than its transit time in the atmosphere.

Oxidation–reduction (redox) potential

Water gains energy as it evaporates, so the energy content of water vapor and its capacity to release energy exceed those of liquid water. The chemical energy of a compound reflects its ability to reduce other compounds and, conversely, its ability to become oxidized. In chemical reactions, an atom is oxidized when it donates an electron to another atom, which becomes reduced by accepting the electron (Box 7.1). An oxidizing agent is therefore one that accepts electrons. The more readily a substance accepts electrons, the greater its oxidizing potential. Molecular oxygen (O_2) is a strong oxidizer. A reducing agent donates electrons, and its strength depends on how readily it does so. In its organic form, carbon is a strong reducer. Giving electrons and taking them are opposite sides of the same coin; each substance may serve as an oxidizing or a reducing agent. But because strong oxidizers hold onto electrons tightly, they always perform as weak reducers, and vice versa. To put it another way, a strong oxidizer can always oxidize (be reduced by) a weaker oxidizer.

Oxygen and carbon are only two of the electron acceptors and donors that are important to biological reactions. Nitrate (NO_3^-) is a powerful oxidizer; hydrogen sulfide (H_2S) and ammonium (NH_3) are good reducers. Indeed, the synthesis and metabolism of many organic compounds, including amino acids, involve assimilatory and dissimilatory redox reactions of nitrogen and sulfur. In general, oxygen is the oxidizer of choice because of its ubiquity and the fact that its common reduced form (H_2O) is innocuous. In contrast, the reduction of nitrate produces nitrite (NO_2^-), which most organisms can tolerate only in minute quantities. Oxidizers such as nitrate and ferric iron can nonetheless become important in soils and aquatic sediments when oxygen has been depleted (anaerobic conditions), as we shall soon see.

The carbon cycle

Three major classes of processes cause the cycling of carbon in aquatic and terrestrial systems (Figure 7.5). The first includes the assimilatory and dissimilatory reactions of carbon in photosynthesis and respiration. These are the major energy-transforming reactions of life. Approximately 10^{17} g (10^{11} metric tons) of carbon enter into such reactions worldwide each year.

The second class includes the physical exchange of carbon dioxide between the atmosphere and oceans, lakes, and streams. Carbon dioxide dissolves readily in water; indeed, the oceans contain about 50 times as much CO_2 as the atmosphere. Exchange across the air–water boundary links the carbon cycles of terrestrial and aquatic systems.

The third type of process that drives the cycling of carbon consists of the dissolution and precipitation (deposition) of carbonate compounds as sediments, particularly limestone and dolomite. On a global scale, dissolution and precipitation approximately balance each other, although certain conditions favoring precipitation have led to the deposition of extensive layers of calcium carbonate (limestone) sediments in the past. In aquatic systems, dissolution

FIGURE 7.5 The global carbon cycle. Pools and fluxes (shown in parentheses) are in billions of metric tons (10^{15} g). After T. Fenchel and T. H. Blackburn, *Bacteria and Mineral Cycling*. Academic Press, New York (1979); W. D. Grant and P. E. Long, *Environmental Microbiology*. Wiley, New York (1981).

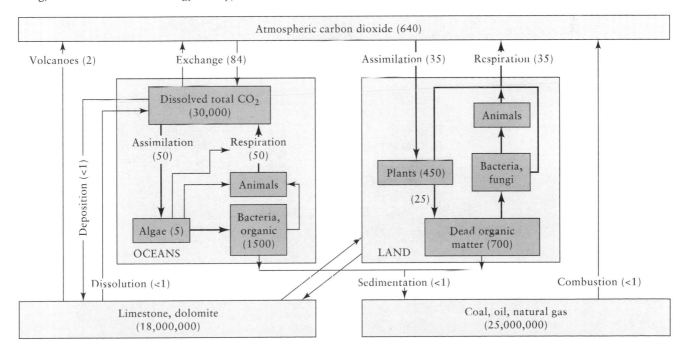

Box 7.1   *An introduction to oxidation and reduction reactions*

Every oxidation–reduction (redox) reaction can be characterized by a pair of half-reactions, one for the reduction (electron-accepting) step and the other for the oxidation (electron-donating) step. For example, when molecular oxygen acts as an oxidizing agent, each atom accepts 2 electrons (an electron is designated e^-). That is, $O_2 + 4e^- \rightarrow 2O^{2-}$. In this reduced form, oxygen readily combines with any of a variety of positive ions, such as H^+, giving

$$O^2 + 4e^- + 4H^+ \rightarrow 2H_2O$$

or C^{4+}, giving

$$O_2 + 4e^- + C^{4+} \rightarrow CO_2$$

These half-reactions include only the reduction step. The electrons must come from an oxidation half-reaction. An example is

$$CH_2O \rightarrow C^{4+} + H_2O + 4e^-$$

The carbon atom donates electrons (it is oxidized), and the electrons are available to reduce an oxidizer. The overall reaction combining the last two half-reactions has the form

$$CH_2O + O_2 \rightarrow CO_2 + H_2O$$

which is the familiar equation for respiration.

The relative strengths of oxidizers and reducers are quantified by their electrical potentials (Eh) measured in volts (V) (see table below). (An electrical current represents the movement of electrons; a chemical battery generates an electrical potential by a pair of redox reactions, one at each electrode.) Because oxygen has a high redox potential (0.81 V at pH 7 and 25 °C), it is an excellent oxidizer. So are nitrate (NO^{3-}), ferric iron (Fe^{3+}), and, to a lesser extent, sulfate (SO_4^{2-}). Near the bottom of the redox scale are carbon dioxide (CO_2) and molecular nitrogen (N_2). The electrical potentials of the half-reactions that reduce CO_2-carbon (C^{4+}) to organic carbon (C^0, -0.42 V) and N_2 to organic (ammonia)-nitrogen (N^{3-}; -0.25 V) are unfavorable thermodynamically.

Any oxidization reaction having a high redox potential coupled with a reduction reaction having a low redox potential can proceed with the release of energy. Thus the oxidation of carbon by oxygen (O_2) releases energy, because oxygen is a stronger oxidizing agent than carbon dioxide. For the reaction to proceed in the opposite direction (for example, fixation of carbon by photosynthesis), energy must be added.

Box 7.1	*Continued*

Redox potentials of selected half-reactions at 25°C and neutral pH

Reaction	Eh (V)
$O_2 + 4H^+ + 4e^- \rightarrow 2H_2O$	0.81
$NO_3^- + 6H^+ + 6e^- \rightarrow 1/2N_2 + 3H_2O$	0.75
$NO_3^- + 2H^+ + e^- \rightarrow NO_2^- + H_2O$	0.42
$NO_3^- + 10H^+ + 8e^- + NH_4^+ + 3H_2O$	0.36
$Fe^{3+} + e^- \rightarrow Fe^{2+}$	0.36
$NO_2^- + 8H^+ + 6e^- \rightarrow NH_4^+ + 2H_2O$	0.34
$CH_3OH + 2H^+ + 2e^- \rightarrow CH_4 + H_2O$	0.17
$CH_2O + 2H^+ + 2e^- \rightarrow CH_3OH$	−0.18
$SO_4^{2-} + 8H^+ + 6e^- \rightarrow S + 4H_2O$	−0.20
$SO_4^{2-} + 10H^+ + 8e^- \rightarrow H_2S + 4H_2O$	−0.21
$CO_2 + 8H^+ + 8e^- \rightarrow CH_4 + 2H_2O$	−0.24
$N_2 + 8H^+ + 6e^- \rightarrow 2NH_4^+$	−0.28
$H^+ + e^- \rightarrow 1/2H_2$	−0.41
$CO_2 + 4H^+ + 4e^- \rightarrow 1/6C_6H_{12}O_6$ (glucose) $+ H_2O$	−0.43
$CO_2 + 4H^+ + 4e^- \rightarrow CH_2O + H_2O$	−0.48
$Fe^{2+} + 2e^- \rightarrow Fe$	−0.85

Source: W. Stumm and J. J. Morgan, *Aquatic Chemistry* (2nd ed.). Wiley, New York (1981).

and deposition occur about two orders of magnitude more slowly than assimilation and dissimilation. Thus the exchange between sediments and the water column is relatively unimportant to the short-term cycling of carbon in the ecosystem. Locally, and over long periods, it can assume much greater importance; in fact, most of the ecosystem's carbon resides in sedimentary rocks (Figure 7.6).

When carbon dioxide dissolves in water, it forms carbonic acid,

$$CO_2 + H_2O \rightleftharpoons H_2CO_3$$

which readily dissociates into hydrogen, bicarbonate, and carbonate ions:

$$H_2CO_3 \rightleftharpoons H^+ + HCO_3^- \rightleftharpoons 2H^+ + CO_3^{2-}$$

Calcium, when present, also equilibrates with the carbonate and bicarbonate ions:

$$Ca^{2+} + CO_3^{2-} \rightleftharpoons CaCO_3$$

FIGURE 7.6 Sedimentary deposits of limestone in mountains of southern Texas represent calcium precipitated out of solution in the shallow seas that once covered the area. Courtesy of the U.S. National Park Service.

Calcium carbonate ($CaCO_3$) has low solubility under most conditions and readily precipitates out of the water column.

Under acid conditions, however, the formation of carbonic acid removes carbonate ions from the system. This removal reduces the amount of dissolved calcium carbonate in the water — which, in turn, enhances the dissolution of calcium carbonate rocks and sediments that the water passes over. By this process, slightly acid streams and groundwater dissolve limestone sediments (and acid rain erodes marble statuary). As these calcium-laden fresh waters enter the oceans and mix with water of nearly neutral pH, calcium carbonate becomes less soluble and precipitates out of the water column to form sediments.

Dissolution and dissociation may be affected locally by the activities of organisms. In the marine system, under approximately neutral conditions, the carbonate system has the overall equilibrium

$$CaCO_3 \text{ (insoluble)} + H_2O + CO_2 \rightleftharpoons Ca(HCO_3)_2 \text{ (soluble)}$$

Removal of CO_2 by photosynthesis shifts the equilibrium to the left, resulting in the formation and precipitation of calcium carbonate. Many algae excrete this calcium carbonate to surrounding water, but reef-building algae and coralline algae incorporate the substance as hard structure (Figure 7.7). In the system as a whole, when photosynthesis exceeds respiration (as it does during algal blooms), calcium tends to precipitate out of the system.

The nitrogen cycle

The ultimate source of nitrogen for the ecosystem is molecular nitrogen (N_2) in the atmosphere. This form of nitrogen dissolves to some extent in water, but no nitrogen of any form is found in native rock. Were it not for biological

FIGURE 7.7 The "skeleton" of coralline algae is made of calcium carbonate precipitated in conjunction with the removal of carbon dioxide from solution in tissues during photosynthesis.

processes, under the present oxidizing conditions at the earth's surface, virtually all nitrogen would occur in its molecular form.

Molecular nitrogen enters biological pathways of the nitrogen cycle (Figure 7.8) through its assimilation (nitrogen fixation) by certain microorganisms. Although this pathway ($N_2 \rightarrow NH_3$) constitutes a small fraction of the earth's annual nitrogen flux, all biologically cycled nitrogen can be traced back to **nitrogen fixation.** Once in the biological realm, nitrogen follows more complicated pathways than those of carbon because of nitrogen's numerous oxidation states. Beginning with reduced (organic) nitrogen, the first step in the nitrogen cycle is **ammonification:** the hydrolysis of protein and the oxidation of amino acids, resulting in the production of ammonia (NH_3), that is accomplished by all organisms. During the initial breakdown of amino acids, some carbon is oxidized, releasing energy, but the energy potential of the nitrogen atom does not change.

Nitrification involves the oxidation of nitrogen, first from ammonia to nitrite, then from nitrite to nitrate, during which transformations the nitrogen atom releases much of its potential chemical energy. Both steps are carried out only by specialized bacteria: $NH_3 \rightarrow NO_2^- \ (N^{3-} \rightarrow N^{3+})$ by *Nitrosomonas* in the soil and by *Nitrosococcus* in marine systems; $NO_2^- \rightarrow NO_3^- \ (N^{5+})$ by *Nitrobacter* in the soil and *Nitrococcus* in the seas.

Because nitrification steps are oxidations, they require the presence of oxygen, which can act as an electron acceptor. In waterlogged, anoxic soils and sediments, nitrate and nitrite can act as electron acceptors (oxidizers), and the nitrification reactions reverse:

$$NO_3^- \rightarrow NO_2^- \rightarrow NO$$

This **denitrification** occurs in soil depleted of oxygen and is accomplished by such bacteria as *Pseudomonas denitrificans*. Additional physical reactions under anaerobic conditions can produce molecular nitrogen; that is,

$$NO \rightarrow N_2O \rightarrow N_2$$

with the consequent loss of nitrogen from biological circulation.

FIGURE 7.8 Schematic diagram of transformations and oxidation states of compounds in the nitrogen cycle. The most reduced state of the atom has the highest chemical energy potential.

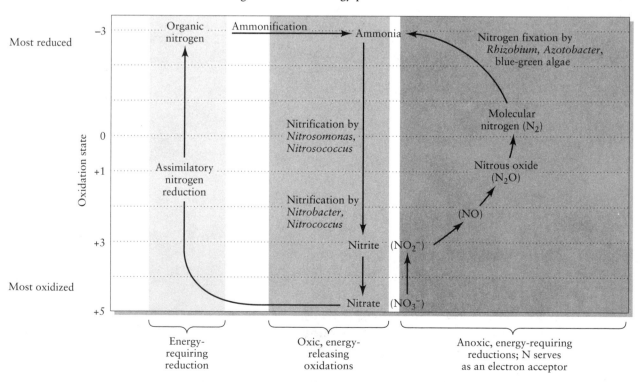

Nitrogen fixation

Denitrification is balanced in terrestrial and aquatic systems by nitrogen fixation. This assimilatory reduction of nitrogen is accomplished by bacteria, such as *Azotobacter*, which is a free-living species; *Rhizobium*, which occurs in symbiotic association with the roots of some legumes (members of the pea family); and the blue-green cyanobacteria.

Nitrogen fixation uses energy, though no more than the conversion of an equivalent amount of nitrate to ammonia by plants. The reduction of one atom of molecular nitrogen (N_2) to ammonia-nitrogen (NH_4^+) requires approximately the amount of energy released by the oxidation of an atom of organic carbon to carbon dioxide.

Nitrogen-fixing microorganisms obtain the energy and reducing power (from organic carbon) they need to reduce N_2 to NH_3 by oxidizing sugars or other organic compounds. Free-living bacteria must obtain these resources by metabolizing organic detritus in the soil, sediments, or water column. More abundant supplies of energy are available to bacteria that enter into symbiotic relationships with plants, which provide them with photosynthate. The best known of these associations is the infection of root nodules of leguminous plants (peas, alfalfa, and their relatives) by the nitrogen-fixing bacterium *Rhizobium*. The nodules are specialized structures of the root cortex, whose development is stimulated by the *Rhizobium* (Figure 7.9). The nodules provide an optimal environment of abundant photosynthate and low oxygen for nitrogen fixation. Nitrogen-fixing symbioses are by no means limited to legumes, and such associations of bacteria have been identified in other terrestrial plants, such as alders and *Casaurina*, and in some marine algae, lichens, and shipworms.

On a global scale, nitrogen fixation approximately balances the production of N_2 by denitrification. The fluxes amount to about 2 percent of the total cycling of nitrogen through the ecosystem. On a local scale, nitrogen fixation can assume much greater importance, especially in nitrogen-poor habitats. When land is exposed to colonization by plants—as, for example, are areas left bare by receding glaciers or newly formed lava—such species with nitrogen-fixing capabilities as the alder *Alnus* dominate newly established vegetation.

The phosphorus cycle

Ecologists have studied the role of phosphorus in the ecosystem intensively, because organisms require phosphorus at a relatively high level (about one-tenth that of nitrogen) as a major constituent of nucleic acids, cell membranes, energy-transfer systems, bones, and teeth. Phosphorus is thought to limit plant productivity in many aquatic habitats, and the influx of phosphorus to rivers and lakes in the form of sewage and runoff from fertilized agricultural lands artificially stimulates production in aquatic habitats, with undesirable consequences.

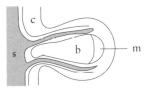

FIGURE 7.9 The root system of an Austrian winter pea plant showing the clusters of nodules that harbor symbiotic nitrogen-fixing bacteria. The diagram shows the arrangement of tissues within a nodule: s = vascular stele, c = root cortex, b = bacteroid-containing tissue, m = nodule meristem. Courtesy of the U.S. Soil Conservation Service. After W. D. Grant and P. E. Long, *Environmental Microbiology*. Wiley, New York (1981).

The phosphorus cycle has fewer steps than the nitrogen cycle: plants assimilate phosphorus as phosphate ion (PO_4^{3-}) directly from the soil or water and incorporate it into various organic compounds in the form of phosphate esters. Animals eliminate excess organic phosphorus in their diets by excreting phosphorus salts in urine; phosphatizing bacteria also convert organic phosphorus in detritus to phosphate ion. Phosphorus does not enter the atmosphere in any form other than dust; thus the phosphorus cycle involves only the soil and aquatic compartments of the ecosystem. Except in very limited microbial transformations, phosphorus does not undergo oxidation–reduction reactions in its cycling through the ecosystem.

The acidity of the environment greatly affects the availability of phosphorus and consequently its uptake and assimilation by plants. In general, at low pH (acid conditions), phosphorus binds tightly to clay particles in soil and forms relatively insoluble compounds with ferric iron and aluminum. At high pH, other insoluble compounds form (for example, with calcium). When both ferric iron or aluminum and calcium are present, the highest concentration of dissolved phosphate—that is, the greatest availability of phosphorus—occurs at a pH of between 6 and 7. Under anoxic conditions in aquatic sediments, phosphorus dissolves and enters the water column when iron is reduced from the ferric (Fe^{3+}) to the ferrous (Fe^{2+}) state, because the iron then forms sulfides rather than phosphate compounds.

The sulfur cycle

Sulfur occurs in the amino acids cysteine and methionine and hence is required by plants and animals. But the importance of sulfur in the ecosystem goes far beyond its assimilation by plants. Like nitrogen, sulfur has many oxidation states, so it follows complex chemical pathways and affects the cycling of other elements (Figure 7.10).

The most oxidized form of sulfur is sulfate (SO_4^{2-}); the most reduced forms are sulfide (S^{2-}) and the organic (thiol) form of sulfur (also S^{2-}). Under oxic conditions, assimilatory sulfur reduction ($SO_4^{2-} \rightarrow$ organic S) balances the oxidation of organic sulfur back to sulfate, either directly or with sulfite (SO_3^{2-}) as an intermediate step. This oxidation occurs when animals excrete excess dietary organic sulfur and when microorganisms decompose plant and animal detritus.

Under anoxic conditions, sulfate may function as an oxidizer. The bacteria *Desulfovibrio* and *Desulfomonas* both couple dissimilatory sulfate reduction ($SO_4^{2-} \rightarrow S^{2-}$) to the oxidation of organic carbon to make energy available. The reduced S^{2-} may then be used as a reducing agent ($S^{2-} \rightarrow S^0 + 2e^-$) by photoautotrophic bacteria, which assimilate carbon by pathways analogous to photosynthesis in green plants. Sulfur takes the place of the oxygen atom in water as an electron donor. The elemental sulfur (S) accumulates unless the sediments are exposed to aeration or oxygenated water, at which point the sulfur may be further oxidized to sulfite (SO_3^{2-}) and sulfate (SO_4^{2-}).

The fate of thiol sulfur (S^{2-}) produced under oxic conditions depends on the availability of positive ions. Frequently hydrogen sulfide forms; it escapes from shallow sediments and mucky soils as a gas having the familiar smell of rotten eggs. Because oxic conditions generally favor the reduction of ferric iron (Fe^{3+}) to ferrous iron (Fe^{2+}), the presence of iron in sediments leads to the formation of iron sulfide (FeS). Sulfides are commonly associated with coal and oil. When exposed to the atmosphere in mine wastes or burned for energy, the reduced sulfur oxidizes (with the help of *Thiobacillus* bacteria in mine wastes), and the oxidized forms combine with water to produce the sulfuric acid of acid mine drainage and acid rain.

The special roles of microorganisms in element cycles

Many of the transformations discussed in this chapter are accomplished mainly or entirely by microorganisms. In fact, were it not for the activities of bacteria and fungi, many element cycles would be drastically altered and the productivity of the ecosystem much reduced. For example, wood breaks down largely because of microorganisms, especially fungi. In their absence, organic carbon would accumulate as cellulose and lignin rather than passing through detritus-based food chains. Over long periods, primary production would drop as organic carbon accumulated in the soil and as levels of carbon dioxide in the atmosphere decreased. For a second example, without the ca-

FIGURE 7.10 Schematic diagram of transformations and oxidation states of compounds in the sulfur cycle.

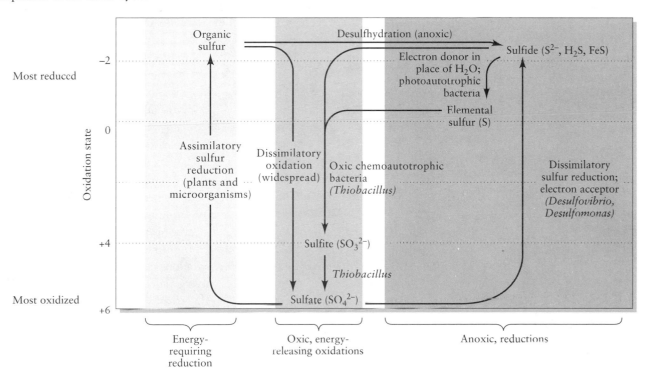

pacity of some microbes to use sulfur and iron as electron acceptors, little decomposition would occur in anoxic organic sediments, and their resulting accumulation would further reduce the amount of inorganic carbon in the ecosystem.

Nitrogen fixation, the sole source of biological nitrogen, depends entirely on microorganisms. Without them, life could not exist. Most plants can assimilate ammonia-nitrogen as easily as, and at less energetic expense than, nitrate-nitrogen. Thus nitrifying bacteria may reduce the primary productivity of terrestrial and aquatic plants under some circumstances by competing for nitrogen substrates in the soil and water.

Many of the transformations carried out by microorganisms, such as the metabolism of sugars and other organic molecules, are accomplished in similar ways by plants and animals. The bacteria and cyanobacteria are distinguished physiologically primarily by the ability of many species to metabolize substrates under anoxic conditions and to use substrates other than organic carbon as energy sources. The requirements of fungi for inorganic substrates as sources of nitrogen, phosphorus, potassium, sulfur, and other elements resemble those of higher plants. What distinguishes the fungi from the bacteria is their greater ability to break down such complex polysaccharides as cellulose and lignin, which make up a large proportion of terrestrial plant litter.

Every organism needs, above all, a source of carbon for building organic structure and a source of energy to fuel the life processes. We may distinguish organisms in terms of their source of carbon. **Heterotrophs,** or organotrophs, obtain carbon in reduced (organic) form by consuming other organisms or organic detritus. All animals and fungi, and many bacteria, are heterotrophs. **Autotrophs** assimilate carbon as carbon dioxide and expend energy to reduce it to an organic form. **Photoautotrophs** use sunlight as their source of energy for photosynthesis. These include all green plants and the cyanobacteria, which use H_2O as an electron donor (reducing agent) and are aerobic, and the purple and green bacteria, which have light-absorbing pigments different from those of green plants, use H_2S or organic compounds as electron donors, and are anaerobic.

Organisms may obtain energy from three sources: sunlight (photoautotrophs), the oxidation of organic compounds (all heterotrophs), and the oxidation of inorganic compounds (chemoautotrophs). Oxidation of glucose combines the oxidation of carbon, which releases energy, and the reduction of oxygen. Heterotrophs all use organic carbon as reduced substrate. Under oxic conditions, they may use O_2 as the oxidizing agent. But many bacteria (such as sulfate-reducing bacteria and methanogens) can use sulfate (SO_4^{2-}), ferric iron (Fe^{3+}), or carbon dioxide (CO_2) as an oxidizing agent under anoxic conditions.

Chemoautotrophs all use CO_2 as a carbon source, but they obtain energy for its reduction by the oxic oxidation of inorganic substrates: methane (for example, *Methanosomonas* and *Methylomonas*); hydrogen (*Hydrogenomonas* and *Micrococcus*); ammonia (the nitrifying bacteria *Nitrosomonas* and *Nitrosococcus*); nitrite (the nitrifying bacteria *Nitrobacter* and *Nitrococcus*); hydrogen sulfide, sulfur, and sulfite (*Thiobacillus*); and ferrous iron salts (*Ferrobacillus* and *Gallionella*). The chemoautotrophs are almost exclusively bacteria, which apparently are the only organisms that can become so specialized biochemically as to make efficient use of inorganic substrates and efficiently dispose of the products of chemoautotrophic metabolism.

In this chapter we have examined the cycling of several important elements from the standpoint of their chemical and biochemical reactions. Each type of habitat presents a different chemical environment, however, and it stands to reason that the patterns of element fluxes differ greatly among them. In the next chapter, we shall contrast element cycling in aquatic and terrestrial habitats by focusing on how some of the unique physical features of each of these environments affect the chemical and biochemical transformations involved in organic production and recycling of elements.

🦅 *SUMMARY*

Unlike energy, nutrients are retained within the ecosystem and are cycled between its physical and biotic components. The paths that elements follow through the ecosystem depend on chemical and biological transformations, which are functions of the chemical characteristics of each element and the need of organisms for them.

1. The movement of energy through the ecosystem parallels the paths of several elements, particularly carbon, whose transformations either require or release energy.

2. The cycling of each element may be thought of as movement between compartments of the ecosystem, the major compartments being living organisms, organic detritus, immediately available inorganic forms, and unavailable organic and inorganic forms, generally in sediments.

3. The water cycle, or hydrological cycle, provides a physical analogy for element cycling. Energy is required to transform water from its liquid phase to atmospheric vapor. Upon condensation and precipitation, that energy is released as heat.

4. Energy transformations in biological systems occur primarily during oxidation–reduction (redox) reactions. An oxidizer is a substance that readily accepts electrons (O_2, NO_3^-); a reducer is one that readily donates electrons (H_2, organic carbon). Upon being reduced, oxidizers gain energy; upon being oxidized, reducers release it. Elements tend to occur in organic compounds in reduced forms, so the biological assimilation of many elements (including carbon, nitrogen, and sulfur) requires energy.

5. All organisms require organic carbon as the primary substance of life. Organic carbon is also the major source of energy for most animals and microorganisms. Carbon shuttles between living forms and the carbon dioxide compartment of the ecosystem by way of photosynthesis and respiration. Precipitation of calcium carbonate in the oceans accounts for the accumulation of limestone sediments.

6. Nitrogen has many oxidation states and consequently follows many pathways through the ecosystem. Quantitatively, most of the flux follows the cycle nitrate → organic nitrogen → ammonia → nitrite → nitrate. The last two steps, referred to as nitrification, are accomplished by certain bacteria in the presence of oxygen. Under anoxic conditions in soils and sediments, nitrate replaces oxygen as an oxidizing agent (denitrification) and the reactions reverse: nitrate → nitrite → (eventually) molecular nitrogen (N_2). This loss of nitrogen from the general biological cycling is balanced by nitrogen fixation by some microorganisms.

7. Plants assimilate phosphorus in the form of phosphate (PO_4^{2-}), whose availability varies with the acidity and oxidation level of the soil or water. The energy potential of the phosphorus atom does not change during its cycling through the ecosystem.

8. Sulfur is an important redox element in anoxic habitats, where it may serve as an oxidizer in the form of sulfate (SO_4^{2-}) or as a reducing agent (for photoautotrophic bacteria) in the forms of elemental sulfur (S^0) and sulfide (S^{2-}).

9. Many elemental transformations, particularly under anoxic conditions, are accomplished by biochemically specialized microorganisms (bacteria, cyanobacteria). The activities of these organisms therefore play important roles in the cycling of elements through the ecosystem.

❧ SUGGESTED READINGS

Berner, R. A., and A. C. Lasaga. 1989. Modeling the geochemical carbon cycle. *Scientific American* 260:74–81.

Coleman, D. C., C. P. P. Reid, and C. V. Cole. 1983. Biological strategies of nutrient cycling in soil systems. *Advances in Ecological Research* 13:1–55.

Fenchel, T., and T. H. Blackburn. 1979. *Bacteria and Mineral Cycling.* Academic Press, New York.

Grant, W. D., and P. E. Long, 1981. *Environmental Microbiology.* Wiley (Halsted Press), New York.

Schlesinger, W. H. 1991. *Biogeochemistry. An Analysis of Global Change.* Academic Press, San Diego.

Sprent, J. I. 1987. *The Ecology of the Nitrogen Cycle.* Cambridge Univ. Press, New York.

Stacey, G., R. H. Burris, and H. J. Evans (eds.). 1992. *Biological Nitrogen Fixation.* Chapman & Hall, New York.

8

NUTRIENT REGENERATION IN TERRESTRIAL AND AQUATIC ECOSYSTEMS

The elements cycle through the ecosystem along paths established by their chemical properties, which in turn determine chemical and biochemical reactions in the biosphere. The regeneration of elements in forms that can be used by other organisms provides a key to understanding the regulation of ecosystem function. These processes of regeneration are different in terrestrial and aquatic systems. To be sure, both systems exhibit similar chemical and biochemical transformations: oxidation of carbohydrates, nitrification, and chemoautotrophic oxidation of sulfur, among many others. But terrestrial and aquatic systems differ in the material basis for nutrient regeneration. In terrestrial habitats, most elements cycle through *detritus* at the soil surface. In aquatic habitats, particularly in lakes and oceans, *sediments* are the ultimate source of regenerated nutrients; these sediments are often far removed from the sites of primary production. In this chapter, we shall discuss how biochemical processes in soils, water, and sediments influence the productivity of the ecosystem and the cycling of elements within it.

Soil formation and the nutrient dynamics of ecosystems

One of the major sources of new nutrients in terrestrial systems is the formation of soil by the weathering of bedrock and other parent materials. How rapidly does this process occur? Weathering normally takes place under deep layers of soil where it is inaccessible to direct measurement. Soil scientists estimate the rate of weathering indirectly, by measuring the net efflux (loss) of certain elements from the system. When a soil attains a steady state, as it might in undisturbed areas, the loss of an element from a system equals the weathering input of that element from the parent material plus any gains from other sources, such as precipitation. Thus it is possible to estimate weathering input from knowledge of precipitation input and total loss. The basic cations—calcium, potassium, sodium, and magnesium—are good candidates for the study of such balances, because they readily leach out of the soil and leave the system as dissolved ions in stream water, where they can be measured easily.

It is easiest to study the cation budgets of large areas in a small **watershed,** from which all surface water and groundwater leave at a single point. The best known of such studies comes from the Hubbard Brook Forest of New Hampshire. Scientists obtained detailed cation budgets for several small watersheds in the area by measuring the inputs in rainwater collected at various locations in the watershed (Figure 8.1) and the outputs in water leaving the watershed by way of the stream that drains it (Figure 8.2). Assuming that the soil is in a steady state, the net loss (precipitation input minus stream outflow) equals the input to the soil from weathering.

FIGURE 8.1 Rain gauges installed in a ponderosa pine stand in California to intercept precipitation falling through the canopy of the forest and running down the trunks of trees. Analysis of the nutrient content of water collected in sampling programs like this helps determine the overall cation budget of the forest and the specific routes of mineral cycles. Courtesy of the U.S. Forest Service.

FIGURE 8.2 A stream gauge at the lower end of a watershed at the Coweeta
Hydrological Laboratory, North Carolina. The V-shaped notch regulates the flow
of water through the weir so that the flow rate is proportional to the water level
in the basin. Courtesy of the U.S. Forest Service.

At Hubbard Brook, annual input of calcium in precipitation averaged 2
kilograms per hectare (kg ha^{-1}), while loss of dissolved Ca^{2+} in stream flow was
14 kg ha^{-1}. Therefore, the net loss to the system equaled 12 kg ha^{-1}. The investigators also determined that living and dead biomass increased in the
watershed during the study period, because the forest was recovering from
earlier clearing; net assimilation of calcium in vegetation and detritus brought
its overall removal from the mineral soil to 21 kg ha^{-1} yr^{-1}. Because calcium
constitutes about 1.4 percent of the weight of the bedrock in the area, its annual loss equaled the weathering of about 1500 kg (21/0.014) of bedrock per
hectare, or approximately 1 mm of depth per year. The major lesson of this and
other studies has been how small the weathering input is relative to the recycling of nutrients by vegetation. The bulk of these nutrients are made available
to plants by regenerative processes within the soil profile; typically, only 10
percent are provided by weathering of bedrock.

Detritus and nutrient regeneration in
terrestrial ecosystems

Plants assimilate elements from the soil far more rapidly than weathering generates them from the parent material. Most of the basic cations (Ca^{2+}, Mg^{2+},
K^+, and Na^+) do not figure prominently in biochemical transformations, and,

for the most part, plants merely assimilate these cations with the water they need in quantity. Plants either sequester or excrete excesses of these elements, whose depletion from the soil probably has little direct effect on plant production. This is not so with supplies of such important nutrients as nitrogen, phosphorus, and sulfur, which are poorly represented in parent material. Igneous rocks — granite and basalt, for example — contain no nitrogen and are only 0.3 percent phosphate (P_2O_5) and 0.1 percent sulfate (SO_3) by mass. Hence weathering adds little of these nutrients to the soil; inputs from precipitation and nitrogen fixation are also small. Plant production therefore depends on the rapid regeneration of nutrients from detritus and their retention within the ecosystem.

Organic detritus occurs everywhere, most conspicuously in terrestrial habitats where the resistance of woody plant parts to herbivory results in the accumulation of abundant plant remains. Regardless of the habitat, however, this reservoir of nutrients is regenerated by the activities of the countless worms, snails, insects, mites, bacteria, and fungi that consume detritus — their primary source of carbon and energy — for food.

The breakdown of leaf litter on the forest floor occurs in three ways: (1) **leaching** of soluble minerals and small organic compounds by water; (2) consumption by large detritus-feeding organisms (millipedes, earthworms, woodlice, and other invertebrates); and (3) further attack by fungi and the eventual **mineralization** of phosphorus, nitrogen, and sulfur by bacteria. Between 10 and 30 percent of the substances in newly fallen leaves dissolve in cold water. Leaching rapidly removes most of these (salts, sugars, amino acids) from the litter; complex carbohydrates and other organic compounds remain behind. Although large detritus feeders assimilate no more than 30 to 45 percent of the energy available in leaf litter, and even less from wood, they nonetheless speed the decay of litter because they macerate leaves (break them down by soaking) in their digestive tracts, and the finer particles in their egested wastes expose new surfaces to microbial feeding.

Leaves of different species of trees decompose at different rates, depending on their composition. For example, in eastern Tennessee, the weight loss of leaves during the first year after leaf fall was found to range from 64 percent for mulberry to 39 percent for oak, 32 percent for sugar maple, and 21 percent for beech. The needles of pines and other conifers also decompose slowly. Differences between species depend to a large extent on the lignin content of the leaves. **Lignins** are a heterogeneous class of **phenolic polymers** (long chains of phenolic subunits). They lend wood many of its structural qualities and are even more difficult to digest than cellulose. In fact, only the "white rot" fungi can break down lignin.

The resistance of some types of litter to degradation points up the unique role of fungi in the regeneration of nutrients. The familiar mushrooms and shelf fungi are merely fruiting structures produced by the mass of the fungal organism deep within the litter or wood (Figure 8.3). Most fungi consist of a network, or mycelium, of **hyphae,** threadlike elements that can penetrate the woody cells of plant litter that bacteria cannot reach. Like bacteria, fungi secrete enzymes into the substrate itself and absorb the simple sugar and amino acid breakdown products of this exocellular digestion. Fungi differ from bacteria in being able to digest cellulose (which a few bacteria, protozoa, and snails also can accomplish) and, especially, lignin. Cellulose digestion begins by the hydrolyzing of polysaccharides (long chains of sugar subunits) into simple sugars that can be absorbed into the hyphae. The breakdown of lignin by fungi apparently is initiated by an oxidation that cleaves the aromatic (phenolic) ring structure. These processes once again emphasize the crucial roles of microorganisms in nutrient cycles.

FIGURE 8.3 Shelf fungi speed the decomposition of a fallen log. The fruiting structures are produced by the fungal hyphae, together called the mycelium, that grow throughout the interior of the log, slowly digesting its structure. Courtesy of the U.S. National Park Service.

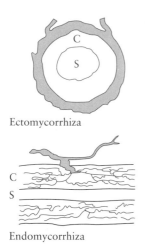

Ectomycorrhiza

Endomycorrhiza

FIGURE 8.4 Generalized structure of ectomycorrhiza and endomycorrhiza; C = root cortex, S = vascular tissue of stele. After W. D. Grant and P. E. Long, *Environmental Microbiology*. Wiley, New York (1981).

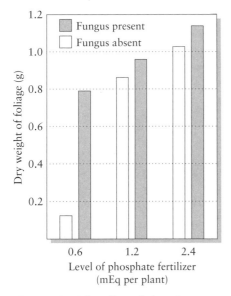

FIGURE 8.5 The effect of phosphate fertilizer and inoculation with the mycorrhizal fungus *Endogone macrocarpa* on the growth of tomato (*Lycopersicon esculentum*). After J. L. Harley and S. E. Smith, *Mycorrhizal Symbiosis*. Academic Press, London (1983).

Mycorrhizae

In addition to their role in decomposing detritus, some kinds of fungi grow on the surface of, or inside, the roots of many types of plants, especially woody species. This association, which is called a **mycorrhiza** (literally, "fungus root"), enhances the plant's ability to extract mineral nutrients from the soil. Although many forms of the association are recognized, they are classified as **endomycorrhizae** when the fungus penetrates the root tissue and as **ectomycorrhizae** when it forms a sheath over the root surface (Figure 8.4).

Mycorrhizae occur everywhere, but their importance is best demonstrated experimentally by growing plants on sterile soil to which spores of mycorrhizae-forming fungi are either added or not added. For example, in an experiment with *Pinus strobus* seedlings, ectomycorrhizae increased by two to three times the uptake of nitrogen, phosphorus, and potassium per unit of root mass and greatly improved the growth of the plant. Infections promote growth more in soils that are relatively depleted of nutrients, as shown in the experiments with tomatoes depicted in Figure 8.5.

Mycorrhizae increase a plant's uptake of minerals by penetrating a greater volume of soil and increasing the total surface area available for nutrient assimilation. Mycorrhizae, especially external forms, may also protect the plant root from infection by pathogens by physically excluding the pathogens or by producing antibiotics (antibacterial toxins). What do the fungi derive from the association? The main advantage appears to be a reliable source of carbon in the form of photosynthates transported to the roots.

The commonest type of endomycorrhiza is the **vesicular–arbuscular** type, so called because of structures developed within the host tissue. These do not grow freely in the soil but rather infect plant roots from spores left behind from dead, previously infected roots. (Presumably, spores blow into virgin soils along with other dust.) Fungi that form vesicular–arbuscular mycorrhizae derive their carbon exclusively from the plant root. Mineral nutrients are derived from the soil, into which the fungi send long hyphae. Apparently, they can use sources of phosphorus not available to plants, such as the highly insoluble "rock phosphorus" $Ca_3(PO_4)_2$; plants can use only the more soluble forms $CaHPO_4$ and KH_2PO_4. The unique ability of fungi to exploit soil nutrients that are otherwise difficult to extract may lie in the secretion of hydrogen ions or organic acids by the hyphae or in their growth in association with phosphate-solubilizing bacteria.

Ectomycorrhizae are widespread, especially among trees and shrubs. In addition to the hyphae, which extend out into the soil, the fungus forms a tough sheath around the root that may account for as much as 40 percent of the weight of the mycorrhizal association. The sheath stores large quantities of soil-derived nutrients and carbon compounds, which may be one of the chief advantages that mycorrhizae offer to the plant. Regardless of how they function, they account for a major part of the energy and nutrient budgets in some habitats. For example, the carbon assimilated by mycorrhizae in a fir (*Abies*) forest accounted for 15 percent of the total net primary production. The mycorrhizal fungus is undoubtedly responsible for a much greater share of the mineral nutrient uptake.

Nutrient regeneration in tropical and temperate forests

Nutrient cycling differs in tropical and temperate ecosystems because of the effects of climate on weathering and on the regeneration of detritus. The deep weathering and poor cation exchange capacity of many tropical soils result in relatively low fertility. But the rapid regeneration of nutrients from detritus under warm, humid conditions and efficient nutrient retention support high productivity. Ecologists have concluded that in typical tropical ecosystems, most of the nutrients occur in the living biomass, and elements are regenerated and assimilated very rapidly. This has important implications for tropical agriculture and conservation.

Many rich and fertile tropical soils—particularly the young soils of mountainous regions—are well suited to the removal-type agriculture commonly practiced in temperate regions. Even with the clearing of forests and the frequent removal of crops, such soils maintain their fertility. In contrast, however, over extensive regions of older, more deeply weathered soil in the tropics, planting such crops as corn on clear-cut land has had disastrous consequences (Figure 8.6). The practice of cutting and burning vegetation releases many mineral nutrients that may support a year or two of crop growth but readily leach out of the soil without the natural tropical vegetation to assimilate them, and soil fertility declines rapidly. Furthermore, as the exposed soil dries, the upward movement of water draws iron and aluminum oxides to the surface, where they form a concretelike substance called **laterite**. Surface run-

FIGURE 8.6 An area of about 2 ha on a steep slope in Costa Rica that has been cut and burned for agriculture. The closeup reveals small corn plants emerging among the debris. This practice exposes soil to erosion and promotes leaching of nutrients and laterization. Such clearings produce crops for 2 or 3 years at most; decades of forest regeneration are required to restore the fertility of the soil. Courtesy of D. H. Janzen.

off over the impenetrable laterite accelerates erosion, further depleting nutrients and choking streams with sediment.

Experience has taught us that vegetation is critical to the development and maintenance of soil fertility in many tropical systems. Even in temperate zones, the removal of vegetation reveals its important role in the retention of soil nutrients (Figure 8.7). Clear-cutting of small watersheds in the Hubbard Brook Forest increased stream flow several times because transpiring leaf surfaces were removed; losses of cations increased 3 to 20 times over cation losses in comparable undisturbed systems. The nitrogen budgets of the clear-cut watersheds sustained the most striking change. Plants assimilate available soil nitrogen so rapidly that undisturbed forest gained nitrogen at the rate of 1 to 3 $kg\ ha^{-1}\ yr^{-1}$ from precipitation and nitrogen fixation. In the clear-cut watershed, net loss of nitrogen as nitrate soared to 54 $kg\ ha^{-1}\ yr^{-1}$, a value comparable to the annual assimilation of nitrogen by vegetation and many times the precipitation input (7 $kg\ ha^{-1}\ yr^{-1}$). The loss of nitrate resulted from nitrification of organic nitrogen by soil microorganisms at the normal annual rate, without simultaneous rapid uptake by plants. Remember that the negatively charged nitrate ion (NO_3^-) does not bind well to soil particles.

FIGURE 8.7 Clear-cut watershed at the Coweeta Hydrological Laboratory, North Carolina, employed in studies of evapotranspiration and runoff in forest ecosystems. Courtesy of the U.S. Forest Service.

Comparative studies of element dynamics in temperate and tropical forests shed further light on their differences. Litter on the forest floor constitutes an average of about 20 percent of the total biomass of vegetation and detritus in temperate coniferous forests, 5 percent in temperate hardwood forests, and only 1 to 2 percent in tropical rain forests. The ratio of litter to the biomass of living leaves is between 5 and 10 to 1 in temperate forests but less than 1 to 1 in tropical forests. Of the total organic carbon, more than 50 percent occurs in soil and litter in northern forests, but less than 25 percent in tropical rain forests. Clearly, dead organic material decomposes rapidly in the tropics and does not form a substantial nutrient reservoir, as it does in temperate regions.

Fewer data exist on the relative proportions of nutrient elements in the soil and in the living vegetation. The distributions of potassium, phosphorus, and nitrogen in a temperate and in a tropical forest with similar living biomass are compared in Table 8.1. Two points stand out. First, the accumulation of nutrients in vegetation, on a weight-for-weight basis, was somewhat greater in the tropical forest. For example, the total dry weight of living vegetation in the Belgian ash–oak forest exceeded that in the tropical deciduous forest of Ghana by 14 percent, but the accumulation of the three elements per gram of dried vegetation was 32 to 38 percent lower in the temperate forest. Second, the ratio of each element in soil to its level in biomass was much lower in the tropics. In the temperate forest, 96 percent of the phosphorus occurred in the soil; in the tropical forest, more than 90 percent of the much smaller amount of phosphorus was found in the living biomass.

TABLE 8.1	Distribution of mineral nutrients in the soil and living biomass of a temperate and a tropical ecosystem

Forest (locality)	Biomass (T ha^{-1})*	NUTRIENTS (kg ha^{-1})		
		Potassium	Phosphorus	Nitrogen
Ash and oak (Belgium)	380			
Living vegetation		624	95	1,260
Soil		767	2,200	14,000
Ratio of soil to living		1.2	23.1	11.1
Tropical deciduous (Ghana)	333			
Living vegetation		808	124	1,794
Soil		649	13	4,587
Ratio of soil to living		0.8	0.1	2.0

*T = metric tons.

Source: P. Duvigneaud and S. Denayer-de-Smet, in D. E. Reichle (ed.), *Analysis of Tropical Forest Ecosystems*, Springer-Verlag, New York (1970), pp. 199–225; D. J. Greenland and J. M. Kowal, *Plant Soil* 12:154–174 (1960); J. D. Ovington, *J. Biol. Rev.* 40:295–336 (1965).

While recognizing the general nutrient poverty of many tropical soils, we must also distinguish between nutrient-rich and nutrient-poor soils within the tropics. **Eutrophic** ("well-nourished") soils develop in geologically active areas where natural erosion is high and soils are relatively young. With the bedrock closer to the surface, weathering adds nutrients more rapidly, and soils retain nutrients more effectively. In the Western Hemisphere, such eutrophic soils occur widely in the Andes Mountains, in Central America, and in the Caribbean. By contrast, nutrient-poor (**oligotrophic**) soils develop in old, geologically stable areas, particularly on sandy alluvial deposits (as in much of the Amazon Basin), where intense weathering removes clay and reduces nutrient retention.

Comparison of an oligotrophic forest (Amazon Basin) and a eutrophic forest (Puerto Rico) illustrates that although production and nutrient flux are similar in the two areas, the distribution of nutrients (calcium, in this case) in the oligotrophic forest favors biomass over soil compared to their distribution in the eutrophic forest (Table 8.2). Moreover, the oligotrophic forest holds nutrients more tightly; that is, loss of calcium from the eutrophic forest in Puerto Rico is equivalent to about half the annual flux through vegetation, whereas the oligotrophic system appears to gain calcium through precipitation input.

Especially in nutrient-poor areas, nutrient retention by vegetation is crucial to the high productivity of tropical ecosystems. Plants retain nutrients by

TABLE 8.2	*Standing crops and fluxes of dry biomass and calcium in a nutrient-rich and a nutrient-poor tropical rain forest*	
Characteristic	Eutrophic*	Oligotrophic*
Soil		
Exchangeable calcium[†] (kg ha^{-1})	1900	306
Standing crop		
Living vegetation (T ha^{-1})	263	298
Production		
Living vegetation (g m^{-2} yr^{-1})	1033	1012
Calcium standing crop		
Living vegetation (kg ha^{-1})	760	529
Calcium flux		
Precipitation (kg ha^{-1} yr^{-1})	21.8	16.0
Subsurface runoff (kg ha^{-1} yr^{-1})	43.1	13.2
Net change (kg ha^{-1} yr^{-1})	−21.3	+2.8

*Eutrophic: montane tropical rain forest, Puerto Rico; oligotrophic: Amazonian rain forest, Venezuela.

[†]Soil calcium measured to a depth of 40 cm.

Source: C. F. Jordan and R. Herrera, *Amer. Nat.* 117:167–180 (1981).

growing a very dense mat of roots (and associated fungi) that remains close to the surface and even extends up the trunks of trees to intercept nutrients washing down from the canopy. Data from Africa reveal that between 68 and 85 percent of the root biomass of forests is concentrated within the top 25 to 30 cm of the soil. In other tropical areas, the application of radioactively labeled compounds has shown that root mats intercept nutrients regenerated by the leaching and decomposition of detritus before they can penetrate into the mineral soil and be washed out of the system.

Nutrient regeneration in aquatic systems

Because most chemical and biochemical processes involved in the cycling of elements take place in an aqueous medium, the processes themselves do not differ markedly in terrestrial and aquatic systems. What is distinctive about most rivers, lakes, and oceans is the **sedimentation** of nutrients into bottom deposits from which they are regenerated and returned to zones of productivity very slowly.

The sediments in aquatic systems are the counterparts of terrestrial soils, but they differ from such soils in two important ways. First, the regeneration of nutrients from terrestrial detritus takes place near plant roots. In contrast, aquatic plants assimilate nutrients directly from the water column in the uppermost sunlit (**photic**) zones, often far removed from sediments at the bottom. Second, decomposition of terrestrial detritus occurs, for the most part, aerobically and hence relatively rapidly. Aquatic sediments often become **anoxic;** the lack of oxygen greatly slows most biochemical transformations and changes the pathways of some nutrient cycles.

The maintenance of high aquatic production depends in part on the proximity of bottom sediments to the photic zone at the surface or on the presence of **upwelling** currents bringing nutrients regenerated from sediments back to the surface. A map of productivity of the oceans (Figure 8.8) reveals high rates of carbon fixation in shallow seas — in both the tropics (for example, the Coral Sea and the waters surrounding Indonesia) and at high latitudes (the Baltic Sea, the Sea of Japan) — and in zones of upwelling. These zones occur along the western coasts of Africa and the Americas, where winds blow surface waters away from shore, thus establishing a vertical current to replace them.

Excretion and microbial decomposition regenerate some nutrients in the water column, where assimilation and production take place, just as they do in terrestrial soils. For some elements, rates of regeneration in the water column itself match rates of production. For example, in a study conducted off the western coast of North America, the uptake of ammonia-nitrogen by phytoplankton closely paralleled its production by zooplankton. Phytoplankton assimilated about half the ammonia directly, and bacteria nitrified ($NH_4^+ \rightarrow NO_3^-$) the other half, regardless of the overall flux. Studies in temperate freshwater systems indicate turnover rates of organic nitrogen on the order of 0.2 to 2 days; in temperate marine systems, turnover varies between 1 and 10 days, suggesting lower productivity overall.

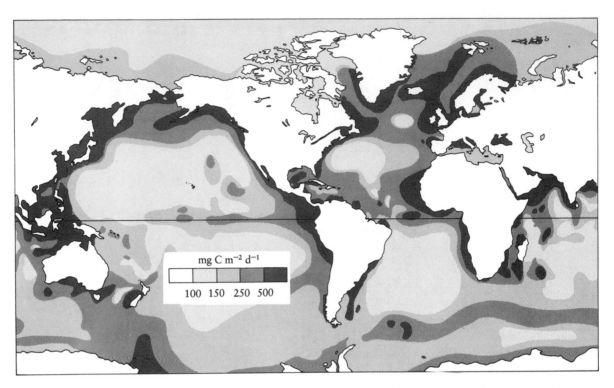

FIGURE 8.8 Primary production in the world's oceans, in milligrams of carbon fixed per square meter per day. Productivity is greatest on the continental shelves and in regions of upwelling on the west coasts of Africa and South and Central America. After R. K. Barnes and K. H. Mann, *Fundamentals of Aquatic Ecosystems.* Blackwell, Oxford, England (1980).

Plants assimilate regenerated nitrogen rapidly, especially in nutrient-depleted waters. Under conditions of high nutrient concentration, phytoplankton can take up more nitrogen than they need for growth. This ability to store nitrogen, called **luxury consumption,** enables algae to take advantage of short-term abundances or local "pockets" of nutrients made available by discrete bursts of excretion or decomposition. We may think of water as homogeneous, but the concentration of nutrients in organisms makes the distribution of elements such as nitrogen and phosphorus extremely heterogeneous. Excretion and decomposition produce transient local abundances of nutrients that plants rely on. Once turbulent mixing disperses these concentrations, the nutrients may become too sparse to be assimilated.

Productivity in most aquatic systems depends on at least occasional interchange between bottom layers and the surface. Nitrogen budgets measured in

TABLE 8.3	Estimates of release and uptake of nitrogen in limnocorrals in the Bay of Quinte, Ontario	
Characteristic	June 5, 1974	September 4, 1974
Concentration (μg N L^{-1} d^{-1})		
Ammonia (NH$_4^-$)	120	176
Nitrate (NO$_3^-$)	84	72
Primary production (μg C L^{-1} d^{-1})		
Gross	185	1281
Net	−139	+807
Uptake (μg N L^{-1} d^{-1})		
Ammonia	20	117
Nitrate	4.5	4.5
Nitrogen fixation	1.2	2.7
Total	26	124
Release (μg N L^{-1} d^{-1})		
Zooplankton grazing	9.2	27
Sedimentation	3.7	63
Total	13	90

Source: C. F. H. Liao and D. R. S. Lean, *J. Fish. Res. Bd. Canad.* 35:1102–1108 (1978).

the Bay of Quinte, Lake Huron, Ontario, illustrate the relative magnitudes of assimilation and regeneration. The studies were conducted within columns of water enclosed by "limnocorrals," which are triangular or circular in cross section and are formed by sheets of plastic suspended by floats at the surface and entrenched in sediment at the bottom (in this case, 4 m beneath the surface. Such enclosures make it possible to study the fluxes of elements by adding isotopically labeled compounds.

Table 8.3 shows that although nitrate accounted for 30 to 40 percent of the regenerated nitrogen, plants preferentially assimilated ammonia by ratios of 4 to 1 early in the growing season (June 5) and 30 to 1 late in the growing season (September 4). Although the levels of nitrogen were similar, gross production in September exceeded that in June by nearly 7 times, probably because of a combination of differences in water temperature and some other, limiting nutrient. Short-term uptake of nitrogen by plants amounted to about one-tenth the level of carbon fixation. Sedimentation of nitrogen in sinking particulate matter averaged 14 percent as much as uptake in June and 28 percent as much as uptake in September. Accordingly, during the September sample period when the total nitrogen in the system (NH$_4^+$, NO$_3^-$, and particulate) was 586 μg L^{-1} and sedimentation was 36 μg L^{-1} d^{-1}, physical removal of nitrogen from the system could have depleted the resource quickly. But sedimentation of particulate nitrogen was approximately balanced throughout the season as vertical mixing of the water column returned ammonia from bottom sediments; thus total nitrogen in the water column varied little relative to internal cycling.

Vertical mixing and thermal stratification

The vertical mixing of water requires an input of energy to accelerate water masses and keep them moving. Winds supply most of this energy, causing turbulent mixing of shallow water. Vertical movement of water can be prevented in such systems when fresh water floats over more dense salt water, as in an estuary, or when the sun heats surface water, establishing a warmer layer of lower density overlying a cooler layer of higher density. Other processes promote vertical mixing. In marine systems, when evaporation exceeds freshwater input, surface layers become more saline, hence denser, and literally fall through the lighter water below. Similarly, when surface layers of water cool during the autumn, the denser surface water tends to fall down through the warmer, less dense layers below.

The vertical mixing of water affects production in two opposing ways. On the one hand, mixing can bring nutrient-rich water from the depths to the sunlit surface and thereby promote production. On the other hand, mixing can carry phytoplankton below the zone of photosynthesis and thereby reduce production. Indeed, when vertical mixing extends far below the photic zone, phytoplankton cannot maintain themselves, much less reproduce. Under such conditions, primary production may shut down altogether, resulting in the seeming contradiction of nutrient-rich water without primary production.

The more typical situation in many temperate-zone lakes and ponds is one in which **thermal stratification** during the summer prevents vertical mixing; then, as sedimentation removes nutrients from the surface layers, production decreases. Nutrients may regenerate themselves in the deeper layers of the lake, but they cannot reach the surface until stratification breaks down and vertical mixing ensues with cool autumn temperatures.

Thermal stratification develops only weakly, if at all, at both high and low latitudes (Figure 8.9). In arctic and subarctic regions, the low heat input into lakes fails to counter turbulent mixing, and the water column warms uniformly to the extent that water temperature rises at all. In the tropics, the lack of a pronounced seasonal temperature cycle reduces the sharpness of thermal stratification, because the sun and constant high air temperatures warm the water uniformly to the bottom of the lake.

In marine systems, two very different water masses may meet at a **front,** and here intermixing may create special conditions for high production. Sometimes at the boundary of a shallow-water system and a deep-water system, mixed (deep) and stratified (shallow) water masses are brought together felicitously. On the mixed side, nutrients may be abundant but phytoplankters do not remain within the photic zone. On the stratified side, nutrients may have been depleted from the surface waters. Where the two meet, some of the nutrient-laden mixed water enters the stratified layer, creating ideal conditions for photosynthesis and nutrient assimilation (Figure 8.10).

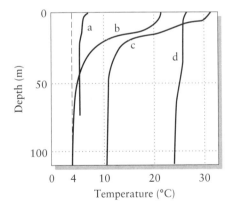

Figure 8.9 Temperature profiles of lakes at the height of summer stratification; a = Flakevatn, Norway; b = Cayuga, New York; c = Ikedako, Japan; d = Edward, Uganda. At high latitude (Norway), the summer sun lacks the intensity to warm the surface layers. At the equator (Uganda), the strong sun and year-round high temperatures warm lakes uniformly to their bottoms. From G. E. Hutchinson, *A Treatise on Limnology,* Vol. 1. Wiley, New York (1957).

Regeneration of nutrients in the hypolimnion of a temperate lake

The layer of water lying above the sharp change (**thermocline**) in a stratified temperature profile is the **epilimnion**; the zone below the thermocline is the **hypolimnion**. During prolonged periods of stratification, bacterial respiration in the hypolimnion depletes the oxygen supply (Figure 8.11), provided that abundant organic matter exists for the bacteria to oxidize.

In the oxygen-depleted environment of bottom sediments and waters immediately over them, such elements as iron, manganese, and nitrogen shift from oxidized to reduced forms, which greatly affects their solubility. In particular, as ferric iron (Fe^{3+}) is reduced to ferrous iron (Fe^{2+}), insoluble iron–phosphate complexes become solubilized, and both elements tend to move into the water column.

Changes observed in the water chemistry of the hypolimnion of an English lake, Esthwaite Water, during the course of a single season show the effects of anoxic conditions (Figure 8.12). After stratification becomes established in June, oxygen at the deepest level of the lake decreases gradually while dissolved carbon dioxide increases. The water becomes anoxic by early July and remains so until the end of stratification and the onset of vertical mixing in late September. During the period of anoxia, levels of ferrous iron, phosphate, and ammonia increase dramatically as reduced forms in the sediments and at the sediment–water boundary solubilize and enter the water column. The return of oxidizing conditions in the autumn reverses the chemistry of the bottom water, initially because of the replacement of bottom water by surface water but ultimately because oxidized forms of several redox elements produce insoluble compounds, which precipitate out of the water column. Nitrogen is a conspicuous exception; under oxic conditions, nitrifying bacteria convert ammonia to nitrate, which generally remains in solution.

Phosphorus and eutrophication

Phosphorus is often scarce in the well-oxygenated surface waters of lakes, and low levels of phosphorus limit the production of freshwater systems. In small lakes on the Canadian Shield, productivity increased dramatically in response to the addition of phosphorus but not nitrogen or carbon (Figure 8.13). Natural lakes exhibit a wide range of productivity, depending on inputs of nutrients from outside (**external loading:** rainfall, streams) and the regeneration of nutrients within the lake (**internal loading**). In shallow lakes lacking a hypolimnion, internal loading occurs continuously through resuspension of bottom sediments. In somewhat deeper lakes, where the thermocline develops only weakly, vertical mixing may occur periodically as a result of occasional strong winds or during unusual periods of summer cold. Such mixing returns regenerated nutrients to the surface and stimulates production. In very deep lakes, bottom waters rarely mix with the surface, and production depends almost entirely on external loading.

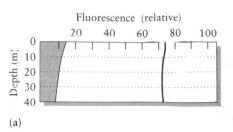

(a)

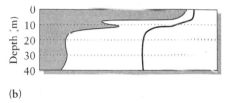

(b)

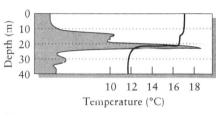

(c)

FIGURE 8.10 Vertical profiles of chlorophyll concentration (fluorescence, represented by the shaded areas) and temperature (represented by the lines) in the western English Channel in July 1975. (a) A well-mixed water mass. (c) A stratified water mass. (b) The "front" in the region of mixing between the two water masses. After R. K. Barnes and K. H. Mann, *Fundamentals of Aquatic Ecosystems.* Blackwell, Oxford (1980).

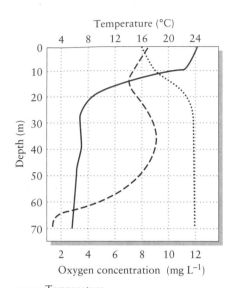

FIGURE 8.11 Profiles of temperature and oxygen concentration in Green Lake, Wisconsin, illustrating oxygen depletion in the hypolimnion during summer stratification. From G. E. Hutchinson, *A Treatise on Limnology*, Vol. 1. Wiley, New York (1957).

Aquatic ecologists classify lakes on a continuum ranging from poorly nourished (oligotrophic) to well nourished (eutrophic), depending on their nutrient status and production. Naturally eutrophic lakes have characteristic temporal patterns of production and nutrient cycling that maintain the system in a steady state. Addition of nutrients in sewage and drainage from fertilized agricultural lands can cause inappropriate nutrient loading and greatly alter natural cycles.

Increased production is not bad in and of itself; indeed, many lakes and ponds are artificially fertilized to increase commercial fish production. But overproduction can lead to imbalance when natural regeneration processes cannot handle the increased demands on cycling. Heavy organic pollution, such as that which results from dumping raw sewage into rivers and lakes, creates **biological oxygen demand (BOD)** resulting from the oxidative breakdown of the detritus by microorganisms. Inorganic nutrients, including runoff from fertilized agricultural land, stimulate the production of organic detritus, adding to the BOD. In its worst manifestations, this type of pollution can deplete the surface water of oxygen, leading to the suffocation of fish and other obligately aerobic organisms.

Despite the frequently devastating effects of external nutrient loading, culturally eutrophied lakes can recover their original condition when inputs are shut off. Eventually, oxic conditions return, phosphorus precipitates out of the water column, and the normal cycles of assimilation and regeneration are restored. A spectacular and convincing demonstration of such a recovery occurred after diversion of sewage from Lake Washington, in Seattle, where an advanced and rather ugly case of eutrophication quickly reversed itself.

Estuaries and marshes

Shallow **estuaries** — semienclosed coastal regions subject to both freshwater inputs from rivers and tidal inputs from the sea — are among the most productive ecosystems on earth. Salt marshes, which are intertidal areas with emer-

FIGURE 8.12 Seasonal course of water chemistry of the hypolimnion of Esthwaite Water, England, showing the solubilizing of reduced phosphorus and iron compounds during the period of summer anoxia. From G. E. Hutchinson, *A Treatise on Limnology*, Vol. 1. Wiley, New York (1957).

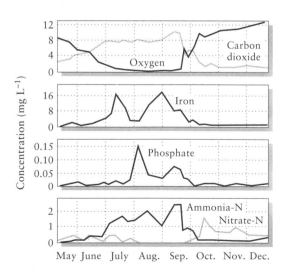

FIGURE 8.13 Experimental lake demonstrating the crucial role of phosphorus in eutrophication. The near basin, fertilized with carbon (in sucrose) and nitrogen (in nitrates), exhibited no change in organic production. The far basin, separated from the first by a plastic curtain, received phosphate in addition to carbon and nitrogen and was covered by a heavy bloom of blue-green algae within 2 months. From D. W. Schindler, *Science* 184:897–899 (1974).

gent vegetation (Figure 8.14), combine the most favorable attributes of aquatic and terrestrial systems, resulting in similarly high production. In addition to these local attributes of estuaries and coastal marshes, their significance for marine systems extends seaward through net export of production. A Georgia salt marsh exports nearly 10 percent of its gross primary production and almost half of its net primary production into surrounding marine systems in the form of organisms, particulate detritus, and dissolved organic material carried out with the tides (Figure 8.15). Because of their high productivity and the hiding places they offer prey organisms, coastal marshes and estuaries are important feeding areas for the larvae and immature stages of many fish and invertebrates that later complete their life cycles in the sea.

FIGURE 8.14 Salt marshes are a common feature of protected bays along most temperate coasts.

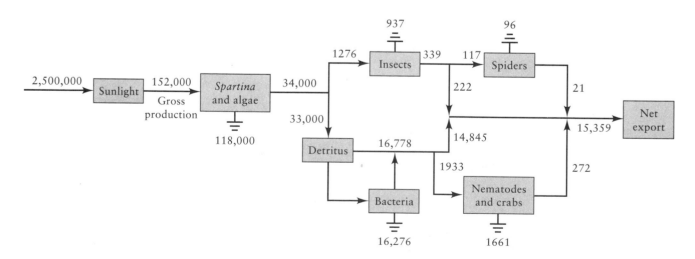

FIGURE 8.15 Energy-flow diagram for a Georgia salt marsh; units are kJ m⁻² yr⁻¹. The striated arrow represents respired energy lost from the system. From D. S. McLusky, *The Estuarine Ecosystem*. Wiley, New York (1981).

The high production of coastal systems reflects high nutrient levels. Because a large fraction of the production of the coastal habitat is carried out to sea, exported nutrients must be replaced by imports. Studies of the nitrogen budget of Great Sippewissett Marsh on Cape Cod have revealed inputs to the marsh through precipitation (minor), groundwater flow from surrounding terrestrial systems, and local fixation of atmospheric nitrogen. These inputs approximately balance losses through denitrification, local accumulation of sediments, and net export in tidal water. A high rate of denitrification, which occurs primarily in the creek bottom draining the salt marsh, underscores the role of anoxic processes in salt marsh metabolism. Rich organic sediments are anoxic just below their surfaces owing to rapid microbial decomposition of organic matter. As a result, oxidations based on denitrification ($N^{5+} \rightarrow N^0$) and sulfate reduction become important for the regeneration of nutrients.

As we have seen, the basic chemical and biochemical transformations of the element cycles are uniquely modified by the physical and chemical conditions created in each type of terrestrial and aquatic system. The paths of elements discussed in this chapter describe patterns of nutrient cycling. In the next chapter, we shall examine the regulation of the fluxes of nutrients through their cycles—and hence the control of the overall productivity of the ecosystem.

❧ SUMMARY

Nutrient cycles in terrestrial and aquatic systems express similar chemical and biochemical reactions but differ with the physical configurations of the habitats.

1. Nutrient regeneration in terrestrial systems takes place in the soil. The weathering of bedrock and the associated release of new nutrients proceed very slowly compared to the assimilation of nutrients from the soil by plants. Therefore, the productivity of vegetation depends on the regeneration of nutrients from plant litter.

2. Nutrients are regenerated from litter by the leaching of soluble substances; consumption by large detritus feeders (millipedes, earthworms); further attack by fungi to break down cellulose and lignin; and the eventual mineralization of phosphorus, nitrogen, and sulfur by bacteria.

3. Mycorrhizae are a symbiotic association of certain types of fungi with the roots of plants. The fungi, which may either penetrate the root tissue or form a dense sheath around the root, enhance the plants' uptake of soil nutrients. They do this primarily by enlarging the volume of soil accessible to the roots. In return, they obtain a reliable source of carbon from the plant.

4. In many lowland tropical habitats, deeply weathered soils retain nutrients poorly. In such habitats, regeneration and assimilation of nutrients proceed rapidly, and most of the nutrients, especially

phosphorus, occur in the living vegetation. When such soils are clear-cut for agriculture, they soon lose their fertility because nutrients are removed along with the native vegetation and crops, as well as because of physical and chemical changes in the soil.

5. The sediments at the bottoms of lakes and oceans resemble terrestrial soils but differ in two important respects. Aquatic sediments are spatially removed from the site of assimilation by aquatic plants. And sediments often develop anoxic conditions that retard the regeneration of some nutrients.

6. The productivity of aquatic systems is maintained either by the transport of nutrients from bottom sediments to the surface, as in shallow waters and areas of upwelling, or by the recycling of nutrients regenerated within the photic zone.

7. Vertical mixing is inhibited by stratification, the formation of layers of water that differ in density because they differ in temperature or salinity. Stratification enhances aquatic production by retaining phytoplankton within the photic zone, but it diminishes production to the extent that the sedimentation of detritus carries nutrients below the depth where light is sufficient for photosynthesis.

8. The annual cycles of temperate-zone lakes include a period of temperature stratification during the summer intervening between spring and fall periods of vertical mixing. Nutrients are brought to the photic zone near the surface during the periods of mixing. During the summer, nutrients are depleted by the sedimentation of organic material.

9. Nutrients are regenerated in aquatic sediments by bacterial decomposition. Anaerobic conditions that develop beneath the thermocline because of the consumption of oxygen by bacteria result in the chemical reduction of iron and magnesium and the solubilizing of phosphate compounds. Thus phosphorus is released from anoxic sediments and may reach the photic zone by vertical mixing.

10. Because phosphorus forms insoluble compounds with iron and precipitates readily under the oxic conditions of surface waters, it frequently is in short supply and limits aquatic production. Sewage and agricultural runoff add phosphorus and other nutrients to streams and lakes and may greatly alter natural patterns of production and nutrient cycling, upsetting natural balances in aquatic ecosystems. One consequence of increased production, augmented by sewage inputs of organic compounds, is depletion of oxygen from all levels of the lake when bacterial consumption exceeds the production of oxygen by photosynthesis and the diffusion of oxygen from the atmosphere. Fish and other animals may suffocate under such conditions.

11. Shallow-water communities, particularly estuaries and salt marshes, are extremely productive because of rapid and local regeneration of nutrients and the external loading of additional nutrients from

nearby terrestrial habitats. Studies of nutrient budgets indicate that marshes and estuaries are major exporters of both organic carbon and mineral nutrients to surrounding marine systems and are therefore an indispensable component of marine production in some areas.

❦ SUGGESTED READINGS

Barnes, R. K., and K. H. Mann (eds.) 1980. *Fundamentals of Aquatic Ecosystems.* Blackwell, Oxford.

Binkley, D., and D. Richter. 1987. Nutrient cycles and H^+ budgets of forested ecosystems. *Advances in Ecological Research* 16:1–51.

Coleman, D. C., C. P. P. Reid, and C. V. Cole. 1983. Biological strategies of nutrient cycling in soil systems. *Advances in Ecological Research* 13:1–55.

Edmondson, W. T. 1970. Phosphorus, nitrogen, and algae in Lake Washington after diversion of sewage. *Science* 169:690–691.

Jordan, C. F. 1982. Amazon rain forests. *American Scientist* 70:394–401.

Jordan, C. F., and R. Herrera. 1981. Tropical rain forests: are nutrients really critical? *American Naturalist* 117:167–180.

Libes, S. M. 1992. *An Introduction to Marine Biogeochemistry.* Wiley, New York.

Mann, K. H., and J. R. N. Lazier. 1991. *Dynamics of Marine Ecosystems. Biological-Physical Interactions in the Oceans.* Blackwell Scientific Publications, Boston.

McLusky, D. S. 1989. *The Estuarine Ecosystem* (2nd ed.). Chapman & Hall, New York.

McNaughton, S. J., R. W. Ruess, and S. W. Seagle. 1988. Large mammals and process dynamics in African ecosystems. *BioScience* 38:794–800.

Richards, B. N. 1987. *The Microbiology of Terrestrial Ecosystems.* Wiley, New York.

Stevenson, F. J. 1986. *Cycles of Soil. Carbon, Nitrogen, Phosphorus, Sulfur, Micronutrients.* Wiley, New York.

Tunnicliffe, V. 1992. Hydrothermal-vent communities of the deep sea. *American Scientist* 80:336–349.

Van Cleve, K., F. S. Chapin III, C. T. Dyrness, and L. A. Viereck. 1991. Element cycling in taiga forest: state-factor control. *BioScience* 41:78–83.

Warning, R. H., and W. H. Schlesinger. 1985. *Forest Ecosystems. Concepts and Management.* Academic Press, Orlando, Fla.

REGULATION OF ECOSYSTEM FUNCTION

Primary production drives the energy flux and cycling of elements within ecosystems. The most productive systems are tropical rain forests, coral reefs, and estuaries, where favorable combinations of high temperature, abundant water, and intense sunlight promote rapid photosynthesis and assimilation. But plant growth also depends on the regeneration of nutrients through biological processes, which themselves respond to variations in temperature, moisture, and other conditions.

Ecologists do not fully understand the control of productivity in most ecosystems. Their data allow broad comparisons of ecosystem function; simple correlations among resources, physical conditions, and production show that temperature, precipitation, and other external factors regulate productivity, but do not show how or where they act.

Experimentation provides another approach to studying the regulation of ecosystem function, but rarely can ecologists conduct experiments on the scale of ecosystems. How, for example, could one eliminate nitrifying bacteria from soil without changing related conditions and resources? Where experiments have been performed (for example, by adding water or nutrients to a system), scientists have had difficulty ascertaining which step or steps in the cycling of critical elements responded to the treatment and whether observed responses to artificial manipulations yield any information about differences between natural systems or temporal changes in them.

Ecologists can also model ecosystem function by mathematical equations. Modeling requires detailed knowledge of the processes that are critical to regulating ecosystem function (which could be interpreted as everything that happens in the system!) and validation of each equation by field observation and experiment. Despite these difficulties, modeling does enable investigators to simulate the effects of environmental changes on the dynamics of a system. Modeling efforts are central to the discipline of **systems ecology**.

In this chapter, we shall develop simple models of systems, with the modest goal of gaining some insights into the control of ecosystem function. We shall then use these insights to interpret particular observations and experimental results. The approach is mathematical but requires only simple algebra. It demonstrates that mathematical modeling can yield new insights about ecosystem function that, in turn, may guide further observation and experimental work.

External forcing functions and internal control feedbacks

In systems terminology, understanding ecosystem function requires knowing how external forcing functions and internal control feedbacks are integrated. **External forcing functions** are material inputs from outside the system and physical conditions of the environment that influence the system's structure and function. The external forcing functions that affect ecosystems include light, temperature, weathering, and precipitation inputs; salinity; and other kindred factors.

Internal control feedbacks result from the chemical behavior of elements in the physical part of the ecosystem and the responses of organisms to the physical environment and to each other. As we saw in Chapter 8, biological production depends on the regeneration of nutrients by microorganisms and other entities; their activities form an internal control feedback in the sense that they determine the availability of nutrients to plants. Because external forcing functions can act on both the assimilatory and the regenerative phases of ecosystem processes, it is impossible to understand the control of ecosystem function without a detailed appreciation of the influence of external forcing functions on all parts of the feedback controls within the system.

An analogy may clarify the action of external and internal controls. The rate at which a bucket fills with water depends on an external forcing function: the flow of water from the tap. If the bottom of the bucket has a hole in it, the hole, as a property of the system, imposes an internal control feedback. When the system achieves a steady state — water entering the bucket equals the amount leaving through the hole — the flow of water through the bucket depends only on the flow from the tap (external control), although the level of water in the bucket depends on both the flow from the tap and the size of the hole (external control and internal feedback).

The regulation of ecosystem function raises two issues. The first is the degree to which variations in ecosystem structure and function result from variations in external forcing variables, as opposed to unique internal properties of ecosystems — essentially, the difference between variation in the flow of water from the tap and variation in the number and size of holes in the bucket. The second issue concerns the particular means by which external forcing variables exert their influence on internal feedback controls.

Variation in primary production

Ecologists agree that basic aspects of ecosystem structure and function respond to external forcing variables, particularly temperature and precipitation in the case of terrestrial ecosystems. We can characterize such relationships by comparing annual primary production to temperature and precipitation. Among data tabulated for 53 localities distributed throughout the world, estimates of annual production varied 50-fold, from 70 to over 3500 grams of dry matter per square meter (g m^{-2}); precipitation varied 48-fold, from 94 to 4500 mm annually; and average temperature ranged from −14.2 to 27.1°C.

We can use these data to develop a predictive equation relating annual production to temperature and precipitation:

$$P = a + bW + cW^2 + dT + eT^2 + fWT$$

where P represents production, W precipitation, T temperature, and a through e the coefficients indicating the strength of each of the terms. In this analysis, production and precipitation are transformed to logarithmic scales because factorial differences between localities are more meaningful than absolute differences. For example, a difference between 100 and 200 g m^{-2} production (a factor of 2) is more significant biologically than a difference of the same absolute magnitude between 1000 and 1100 g m^{-2} (a factor of 1.1).

The presence of squared (quadratic) terms in the foregoing equation indicates nonlinearity in the relationship (Figure 9.1). The presence of a term that is the product of W and T indicates a synergism between temperature and precipitation (Figure 9.2). The significance of any of these terms is revealed by a statistical analysis of their relationship to the dependent variable, annual production P.

Variation in terrestrial production clearly shows the influence of the physical environment. All the terms of the predictive model made significant, unique contributions to variation in production, yielding the equation

$$P = 4.51W - 0.82W^2 - 0.11T - 0.0015T^2 + 0.053WT$$

The signs of the coefficients b (4.51) and c (−0.82) show that productivity increases in direct proportion to precipitation over the lower part of the range (b positive) and then levels off (c negative), presumably because some other factor becomes more limiting. When the contribution of precipitation to the

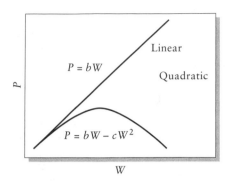

FIGURE 9.1 Illustration of linear and quadratic relationships of productivity (P) to precipitation (W).

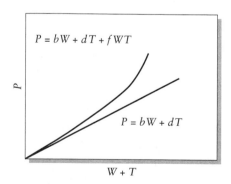

FIGURE 9.2 Illustration of additive and interactive (synergistic) relationships of productivity (P) to the sum of precipitation and temperature ($W + T$).

variation in production is accounted for statistically, as it is in the preceding equation, we find that higher temperatures lead to decreased production (*d* and *e* negative); this confirms the notion that water goes farther in cold climates than in warm ones. A positive synergism between temperature and rainfall (*f* positive) partially balances this effect. Thus hot, humid climates have higher productivity than we would predict on the basis of the separate relationships of production to precipitation and to temperature.

This analysis suggests that external forcing variables in some way cause much of the variation in terrestrial production. But how much? Statistically, their effect is estimated by the coefficient of determination (R^2), which is the proportion of the variation in production that is related to variation in temperature and precipitation. For these data, $R^2 = 0.73$; that is, 73 percent of the variation in productivity is explained by the equation, and 27 percent is not.

Temperature and precipitation clearly "control" terrestrial production to a large extent. The unexplained variation could be attributable to measurement error, inappropriate choice of mathematical terms for the equation, external factors not included in the equation (such as seasonal variations in temperature and precipitation, solar radiation, and nutrient status of the soil), and internal control feedbacks unique to each locality. These unique feedbacks may derive from particular species, perhaps herbivores, that are present in one locality but not in others. It is evident, however, that one index of ecosystem function—net aboveground primary productivity—responds to external forcing functions.

Energy and element fluxes in ecosystem function

Energy provides the most generalizable currency of ecosystem function, but the availability of energy (the external forcing function is light) appears to cause little of the variation among systems. As we have seen, most of this variation can be related to variations in temperature and precipitation. How does one identify these regulatory factors and the component or components of the system on which these regulators act?

Peter Vitousek, a plant ecologist at Stanford University, has attempted to pinpoint the influence of external forcing functions by comparing the fluxes of individual elements to the flux of energy through the system as a whole. Vitousek supposed that the element whose flux shows the strongest correlation with primary production exercises predominant control.

When a forest achieves a steady state, net aboveground primary production approximately equals litter production. Similarly, the flux of each element approximately equals the amount of the element that falls each year in litter. Vitousek compared the flux of total dry matter in the litter (which is proportional to carbon and hence to energy content) to the fluxes of several elements; nitrogen and phosphorus fluxes are portrayed in Figure 9.3. As can be seen in the figure, production more closely parallels the flux of nitrogen than that of phosphorus (or that of calcium, which is not shown). Vitousek concluded that factors regulating the cycling of nitrogen predominantly control primary production in forests.

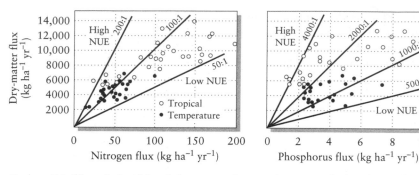

FIGURE 9.3 The relationship of dry-matter flux to nitrogen and phosphorus fluxes in the litter fall of temperate and tropical forests. NUE = nutrient use efficiency. After P. M. Vitousek, *Amer. Nat.* 119:553–572 (1982).

Vitousek's study also showed, however, that production does not vary in direct proportion to nutrient flux. Rather, forests with low fluxes exhibit relatively *greater* production per unit of nutrient cycled (**nutrient use efficiency, NUE**) than do those with high fluxes. Two factors result in higher nutrient use efficiency. First, trees may assimilate more energy per unit of nutrient assimilated; second, trees may retain nutrients for reuse by drawing them back into the stem before they drop their leaves. For both nitrogen and phosphorus, nutrient use efficiency clearly decreases as nutrient flux increases, and, for phosphorus, NUE in the tropics exceeds that in temperate latitudes; tropical trees evidently retain phosphorus to a greater extent than do temperate trees.

Nutrient use efficiency may express certain adaptations of the plant that enable it to retain nutrients or reduce nutrient requirements for production. Thus indications of a regulatory role based on the relationship between production and nutrient flux may be obscured by adaptations of plants to manage nutrient use. In an attempt to sort out the factors responsible for the regulation of ecosystem function, we shall develop simple systems models that incorporate mathematical expressions for the effects of these factors and then examine the behavior of these models. In particular, we wish to determine whether variations in external forcing functions and internal feedback controls produce diagnostic, measurable variations in ecosystem properties, from which we can infer the mechanisms that regulate ecosystem function. First, however, some basics.

Development of a systems model

Each form of a nutrient or of the energy within a system defines a distinct compartment, which we shall designate X_i for the ith form (organic-N, ammonia, or nitrate, for example). Each compartment has inflows and outflows, which we shall designate by Js. A schematic diagram of a single compartment (X_1) with one input (J_0) and one output (J_1) is shown in Figure 9.4. If the compartment represented the water in our bucket, then J_0 would be the flow from the tap and J_1 would be what leaks out through the hole in the bottom. The

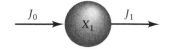

FIGURE 9.4 Representation of a single compartment in an ecosystem model, with input (J_0), output (J_1), and compartment size (X_1).

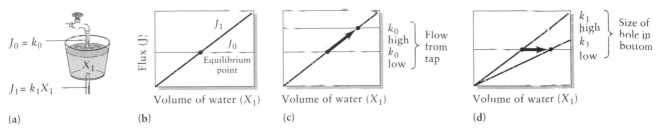

(a) (b) (c) (d)

FIGURE 9.5 The effect of changing input (J_0) and output (J_1) rates on the volume of water (X_1) in a bucket. (a) Because water leaves the bucket through a hole in the bottom, the rate of output is proportional to the water pressure at the bottom and thus to the depth and volume of water. (b, c) Water depth and flux increase as water input (J_0) increases. (d) Water depth (X_1) increases, but flux (J) does not change, as rate of emptying (k_1) decreases.

rate of change in the amount of water in the bucket (dX_1/dt) is equal to the difference between the input and the output — algebraically,

$$\frac{dX_1}{dt} = J_0 - J_1$$

The amount of water in the bucket achieves a **steady state** (that is, $dX_1/dt = 0$) when inflow equals outflow ($J_1 = J_0$).

Each flux J may be a constant, or it may vary depending on the state of other factors, including the value of X and the values of external forcing functions. We may express such variable fluxes in symbolic notation as $J = f(X, S, P)$, where f denotes that J is a function of the values inside the parentheses; S represents the state of the system (for example, the number and size of holes in the bucket, the leaf area available for photosynthesis, the population density of nitrifying bacteria); and P stands for various parameters that are the external forcing functions (such as the flow of water from the tap, the temperature of the environment, the amount of precipitation).

In the bucket example, suppose that J_0 is a constant ($J_0 = k_0$) but that J_1 increases in direct proportion to the volume of water in the bucket (that is, $J_1 = k_1 X_1$). As the bucket fills, the water pressure at the bottom increases the flow of water through the hole. The amount of water in the bucket (X_1) reaches a steady state when $J_0 = J_1$, and therefore $k_0 = k_1 X_1$. We may rearrange this equation to show that the steady-state amount of water (X_1) equals k_0/k_1. Thus, when the value of the forcing function (k_0) increases, water reaches a higher level in the bucket. Reducing the size of the hole in the bucket (the variable k_1) has the same effect on X_1 (Figure 9.5), but note that the flux through the system is always controlled by k_0 alone.

We need at least two compartments to depict the internal cycling of elements within ecosystems. Realistic systems models can become much more complicated, but we shall first consider the simple case of two compartments that cycle an element between them (Figure 9.6). Compartment sizes are X_1 and X_2, and fluxes are J_1 and J_2. As we have seen, the Js represent functions. These may be of zero order, in which case J is a constant ($J = c$); of first order,

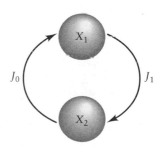

FIGURE 9.6 Diagram of a closed, two-compartment system.

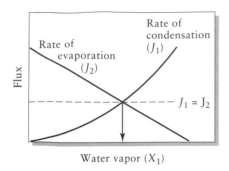

FIGURE 9.7 Rate of evaporation decreases and rate of condensation increases with increasing water vapor in the atmosphere. This creates an equilibrium level (arrow) of water vapor in the atmosphere, at which point evaporation equals condensation.

in which case J_1 is a function of X_1; or of second order, J_1 being a function of both X_1 and X_2.

We can say little about the functioning of a particular system without knowing details of the flux functions. Simple models can, however, lead to a general understanding of some broad features of system function. The global water cycle, for example, can be thought of as two compartments, vapor (X_1) and liquid (X_2). Knowledge of the physics of the state change of water enables us to describe functions for the fluxes in very general terms. Precipitation (J_1) is a first-order equation depending only on the water vapor of the atmosphere (X_1) and on various external forcing functions such as air temperature; the amount of water at the earth's surface (X_2) has no direct effect on precipitation. Air has a limited capacity to hold water vapor at a given temperature, so condensation increases disproportionately as water vapor increases toward this limit.

Evaporation (J_2) is a second-order equation depending on the surface area of water (some function of X_2), the vapor pressure of water in the atmosphere (proportional to X_1), and external forcing functions: temperature and insolation (intensity of sunlight). On a global scale, precipitation probably has a smaller effect on the surface area of water than evaporation has on atmospheric water vapor, so we can ignore X_2 in the function J_2. Thus we may portray the general features of the hydrological cycle on a graph relating J_1 and J_2 to X_1 (Figure 9.7). The model shows that the system assumes a steady state with $J_1 = J_2$ at some intermediate value of X_1; changes in the forcing functions affecting J_1 or J_2 (change in temperature, for example) would adjust the equilibrium point for the entire system. The model can't be used to predict the local weather.

Lotka's model of ecosystem function

In his book *The Elements of Physical Biology,* published in 1925, Alfred J. Lotka investigated the behavior of biological systems by applying the insights of thermodynamics and the tools of mathematical modeling borrowed from the study of chemical equilibria and other physical phenomena. In the space of a few pages, he outlined the application of systems modeling to nutrient cycling in ecosystems. Lotka treated the specific case of three compartments X_1, X_2, X_3 through which some element or other material cycles with fluxes J_1, J_2, J_3 (Figure 9.8).

Change in any compartment—for example, dX_1/dt—equals the difference between the fluxes into and out of the compartment (J_3 and J_1 in the case of compartment X_1). To illustrate the behavior of such a system, Lotka described each flux as a first-order term $f_i(X_i)$ arbitrarily having the form g_iX_i; g_i is the rate at which material in compartment X_i is transferred to the next compartment. Thus $dX_1/dt = g_3X_3 - g_1X_1$. When the system is in a steady state, all the fluxes are equal; hence $J_1 = J_2 = J_3$, or $g_1X_1 = g_2X_2 = g_3X_3$. Because $X_i = J_i/g_i$ and all the Js are equal, the sizes of the compartments are in the relative proportions

$$X_1 : X_2 : X_3 = \frac{1}{g_1} : \frac{1}{g_2} : \frac{1}{g_3}$$

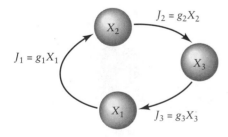

FIGURE 9.8 A diagram of Lotka's system of three compartments in which the flux between one compartment and the next is a function only of the first compartment.

Now, all the compartments of the system contain a total $M (= X_1 + X_2 + X_3)$ of the cycling material. Lotka showed that the compartment sizes may be described by equations of the form

$$X_i = \frac{M}{g_i} \left(\frac{g_1 g_2 g_3}{g_1 + g_2 + g_3} \right)$$

and the fluxes by

$$J_i = g_i X_i = M \left(\frac{g_1 g_2 g_3}{g_1 + g_2 + g_3} \right)$$

These equations tell us that the structure (Xs) and function (Js) of the system are defined completely by the transfer functions (gs), which incorporate the external forcing variables and internal feedback controls. In particular, the size of compartment X_i is inversely related to transfer rate g_i.

Lotka further showed that differences in structure and function between systems probably derive from differences in the lowest transfer rates. Low transfer rates place bottlenecks in the path of material flow through a system, causing material to accumulate in the immediately preceding compartment. Therefore, the underlying cause of variation in flux through a system should be apparent in shifts of materials among compartments in the system. For example, if an increase in J is accompanied by a shift of material from compartment X_1 to compartments X_2 and X_3, we may infer that an increase in the function g_1 was responsible. Distinguishing among the roles of state variables, internal control feedbacks, and external forcing functions in this change would require additional study of the system, but research efforts are greatly focused by heeding the insights of the systems model. The practical lesson to be learned is that a change in one segment of a nutrient cycle can alter the function of the entire system: a chain is only as strong as its weakest link.

An aquatic systems model

The cycling of nitrogen within a water column by and large follows the path (ammonia → nitrate) → particulate-N (organisms + detritus) → dissolved organic nitrogen (DON) → (ammonia → nitrate). Algae may assimilate either ammonia or nitrate. In the study of nitrogen transformations in the water column of the Bay of Quinte, Ontario, the sizes of some compartments were found to change dramatically with season (Figure 9.9). In particular, between winter and summer, particulate and dissolved organic nitrogen increased while nitrate decreased. This shift implies that the primary difference between the cycling of nitrogen in summer and in winter was the rate of uptake of nitrate by phytoplankton. A simple first-order systems model will illustrate how we can elaborate on this idea.

When the system in Figure 9.9 has achieved a steady state, fluxes into and out of each compartment must be balanced. For example, $J_2 = J_3, J_3 = J_4 + J_5$, and so on. Hence, under steady state conditions,

$$g_2 X_2 = g_3 X_3 = (g_4 + g_5) X_4 = (g_1 X_1 + g_4 X_4)$$

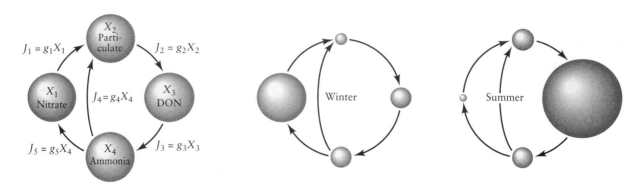

FIGURE 9.9 A first-order systems model of nitrogen cycling in the Bay of Quinte, Ontario, showing the changes in compartment sizes between summer and winter.

From the last two quantities, we can show algebraically that $X_1/X_4 = g_5/g_1$, and by making the appropriate substitutions, we obtain the relationship

$$X_1 : X_2 : X_3 : X_4 = \frac{g_5}{g_1(g_4 + g_5)} : \frac{1}{g_2} : \frac{1}{g_3} : \frac{1}{(g_4 + g_5)}$$

Setting each X_i equal to its proportion of the total nitrogen enables us to estimate relative values for the g_is during the winter and summer (Table 9.1).

TABLE 9.1	*Transfer rates for nitrogen in the Bay of Quinte, Ontario*		
Transfer rate	Description	Winter	Summer
g_2	Production of dissolved organic nitrogen by grazers, excretion, and leakage	8.0	4.1
g_3	Ammonification	4.4	2.0
$g_4 + g_5$	Assimilation of ammonium plus nitrification	4.2	4.7
$\dfrac{g_1}{g_5}$	Ratio of nitrate assimilation to nitrification	0.6	5.1
$\dfrac{g_1}{g_5}(g_4 + g_5)$		2.5	24

Source: From data in C. F. H. Liao and D. R. S. Lean, *J. Fish. Res. Board. Canad.* 35:1102–1108 (1978).

These indicate, quite dramatically, that the ratio g_1/g_5 (the ratio of plant assimilation to nitrification) is an order of magnitude lower during the winter than it is during the summer. The sum $g_4 + g_5$ differs little between the seasons, so we may assume that g_5 is relatively constant. Therefore, it is evidently the value of g_1 that decreases so much between summer and winter (g_1 describes the rate of assimilation of nitrate-N by plants). During the winter, assimilation decreases markedly compared to nitrification (g_5) and ammonification (g_3), suggesting that some factor, such as light, might limit production. Alternatively, g_1 and g_4 (assimilation of nitrate and ammonia, respectively) might both decrease in winter, in which case nitrification rate (g_5) would have to increase at the same time to maintain the sum $g_4 + g_5$ constant.

The preceding model may oversimplify or even misrepresent nitrogen transformations in the water column. It was meant to illustrate an approach. Most systems models are much more complicated, often including dozens of compartments and equations of much higher order than one. In our simple model, for example, fluxes J_1 and J_4 almost certainly should have been represented by second-order equations involving X_2 (the populations of phytoplankton that accomplish the assimilation; no nitrogen would be assimilated in the absence of plants). More realistic equations could be developed for particular systems, but simplified models, such as this one and the one that follows, may also serve to direct inquiry into more general comparisons of function between ecosystems.

A terrestrial systems model

In broad comparisons among terrestrial ecosystems, measurements of annual net primary production (in grams of dry matter per square meter) vary almost 30-fold between desert shrub (70) and tropical rain forest (2000). This variation is clearly related to climate, yet ecologists have not determined where external forcing variables exert their influence within the system. A systems approach can provide clues in this case.

Let us consider the cycling of nitrogen again. We shall represent an ecosystem as three compartments — mineral soil (X_1), living plant biomass (X_2), and organic detritus (X_3) — with fluxes $J_1, J_2,$ and J_3 (Figure 9.10). (Animals are trivial in this model; the activities of microorganisms are implicit in the flux J_3). J_2 (the annual dropping of leaves and other detritus) and J_3 (the mineralization of detritus by microorganisms) can be considered first-order processes. The rate of nitrogen assimilation (J_1) is modeled as a second-order process, incorporating both nutrient availability and plant activity. Under steady-state conditions, we obtain the relationships

$$g_1X_1X_2 = g_2X_2 = g_3X_3$$

and

$$X_1X_2 : X_2 : X_3 = \frac{1}{g_1} : \frac{1}{g_2} : \frac{1}{g_3}$$

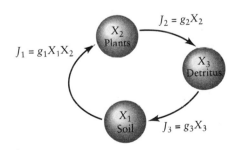

FIGURE 9.10 A second-order systems model of nitrogen cycling in a forest ecosystem. Note that the assimilation of nitrogen by plants (J_1) is a function of both the availability of nitrogen in the soil and the amount of living plant biomass.

This simple model offers some surprises. For example, the level of inorganic nitrogen in the soil, mostly nitrate (X_1), equals the ratio of g_2 (rate of detritus production) to g_1 (rate of nitrogen assimilation) and is independent of g_3 (rate of microbial regeneration of inorganic nitrogen). Nitrogen flux does depend on g_3, of course. Algebra shows that the relationship of flux to the rates g_i follows

$$J = \left[\frac{g_2 g_3 (g_1 - g_2)}{g_1 (g_2 + g_3)}\right] M$$

where M equals the total nitrogen in the system. Fluxes of elements are difficult to measure directly, but we may estimate relative values of g_i from the sizes of compartments. Finally, differences in the relative rates of transfers between compartments in different systems can provide insights into points of feedback control.

To illustrate, we shall compare the cycles of nitrogen and phosphorus in a 47-year-old Scots pine (*Pinus sylvestris*) plantation in England and a 50-year-old mixed tropical forest in Ghana (Figure 9.11). The differences are striking. In England, for both nitrogen and phosphorus, the values of $g_1, g_2,$ and g_3 are of the same magnitude. In the nitrogen cycle in Ghana, g_1 and g_2 are similar, but g_3 is almost two orders of magnitude greater. Thus the regeneration of mineral nutrients from organic detritus apparently proceeds much more rapidly in the tropics than in temperate areas, confirming direct measurement of litter decomposition. External forcing functions appear to affect soil microorganisms rather than plants.

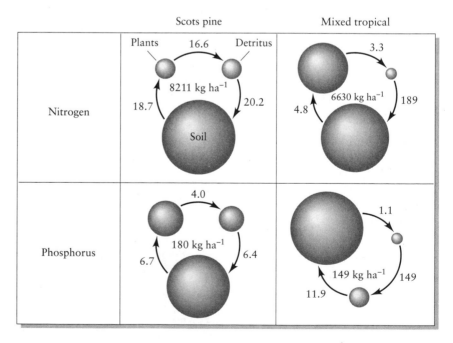

FIGURE 9.11 Compartment sizes and values of g_i for nitrogen and phosphorus estimated from compartment sizes in a Scots pine plantation in England and a tropical forest in Ghana. After data in J. D. Ovington, *Adv. Ecol. Res.* 1:103–192 (1962); D. J. Greenland and J. M. L. Kowal, *Plant Soil* 12:154–174 (1960).

Relative transfer rates of phosphorus behave similarly, but they reveal another difference between the tropical and temperate forests. In Ghana, the assimilation coefficient ($g_1 = 11.9$) greatly exceeds the rate at which vegetation gives up nutrients ($g_2 = 1.1$); the values of g_1 and g_2 in England are 6.7 and 4.0. These values of g imply that, compared with *Pinus sylvestris*, tropical forest trees assimilate phosphorus more efficiently and hold onto it more tightly (for example, by withdrawing phosphorus into the stem before shedding leaves). Therefore, the systems model is consistent with the higher nutrient use efficiency observed in tropical vegetation.

Energy and materials flow through the ecosystem largely because of the activities of organisms. Understanding the regulation of ecosystem function therefore depends largely on understanding how organisms respond to physical conditions and how they interact with others of the same and different species. These interactions form the basis for the study of organism and population processes, which are discussed in Parts 3 and 4.

🦅 *SUMMARY*

1. Ecologists use three methods to study the regulation of ecosystem function: comparison, experiment, and modeling. The results of comparisons and experiments are most readily interpretable when they are compared to the predictions of models based on mechanisms of ecosystem function.

2. Systems models incorporate external forcing functions, which are factors that affect the system from outside, and internal control feedbacks, which result from interactions among the components of a system.

3. Examination of terrestrial systems shows that primary production correlates with the external forcing functions of temperature and precipitation. Hence, ecosystem function responds to the physical environment.

4. We can understand the regulation of terrestrial production by examining the correlation between production and element cycling. The concept of nutrient use efficiency (NUE) describes the relationship between production and nutrient assimilation.

5. Systems models based on a particular element or on energy consist of compartments (Xs) and fluxes (Js) between compartments. When compartments achieve a steady state, input equals output. For an element cycling in a steady state, fluxes through each segment of the cycle are equal. Each flux out of a compartment equals a transfer rate g times the compartment size. Each transfer rate may be a function of the attributes of one or more compartments, other properties of the system, and factors external to the system.

6. A. J. Lotka first applied systems models to ecological systems. He showed that the flux of a cycling element responds most sensitively to variation in the smallest transfer rate.

7. When a system exists in a steady state, relative transfer rates can be estimated from compartment sizes. For a simple model of nitrogen cycling within the water column of a lake, seasonal changes in nitrogen compartments indicated that the rate of assimilation of nitrate-N decreased dramatically during the winter.

8. A three-compartment model (soil, plant, and detritus) was developed for forest ecosystems. Comparisons of compartment sizes in temperate and tropical forests indicate that nitrogen and phosphorus are regenerated from litter much more rapidly in the tropics and that tropical trees, to a greater extent than temperate trees, withdraw phosphorus from leaves about to be shed.

❧ SUGGESTED READINGS

Carpenter, S. R., and J. F. Kitchell. 1988. Consumer control of lake productivity. *BioScience* 38:764–769.

Jeffers, J. N. R. 1978. *An Introduction to Systems Analysis. With Ecological Applications.* Edward Arnold, London, and Univ. Park Press, Baltimore.

Kitching, R. L. 1983. *Systems Ecology. An Introduction to Ecological Modelling.* Univ. of Queensland Press, St. Lucia, London, and New York.

Miller, P. C., W. A. Stoner, and L. L. Tieszen. 1976. A model of stand photosynthesis for the wet meadow tundra at Barrow, Alaska. *Ecology* 57:411–430.

Odum, H. T. 1983. *Systems Ecology. An Introduction.* Wiley, New York.

Perry, D. A., M. P. Amaranthus, J. G. Borchers, S. L. Borchers, and R. E. Brainerd. 1989. Bootstrapping in ecosystems. *BioScience* 39:230–237.

Vitousek, P. M. 1982. Nutrient cycling and nutrient use efficiency. *American Naturalist* 119:553–572.

Vitousek, P. M. 1984. Litterfall, nutrient cycling, and nutrient limitation in tropical forests. *Ecology* 65:285–298.

3

Organisms

10

HOMEOSTASIS, ACCLIMATION, AND DEVELOPMENTAL RESPONSE

Earlier, we discussed the physical factors in the environment that are important to the well-being of organisms and the ways in which plants and animals handle these factors. In general, organisms maintain themselves out of equilibrium with the physical environment with respect to heat, water, salts, and chemical energy. Keeping this state of unbalance — upholding gradients in the concentrations of energy and substances between the body and the surroundings — requires the expenditure of considerable energy. We have also seen that organisms are specialized to function most efficiently and productively within a narrow range of environmental conditions, and each form has its optimum. But the environment also varies, forcing individuals to adjust continuously to their surroundings. In this chapter, we shall explore some of the problems created by change in the environment and the mechanisms that plants and animals employ to counter them.

Change pervades an organism's surroundings: the annual cycle of the seasons, daily periods of light and dark, frequent unpredictable turns of climate. The survival of each individual depends on its ability to cope with variations in the environment. Humans can be conscious of their own responses to change. When we step from a warm room into the outdoors on a cold day, we soon shiver to generate heat. A few weeks on the beach and our skin darkens, blocking some of the damaging radiation from the sun. Like other organisms, we respond to environmental change in such a way as to maintain our internal conditions within a suitable range for proper functioning. But what determines the best internal condition? At what rate should the individual function? How can the organism respond to environmental change most effectively?

Responses can be evaluated, like a problem in economics, in terms of costs and benefits. When the environment cools, shivering increases the probability of survival of a warm-blooded organism. But shivering also requires energy, which in turn may deplete fat reserves and render life more precarious in the face of a sudden food shortage. The organism may reduce the cost of temperature regulation by lowering its internal temperature, just as we turn down the thermostat to save fuel. But turning down the fire of an organism's life also reduces its rate of activity and hence its food-gathering and predator-avoiding abilities.

When we consider the many factors that affect the costs and benefits of a particular response, we begin to understand why different organisms regulate their internal conditions at different levels, or not at all, and why they employ different means of response to environmental change.

Homeostasis

Homeostasis is the ability of the individual to maintain constant internal conditions in the face of a varying external environment. All organisms exhibit homeostasis to some degree with respect to some environmental conditions, though the occurrence and effectiveness of homeostatic mechanisms vary.

Temperature regulation

Most mammals and birds closely maintain their body temperature between 36 and 41°C, even though the temperature of their surroundings may vary from −50 to +50°C. Such regulation, referred to as **homeothermy** (the Greek root *homos* means "same"), guarantees that biochemical processes within cells can occur under constant-temperature (homeothermic) conditions. The internal environments of cold-blooded (**poikilothermic**) organisms, such as frogs and grasshoppers, conform to the external temperature (the root *poikilo* means "varying"). Of course, frogs cannot function at either high or low temperature extremes, so they are active only within a narrow part of the range of environmental conditions over which mammals and birds thrive.

How is body temperature regulated? Most homeotherms have a sensitive thermostat in the brain that responds to changes in the temperature of the blood by secreting hormones into the bloodstream to slow down or accelerate the generation of heat in body tissues. In addition, most homeotherms partly regulate body temperature by altering gains and losses of heat from the environment. For example, humans put on heavy clothes in cold weather and avoid standing in the sun on a hot day; birds fluff up their feathers, increasing insulation against the cold.

Because the so-called cold-blooded organisms can generate heat only when they are active, many adjust their heat balance behaviorally by simply moving into or out of the shade or by changing the orientation of the body with respect to the sun. When horned lizards are cold, they increase the profile of their bodies exposed to the sun by lying flat against the ground; when hot, they decrease their exposure by standing erect upon their legs. By lying flat against the ground, horned lizards also gain heat by conduction from the sun-warmed surface. Such behavior, widespread among reptiles, effectively regulates body temperature within a narrow range; their temperature may rise considerably above that of the surrounding air. So long as lizards and snakes can gain heat from the sun, they act as homeotherms. But because their source of heat lies outside the body, biologists refer to them as **ectotherms** (external heat) rather than **endotherms** (internal heat).

Negative feedback

Regardless of the particular mechanism of regulation, all homeostasis exhibits properties of a **negative feedback** system, exemplified by the working of a thermostat. When the temperature of a room exceeds the desired temperature, a temperature-sensitive switch turns off the heater; when the temperature drops too low, the switch turns the heater on. In accordance with the same general principle, when we walk from a dark room into bright sunlight, the pupils of our eyes contract rapidly, restricting the amount of light entering the eye. Similarly though less rapidly, a sudden exposure to heat brings on sweating, which increases evaporative heat loss from the skin and helps to maintain body temperature at its normal level.

Such patterns of response represent negative feedback: when external influences cause a system to deviate from its norm, or desired state, internal response mechanisms act to restore that state. The essential elements of a negative feedback system are (1) a mechanism that senses the internal condition of the organism; (2) a means of comparing the actual internal state with the desired state; and (3) an apparatus that alters the organism's internal condition in the direction of the desired condition. A schematized and simplified negative feedback mechanism for regulating body temperature is illustrated in Figure 10.1. Such negative feedback mechanisms characterize the regulation of ions, acidity, blood gases, enzyme concentrations, and many other conditions and substances within the body.

Energy costs of endothermy

To sustain internal conditions that differ significantly from the external environment requires work and energy. The maintaining of constant high body temperatures by birds and mammals in cold environments illustrates the meta-

Feedback control loop

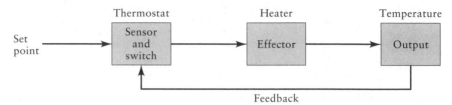

FIGURE 10.1 The major attributes of a negative feedback system, with particular components indicated for the regulation of body temperature.

bolic cost of homeostasis. As air temperature decreases, the gradient between the internal and external environments increases, and the body surface loses heat proportionately more rapidly. An animal that maintains its body temperature at 40°C loses heat twice as fast at an **ambient temperature** (surrounding temperature) of 20°C (a gradient of 20°C) as at an ambient temperature of 30°C (a gradient of only 10°C). The principle that heat loss varies in direct proportion to the gradient between body and ambient temperature is called **Newton's law of cooling.** To maintain a constant body temperature, endothermic organisms replace heat lost to the environment by generating heat metabolically. Thus the rate of metabolism required to maintain body temperature increases in direct proportion to the difference between body and ambient temperature, all other things being equal (Figure 10.2).

If the only purpose of metabolism were temperature regulation, the individual would require no metabolic heat production when body temperature equaled ambient temperature, at which point no heat would flow between the organism and its surroundings. But organisms generate metabolic heat in sustaining heartbeat, breathing, muscle tone, and kidney function, regardless of the ambient temperature. The metabolism of an organism resting quietly and without food in its digestive tract (postabsorptive) achieves a **basal** or **resting metabolic rate** (**BMR** or **RMR**). The BMR represents the lowest level of energy release under normal conditions. At this basal rate, the individual produces sufficient heat to maintain its body temperature when ambient temperatures exceed a certain level, the **lower critical temperature** (T_{lc}). This tem-

FIGURE 10.2 Relationship between energy metabolism and ambient temperature for a homeothermic bird or mammal whose body temperature is maintained at T_b. T_{lc} is the lower critical temperature, below which metabolism must increase to maintain body temperature. Point c is the lower lethal temperature, and point b is the lowest temperature at which the organism can maintain itself indefinitely.

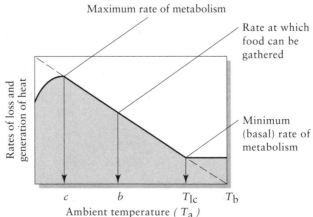

Environments without life

The Antarctic ice cap is too cold for life. Water freezes and life processes come to a standstill. In contrast, new land is formed when lava reaches the surface in volcanic eruptions, as at Kilauea volcano in Hawaii. Several decades will pass after the lava has cooled before the first plants begin to colonize. Hot temperatures and noxious gases keep plants from establishing in the crater of Poas volcano in Costa Rica.

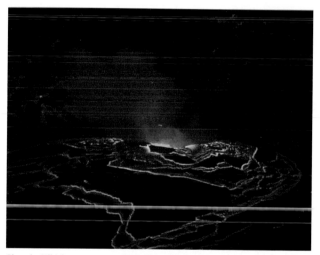

Photo by Bill Johnson, courtesy of U.S. Dept. of Agriculture, Soil Conservation Service.

Photo by R. E. Ricklefs.

Photo by R. E. Ricklefs.

The tropical rain forest

Warm temperatures and abundant rainfall in
many parts of the tropics support the highest
levels of biological production and diversity of
life on earth.

Photo by R. E. Ricklefs.

The unbroken canopy of a Panamanian rain forest may include 200 species of trees. Some of these may reach 50 to 60 meters in height. In the monotonous green of the forest, plants use bright colors, especially red, to attract pollinators and seed dispersers.

Photo by Carl C. Hansen, courtesy of Smithsonian Tropical Research Institute.

Photo by Carl C. Hansen, courtesy of Smithsonian Tropical Research Institute.

Photo by Marcos A. Guerra, courtesy of Smithsonian Tropical Research Institute.

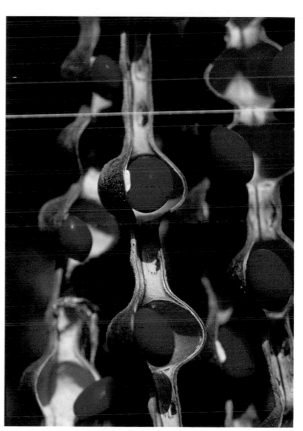

Photo by Carl C. Hansen, courtesy of Smithsonian Tropical Research Institute.

Many tropical animals have bright colors that attract mates or warn predators that they are unpalatable. The predominantly green color of some tree frogs is cryptic, but bright colors on the legs and ventral surfaces may be displayed during behavioral encounters. Conspicuous banding in a caterpillar often warns of noxious chemicals. The coloration and antenna-like tails on the wings of this tropical butterfly confuse predators into mistaking its hind end for its head.

Photo by Marcos A. Guerra, courtesy of Smithsonian Tropical Research Institute.

Photo by J. Burgett, courtesy of Smithsonian Tropical Research Institute.

Photo by Marcos A. Guerra, courtesy of Smithsonian Tropical Research Institute.

Photo by Carl C. Hansen, courtesy of Smithsonian Tropical Research Institute.

Conspicuously marked noxious animals often aggregate to enhance the effect of their coloration, as in the case of these Panamanian caterpillars and true bugs (hemipterans). Alternating yellow and dark bands are a common theme, as illustrated by these beetles and a saturnid moth, which curls up when disturbed to display its banded abdomen. In the tropics, enemies are so numerous that escape and defense tactics are conspicuous; a coccinclid beetle guards her eggs against ants and wasps.

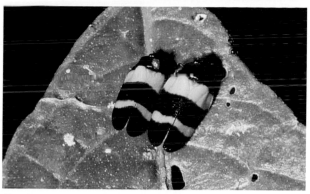

Photo by Marcos A. Guerra, courtesy of Smithsonian Tropical Research Institute.

Photo by Carl C. Hansen, courtesy of Smithsonian Tropical Research Institute.

Photo by Marcos A. Guerra, courtesy of Smithsonian Tropical Research Institute.

Photo by Carl C. Hansen, courtesy of Smithsonian Tropical Research Institute.

Photo by Carl C. Hansen, courtesy of Smithsonian Tropical Research Institute.

With the high productivity of rain forest plants, herbivores are also abundant and varied, including thousands of species of insects. Among vertebrate herbivores, the sloth may by thought of as the sheep of the rain forest. Even reptiles get into the act: the green iguana is a major consumer of leaves. Because birds have high energy requirements and could not fly with their stomachs filled with fermenting vegetation, they concentrate on nectar and fruits.

Photo by Marcos A. Guerra, courtesy of Smithsonian Tropical Research Institute.

Photo by Carl C. Hansen, courtesy of Smithsonian Tropical Research Institute.

Photo by Marcos A. Guerra, courtesy of Smithsonian Tropical Research Institute.

Photo by Carl C. Hansen, courtesy of Smithsonian Tropical Research Institute.

Within the tropics, high mountains and other features create rain shadows and seasonally dry conditions. The false-color satellite image of western Panama shows heavy forest (brown) to the north of the continental divide where the prevailing winds blow humid air from the Caribbean Sea. On the Pacific side of the isthmus, the green color in this dry-season image indicates pasture and dry forest, illustrated in the accompanying scene from western Costa Rica. Over much of the tropics, pastures are burned during the dry season to kill tree seedlings. Forest clearing is extending the acreage under pasture and cultivation in the tropics, for which most forest soils are poorly suited.

Photo by R. E. Ricklefs.

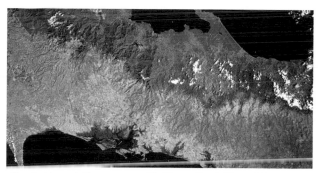

Photo by Marcos A. Guerra, courtesy of Smithsonian Tropical Research Institute.

Photo by Marcos A. Guerra, courtesy of Smithsonian Tropical Research Institute.

Photo by Marcos A. Guerra, courtesy of Smithsonian Tropical Research Institute.

Throughout much of tropical Africa, seasonal rainfall supports mixed grassland and woodland vegetation, called savannah, with vast herds of grazing herbivores; Amboseli National Park, Kenya, with Mt. Kilimanjaro in the background.

Photo by R. E. Ricklefs.

perature depends on BMR and thermal conductance. As body size increases among species, basal metabolic rate increases more rapidly than body surface area, and thicker fur and feathers tend to reduce thermal conductances. As a result, T_{lc} decreases with increasing size — for example, from about 30°C in sparrow-sized birds to below 0°C in penguins and other large species.

An organism's ability to sustain a high body temperature while exposed to extremely low ambient temperatures reaches limits set over the short term by its physiological capacity to generate heat and over the long term by its ability to gather food or metabolize nutrients to satisfy the energy requirements of generating heat. The maximum rate at which an organism can perform work generally does not exceed 10 to 15 times its basal metabolism. Over the course of a day, few organisms expend energy at a rate exceeding 4 times BMR.

When the environment becomes so cold that heat loss exceeds the organism's physiological capacity to produce heat (point c in Figure 10.2), body temperature begins to drop, a condition that is fatal to most homeotherms. The lowest environmental temperatures that homeotherms can survive for long periods often depend on their ability to gather food (point b) rather than on their ability to assimilate and metabolize the energy in food. Animals can quite literally starve to death at low temperatures because they metabolize food energy more rapidly than they can gather food. In temperate and arctic climates, the coldest temperatures often arrive with storms and snow accumulation, which may reduce access to food, thereby doubly taxing the physiological and ecological sources of heat energy.

When homeothermy requires more energy than the individual can provide, certain "economy measures" are available. For example, the regulated temperature of portions of the body may be lowered, which reduces the temperature difference between air and body. Because the legs and feet of most birds do not have feathers, they would act as a major avenue of heat loss in cold regions were they not held at a lower temperature than the rest of the body (Figure 10.3). Gulls conserve heat via a countercurrent heat exchanger in which warm blood in the arteries leading to the feet cools as it passes close to the veins that return cold blood to the body. In this way, heat is transferred from arterial to venous blood and transported back into the body rather than being lost to the environment.

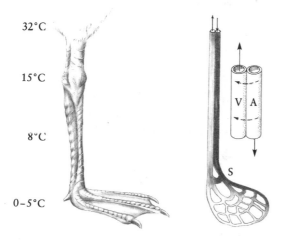

32°C

15°C

8°C

0–5°C

V A

S

FIGURE 10.3 Skin temperatures of the leg and foot of a gull standing on ice. The anatomical arrangement of blood vessels and the countercurrent heat exchange between arterial blood (A) and venous blood (V) are diagrammed at right. Arrows indicate direction of blood flow, and dashed arrows indicate heat transfer. A shunt at point S allows the blood vessels in the feet to constrict, thereby reducing blood flow and heat loss further, with no increase in blood pressure. After L. Irving, *Sci. Amer.* 214:93–101 (1966); K. Schmidt-Nielsen, *Animal Physiology.* Cambridge Univ. Press (1983).

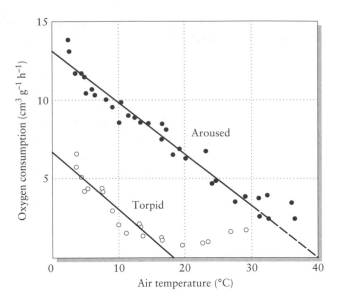

FIGURE 10.4 Relationship between energy metabolism and air temperature in the hummingbird *Eulampis jugularis* during periods of torpor and normal arousal, illustrating endothermy in each case but with body temperature regulated at different set points. After F. R. Hainsworth and L. L. Wolf, *Science* 168:368–369 (1970).

Because they are small, hummingbirds expose a large surface area relative to their weight and consequently lose heat rapidly compared with their ability to produce it. As a result, hummingbirds sustain very high metabolic rates to maintain their at-rest body temperature near 40°C. Species inhabiting cool climates would risk starving overnight if they did not become **torpid,** entering a condition of lowered body temperature and inactivity that resembles hibernation. The temperature of the West Indian hummingbird, *Eulampis jugularis,* drops to between 18 and 20°C when the animal is resting at night. It does not cease to regulate body temperature; it merely changes the setting on its thermostat to reduce the difference between ambient and body temperature (Figure 10.4).

Regulators and conformers

Regulators maintain constant internal environments; **conformers** allow their internal environments to follow external changes. Few organisms fit either ideal. Frogs regulate the salt concentration of their blood but conform to external temperature. Even warm-blooded animals conform partially to ambient temperature: in cold weather our hands, feet, noses, and ears—our exposed extremities—become noticeably cool.

Organisms sometimes regulate their internal environments over moderate ranges of external conditions but conform under extremes. Small aquatic amphipods of the genus *Gammarus* control the salt concentrations of their body fluids when placed in water with less concentrated salt than their blood, but not when placed in water with more concentrated salt (Figure 10.5). The freshwater species *G. fasciatus* regulates the salt concentration of its blood at a lower level than the saltwater species *G. oceanicus* and thus begins to conform to concentrated salt solutions at a lower level. In their natural habitat, however, neither the freshwater species nor the saltwater species encounters salt more concentrated than that in its blood. Animals that inhabit salt lakes and brine pools, by contrast, actively maintain the salt concentration of their

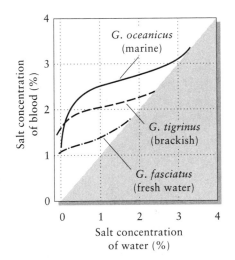

FIGURE 10.5 Salt concentration in the blood of three gammarid crustaceans from different habitats as a function of the salt concentration of their external environment. The normal salt concentration of seawater is 3.5 percent. After C. L. Prosser and F. A. Brown, *Comparative Animal Physiology* (2nd ed.). Saunders, Philadelphia (1961).

blood below that of the surrounding water. The brine shrimp *Artemia* maintains an internal salt concentration below 3 percent even when placed in a 30 percent salt solution.

All birds and mammals generate heat metabolically to regulate body temperature, but many cold-blooded species also become endothermic or partially endothermic at times. For example, pythons maintain high body temperatures while incubating eggs. Some large fish, such as the tuna, use a countercurrent arrangement of blood vessels to maintain temperatures up to 40°C in the center of their metabolically active muscle masses; swordfish employ specialized metabolic heaters, derived from muscle tissue, to keep their brains hot. Large moths and bees often require a preflight warm-up period during which flight muscles shiver to generate heat. Even among plants, temperature regulation based on metabolic heat production has been discovered in the floral structures of philodendron and skunk cabbage.

Clearly, organisms other than birds and mammals can generate heat and maintain elevated body temperatures, and many do so under certain conditions. Why, then, is endothermy so rare throughout the animal and plant kingdoms? Part of the answer certainly lies in a consideration of body size. Birds and mammals are relatively large as animals go. As body size increases, volume increases relatively more rapidly than surface area, across which heat leaves the body. In general, the lower the ratio of surface area to volume, the more comprehensive and precise regulation can be.

Although body size may explain why mammals are hot and insects are not —large moths that exhibit preflight warm-up approach the size of small mammals — large fish and reptiles also have not generally made the shift to homeothermy. Water contains too little oxygen and conducts heat away from the body too rapidly for endothermy to be practical for most fish. Moreover, the metabolic rates of resting birds and mammals may be 10 times those of fish, amphibians, and reptiles of similar size.

Reptiles have evolved endothermy several times, but we now call the resulting organisms birds and mammals. Other reptiles are effective ectothermic homeotherms. Why wasn't the transition to homeothermy complete? After all, contemporary reptiles are comparable in body size to small mammals, and they breathe air. The major innovation leading to endothermy appears to have been the insulation provided by fur and feathers. But even though birds and mammals evolved from reptile lineages, they did not displace their ancestors completely, probably because ectothermy confers advantages under certain environmental conditions. For example, although reptiles lack the high activity levels of warm-blooded animals, their low energy requirements equip them to cope with highly seasonal and erratic food supplies. The lifestyles of reptiles differ from those of birds and mammals but provide equally effective alternatives.

Activity space

Each organism functions best within a limited range of conditions, which we may refer to as its **activity space**. As conditions change, animals move to more favorable parts of the environment. Plants cannot uproot themselves and move, so most of them become inactive during periods of unsuitable conditions.

For proper germination, seeds require specific combinations of light, temperature, and moisture, which may vary even among closely related species. When seeds arrive at suitable sites, the seedlings survive and grow; elsewhere they die. Irregularities in the surfaces of natural soils provide the variety of conditions needed to allow the germination of many species. An experiment in which investigators created a heterogeneous soil environment dramatized the differences in germination requirements between closely related species. Plantains are common lawn and roadside weeds in the genus *Plantago*. Three species sown in seedbeds responded differently to the modifications in environment produced by slight depressions, by squares of glass placed on the soil surface, and by vertical walls of glass or wood (Figure 10.6). Relatively few seeds germinated on the smooth surface of soil that had not been disturbed.

Unlike plants, most animals can choose where they live. In many environments, the diurnal behavior cycle of lizards responds to the varying temperatures of habitat patches. Although few lizards regulate their temperature by generating heat metabolically, they do take advantage of solar radiation and warm surfaces to maintain their body temperatures within the optimum range.

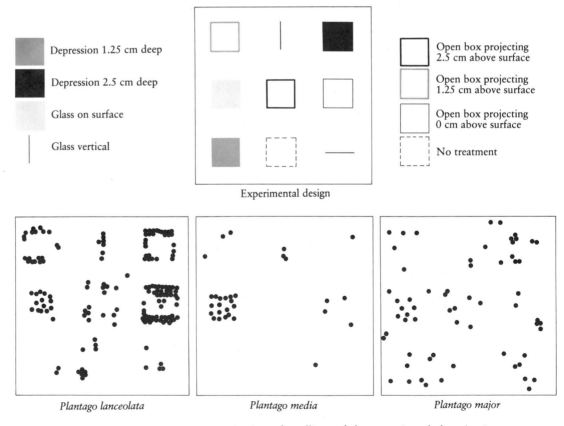

FIGURE 10.6 Germination of seedlings of three species of plantains (genus *Plantago*) with respect to artificially produced variation in the conditions at the soil surface. After J. L. Harper, J. T. Williams, and G. R. Sagar, *J. Ecol.* 51:273–286 (1965).

At night, these sources of heat disappear, and the lizard's body temperature gradually drops to that of the surrounding air. The mallee dragon (*Amphibolurus fordi*), a lizard of Australia, is fully active only when its body temperature ranges between 33 and 39°C. In the early morning, before its body temperature has risen above 25°C and when it still moves sluggishly, the mallee dragon basks within large clumps of the grass of the genus *Triodia*, where it enjoys protection from predators (Figure 10.7). When the temperature of an individual rises above 25°C, it leaves the *Triodia* clump and basks in the sunshine nearby, with its head and body in direct contact with the ground surface, from which it absorbs additional heat. When its body temperature enters the range for normal activity, the dragon ventures farther from *Triodia* clumps to forage, its head and body normally raised above the ground as it moves. When its body temperature exceeds 39°C, it moves less rapidly and seeks the shade of small *Triodia* clumps; above 41°C, it reenters large *Triodia* clumps, at whose centers it finds cooler temperatures and deeper shade. It may also pant to dis-

FIGURE 10.7 The mallee dragon (*Amphibolurus fordi*) at different times during its activity cycle. *Top right:* Early morning basking in *Triodia* grass clump. *Left:* Midmorning basking on ground (note that the body is flattened against the surface to increase the exposed profile and the animal's contact with warm soil). *Bottom right:* Normal foraging attitude. Courtesy of H. Cogger. From H. Cogger, *Austral. J. Zool.* 22:319–339 (1974).

sipate heat by evaporation. If heat stress continues unabated, it loses locomotor ability above 44°C; it dies if its body temperature exceeds 46°C.

On a typical summer day, during which air temperature varies from about 23°C at dawn to 34°C at midday, the mallee dragon does not begin to forage until about 8:30 A.M. By 11:30 the habitat has become too hot for normal activity, and most individuals seek shade and become inactive. By 2:30 P.M. the habitat has cooled off enough for the dragons to resume foraging, but by 6:00, colder evening temperatures force them back into *Triodia* clumps before their bodies cool. Any individuals that remain in the open after this time of day move sluggishly and are easily caught by warm-blooded predators.

The desert iguana (*Dipsosaurus dorsalis*) of the southwestern United States faces a more severe environment, with greater annual temperature fluctuation. Shade temperatures can reach 45°C in summer and plunge below freezing in winter. During mid-July, the thermal environment fluctuates so rapidly between extremes that the desert iguana can move about within its preferred body temperature range of 39 to 43°C for only about 45 minutes in mid-morning and a similar period in the early evening (Figure 10.8). During the remainder of the day, it seeks the shade of plants or the coolness of its burrow, where the temperature rarely rises above the preferred range. At night the desert iguana enters its burrow, where it finds safety from predators; at dawn the burrow is warmer than the desert surface, so the early morning warm-up period is correspondingly brief.

Whereas in summer the desert iguana restricts its activity to two brief bouts separated by inactivity through midday to avoid heat stress, spring offers more favorable temperatures for activity. The thermal environment in May does not exceed its preferred range, and individuals forage actively aboveground from 9:00 A.M. to 5:00 P.M., only occasionally seeking the cool shade of plants. Winter cold restricts *Dipsosaurus* to brief periods of activity in the middle of

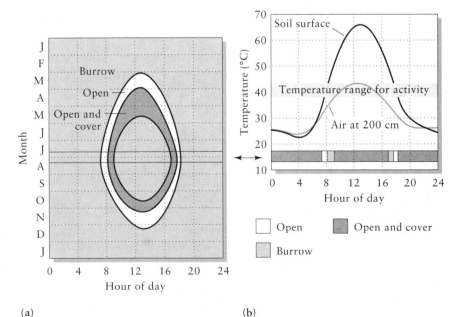

FIGURE 10.8 Seasonal activity space of the desert iguana (*Dipsosaurus dorsalis*) in southern California. (a) The daily activity budget for an entire seasonal cycle. (b) The activity budget for July 15 is shown with the time course of environmental temperature. After W. A. Beckman, J. W. Mitchell, and W. P. Porter, *J. Heat Transfer* (May 1973): 257–262.

(a)

(b)

the day when body temperature rises to the point where individuals can come aboveground and forage. Between early December and the end of February, most days are so cold that the desert iguana cannot venture from its burrow.

Microhabitat selection

Subdivisions of the habitat that exhibit distinctive attributes of structure, temperature, salinity, and so on, are referred to as **microhabitats.** In the desert, the shaded ground under a shrub offers cooler, moister conditions than surrounding patches of desert exposed to direct sunlight. Perches in vegetation a meter or more off the ground lie above the searing boundary layer of hot air at the surface but don't offer the escape from summer heat and winter cold provided by subterranean burrows and niches in the soil.

The cactus wren, a desert insectivore, actively seeks favorable conditions within which to feed (**microhabitat selection**) and modifies the habitat to provide suitable conditions for its young. In the cool temperatures of the early morning, wrens forage throughout most of the environment, searching for food among foliage and on the ground. As the day brings warmer temperatures, they select cooler parts of their habitat, particularly the shade of small trees and large shrubs, always managing to avoid feeding where the temperature of the microhabitat exceeds 35°C (Figure 10.9). When the minimum tem-

FIGURE 10.9 Microhabitat use by cactus wrens in southeastern Arizona during the course of a day in late spring. Microhabitats vary in degree of thermal stress between exposed ground (a) and the deep shade of trees (e). Wrens distribute their activity among all microhabitats in the cool hours of early morning (7:00 A.M.) but restrict their activity to cool shade (e) during the hottest part of the day (2:30 P.M.), when other microhabitats are above 40°C. From R. E. Ricklefs and F. R. Hainsworth, *Ecology* 49:227–233 (1968).

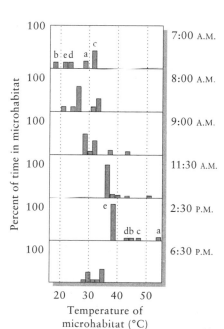

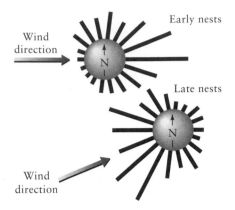

Wind direction

Early nests

Late nests

Wind direction

FIGURE 10.10 Orientation of nest entrances of the verdin (*Auriparus flaviceps*) during the early (cool) and late (hot) part of the breeding season in Nevada and Arizona. Orientation of cactus wren nest entrances is similar. Data courtesy of G. T. Austin.

perature in the environment rises above 35°C, at which point wrens must use evaporative cooling to maintain their body temperatures even when inactive, the birds stop feeding.

Many desert birds build enclosed nests or place their nests in holes in the stems of large cacti, where the young enjoy protection from the sun and from extremes of temperature. Cactus wrens build untidy nests, bulky and somewhat haphazard balls of grass, with side entrances. Once a pair of wrens has built a nest, its position or orientation cannot be changed. For a month and a half, from the laying of the first egg until the young leave the nest, the nest must provide a suitable environment day and night, in hot and cool weather. Cactus wrens usually rear several broods of young, in different nests, during the breeding period (March through September) in southern Arizona. They build early nests so that the entrances face away from the direction of the cold winds of early spring; during the hot summer months, they orient the nests to face prevailing afternoon breezes, which circulate air through the nest chamber and facilitate heat loss (Figure 10.10). It makes a difference! Nests oriented properly for the season are consistently more successful (82 percent produce viable offspring) than nests facing the wrong direction (45 percent do so).

The time course of response

The suitability of the environment for each organism varies over time and space. At any given moment, the cactus wren may feed within any of several parts of the habitat, each offering different thermal characteristics and access to food. These distinctive patches of microhabitat shift continuously throughout the daily cycle of solar radiation and temperature, and they change seasonally as both the physical environment and populations of suitable prey vary. Like many other animals, the cactus wren must adjust behaviorally and physiologically to these ever-changing conditions.

Appropriate responses to temporal changes in the environment depend on the duration of the change. The choices presented to organisms by spatial variation, on the other hand, depend on the distance between patches of different habitat type. But what seems long or brief, and what seems near or far, depends on the life span, response time, and mobility of the individual. Humans do not distinguish the upper and lower surfaces of a leaf as we push aside the branch of a tree, but these surfaces present different worlds to the aphid that clings to the undersurface and sucks plant juices. A tropical storm or passing cold front may mean little more to us than spoiled plans for a picnic, but it may completely encompass the adult life of a mayfly, during which it must mate and lay eggs.

Suitable responses to changing conditions must take substantially less time than the period of environmental change. Otherwise, today's form and function may reflect yesterday's conditions. The most rapid responses involve changes only in the *rates* of physiological processes and behavioral modifications. Examples include the homeotherm's elevation of metabolism in response to cold stress and its shade-seeking behavior under heat stress. Such responses, which we have already discussed, do not require modification of existing morphology or biochemical pathways.

Acclimatory response (or **acclimation**) involves more substantial changes of structure, such as the thickening of fur in winter, an increase in the number of red cells in the blood at high altitude, and the production of enzymes with different temperature optima. These changes may be thought of as shifts in the *ranges* of the behavioral and physiological responses of the individual. Acclimatory responses are reversible, as they must be to follow the ups and downs of the environment.

When the environment changes slowly, a given set of conditions may persist throughout the adult life span, and the responses of an individual may alter its development to produce the phenotype most suitable to the prevailing conditions. Such **developmental responses** require long periods and generally do not reverse themselves.

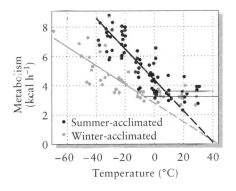

FIGURE 10.11 Metabolic responses of willow ptarmigan acclimated to summer and winter temperatures. Winter-acclimated birds have thicker plumages, providing better insulation than summer plumages. Hence their metabolic rates are lower at any given temperature, and their lower critical temperatures are also lower. After G. C. West, *Comp. Biochem. Physiol.* 42A:867–876 (1972).

Acclimation

During winter, many birds don a heavier plumage than they sport during the hot summer months. These species replace their body feathers in spring and autumn; each plumage suits them to the environmental conditions that typically occur between molts. The willow ptarmigan, a ground-feeding arctic bird, trades its lightweight, brown summer plumage in the fall for a thick, white winter plumage, which provides insulation as well as camouflage against a background of snow. With increased insulation, the ptarmigan expends less energy to maintain its body temperatures during the winter (Figure 10.11). Seasonal change in plumage thickness effectively shifts the regulatory response range to match the prevalent temperature range of the season. The ptarmigan needs a similar metabolic rate to maintain its body temperature when the surroundings are −40°C in winter and when they are 10°C in summer (a conceivable temperature in its arctic home). Although winter-acclimated individuals would seem well adapted for both winter and summer climates when at rest, summer activity combined with a winter plumage would quickly produce heat prostration. By adjusting insulation to enhance heat conservation in winter and to facilitate heat dissipation in summer, the organism can maintain a constant body temperature at the least possible cost.

Temperature-conforming animals and plants also acclimate to seasonal changes in their environments. By switching between enzymes and other biochemical systems with different temperature optima, cold-blooded animals adjust their tolerance ranges in response to prevailing environmental conditions. The relationship between the swimming speed of fish and the water temperature shows the advantages and the limitations of acclimation. Goldfish swim most rapidly when acclimated to 25°C and placed in water between 25 and 30°C, conditions that closely resemble their natural habitat (Figure 10.12). Lowering the acclimation temperature to 5°C increases the swimming speed at 15°C but reduces it at 25°C. Increased tolerance of one extreme often brings reduced tolerance of the other.

When goldfish are placed at temperatures approaching their upper lethal limits, they exhibit progressive stages of behavioral abnormality, passing from hyperexcitability to loss of equilibrium and coma. The temperatures at which these syndromes appear can be shifted by acclimation over periods of 2 to 6

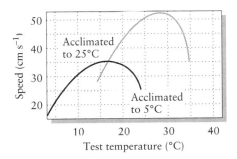

FIGURE 10.12 Swimming speed of goldfish as a function of temperature. Separate curves are graphed for individuals acclimated to 5°C and to 25°C. After F. E. J. Fry and J. S. Hart, *J. Fish. Res. Bd. Canad.* 7:169–175 (1948).

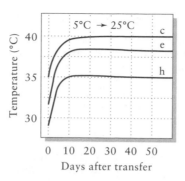
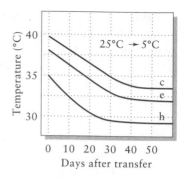

FIGURE 10.13 Time courses of acclimation to high-temperature stress for three behavioral indices—hyperexcitability (h), loss of equilibrium (e), and coma (c)—in goldfish. Individuals were first acclimated to either (a) 5°C or (b) 25°C and then transferred to the other temperature. From A. R. Cossins, M. J. Friedlander, and C. L. Prosser, *J. Comp. Physiol.* 120:109–121 (1977).

weeks, as shown in Figure 10.13. These neurological traits are strongly correlated with the fluidity of lipids in the membranes of nerve synapses, which apparently affects the transmission of nerve impulses to the muscles. Acclimation depends on the turnover of lipid components of the membrane, which, as one would expect, requires longer at 5°C than at the higher acclimation temperature of 25°C.

Acclimation of photosynthetic rate to temperature shows that the capacity for acclimation often reflects the range of temperatures experienced in the natural environment (Figure 10.14). *Atriplex glabriuscula*, a species of saltbush

FIGURE 10.14 Light-saturated photosynthetic rate as a function of leaf temperature in three species of plants (genera *Atriplex, Tidestromia,* and *Larrea*) grown under moderate and hot temperatures. From P. W. Hochachka and G. N. Somero, *Biochemical Adaptation,* Princeton Univ. Press, Princeton (1984); after O. Bjorkman, M. R. Badger, and P. A. Arnold, in N. C. Turner and P. J. Kramer (eds.), *Adaptation of Plants to Water and High Temperature Stress.* Wiley, New York (1980), pp. 231–249.

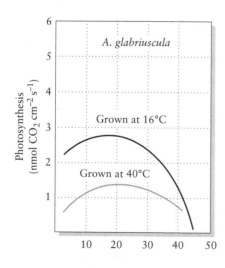
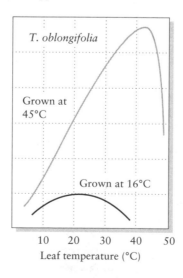
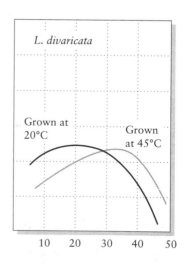

native to cool coastal regions of California, does not increase its photosynthetic rate at high temperature when acclimated to 40°C; plants acclimated to 16°C, within the range of temperature the species normally experiences, are uniformly more productive at all temperatures. The heat-loving (**thermophilic**) species *Tidestromia oblongifolia* cannot acclimate to cool temperatures. Its temperature optimum shifts to a lower temperature, but photosynthesis declines over a wide range of leaf temperatures from 10 to 40°C. *Larrea divaricata* (creosote bush), a species that inhabits interior deserts but maintains photosynthetic activity during the cool winters as well as the hot summers, shows the classic shift in temperature optima characteristic of thermal acclimation. The basis for this acclimation appears to be associated with changes in the viscosity of membranes directly related to photosynthetic pathways. That some species lack this capability suggests that maintaining mechanisms of acclimatory change entails a cost to the individual.

Developmental response

In many organisms, growth and differentiation are sensitive to environmental variation. Such developmental responses generally do not reverse themselves; once fixed during development, they remain unchanged for the rest of the organism's life. Because of their long response time and their irreversibility, developmental responses cannot accommodate short-term environmental changes. As a general rule, therefore, plants and animals exhibit developmental flexibility only in environments with persistent variation for the individual. When environmental changes occur slowly compared to the life span of an organism, as do seasonal conditions for short-lived animals, developmental responses are often appropriate. The strategy of developmental flexibility also makes good sense for plant species whose seeds may settle in many different kinds of habitats.

Light intensity, among many other factors, influences the course of development in plants. Loblolly pine seedlings grown in shade have smaller root systems and more foliage than seedlings grown in full sunlight. Because the shaded environment taxes the plant's water economy less, shade-grown seedlings can allocate more of their production to stem and needles; sun-grown seedlings must develop more extensive root systems to obtain sufficient water. The larger proportion of foliage of the shade-grown seedlings results in a higher rate of photosynthesis per unit of plant mass under given light conditions, particularly under low light intensity (Table 10.1).

A striking example of developmental response is the coloration of several species of locusts and grasshoppers. In tropical habitats with seasonal precipitation, the onset of the wet season brings growth of lush, green vegetation. During the early part of the dry season this vegetation browns and dies, often exposing red-brown earth. As the seasonal drought intensifies, natural fires and those that humans set blacken the ground over vast areas. As a result, there is a regular seasonal progression of color in the dry tropics from green to brown to black and back to green again. Grasshoppers develop so rapidly that the life span of each individual may coincide largely with one background

TABLE 10.1	*Distribution of dry matter and rates of photosynthesis in loblolly pine (Pinus taeda) seedlings grown under shade and in full sunlight*	
Characteristic	Shade-grown	Sun-grown
Percentage of dry weight, 6-month-old seedlings		
Roots	35	52
Needles	47	37
Stems	18	11
Photosynthetic rate, 4-month-old seedlings (mg CO_2 h^{-1} g^{-1})		
Low light intensity (500 fc)*	1.9	1.0
Moderate light intensity (1500 fc)	4.6	4.0
High light intensity (4500 fc)	7.2	6.6

*fc = footcandles.

Source: After F. H. Bormann, in D. V. Thimann (ed.), *The Physiology of Forest Trees*. Ronald Press, New York (1958), pp. 197–215.

color. And many species of grasshoppers do, in fact, match the background coloration of their environment closely, regardless of season.

The coloration of *Gastrimargus africanus,* for example, responds to environmental conditions, particularly the quality and intensity of light and the humidity. The epidermis (outer skin) of the grasshopper has a pigment system that permits any given area of the skin to be either green or brown; both colors may occur on a single animal but not in the same area of the body. The green and brown colors represent small biochemical variations on a single pigment molecule. Where brown pigment occurs in the epidermis, additional pigments may produce colors ranging from yellow through orange and red to black. Furthermore, black pigment (melanin) may be deposited in the cuticle that covers the epidermis.

The color response in *Gastrimargus* is most pronounced in the early developmental stages; adults have a more characteristic and unvarying pattern of color distribution. Between developmental stages the grasshopper sheds its epidermis, discarding its pattern of camouflaging coloration with it. A new layer of epidermis develops underneath, and thus the grasshopper may change its color with each molt, just as a ptarmigan does its feathers. High-intensity light, a characteristic of the dry season during which much of the vegetation has died back, leads to predominance of the brown and black pigment systems. The low-intensity sunlight and high humidity that prevail during the rainy season increase the proportion of green coloration. When black ash covers a recently burned area, the soil reflects little incident light, a condition that stimulates the black pigment system in grasshopper nymphs feeding on newly sprouting vegetation. These environmental signals are perceived by the eye and transmitted to the epidermis by hormones produced in the brain. The sensitivity of pigment systems in the developing epidermis to prevailing conditions of light and humidity thus enables the grasshopper to match its coloration to the changing conditions experienced during its growth period.

❧ SUMMARY

In varying environments, organisms must be able to respond to changing external conditions in order to maintain suitable internal environments, feed on appropriate food items, and avoid predation. Responses include a variety of physiological changes and structural modifications, and each has a characteristic time period.

1. Maintenance of constant internal conditions, called homeostasis, depends on negative feedback responses. Organisms sense changes in their internal environments and respond in such a manner as to return those conditions toward their optimum.

2. Homeostasis requires energy when a gradient must be maintained between internal and external conditions. This is illustrated by endotherms (birds and mammals), which maintain elevated body temperatures by generating heat metabolically to balance loss of heat to their cooler surroundings.

3. The magnitude of the body–environment gradient and the constancy of internal conditions that organisms adopt affect the balance the organism strikes between the costs of homeostasis (maintaining the physiological apparatus and sustaining gradients) and the benefits of closely regulated internal conditions.

4. A major component of homeostasis is the selection of microhabitats that minimize the body–environment gradient. This is illustrated by the temperature-dependent foraging behavior of desert birds and by their construction of nests in such a way as to protect their chicks from environmental extremes.

5. The activity space of an individual—that is, its selection of habitat and microhabitat—depends on the suitability of conditions and the availability of resources at any given time. This is illustrated by the influence of changing thermal conditions on habitat utilization by desert lizards.

6. Behavioral selection of habitats cannot fully compensate for environmental change, and organisms must bring a variety of homeostatic responses into play. The most rapid of these involve changes in the rates of processes and the orientation of the organism, but not changes in morphological structures or biochemical pathways. Shivering in response to cold stress and contraction of the pupil in response to bright light are examples. Such responses are rapid and reversible.

7. Acclimatory responses involve reversible changes in structure (for example, fur thickness) or biochemical pathway (induction of enzymes

or changes in the amounts of enzymes and their products). Such changes require longer periods (usually days or weeks). Acclimation plays a prominent role in long-lived organisms' responses to seasonal change.

8. Developmental responses express the interaction between an organism and its environment during the growth period. Different environments lead to characteristic structures and appearances that generally do not reverse themselves. Developmental responses include those of plant seedlings to sun or shade and the expression of pigment systems in grasshoppers depending on light intensity.

🦅 SUGGESTED READINGS

Gordon, M. S. 1968. *Animal Function. Principles and Adaptations.* Macmillan, New York.

Hardy, R. N. 1983. *Homeostasis* (2nd ed.). Edward Arnold, London.

Heinrich, B. 1979. *Bumblebee Economics.* Harvard Univ. Press, Cambridge, Mass.

Porter, W. P., J. W. Mitchell, W. A. Beckman, and C. B. Dewitt. 1973. Behavioral implications of mechanistic ecology. Thermal and behavioral modeling of desert ectotherms and their microenvironment. *Oecologia* 13:1–54.

Pough, F. H. 1980. The advantages of ectothermy for tetrapods. *American Naturalist* 115:92–112.

Prosser, C. L. (ed.). 1991. *Environmental and Metabolic Animal Physiology.* Wiley-Liss, New York.

Ricklefs, R. E., and F. R. Hainsworth. 1968. Temperature dependent behavior of the cactus wren. *Ecology* 49:227–233.

Schmidt-Nielsen, K. 1972. *How Animals Work.* Cambridge Univ. Press, London and New York.

Schmidt-Nielsen, K. 1990. *Animal Physiology. Adaptations and Environment* (4th ed.). Cambridge Univ. Press, London and New York.

11

BEHAVIOR IN HETEROGENEOUS ENVIRONMENTS

The world varies in time and space. No organism can handle all types of habitats, prey, or physical conditions equally well. Thus **heterogeneous environments**—so called because they comprise a range of varying characteristics—require individuals to make choices concerning habitat use, microhabitat selection, prey, and so on. Even plants can make such choices by controlling germination and flowering times in response to changing conditions and by variously directing root growth in heterogeneous soils. The behaviors that govern these choices follow a set of rules that optimize the individual's relationship to its surroundings. Environmental variation, temporal or spatial, may be of large or small scale in relation to a given individual and the activities it undertakes to supply its needs. The most successful response will reflect this relationship.

Environmental grain

Imagine for a moment that environmental alternatives consist of patches of different conditions distributed as a mosaic in time and space, resembling a patchwork quilt or the pattern made by adjacent fields of different crops. Patches of different characteristics have different typical sizes, and patterns of various kinds of patches can be superimposed. For example, patches of uniform air temperature tend to be larger than patches of similar soil moisture, because topographic influences subdivide the soil moisture landscape more finely.

The concept of **grain** relates the size of a patch to the activity space of the organism (the range of environmental conditions that are suitable for the activity of the individual). We define a **coarse-grained** environment as one in which the patches are large enough for the individual to distinguish and choose among them. A **fine-grained** environment is one in which the patches are so small that the individual cannot usefully distinguish among them, and the environment appears essentially uniform. To anyone but a trained botanist, for example, a field looks like a uniform carpet of plants, a type of patch readily distinguished from patches of forest, marsh, beach, and so on. But to the caterpillar, a single plant within the field may be its home for the duration of its larval life. Grasshoppers, which can fly from plant to plant, choose carefully among the various species in the field, since some provide better feeding than others. Therefore, whereas individual plants in a field appear as fine grains to us, they are coarse grains to small insects. Grain also depends on activity. If we set out to pick flowers of a particular kind, we, like grasshoppers, perceive individual plants as coarse grains, readily distinguishable from one another.

Patches occur in time as well as space. At one extreme, conditions that fluctuate through a daily cycle or over a shorter period may be thought of as fine-grained to most organisms, inasmuch as such rapid changes offer too little time for them to respond. At the other extreme, seasonal changes and longer trends are coarse-grained for most organisms. Given persistent changes of several months' duration, plants and animals can undertake morphological changes, such as dropping their leaves or increasing the thickness of their fur; corresponding changes in physiological mechanisms; and, in the case of animals, migrations to areas with more favorable conditions.

Optimal foraging

When an animal moves through a habitat in search of food, it encounters potential prey in sequence. At each encounter, the individual chooses whether or not to pursue and eat the prey. Pursuit requires time and the expenditure of energy and, if successful, results in the acquisition of food energy. The individual may, however, pass up prey of low quality in favor of continued search for better prey. When a predator meets potential prey infrequently, compared to the time required to subdue and consume them, it will eat most kinds of potential prey. But when a predator encounters prey often, it will pass by less desirable types because it will soon find better ones. The rules that govern the foraging behavior of organisms tend to maximize their rate of food intake or minimize the time required to obtain a given amount of food. These rules are referred to as **optimal foraging** rules.

Let us assume that each predator selects a diet that minimizes the average time it takes to find and consume an individual prey item. When a predator increases the breadth of its diet by adding a new kind of prey, its rate of food consumption is affected in two ways. A broader diet means that more potential prey individuals are available, and therefore the predator encounters prey more frequently. Assuming, however, that the predator always includes the most suitable prey species in its diet, the average ease of pursuit, capture, and consumption *decreases* through the addition of less suitable species. Thus a predator should broaden its diet until the resulting decrease in the average quality of its prey more than offsets the advantage of the decreased searching time that results from the greater abundance of prey included within the diet.

The solution to this problem can be very complex, and many theoretical models have been developed to explore the factors that can influence diet breadth. These models generally agree in predicting that increased abundance of resources favors increased diet specialization and leads to increased rate of prey consumption.

Optimal patch use

Most prey occur in patches separated by unsuitable habitats. Because predators must travel between patches, they must make choices about what prey to use within patches and when to leave a patch to search elsewhere. In general, the quality of a patch decreases as the predator captures prey and reduces the resources it contains. Eventually, the predator receives greater reward by moving to a new patch than by continuing to search in a depleted one. The amount of time a predator spends in a patch before leaving is called the **giving-up time** (**GUT,** certainly a happy acronym).

As a predator initially searches within a patch, it encounters and consumes food at a high rate. As it depletes the resource, its return per unit of time decreases, and the total gain from the patch begins to level off. The rate of gain sinks to zero as the resource is exhausted. During a period of search, the average rate of gain from a patch is the amount of food consumed divided by the sum of the search time and the time required to travel between patches. Assuming that travel time is determined by the spatial distribution of patches, the optimal giving-up time is the period of search that maximizes the average rate of gain from a patch. This model predicts that optimal giving-up time decreases as travel time between patches decreases. Intuition confirms that when a new patch with a high resource level can be reached quickly, a predator cannot profit from staying long in a patch of diminishing quality. And conversely, the greater the distance between patches, the longer the consumer should stay in each.

The ideal free distribution

The models of optimal choice that we have discussed so far address the behavior of a single individual confronted with environmental heterogeneity. Frequently, however, many individuals face the same options, and a given individual's optimal behavior may be influenced by the decisions of other members of the population. For example, a patch becomes less attractive to newcomers as the number of predators exploiting it increases. Such a patch probably has fewer remaining resources than undiscovered patches; furthermore, competing individuals may precipitate costly behavioral conflicts. We would expect that, presented with many patches of varying intrinsic quality, consumers that had complete knowledge and freedom of choice would occupy or exploit patches in direct proportion to their quality. Patches with the highest levels of resources should attract the most consumers, those with the lowest levels the fewest.

Each consumer bases its choice of a patch on criteria that maximize its own rate of gain of resources. Imagine two patches, one having more resources than the other. At first, consumers choose the intrinsically superior patch. But as the population of consumers builds up in that patch, its apparent quality decreases, owing to depletion of resources and antagonistic interactions among the consumers, until the second patch becomes the better choice. At this point, additional consumers choose the second and the first patch alternately as the quality of both continues to decline. As a result, each individual in the population exploits a patch of equal *realized* quality, regardless of the variation in *intrinsic* patch quality in the absence of consumers. This outcome is called the **ideal free distribution.**

Several lines of evidence suggest the operation of ideal free distributions of individuals in nature; among the most compelling is the pattern of habitat selection by birds. In migratory species, individuals arrive on the breeding grounds over a period of several weeks. Early arrivals generally fill certain habitats that confer high breeding success before newcomers begin to establish territories in poorer habitats. When individuals are removed from good habitats, their places are quickly taken by others moving in from poor habitats. When population densities are low, perhaps following a particularly severe winter, the occupancy of poor habitats decreases more than does that of good habitats (Figure 11.1).

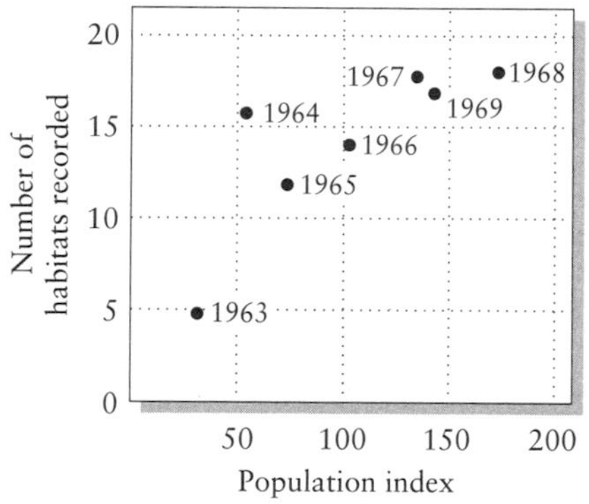

The achievement of an ideal free distribution has also been investigated in the laboratory, where the quality of patches can be controlled. One series of experiments involved stickleback fish that were provided with food (water fleas) at different rates at opposite ends of an aquarium. Each end of the tank could be regarded as a patch, and the system had the following conditions conducive to establishing an ideal free distribution: (1) the two patches differed in profitability, (2) profitability decreased as the number of fish using a patch increased, and (3) the fish were free to move between patches.

Hungry fish were placed in the aquarium about 3 hours before the start of the experiment. During trials, the numbers of fish in each half of the tank were recorded at the end of each 20-second interval. Before the addition of any food to the tank, the fish were distributed equally between the two halves. In one experiment, water fleas were added at a rate of 30 per minute to one end of

Figure 11.1 Relationship between population levels of the kestrel and number of habitats used for nesting in England between 1963 and 1969. The population of kestrels, which are small hawks, declined precipitously between 1962 and 1963 because of a severe intervening winter. By 1969, the population had increased to levels determined by the amount of suitable breeding habitat. After R. J. O'Connor, *Bird Study* 29:17–26 (1982).

the tank and at a rate of 6 per minute to the other, a ratio of 5 to 1. Within 5 minutes, the fish had distributed themselves between the two halves in the same ratio as predicted for an ideal free distribution. In a second experiment, water fleas were provided at rates of 30 and 15 per minute, a 2-to-1 ratio. Again the distribution of fish followed suit. When the better and poorer patches in the tank were reversed, the fish reversed their distribution within about 5 minutes. The behavioral mechanisms used to achieve an ideal free distribution were not determined, but cues for behavioral choices must incorporate both the rate of food provisioning and the number of competitors within the patch. Such experiments demonstrate the considerable sensitivity of organisms to conditions of their environment, as well as their behavioral flexibility in making choices.

Behavior in temporally varying environments

Temporal heterogeneity differs from spatial heterogeneity. An individual can move among patches of space; it cannot move among patches of time. But although the organism cannot choose among temporal environments, it can choose what it does under the conditions it encounters at any particular time, and sometimes that "choice" takes developmental form. Temporal variation requires a flexible response, the nature of which depends on the time course, predictability, and amplitude of variation. We saw in Chapter 10 that some grasshoppers, through a response of skin pigments, match the coloration of the habitat in which the organism develops. Plant growth form also responds to the levels of light, water, and nutrients in the environment.

The case of wing development in water striders illustrates the complexity of response to temporal variation. Water striders are freshwater bugs of the genus *Gerris*. They spend most of their lives moving about the surface of streams and ponds, where they feed on other small insects. They fly only occasionally: to disperse between feeding areas and at the end of the summer, when individuals of many species leave the lakes and ponds to find protected sites in forests to spend the winter. In Europe some species of water striders have short, nonfunctional wings; others have long, functional wings; and still others may produce individuals with either wing type.

The length to which wings develop depends on the seasonal variability and predictability of the habitat (Table 11.1). At one extreme, species inhabiting large, permanent lakes have short wings, or none at all, and do not disperse between lakes. At the other extreme, species living in temporary ponds usually have long, functional wings and disperse to find suitable sites for breeding each year. Between these extremes, species that are characteristic of small ponds, which are more or less persistent from year to year but tend to dry up during the summer, frequently have both long-winged and short-winged forms. In these cases, the length of the wings of each individual is determined genetically and does not respond to environmental conditions. Even where both long-winged and short-winged individuals occur within a single population, genes control the differences between these individuals.

TABLE 11.1	*Wing lengths of water striders* (Gerris) *inhabiting bodies of fresh water with different levels of permanence and predictability*		
Characteristic of habitat	Characteristic wing length*	Mechanism of determination	
Permanent	Short	Genetic	
Fairly persistent but unpredictable	Both short and long	Genetic dimorphism	
Seasonal	Seasonally dimorphic (summer)	Developmental switch	
Very unpredictable	Long	Genetic	

*Short wings are not functional and prevent dispersal.

Source: K. Vepsalainen, *Acta Zool. Fenn.* 141:1–73 (1974).

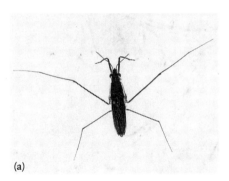

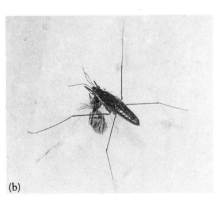

FIGURE 11.2 Alary (wing) polymorphism in the water strider *Gerris odontogaster*. (a) The long-winged form; (b) The short-winged form. Courtesy of K. Vepsalainen.

The life cycle of most *Gerris* species in central Europe, England, and southern Scandinavia includes two generations per year. The first (summer) generation hatches during the spring, reproduces during the summer, and then dies. The second, hatched from eggs laid by females of the summer generation, develops to the adult stage during late summer, then overwinters before breeding in early spring the following year. In species that inhabit seasonal ponds, the summer generation is **dimorphic** (literally, "of two forms"); both long-winged and short-winged forms are represented (Figure 11.2). All the individuals in the winter generation have long wings and can fly. They leave the pond in late summer and move into nearby woodlands for the winter. In the spring, the same individuals return to small bodies of water to lay eggs.

Dimorphism in the summer generation reflects two extreme strategies. The long-winged forms can fly to other habitats if their pond dries up, a particular advantage when this happens early in the season. The short-winged forms gamble that their pond will persist. The advantage of short wings may be higher fecundity, in that nutrients that would have become wings and flight muscles are converted into eggs.

The fact that all members of the winter generation have long wings suggests that seasonal dimorphism must be controlled by responsiveness to particular environmental conditions during development, rather than by genetic factors. Whether an individual grows long or short wings depends on the length of the day during the early development period. When day length increases continuously during the development period and, in southern Finland, exceeds 18 hours during the last nymphal (immature) stage prior to adulthood, individuals grow short wings. When day length begins to decrease before the end of the nymphal stage (as it does when larval development extends beyond June 21, the summer solstice), long-winged adults are produced. Thus summer-generation individuals become short-winged or long-winged depending on hatching date and rate of nymphal development. Because ponds dry up in late summer, late hatching and slow development place the individual in jeopardy more often than not. The switch between long and short wings is also influenced by temperature; high temperatures, which cause ponds to dry quickly, favor the development of long-winged forms.

Migration, storage, and dormancy

In many parts of the world, freezing temperatures, extreme drought, low light, and other adverse conditions prevent animals and plants from pursuing their usual activities. Under such conditions, organisms resort to a number of extreme responses. These include **migration** (moving to another region where conditions are more suitable), **storage** (relying on resources accumulated during bountiful seasons), and **dormancy** (becoming inactive).

Many animals, particularly among those that fly or swim, undertake extensive migrations. The arctic tern probably holds the record for long-distance migration. It makes a yearly round trip of 30,000 km between its North Atlantic breeding grounds and its antarctic wintering grounds. Each fall hundreds of species of land birds leave temperate and arctic North America, Europe, and Asia for the south in anticipation of cold winter weather and dwindling supplies of their invertebrate food. In East Africa, many of the large ungulates, such as the wildebeest, migrate long distances following the geographic pattern of seasonal rainfall and fresh vegetation (Figure 11.3). A few insects, such as the monarch butterfly, perform impressive migratory movements each year, and some others undertake local movements, but most species of temperate and arctic insects overwinter in dormant stages as eggs or pupae.

FIGURE 11.3 Distribution of the wildebeest population of the Serengeti ecosystem of northern Tanzania and southern Kenya throughout the annual cycle during 1969–1972. (a) December to April; (b) May to July; (c) August to November. The size of the dot indicates the relative size of the population in each area. Adapted from L. Pennycuick, in A. R. E. Sinclair and M. Norton-Griffiths (eds.), *Serengeti. Dynamics of an Ecosystem.* Univ. of Chicago Press, Chicago (1979), pp. 65–87.

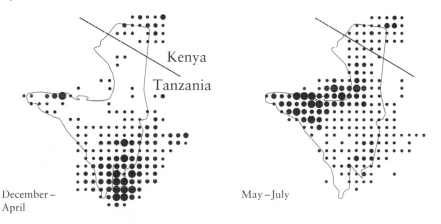

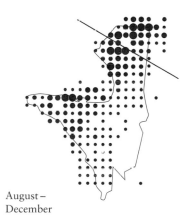

(a)　　　　　　　　　　　(b)　　　　　　　　　　　(c)

Some marine organisms undertake large-scale migrations to reach spawning grounds, to follow a food supply, or to keep within temperature ranges suitable for development. The migration of salmon from the ocean to their spawning grounds at the headwaters of rivers and the reverse migration of adult freshwater eels to their breeding grounds in the Sargasso Sea are striking examples. Many whales undertake seasonal migrations between feeding and breeding areas.

Some populations exhibit irregular or sporadic movements that are tied to food scarcity during particular years rather than to seasonal conditions. The occasional failure of cone crops in the coniferous forests of Canada and the mountains of the western United States forces large numbers of birds that rely on seeds to move to lower elevations or latitudes. Birds of prey that normally feed on rodents disperse widely when their prey populations decline sharply. Even insects are subject to irregular movements. Outbreaks of migratory locusts, from areas of high local density where food has been depleted, can reach immense proportions and cause extensive crop damage over wide areas (Figure 11.4). This behavior is a developmental response to population density. When the locusts occur in sparse populations, they become solitary and sedentary as adults. In dense populations, however, frequent contact with other locusts stimulates young individuals to develop a gregarious, highly mobile behavior, often leading to mass emigration following the local depletion of food resources.

Although homeostasis and migration help maintain function in the face of a changing physical environment, environmental changes often plunge organisms from feast into famine. When an environment supports life only mar-

FIGURE 11.4 A dense swarm of migratory locusts in Somalia, Africa, in 1962. Courtesy of the U.S. Dept. of Agriculture.

ginally, even small fluctuations in food or water supply can be critical; to prevent disaster in such circumstances, many plants and animals store resources during periods of abundance for use in times of scarcity. During infrequent rainy periods, desert cacti store water in their succulent stems. Plants growing on infertile soils absorb, in times of abundance, more nutrients than they require and use them when soil nutrients are depleted. Many temperate and arctic animals accumulate fat during mild weather in winter as a reserve of energy for periods when snow and ice make food sources inaccessible. Some winter-active mammals (beavers and squirrels) and birds (acorn woodpeckers and jays) cache food supplies underground or under the bark of trees for later retrieval. In habitats that frequently burn severely—as in the chaparral of southern California—perennial plants store food reserves in fire-resistant root crowns, which sprout and send up new shoots shortly after a fire has passed (Figure 11.5).

The environment sometimes becomes so extremely cold, dry, or nutrient-poor that animals and plants can no longer function normally. In such circumstances, species that are not capable of migration enter physiologically dormant states. Many tropical and subtropical trees shed their leaves during seasonal periods of drought; temperate and arctic broad-leaved trees shed theirs in the fall, because they cannot obtain from frozen soil the moisture needed to maintain them. For many small invertebrates and cold-blooded vertebrates, freezing temperatures directly curtail activity and lead to dormancy. Many mammals **hibernate** (spend the winter in a dormant state) because they lack food, not because they are physiologically unable to cope with the harsh physical environment.

In most species, conditions requiring dormancy are anticipated by a series of physiological changes in the individual (for example, production of antifreezes, dehydration, and fat storage) that prepare the organism for a complete shutdown. Before the winter, insects enter a resting state known as **diapause,** in which water is chemically bound or reduced in quantity to prevent freezing,

FIGURE 11.5 Root-crown sprouting by chamise (*Adenostoma fasciculatum*) following a fire in the chaparral habitat of southern California. *Left:* Photograph taken on May 4, 1939, 6 months after the burn. *Right:* Photograph taken on July 16, 1940, showing extensive regeneration. Courtesy of the U.S. Forest Service.

and metabolism drops to near nought. In summer diapause, drought-resistant insects respond in one of two ways. Either their bodies dry out and tolerate desiccation, or they secrete an impermeable outer covering to prevent drying. Plant seeds and the spores of bacteria and fungi exhibit similar dormancy mechanisms.

By whatever mechanism it occurs, dormancy reduces exchange between the organism and its environment, enabling animals and plants to "ride out" unfavorable conditions.

Proximate and ultimate factors

What stimulus indicates to birds wintering in the tropics that spring is approaching in northern forests? What urges salmon to leave the seas and migrate upstream to their spawning grounds? How do aquatic invertebrates in the arctic sense that if they delay entering diapause a quick freeze may catch them unprepared for winter? In 1938 J. R. Baker made the important distinction between **proximate factors**—cues, such as day length, by which organisms can assess the state of the environment but that do not directly affect its wellbeing—and **ultimate factors**—features of the environment, such as food, that bear directly on the well-being of the organism.

Virtually all plants and animals sense the length of the day (**photoperiod**) as a proximate factor that indicates season, and many can distinguish periods of lengthening and shortening days. We have seen how wing development in water striders responds to day length. Similarly, the oriental fruit moth enters diapause in the early fall when days become less than 13 hours long (Figure 11.6); at that day length, early frosts may occur.

Similar organisms may differ strikingly in their responses to photoperiod in different areas. Under controlled cycles of light and dark, southern populations (at 30°N) of side oats gramma grass flower when day length is 13 hours, whereas more northerly populations (at 47°N) flower only when the light period exceeds 16 hours each day. In Michigan, at 45°N, populations of small freshwater crustaceans known as water fleas (*Daphnia*) form diapausing broods at photoperiods of 12 hours (mid-September) or less. In Alaska, at 71°N, related species enter diapause when the light period decreases to fewer than 20 hours per day, which happens in mid-August. Warm temperature and low population density tend to shorten the day length that triggers diapause (and hence delay the inception of diapause in the fall), suggesting that these factors portend more favorable environmental conditions for *Daphnia*. Differences in response to day length are under genetic control.

When day length does not accurately predict sporadically changing conditions, animals and plants must take their cues more directly from changes in ultimate factors in the environment. Annual cycles in equatorial regions, where day length is nearly constant, follow upon seasonal cycles of rainfall and their effects on humidity and vegetation. In such highly unpredictable environments as deserts, many organisms adopt a conservative strategy of readiness during the entire period in which sporadic rains are likely to occur. In some desert birds, for example, lengthening photoperiod in the spring stimu-

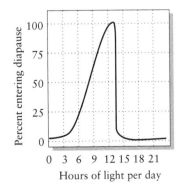

FIGURE 11.6 The influence of day length on entrance into diapause by larvae of the oriental fruit moth (*Grapholitha modesta*). Note that the sharpest break in the response occurs between 13 and 14 hours of day length, corresponding to early fall, which is biologically the most critical point for the inception of diapause. After R. C. Dickson, *Ann. Ent. Soc. Amer.* 42:511–537 (1949).

lates the development of reproductive organs to a point just short of breeding. The gonads maintain this state of readiness throughout the time when rains might occur, but only rainfall itself finally triggers the completion of their physiological development and the initiation of breeding.

✾ SUMMARY

The environment of the individual varies both temporally and spatially and includes a diverse array of resources. Analyses based on the performance of organisms in different environments and on different resources provide insights into the rules governing decisions about diet and habitat selection.

1. We use the concept of environmental grain to express the scale of variations in the environment from the perspective of organisms using resources within it. A fine-grained environment appears uniform, and habitat patches are so small that the individual cannot distinguish among them. In a coarse-grained environment, the individual is able to distinguish and choose among the relatively large habitat patches that occur.

2. Simple models of prey choice, based on considerations of search time, handling time, and reward, predict that predators should specialize more when prey are abundant than when prey are scarce. In other words, beggars can't be choosers.

3. The optimal period during which a consumer should remain within a patch of resources before moving on to another patch is the optimal giving-up time. According to simple models, this period increases as the initial level of resources within patches and the traveling time between patches increase.

4. When many individuals choose among patches of habitats, assessment of habitat quality is influenced by the presence of other, competing individuals. In theory, individuals should distribute themselves among habitat patches so as to equalize their suitabilities, resulting in what is called an ideal free distribution.

5. Temporal variations are encountered in sequence. Response to temporal change depends on the period, predictability, and severity of variation in the environment.

6. When conditions exceed organisms' ranges of tolerance, they may migrate elsewhere, rely on materials stored during periods of abundance, or enter inactive states (such as torpor, hibernation, or diapause).

7. In many cases, the individual must anticipate environmental changes in order to respond successfully. Organisms rely on proximate cues, such as day length and seasonal changes in climate, to predict changes in ultimate factors, such as food supply and temperature, that directly affect their well-being.

❧SUGGESTED READINGS

Baker, R. R. 1978. *The Evolutionary Ecology of Animal Migration.* Holmes & Meier, New York.

French, A. R. 1988. The patterns of mammalian hibernation. *American Scientist* 76:568–575.

Gathreaux, S. A., Jr. (ed.). 1981. *Animal Migration, Orientation and Navigation.* Academic Press, New York.

Gunn, D. L. 1960. The biological background of locust control. *Annual Review of Entomology* 5:279–300.

Krebs, J. R., and N. B. Davies. 1981. *An Introduction to Behavioural Ecology.* Sinauer, Sunderland, Mass.

Lyman, C., et al. 1982. *Hibernation and Torpor in Mammals and Birds.* Academic Press, New York.

Milinski, M. 1979. An evolutionarily stable feeding strategy in sticklebacks. *Zeitschrift für Tierpsychologie* 51:36–40.

Shettleworth, S. J. 1984. Learning and behavioural ecology, in J. R. Krebs and N. B. Davies (eds.), *Behavioural Ecology. An Evolutionary Approach* (2nd ed.). Sinauer, Sunderland, Mass., pp. 170–194.

Stephens, D. W., and J. R. Krebs. 1986. *Foraging Theory.* Princeton Univ. Press, Princeton.

12

LIFE HISTORIES

The attributes of organisms generally suit them to the conditions of their environments. Morphology and physiology vary in parallel with temperature, water availability, salinity, oxygen, and other factors. Homeostatic mechanisms enable the organism to respond to temporal and spatial variations in its environment. Furthermore, these responses appear to follow sets of rules that maximize survival and reproduction. So far, we have considered the features of morphology, physiology, and behavior that handle particular environmental challenges, allowing the organism to maintain suitable internal conditions for proper growth and physiological function and to gather adequate supplies of oxygen, energy, and nutrients. But animals and plants must also reproduce —that is, they must convert the resources available to them into offspring. Reproduction involves choosing among options: when to begin reproducing, how many offspring to have at one time, how much care to bestow upon them. Each of these choices affects other aspects of the individual's life. Because breeding entails risks, investing in offspring generally adversely affects the survival of the parent. In many cases, rearing offspring drains the parent's resources so much that fewer offspring are produced later.

The set of rules and choices pertaining to the individual's schedule of reproduction is referred to as its **life history**. Each life history has many components, the most important of which concern **maturity** (age at first reproduction), **parity** (number of episodes of reproduction), **fecundity** (number of offspring produced per reproductive episode), and termination of life (senescence and programmed death). In most instances, choices regarding one attribute of the life history affect other attributes. The individual has limited time and resources at its disposal. Breeding takes time and resources from other activities, possibly reducing the parent's chances of reproducing again. Thus the life history represents the resolution, to the best advantage of the individual in terms of perpetuating its characteristics, of conflicts between the competing demands of survival and reproduction.

Variation in life histories

During the early 1940s Reginald Moreau, a British ornithologist who had worked in Africa for many years, called attention to the fact that songbirds in the tropics lay fewer eggs (2 or 3, on average) than their counterparts at higher latitudes (generally 4 to 10, depending on the species). But the study of life histories is largely a legacy of one of Moreau's colleagues, David Lack, whose influence is ubiquitous in population biology and evolutionary ecology. Lack recognized that an increase in **clutch size** (number of eggs per nest) would increase the overall reproductive success of the bearer, unless something reduced survival of the offspring in large broods. He presumed that the ability of adults to gather food for their young was limited and, accordingly, that broods with more than a certain critical number of offspring, determined by the availability of food, would be undernourished and the survival of the chicks thereby reduced. Lack further suggested that because of the longer day length at high latitudes (in the season when offspring are reared), temperate-zone and arctic-zone birds could gather more food and therefore rear more offspring than could birds breeding in the tropics, where day length remains close to 12 hours year-round.

Lack made three important points. First, he clearly related life history traits, such as fecundity, to reproductive success and thus to evolutionary fitness. Second, he demonstrated that life histories vary in parallel with factors in the environment, suggesting a responsiveness similar to that exhibited by morphological and physiological traits. Third, he proposed a hypothesis that could be subjected to experiment. In the particular case of clutch size, Lack suggested that food supply limits the number of offspring that parents can rear. To test this idea, an investigator could add eggs to the nest to create an enlarged clutch. According to Lack's hypothesis, the parents should be unable to rear the added chicks because of their limited ability to gather food. We shall return to this problem in a moment.

Life history traits vary with respect to environmental conditions and to one another. Consistent patterns of variation, such as the general decrease in reproductive rate of animals from the poles to the tropics, have aroused curiosity and at the same time suggested relationships between adaptation and environment that may govern the pattern. Lack noted that clutch size and day length vary together; additional explanations for the increase in clutch size with latitude have been based on other environmental conditions that vary with latitude, such as average temperature, seasonal variation in temperature, and rates of predation on bird nests.

Some life history traits, such as fecundity and mortality (Figure 12.1), tend to vary in close association. At one extreme, elephants, albatrosses, giant tortoises, and oak trees exhibit long life, slow development, delayed maturity, high parental investment, and low reproductive rate; at the other extreme are mice, fruit flies, and weedy plants. In broad comparisons within the plant and animal kingdoms, such associations of traits vary with body size and undoubtedly reflect the relative slowness of all life processes in large organisms. But even among organisms of similar size and body plan, different environments produce widely divergent life histories. Storm petrels, which are seabirds the size of thrushes, rear at most a single chick each year, do not begin to reproduce until they are 4 or 5 years of age, and may live 30 or 40 years. Thrushes may produce several broods of 3 or 4 young each year, beginning with their first birthday, but they rarely live beyond 3 or 4 years. Similarly varied life histories may be found even among different populations of the same species — for example, among populations of the fence lizard *Sceloporus undulatus* (Table 12.1).

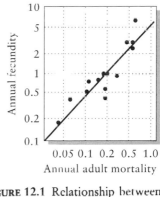

FIGURE 12.1 Relationship between annual fecundity and adult mortality in several populations of birds ranging from albatross (low values) to sparrow (high values). From data in R. E. Ricklefs, *Amer. Nat.* 111:453–478 (1977).

| TABLE 12.1 | *Life history traits of several populations of the fence lizard* Sceloporus undulatus |

	LOCATION							
Trait	Arizona	Utah	Colorado	New Mexico	Kansas	Texas	Ohio	South Carolina
Clutch size	8.3	6.3	7.9	9.9	7.0	9.5	11.8	7.4
Clutches per year	3	3	2	4	1–2	3	2	3
Egg weight (g)	0.29	0.36	0.42	0.24	0.26	0.22	0.35	0.33
Relative clutch mass	0.22	0.21	0.23	0.21	0.28	0.27	0.25	0.23
Age at maturity (months)	12	23	21	12	12	12	20	12
Survival to breeding	0.07	0.05	0.11	0.03	0.10	0.06	0.03	0.11
Annual adult survival	0.24	0.48	0.37	0.20	0.27	0.11	0.44	0.49

Source: D. W. Tinkle and A. E. Dunham, *Copeia* (1986):1–18.

Allocation of limited time and resources

Ecologists believe that observed life histories represent the best resolution of conflicting demands on the organism. Thus a critical part of the study of life histories has been the effort to understand the allocation of limited time and resources to competing functions. An increase in one of an organism's functions places demands on it to increase the delivery of energy and nutrients or its allocation of production toward that function. Energy, nutrients, and production occur in limited supply. Hence any increase in one component of demand, such as that created by reproduction, results in a decrease in delivery to some other component of demand.

For example, animals cannot devote time spent searching for food to caring for offspring or watching for predators. They cannot use for growth the energy and nutrients allocated to reproduction. In many species, individuals produce eggs in direct proportion to the size of their gonads or produce seeds in direct proportion to the number of flowers. Photosynthetic rate depends in part on how much of a plant's production ends up in photosynthetic tissue at the expense of root and support tissue (Figure 12.2). Plants also make use of modular body plans. In some species, nodes at points of leaf attachment may produce either lateral branches or flowers, but not both, thus sacrificing growth at that point for the sake of reproduction.

Demonstrating the validity of this assumption of trade-offs has proved difficult. Adding and subtracting eggs in the nests of several species of birds has,

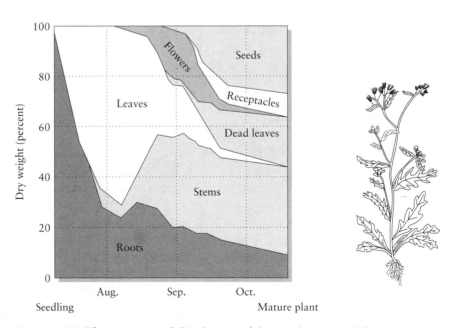

FIGURE 12.2 The proportional distribution of dry weight among different plant parts of the groundsel, *Senecio vulgaris* (Compositae), during its life cycle. Note the development of reproductive parts at the expense of leaves and roots toward the end of the growing season. Based on J. Ogden, in J. L. Harper, *J. Ecol.* 55:247–270 (1967).

in many cases, revealed an inverse relationship between the number of chicks in the nest and their survival. As a result, production of offspring is often greatest from clutches of intermediate size, as predicted by David Lack. For example, magpies (relatives of crows) lay a clutch that corresponds to the maximum number of offspring that a pair can rear. Either adding or subtracting eggs results in fewer offspring fledged (Figure 12.3). Similar experiments with the blue tit and the European kestrel (a small hawk) revealed that fewer parents that were rearing enlarged broods survived to the following breeding season, and those that did survive produced fewer offspring.

Experimental manipulations have not always demonstrated trade-offs between life history traits. For example, in one experiment with guppies, the amount of resource allocated to reproduction was reduced by preventing females from mating with males. Now if growth and reproduction *competed* for allocation of assimilated resources, the experimental fish should have attained a larger size than controls by the end of the study period. But, in fact, little of the difference in accumulated reproductive tissue between mated and unmated females was converted to growth (Figure 12.4). These results suggest several interpretations. Possibly, food availability does not limit reproduction in guppies. Alternatively, genetic factors may adjust the growth rate of females to correspond to the expected availability of resources (guppies normally do not abstain from mating). Finally, growth rate may respond directly to rate of food intake, as it does in immature fish, and not to the relative demand of producing eggs. That is, the experimenters may not have tested their primary hypothesis because they provided an inappropriate cue. Does failure to demonstrate a response to a particular experimental manipulation carry the same weight as confirming a more general role of food in limiting growth and reproduction?

Irrespective of the difficulties involved in experimentation, most issues concerning life histories can be phrased in terms of three questions: When should an individual begin to produce offspring? How often should it breed? How many offspring should it attempt to produce in each breeding episode? The answers that each species has provided to these questions express in a different way the fundamental trade-off between fecundity and adult survival.

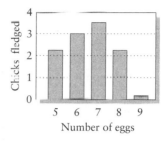

FIGURE 12.3 Number of chicks fledged from nests of European magpies in which 7 eggs were laid but the experimenter added or removed eggs to make up manipulated clutches of 5 to 9 eggs. The most productive clutch size was 7. After G. Hogsted, *Science* 210:1148–1150 (1980).

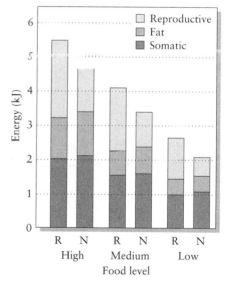

FIGURE 12.4 Contents of energy in somatic tissues, fat deposits, and reproductive tissues (including eggs) in female guppies that were raised at three food levels and either were permitted access to males (R) or were not (N). After D. Reznick, *Ecology* 64:862–873 (1983).

Age at first reproduction

When should an animal or plant begin to breed? Long-lived organisms typically begin to reproduce at an older age than short-lived ones (Figure 12.5). Why should this be so? At every age, an individual must choose between attempting to reproduce and abstaining from breeding. When young individuals resolve this choice in favor of abstention, they effectively delay the onset of sexual maturity. Thus we may understand age at first reproduction in terms of the benefits and costs of breeding at a particular age. The benefit appears as an increase in fecundity at that age. The cost may appear as reduced survival to older ages, reduced fecundity at older ages, or both.

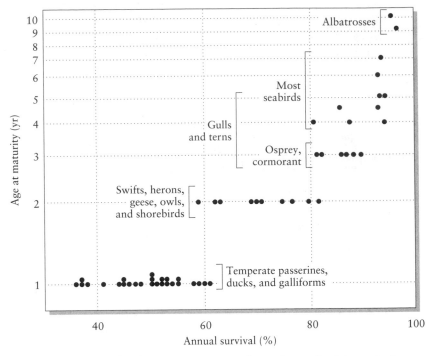

FIGURE 12.5 Relationship between age at maturity and annual adult survival rate, which is directly proportional to life span, in a variety of birds. From R. E. Ricklefs, in D. S. Farner (ed.), *Breeding Biology of Birds*. National Academy of Sciences, Washington, D.C. (1973), pp. 366–435.

Consider the following example. A type of fish continues to grow only until it reaches sexual maturity. Its fecundity varies in direct proportion to body size. Suppose that the number of eggs laid per year increases by 10 for each year that the individual delays reproduction, so that individuals first breeding in their first year produce 10 eggs that year and the same number each year thereafter, individuals first breeding in their second year produce 20 eggs per year, individuals first breeding in their third year produce 30 eggs, and so on. Comparing the cumulative egg production of early-maturing and late-maturing individuals (Table 12.2) reveals that the age at maturity that maximizes lifetime reproduction varies in direct proportion to expected life span.

For organisms that do not grow after their first year (most birds, for example), the choice between breeding and not breeding comes down to balancing current reproduction against survival. Nonbreeding individuals avoid the risks of preparing for reproduction: courtship, nest building, migration to breeding areas. Presumably, life experience gained with age also reduces the risks of breeding, increases the realized fecundity of a certain level of parental investment, or both, thereby favoring delayed reproduction. Among birds, age at maturity varies directly with annual survival rates of adults, up to about 10 years in certain long-lived seabirds. Tending to offset the advantages of delayed reproduction are many factors that reduce the expectation of future reproduction. These include high predation rates; encroaching senescence at old

| TABLE 12.2 | *Total eggs produced by individuals in a hypothetical population as a funtion of life span and age at first reproduction* | | | | | | | |

Age at first reproduction (years)	LIFE SPAN (YEARS)							
	1	2	3	4	5	6	7	8
1	10*	20	30	40	50	60	70	80
2	0	20	40	60	80	100	120	140
3	0	0	30	60	90	120	150	180
4	0	0	0	40	80	120	160	200
5	0	0	0	0	50	100	150	200
6	0	0	0	0	0	60	120	180

*Bold type indicates most productive age at first reproduction for a given life span.

age; and, for organisms that live a single year or less in seasonal environments, the end of the productive season.

Annual and perennial life histories

Plants and animals either reproduce during a single season and die (**annual reproduction**) or have the potential to reproduce over a span of many seasons (**perennial reproduction**). Population biologists have pondered the relative advantages of each habit in terms of the trade-off between survival probability and fecundity. To survive the nonreproductive "winter" period, a perennial plant must allocate resources to storing materials in roots and forming freeze-resistant or drought-resistant buds, presumably at the expense of production. At some point, the advantages of the perennial habit must outweigh the costs in reduced fecundity relative to that of annuals. Theory shows that the pertinent consideration involves the ratio of the survival of the adult perennial to that of the immature plant (up to the point of reproduction; see Box 12.1).

When few perennials survive from one breeding season to the next or when perennials produce relatively few seeds, the annual habit is favored. Where individuals survive well once they are established, but seedlings survive poorly (high ratio of adult to immature survival), an annual must have many seeds in order to outproduce a perennial. Thus annuals predominate among the floras of deserts, where few adult plants can survive drought periods, and perennials dominate the floras of the tropics, where competition and predator pressure make the establishment of seedlings difficult.

BOX 12.1 *A theoretical comparison of annual and perennial reproduction*

The following algebraic model comes from E. L. Charnov and W. M. Schaffer, *Amer. Nat.* 107:791–793 (1973). Suppose a population of plants contains some individuals that produce a large number of seeds at the end of the first growing season and then die (annual) and others that produce fewer seeds but survive through the winter to reproduce in subsequent growing seasons (perennial). Which has the greater fitness? For simplicity, let us assume that annual and perennial plants have the same probability of survival during their first (in the annual's case, only) growing season (S_0) and that perennials have a constant probability of survival thereafter (S).

The factor by which a population of an annual plant grows (λ) equals the number of seeds each individual produces (B_a) times their survival to reproductive age (S_0):

$$\lambda_a = B_a S_0$$

The increase of a population of a perennial plant equals the number of seeds (B_p) times their survival (S_0) plus the probability of survival of the parent (S); hence

$$\lambda_p = B_p S_0 + S$$

The population growth rate of the annual exceeds that of the perennial ($\lambda_a > \lambda_p$) when $B_a S_0 > B_p S_0 + S$. Dividing both sides of the inequality by S_0, we obtain $B_a > B_p + S/S_0$ or, rearranging,

$$B_a - B_p > \frac{S}{S_0}$$

Accordingly, an annual life history leaves more offspring than a perennial life history when the number of seeds produced by the annual exceeds the fecundity of the perennial by the ratio S/S_0.

Reproductive effort

Annual plants have little expectation of life beyond the first breeding season and therefore devote all their resources to current reproduction. Perennials, however, must allocate resources between current reproduction and functions that prolong life or increase size. When a particular life history attribute affects both fecundity and growth or survival, the individual must optimize the trade-off between the two. Intuitively, we would expect a short life span resulting from factors that affect survival independently of the consequences of reproduction to tip the balance in favor of current fecundity. Conversely, individuals with long potential life spans should not increase current fecundity enough to unduly jeopardize future reproduction. This insight has a simple algebraic proof (Box 12.2).

Many plants and invertebrates and some fish, reptiles, and amphibians do not have a characteristic adult size. They grow, at a continually decreasing rate, throughout their adult lives (**indeterminate growth**). Fecundity is directly related to body size in most species with indeterminate growth. Because egg production and growth draw on the same resources of assimilated energy and nutrients, increased fecundity during one year must be weighed against reduced expectation of fecundity in subsequent years. Long life expectancy should favor growth over fecundity during each year. For organisms with less chance of living to reproduce in future years, allocating limited resources to growth instead of to eggs wastes potential fecundity.

Consider two hypothetical fish that both weigh 10 g at sexual maturity but allocate resources to growth and reproduction differently. Both gather enough food each year to reproduce their weight in new tissue or eggs. Fish A allocates 20 percent of its production to growth and 80 percent to eggs, whereas fish B allocates half of its production to growth and and half to eggs. Calculated growth, fecundity, and accumulated fecundity (Table 12.3) show that for fish living 4 or fewer years, on average, high fecundity and slow growth result in greater overall productivity, whereas for fish living longer than 4 years, lower fecundity and more rapid growth are more productive. Adult

---

Box 12.2   *The balance between fecundity and survival*

We use a model similar to the one presented for perennials in Box 12.1, in which the rate of population growth (λ) is equal to $S_0B + S$. However, we partition adult survival into two components, one directly related to reproduction (S_R) and the other independent of reproduction (S). Now, the number of descendants may be expressed as

$$\lambda = S_0B + SS_R$$

Certain reproductive traits that cause small changes in the values of survival (ΔS_R) and fecundity (ΔB) will influence the number of descendants as follows:

$$\Delta\lambda = S_0\Delta B + S\Delta S_R$$

When changes that enhance fecundity (ΔB positive) also reduce survival (ΔS_R negative), their effects on $\Delta\lambda$ depend on the relative values of S and S_0. In general, when S is large compared to S_0, the criterion of number of descendants favors traits that increase adult survival at the expense of fecundity, and vice versa. Thus one expects parental investment in offspring to decrease with increasing adult life span.

TABLE 12.3	*Numerical comparisons of the strategies of slow growth–high fecundity (A) and rapid growth–low fecundity (B) in two hypothetical fish (all weights in grams)*					
	YEAR					
Characteristic	1	2	3	4	5	6
Strategy A						
Body weight*	10	12	14.4	17.3	20.8	25.0
Growth increment†	2	2.4	2.9	3.5	4.2	5.0
Weight of eggs	8	9.6	11.5	13.8	16.6	20.0
Cumulative weight of eggs*	8	17.6	29.1	42.9	59.5	79.5
Strategy B						
Body weight	10	15	22.5	33.8	50.7	76.1
Growth increment	5	7.5	11.3	16.9	25.4	38.1
Weight of eggs	5	7.5	11.3	16.9	25.4	38.1
Cumulative weight of eggs	5	12.5	23.8	40.7	66.1	104.2

*Body weight + growth increment = next year's body weight.

Cumulative weight of eggs to last year + weight of eggs = cumulative weight of eggs to this year.

†Growth increment and weight of eggs in each year are equal to the body weight.

mortality, therefore, determines the optimal allocation of resources between growth and reproduction.

Semelparity

Some species of salmon grow rapidly for several years, then make a single immense reproductive effort during which they convert a large portion of their body tissues to eggs, followed shortly after spawning by death. Because salmon make so great an effort to migrate upriver just to reach their spawning grounds, it may be to their advantage to make the trip just once, at which time they should produce as many eggs as possible, even if this supreme reproductive effort requires the conversion of muscle and digestive tissue to eggs and ensures their death. This pattern is called **programmed death.**

The salmon life history is sometimes called "big-bang" reproduction but is more properly referred to as **semelparity.** This term comes from the Latin *semel* ("once") and *pario* ("to beget"); the opposite of *semelparity* is **iteroparity,** from *itero* ("to repeat"). Semelparity does not equal annual reproduction. For one thing, annuals may have more than one episode of reproduction within a season; for another, like perennials, semelparous individuals must survive one nonbreeding season—and usually many. Semelparity is rare among animals and plants that live for more than 1 or 2 years. Usually, animals and plants allocate so many more resources to survival between growing seasons than they use to prepare for breeding that, once they have adopted a perennial life form, reproduction every year seems the most productive pattern.

The best known cases of semelparous reproduction in plants occur in the agaves and the bamboos, two distinctly different groups. Most bamboos are tropical or warm-temperate-zone plants that form dense stands in disturbed habitats. Reproduction does not appear to require substantial preparation or resources, such as are needed to grow a heavy flowering stalk. But there are probably few opportunities for successful seed germination. Once established, a bamboo plant increases by asexual reproduction, continually sending up new stalks until the habitat in which it germinated is fairly packed with bamboo. Only then, when vegetative growth becomes severely limited, do plants benefit from producing seeds, which may colonize disturbed sites.

The environment and habits of agaves occupy the opposite end of the spectrum from those of bamboos. Most species of agave inhabit arid climates with sparse and erratic rainfall. Plants grow vegetatively for several years (the number varies from species to species) and then send up a gigantic flowering stalk. After producing its seeds, the agave dies (Figure 12.6). One curious fact about agaves is that they frequently live side by side with yuccas, a group of plants that have a similar growth form but flower year after year. The root systems of agaves do, however, differ from those of yuccas. Yucca roots descend deeply to tap persistent sources of groundwater; agaves have shallow, fibrous roots

FIGURE 12.6 Stages in the life cycle of the Kaibab agave (*Agave kaibabensis*) in the Grand Canyon of Arizona. The plant grows as a rosette of thick, fleshy leaves for up to 15 years. Then it rapidly sends up its flowering stalk and sets fruit, after which the entire plant dies (left-hand rosette).

that catch water percolating through the surface layers of desert soils after rain showers but that are left high and dry during drought periods. The erratic water supply of the agave may prevent successful seed production or seedling establishment from occurring every year, and the period between suitable years may be very long. Under these conditions, agaves may benefit from growing and storing nutrients until an unusually wet year — perhaps 1 in 10 or even 1 in 100 (agaves are often called century plants) — and then putting all resources into reproduction.

Senescence

Although few organisms exhibit programmed death associated with reproduction, most do experience a gradual increase in mortality and a decline in fecundity resulting from the deterioration of physiological function known as **senescence**. For example, the rates of most physiological functions in humans decrease in a roughly linear fashion between the ages of 30 and 85 years, to 80 to 85 percent of the value in 30-year-old individuals for nerve conduction and basal metabolism, 40 to 45 percent for the volume of blood circulated through the kidneys, and 37 percent for maximum breathing capacity. Birth defects in offspring and infertility generally occur with increasing prevalence in women progressively older than 30 years (Table 12.4). Reproductive decline and death in old age do not result from abrupt physiological change. Rather, these consequences of senescence follow upon a gradual decrease in physiological function with age. Such changes are found throughout the animal kingdom.

TABLE 12.4	*Number of defects per 10,000 human births according to age of the mother*						
	AGE OF MOTHER (YEARS)						
Birth defect	15–19	20–24	24–29	30–34	35–39	40–44	45+
Anencephalus	10	9	7	8	9	10	6
Spina bifida	12	11	10	10	13	13	16
Hydrocephalus	18	7	7	8	10	15	25
Microcephalus	1	1	1	1	1	13	—
Heart defects	17	18	17	17	24	28	50
Cleft lip, palate	10	11	11	10	12	16	—
Down syndrome	2	2	2	3	11	33	89
Club foot	15	14	11	11	12	24	18
Total	85	73	66	68	92	142	202

Source: S. Milham, Jr., and A. M. Gittelsohn, *Human Biol.* 3:13–22 (1965).

Why does senescence exist, when survival and reproduction presumably confer advantage on an individual at any age? The answer to this question originates in the fact that even in a population without senescence, accidental causes of death that affect the young and old equally allow few individuals to live to old age. With few individuals surviving to old age, progressive physiological deterioration has little effect, on average, on lifetime reproduction. Senescence may reflect the accumulation of molecular defects that fail to be repaired, just as an automobile eventually deteriorates and has to be junked.

Alternatively, senescence may arise from life history traits that enhance reproduction at an early age and reduce it later. Because few individuals live to old age, such traits confer an overall benefit on the individual. The presence of such traits has been demonstrated in experiments in which investigators artificially terminated the life span of fruit flies (*Drosophila*) at an early age and, after several generations, observed an increase in rate of egg production early in life coupled with a decrease in life span among individuals left in peace until they died of natural causes (Figure 12.7). In the experimental situation, reproduction at an earlier age assumed greater value to the individual, thus explaining the shift in life history traits.

☙ *SUMMARY*

1. Life history traits of organisms include age at first reproduction, reproductive rate, and life span. Their expression can be interpreted as resolving conflicts over the allocation of limited time and resources among various activities.

2. Variation and correlation of life history traits among species constitute the phenomenological basis for their study. Delayed reproduction, long life, and low reproductive rate—and their opposites—frequently are associated.

3. Theories of life history variation among species, including correlations among life history traits, are based on adjusting the allocation among competing activities in such a way as to maximize lifetime reproductive success.

4. When survival between breeding seasons is low or requires large sacrifices in fecundity, annual life histories are favored over perennial life histories.

5. When reproduction requires costly preparation, selection may favor a single all-consuming reproductive event followed by death, as in salmon, agaves, and bamboos. This pattern of reproduction is called semelparity, which is the converse of iteroparity, or repeated reproduction.

6. Senescence, the progressive deterioration of physiological function with age, leads to declines in fecundity and probability of survival. Senescence arises because, owing to accidental death, few individuals survive to old age under any circumstances. As a result, traits that enhance early reproduction at the expense of late reproduction and long life benefit the individual overall.

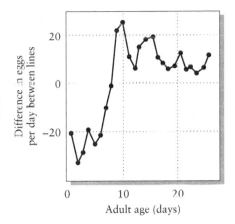

FIGURE 12.7 Difference in daily egg production as a function of age between a control line of fruit flies and a line in which life span was terminated at an early age. Negative values indicate increased rate of egg production among short-lived flies. From M. R. Rose and B. Charlesworth, *Genetics* 97:187–196 (1981).

❦ SUGGESTED READINGS

Bazzaz, F. A., N. R. Chiarello, P. D. Coley, and L. F. Pitelka. 1987. Allocating resources to reproduction and defense. *BioScience* 37:58–67.

Fleming, I. A., and M. R. Gross. 1989. Evolution of adult female life history and morphology in a pacific salmon (Coho: *Oncorhynchus kisutch*). *Evolution* 43:141–157.

Janzen, D. H. 1976. Why bamboos wait so long to flower. *Annual Review of Ecology and Systematics* 7:347–391.

Reznick, D. 1985. Costs of reproduction: an evaluation of the empirical evidence. *Oikos* 44:257–267.

Roff, D. A. 1986. Predicting body size with life history models. *BioScience* 36:316–323.

Rose, M. R. 1991. *Evolutionary Biology of Aging.* Oxford Univ. Press, Oxford.

Schaffer, W. M., and M. V. Schaffer. 1977. The adaptive significance of variations in reproductive habit in the Agavaceae. In B. Stonehouse and C. Perrins (eds.), *Evolutionary Ecology.* Macmillan, London, pp. 261–276.

Shine, R., and J. J. Bull. 1979. The evolution of live-bearing in snakes and lizards. *American Naturalist* 113:905–923.

Sibly, R. M., and P. Calow. 1986. *Physiological Ecology of Animals. An Evolutionary Approach.* Blackwell Scientific Publications, Oxford.

Stearns, S. C. 1976. Life-history tactics: a review of the ideas. *Quarterly Review of Biology* 51:3–47.

Williams, G. C. 1966. Natural selection, the costs of reproduction, and a refinement of Lack's principle. *American Naturalist* 100:687–690.

13

SEX, FAMILY, AND SOCIETY

During the course of its life, each individual interacts with many others of the same species: mates, offspring, nondescendant relatives, and members of social groups. Each interaction requires the individual to perceive the behavior of others and make appropriate responses. In general, interactions between close relatives are mutually supportive because of the shared genetic factors they have inherited from their ancestors. Mates also must cooperate if they are to raise offspring successfully. Nonetheless, social behaviors among unrelated individuals emphasize that all interactions between members of the same species delicately balance the conflicting tendencies of cooperation and competition, altruism and selfishness.

In this chapter, we shall explore some of the consequences for the individual of sexual reproduction, parental care, and social interaction, and we shall describe the various ways in which these relationships are managed behaviorally.

Sex and gender

Sexual reproduction unites two **haploid gametes** to form a **diploid** cell, the **zygote**. It makes possible a mixing of the genetic material of two individuals, resulting in the formation of new combinations of genes in the offspring. This **recombination** of genes guarantees that siblings differ from each other genetically. Thus, in a variable environment, some offspring have a genetic constitution that enables them to survive and reproduce, regardless of the conditions of the environment. Sexual reproduction may also produce new combinations of genetic factors (**genotypes**) previously absent from the population, thereby providing new genetic variation as raw material on which the processes of evolution operate.

As with many issues, botanists and zoologists have divergent perspectives on sex. We are used to thinking in terms of two genders, female and male. But female and male sexual functions may be combined in the same individual, or the individual may change its gender during its lifetime. In most species of plants, individuals have both male and female functions (Figure 13.1). Animals in which sexual function is divided into male and female individuals are referred to as **gonochoristic;** the word derives from the Greek *gonos,* which per-

FIGURE 13.1 Flower of the hibiscus plant showing the receptive structure (stigma) of the female part of the plant (F) and the anthers of the male part of the plant (M), which produce the pollen.

tains to procreation, and *choris,* meaning "apart" or "separate." Botanists refer to plants exhibiting the same condition as **dioecious,** from the Greek *di-* ("two") and *oikos* ("dwelling," also the root of the word *ecology*).

When both sexes occur in the same individual, biologists term the individual a **hermaphrodite,** after the mythological Hermaphroditus, son of Hermes and Aphrodite, who while bathing became joined in one body with a nymph. Hermaphrodites may be **simultaneous,** as in the case of many snails and most worms, or they may be **sequential:** male first in some mollusks and echinoderms, female first in some fishes. Botanists apply an additional term, **monoecious,** to individual plants that bear separate male and female flowers, rather than the so-called **perfect flowers** that include both male and female parts. Though perfect-flowered hermaphrodites are the rule among plants (by one estimate they account for 72 percent of plant species), nearly all imaginable combinations of sexual patterns are known, including populations with hermaphrodites and either male or female plants; populations with male, female, and monoecious plants; and hermaphrodite individuals with perfect flowers that also bear either male or female flowers.

One can measure the benefit of male and female sexual function by the number of sets of genes transmitted to offspring through either male or female gametes. When a female can achieve a certain amount of male function by giving up a smaller amount of her female function, she (subsequently it) should do so. Similarly, males should add female function when this change does not cut deeply into their male productivity. For the flowers of female plants to produce a little bit of pollen, few resources need shift from female to male function. And so hermaphroditism should arise frequently, as it has among plants.

Separate sexes provide the best compromise when the gains from adding function of one sex bring about even greater losses in function of the other. This may occur when establishing new sexual function in an individual entails a substantial fixed cost before any gametes appear. Sexual function requires gonads, ducts, and other structures for transmitting gametes and necessitates secondary sexual characteristics for attracting mates and for competition among individuals of the same sex. In many animals, where maleness requires specializations for mate attraction and antagonistic interaction with other males, or where femaleness requires specializations for egg production or brood care, such fixed costs may put hermaphroditism at a disadvantage compared to sexual specialization. In fact, hermaphroditism occurs rarely among active animal species and those that engage in brood care, but it is much more common among sedentary aquatic species that shed gametes into the water.

Sequential hermaphroditism reflects changes in the costs and benefits of male and female sexual function as an organism grows. In some marine gastropods having internal fertilization, such as the slipper shell *Crepidula* (Figure 13.2), insemination requires the production of only small amounts of sperm. Hence male function consumes few resources and has little effect on growth. As a consequence, individuals of many such species are male when small and female when large and thus are able to produce correspondingly large clutches of eggs. When competition between males for mates influences reproductive success, large size confers advantage. This apparently has led to female-first sequential hermaphroditism in some species of fish that inhabit reefs, where males compete among themselves for breeding territories.

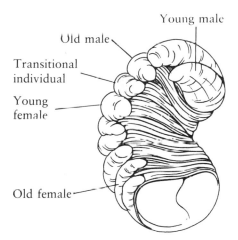

Young male
Old male
Transitional individual
Young female
Old female

FIGURE 13.2 A drawing of stacked individuals of the slipper shell *Crepidula fornicata* displaying sequential hermaphroditism. The bottom of the stack consists of reproductively mature females and the top consists of mature males, with transitional individuals in between. After R. C. Brusca and G. J. Brusca, *Invertebrates.* Sinauer, Sunderland, Mass. (1990).

Sex ratio

Two facts about **sex ratio** stand out. First, many populations have very nearly equal numbers of females and males. Second, many populations present exceptions to this rule. Because females and males occur in the human population in a ratio of approximately 1 : 1, and because roughly equal numbers of females and males characterize populations of most species, we consider the 1 : 1 sex ratio the standard condition and regard deviations from this ratio as special cases.

Biologists explain this predominant 1 : 1 ratio by the following simple reasoning. Every product of a sexual union has exactly one mother and one father. Consequently, individuals of the rarer sex in the population will enjoy greater reproductive success because they compete for matings with fewer others of the same sex. For example, when a population of 5 males and 10 females produces 100 offspring, each male contributes 20 sets of genes, but each female contributes only 10 sets of genes. Thus individuals of the rarer sex produce more offspring. With a sex ratio skewed toward females, any parental genetic tendency resulting in a larger proportion of male offspring would be favored, and the frequency of males in the population would increase. Similarly, if females were the rarer sex, genotypes that increased the proportion of female progeny would be favored, and the frequency of females would increase in the population. When males and females are equally numerous, individuals of both sexes contribute equally to future generations, on average, and the frequencies of males and females among the progeny of an individual are of no consequence to its long-term reproductive success.

Some situations exist, however, in which a parent may benefit from a skewed sex ratio among its progeny. In many species, competition for matings among individuals of one sex (usually males) creates tremendous variation in reproductive success: where competition is keen, some males achieve many matings, others none. In harem-forming species such as many ungulates, for example, a few males control most of the females, and other males lower in social dominance have little access to mates. Generally, the largest males win contests. In mammals, in which the mother cares directly for her offspring through the periods of gestation and lactation, the condition of the mother is likely to influence that of her male offspring. As a result, to maximize their long-term reproductive success, females in good condition should produce male offspring, which will grow large and fare well in male–male competition for mates. Females in poor condition should place their investment in female offspring, which will mate successfully regardless of the parental care they receive.

Experimental confirmation of this idea has emerged from a laboratory study of female wood rats (*Neotoma floridana*). In this species, as in most mammals, females normally invest equally in male and female offspring. But when investigators restricted food intake during the first 3 weeks of the lactation period to below the maintenance level of a nonreproductive female, mothers actively rejected the attempts of male offspring to nurse. As a result, males starved and the sex ratio at 3 weeks shifted to about one male for each two females.

Local mate competition and sex ratio in wasps

When individuals do not disperse far within populations, or when they mate prior to dispersal, mating often takes place among close relatives. At the extreme, mating may occur among the progeny of an individual parent. In this situation, mate competition takes place locally among brothers (hence the term **local mate competition**). From the standpoint of the parent of these siblings, one son would serve as well as many the purpose of fertilizing its sisters. In the case of sib mating, the number of grandchildren depends on the number of daughters, so the female that produces daughters at the expense of sons will have the greater number of grandchildren. Sib mating occurs commonly in certain wasps that parasitize other insects and whose progeny mate on their hosts before dispersing. These wasps can also alter the sex ratio among their progeny, and they do so in a manner predicted by local mate competition.

The hymenopterans (bees, ants, and wasps) have an unusual sex-determining mechanism by which fertilized eggs produce females and unfertilized eggs produce males. As a result, females are diploid and males haploid, giving rise to the term **haplodiploid** sex-determining mechanism. Reproductive females can control the sex of their offspring simply by fertilizing eggs or not.

Many wasps parasitize other insects or complete their larval development within the fruits of certain plants. For some of these species, hosts are so scarce and mates so difficult to find that females mate where they grew up before dispersing to find new hosts on which to lay their own eggs. When a host is parasitized by a single female wasp, females within each brood are limited to mating with their brothers. Under this circumstance, male offspring make a reduced contribution to reproductive success, and wasps skew the sex ratios of their progeny greatly in favor of females — to the point of producing only one male per brood. Such males often lack wings and, in extreme cases, fertilize females as larvae within the host. These observations confirm the idea that when local mate competition occurs only among sibs, the value of males diminishes greatly, and, where possible, mothers limit the number of males among their progeny.

Mating systems

It is a basic asymmetry of life that the fitness contribution of a female depends on her ability to make eggs and otherwise provide for her offspring, whereas the fitness contribution of a male usually depends on the number of matings he can procure. The larger female gamete requires more resources than the male gamete, so the female's ability to gather resources to make eggs determines her fecundity. The males of many species neither contribute resources to their mates (for example, by defending territories for the females to feed in or by feeding the female directly) nor care for their young. In such cases, a male can increase its own fecundity only by mating with additional females.

When a male mates with as many females as he can locate and persuade and provides his offspring nothing more than a set of genes, he is said to be **promiscuous.** Promiscuity generally precludes a lasting pair bond. When adopted by a population, it also tends to increase variance in mating success among males, some individuals obtaining perhaps dozens of matings while others get none. Among animal taxa as a whole, promiscuous mating is by far the commonest system.

When the male can contribute to the fecundity of its mate by procuring resources for the female or caring for their offspring directly, pair bonds may outlast the copulatory rapture of promiscuous species. As males increasingly enhance the realized fecundity of a single mating, mating systems progress from promiscuity (least care) through polygamy to strict monogamy (most care, relative to the female). In **polygamy,** a single individual of one sex forms long-term pair bonds with more than one individual of the opposite sex. Normally males mate with more than one female, in which case the system is referred to as **polygyny** (literally, "many females"). Polygyny expresses itself through the defense of several females against the mating attempts of other males (a harem) or the defense of territory or nesting sites to which more than one female gravitate to breed as well as mate. Thus polygyny may arise through the ability of a male to control matings by defending females, in which case the contribution of the male to its progeny may be primarily genetic, or through the ability of a male to control or provide resources that the female needs to reproduce.

Monogamy is the formation of a pair bond between one male and one female that persists through the period required to rear offspring and may endure until one of the pair dies. Monogamy arises primarily when males can contribute to the number and survival of offspring. Hence it is most common in species with prolonged dependence of offspring, in which both sexes can provide for the young. Monogamy is not common in mammals because males cannot provide milk. But it is common among birds, especially those in which parents feed their offspring, a duty both sexes can manage equally well.

Mating system and ecology go hand in hand. In African weaverbirds, for example, monogamy predominates among insectivorous species inhabiting forest and savanna, whereas polygyny is the rule among seed-eating species of open savannas and grasslands. Furthermore, most of the monogamous species breed as solitary pairs on widely dispersed territories, whereas polygynous species are invariably colonial. But why do some species adopt polygynous habits and others do not? And why should mating system be related to habitat?

The polygyny threshold model

As long as his territory holds sufficient resources, a male gains by increasing the number of his mates. A female benefits from choosing a territory or a mate of high quality. Thus polygyny arises when a female will experience greater reproductive success by sharing a male with one or more other females than by forming a monogamous relationship. Suppose that the quality of two males' territories differs so much that a female could rear as many offspring on the better territory, sharing it with other females and having little or no help from her mate, as she could on the poorer territory with help from her monogamous mate. The difference in territory quality at which polygynous and monogamous females do equally well is referred to as the **polygyny threshold** (Figure 13.3).

According to the polygyny threshold model, then, polygyny will occur only when the quality of male territories varies so much that some females will have higher fitness mated to a polygynous male on a territory of high quality than they would mated monogamously to a male on a territory of poor quality. If females can assess the quality of territories, the first individuals to pair should mate monogamously, latecomers choosing mates with territories of progressively lower quality until the polygyny threshold is reached. At this point, females should be ambivalent about pairing with unmated or previously mated males.

Promiscuous mating

Although promiscuous mating is the rule in the animal and plant kingdoms, it appears rarely and sporadically among birds, being predominant only among grouse and a few groups of tropical, frugivorous species — the manakins, cotingas, and birds of paradise. Representatives of all these groups have communal mating areas, called **leks,** in which congregated males perform elaborate displays to attract females to mate with them. Because such mating systems are so unusual among birds, ecologists have paid considerable attention to the surfacing of promiscuity on the calm sea of avian monogamy.

The males of most species of birds contribute substantially to the fecundity of their mates (and therefore to their own fecundity) either by defending territories or by directly providing for the young. The key to promiscuity therefore has been perceived as the emancipation of the male from parental investment without a substantial reduction in the fecundity of his mates. This can happen in several situations. First, the young of precocial birds feed themselves, so factors other than the provisioning of food by the parents may limit fecundity. Females of such species, including the promiscuously mating grouse and their relatives (Figure 13.4), can rear nearly as many young by themselves as they could with the help of males. Second, in mammals the female alone nurses the young, and potential male contributions (defending resources, protecting the litter, and providing food directly to the female) pale by comparison. Third, the resources of some fruit-eating species may be conspicuous and of fixed quantity, in which case a female alone can gather as much as a male and a female working together. Clearly, when males can desert their mates with little loss of fecundity, they may increase their reproductive success by attempting to attract additional mates.

Sexual selection

Regardless of the mating system, the initial stages of reproduction involve choosing a mate. In polygynous and promiscuous mating systems, males gain by mating indiscriminately with females, and choice usually is the prerogative

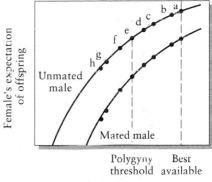

FIGURE 13.3 The polygyny threshold model. Female fitness (expectation of offspring) varies with the intrinsic quality of a male's territory, and the presence of a mated female reduces that quality to subsequent females. Females should select unmated males (a through e) until the quality of the territory of any remaining unmated males (f, g, and h) drops below the "net" quality (presence of another female "factored in") of the best territory controlled by a mated male (a). At this point, called the polygyny threshold, female choice will alternate between mated males (a through e, which then become bigamists) and unmated males (f through h). After G. H. Orians, *Amer. Nat.* 103:589–603 (1969).

FIGURE 13.4 Two male (*left*) sharp-tailed grouse displaying on a communal courting area (lek) in southern Michigan. The female (*right*) has a dull plumage. Courtesy of U.S. Soil Conservation Service.

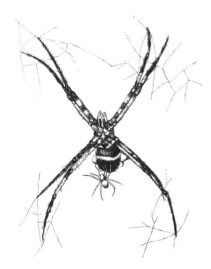

FIGURE 13.5 Extreme sexual dimorphism in size in the garden spider *Argiope argentata*. The male is much smaller than the female, which is portrayed in the normal resting position at the hub of her web.

of the female. When males differ by characteristics that could affect the reproductive success of a female and that her offspring could inherit, she should choose to mate with the male of highest quality. And, of course, males should do everything in their power to advertise their high quality—that is, they should strut their stuff. This phenomenon of mate choice, also known as **sexual selection,** sets the stage for intense competition among males for mates and the development of attributes in the males that are effective in fighting other males or in attracting females. The usual result of sexual selection is strong **sexual dimorphism,** especially of ornamentation, coloration, and courtship behavior. Such traits, which distinguish gender over and above the primary sexual organs, are known as **secondary sexual characteristics.**

Charles Darwin, in his book *Descent of Man and Selection in Relation to Sex* (1871), was the first to propose that sexual dimorphism could be explained by selection applied differentially to one sex. Sexual dimorphism can arise in three different ways. First, the different sexual roles of males and females may place each in a different relationship to the environment. For example, because females produce large gametes, fecundity often increases in direct relation to body size; this may account for females being larger than males in many species (Figure 13.5). In addition, the special nutritional requirements of producing eggs and the duties of protecting the eggs and young, which usually fall upon the female, may lead females to use the environment differently than males. Simply having to find suitable nest sites may require females to use different habitats than males during the nesting season.

Second, sexual dimorphism may result from contests between males, which may select elaborate weapons for combat, such as the antlers of deer (Figure 13.6) and the horns of the mountain sheep.

Figure 13.6 Elk have immense antlers that are used during contests between males to establish control over harems of females. Courtesy of U.S. Department of Interior.

Third, sexual dimorphism may arise through the direct exercise of choice by individuals of the opposite sex. With few exceptions, females do the choosing, and males attempt to influence that choice with magnificent displays of vainglorious courtship ritual. Why females choose, and males compete among themselves for the opportunity to mate, is a consequence of the general asymmetry of parental investment that defines the male and female conditions. Males enhance their fecundity in direct proportion to the number of matings they obtain; females are limited in number of offspring by the number of eggs they can produce, but they stand to gain in quality of offspring by choosing to mate with males bearing superior genotypes.

Female choice is experienced at some level by most males. A particularly compelling demonstration of female choice comes from an experimental study of the tail length of male long-tailed widowbirds (*Euplectes progne*). This polygynous species inhabits open grasslands of central Africa. Females, about the size of a sparrow, are mottled brown, short-tailed, and altogether ordinary in appearance. The males are black, except for a red shoulder patch, and sport a half-meter-long tail that is conspicuously displayed in courtship flights. Males may attract a half-dozen females to nest in their territories, but they provide no care for their offspring. Tremendous variation in male reproductive success provides the classic conditions for sexual selection. The experiment was simple and straightforward. The tail feathers of some males were cut to shorten them, and the clipped feather ends were glued onto the feathers of other males' tails to lengthen them. Length of tail had no effect on a male's ability to maintain a territory, and reproductive success apparently was not influenced by the effect of tail length on contests between males. But, strikingly,

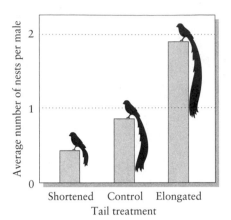

FIGURE 13.7 Relative reproductive success of male long-tailed widowbirds with artificially shortened and artificially elongated tails. After M. Andersson, *Nature* 299:818–820 (1982).

males with experimentally elongated tails attracted significantly more mates than those with shortened or unaltered tails (Figure 13.7). This result strongly suggests choice of mate on the basis of tail length.

How do such outlandish traits as the long tail of the long-tailed widowbird arise? Presumably, such traits burden males by making them more conspicuous to predators and by requiring more energy and resources to maintain. One intriguing possibility is that male secondary sexual characteristics act as handicaps: that a male can survive while bearing such a handicap signals a female that he has an otherwise superior genotype. This idea is known as the **handicap principle.** It may sound crazy, but if you wanted to demonstrate your strength to someone, you might be able to make your point by carrying around a large set of weights. A weaker individual couldn't do it and thus could not falsely advertise strength. Accordingly, the greater the handicap borne, the greater the ability of the individual to offset the handicap by other virtues—and to pass these on to his offspring.

Society

Humans are the most social of all animals. Societies are sustained by role specialization among members, the interdependence attendant upon specialization, and the cooperation that interdependence requires. Yet humans also are competitive, to the point of violence, within this mutually supportive structure. Social life balances contrasting tendencies toward mutual help and conflict. Some animal populations exhibit much of the complexity of human societies. The social insects—ants, bees, termites—are remarkable for their division of labor and the behavioral integration of the hive or nest. Similar subtlety of social interaction, including role specialization and altruistic behavior, is being discovered increasingly among mammals and birds.

Social behavior includes all types of interactions between individuals, from cooperation to antagonism. Sometimes social behavior provides a means of organizing and ritualizing the expression of basic conflicts within groups: outright conflict is averted when it is channeled into the posturing of males for social rank or access to mates. Defense of territories and the establishment of dominance hierarchies also serve this purpose with a minimum of social strife.

Dominance hierarchies and territoriality

Any area defended against the intrusion of others may be regarded as a **territory.** These may be transient or more or less permanent, depending on the stability of the resource and the individual's need of it. At times, dispersion of individuals on territories may not be practical because of the social pressure of high population density, transience of the critical resource, or overriding benefits of living in groups. In such circumstances, conflict among individuals is still resolved by contest, but social rank rather than space is the winner's prize. Once individuals order themselves into a **dominance hierarchy** of social status, subsequent contests between them are resolved quickly in favor of the higher ranking individual. When a social hierarchy is linearly ordered, the first-ranked member of the group dominates all others, the second-ranked dominates all but the first-ranked, and so on down the line to the last-ranked individual, who dominates none.

The position of an individual in a dominance hierarchy is sometimes reflected by its spatial position within a group. In large foraging flocks of wood pigeons, individuals low in the dominance hierarchy tend to be at the periphery, where they are more exposed to predators than the dominant individuals that occupy the center of the flock. Peripheral birds appear nervous, and because they spend much of their time looking up from feeding, they are often undernourished. Birds in the center of the flock remain calm and feed more, because they are protected from the surprise attack of a predator by the vigilance of individuals at the periphery.

Ritualized antagonistic behavior

Most conflict associated with social rank or territorial defense involves **ritualized behaviors** that rarely lead to risky physical struggle. Certain appearances or behaviors appear to signal higher status than others. Why, then, don't less dominant individuals assume the behavior of their betters? Why don't they try to deceive other members of the population into thinking that they are more aggressive than they really are? Part of the answer is that some ritualized behaviors allow contestants to judge each other's size, which is difficult to fake and often determines the outcome of earnest combat. Another part is that signaling status and ability to back it up must go together. Consider the following experiments, which demonstrate this fact.

Harris's sparrows breed in the Canadian arctic and winter in small flocks in the central United States. Social status within flocks is correlated with the amount of dark coloration in the plumage of the throat and upper breast (Figure 13.8). Upon first encounters, lighter birds generally avoid darker individuals, so dark coloration meaningfully signals status. When the plumage of light individuals was dyed dark, they were mercilessly attacked by others who easily saw through the ruse. But dark individuals bleached to a lighter color found themselves constantly having to attack naturally light birds to regain their status. The system works because plumage and aggression go together — there are no cheaters. When lighter birds were implanted with testosterone to raise their level of aggression, such individuals that were also dyed darker rose in the dominance hierarchy whereas those left with their light plumage were persecuted by naturally dark birds and could not rise in status, despite their higher level of aggression. Behavior and plumage are probably affected by common physiological conditions associated with hormone levels. When the two do not match, birds get into trouble because they either don't get the respect that is due them or are treated as imposters.

These results raise a difficult question. If social status depends on the level of a hormone whose production imposes little physiological cost, why don't all members of the population have the highest level of aggression? Hormone implants have turned monogamous song sparrows into bigamists and nonterritorial red grouse into landowners. Presumably, either (1) status is linked to age and experience, on the basis of which hormone-mediated levels of behavior are adjusted to serve other purposes, or (2) variation in rank reflects the fact

FIGURE 13.8 Examples of variation in throat plumage of Harris's sparrows. After D. Morse, *Behavioral Mechanisms in Ecology.* Harvard Univ. Press, Cambridge, Mass. (1980).

that dominance has its costs and all members of the hierarchy have roughly equivalent fitness.

Given that status truly represents an individual's capabilities in behavioral encounters, one might also ask why low-ranking individuals, which are excluded from food and mates and are exposed to higher risks, remain with the flock. The answer must simply be that it is better to be a low-ranking individual in a flock, perhaps rising in rank with age, than to be a loner.

Social groups

Animals get together for a variety of reasons. Sometimes they are independently attracted to suitable habitat or resources and form aggregations, such as those of vultures around a carcass or of dungflies on a cowpat. Within such groups, individuals may interact, usually to contest space, resources, or mates. In other cases, progeny remain with their parents to form family groups, and aggregation results from a simple failure to disperse. True **social groups,** however, arise through the attraction of unrelated individuals to one another—that is, through joining together.

Animals form groups to increase their survival, facilitate feeding, or find mates. In groups, individuals spend more time feeding and less time looking out for predators, as we have seen in the wood pigeon; individuals also may learn the location of food from others or cooperate in obtaining food. Cooperation involves the coordination of individual behavior toward a common goal: defense of the herd by male musk oxen or pack-hunting by wolves and killer whales. Such instances of cooperation are not common, but where they do occur, social behavior fosters mutualism between individuals more than it organizes competition between them: individual behavior becomes subservient to the group, and groups assume a characteristic behavior of their own.

Social behavior

Any social interaction other than mutual display can be dissected into a series of behavioral acts by one individual (the **donor** of the behavior) directed toward another (the **recipient**). One individual delivers food, the other receives it; one threatens, the other is threatened. When one individual attacks another, the attacker may be thought of as the donor of a behavior. The attacked individual (the recipient in this case) may respond by standing its ground or fleeing; in either case, it then becomes the donor of a behavior. The donor–recipient distinction is useful because each behavioral act results in a change in the reproductive success of both the donor and recipient of the behavior. These increments may be positive or negative, depending on the interaction. Four combinations of cost and benefit for donor and recipient organize the outcomes of social interactions into four categories (Figure 13.9).

Cooperation and **selfishness** both benefit the donor of the behavior and therefore should appear frequently as an expression of social behavior. **Spitefulness**—reducing another's reproductive success despite the cost to one's own success—cannot be favored under any circumstance. Selfishness

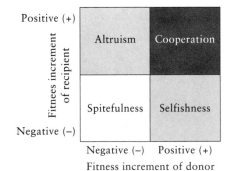

FIGURE 13.9 Four types of behavior classified according to the fitness increments of the donors and recipients of the actions.

should normally be favored to the exclusion of **altruism** because it increases the reproductive success of the donor. Altruism presents a difficult problem because it requires the evolution of behaviors that reduce the fitness of the individual. Yet this appears to have happened in colonies of social insects, in which workers forgo personal reproduction to rear the offspring of the queen.

Kin selection

The dilemma of the apparent altruism of social insects is resolved when one realizes that ant colonies are discrete family units, mostly the offspring of a single female (the queen). Behavioral interactions occur between close relatives, in this case siblings. When an individual directs an altruistic act toward a sibling, it enhances the reproductive success of an individual with which it shares a substantial part of its own genetic makeup. When an individual behaves altruistically because of a single gene inherited from one parent, a brother or sister has a probability of one-half of bearing a copy of the same gene. The likelihood that two individuals share a copy of the same gene is the probability of **identity by descent,** the value of which varies with the degree of relationship (Table 13.1).

Biologists refer to the total fitness of a gene responsible for a particular behavior as its **inclusive fitness,** which equals the change in reproductive success of the donor resulting from its behavior *plus* the product arrived at by multiplying the change in reproductive success of the recipient by the probability that it carries a copy of the same gene. Therefore, the inclusive fitness of the altruistic gene would exceed that of its selfish alternative as long as the cost to the altruist was less than one-half the benefit to the recipient. In general, a single altruistic act between individuals with genetic relationship r increases the

TABLE 13.1	*Probabilities of identity by descent between individuals with various degrees of relationship*
Relationship	Probability of identity by descent
Parent	0.50
Offspring	0.50
Full sibling	0.50
Half-sibling	0.25
Grandparent	0.25
Grandchild	0.25
Uncle or aunt	0.25
Nephew or niece	0.25
First cousin	0.125

fitness of the altruist gene as long as the cost *C* is less than the benefit *B* times the coefficient of relationship; that is, $C < Br$. The evolution of altruistic behaviors with high inclusive fitness is said to occur through the action of **kin selection.**

The maintenance of altruistic behavior by kin selection requires that such behavior be restricted to close relatives. Individuals of many species tend to associate in family groups, and limited dispersal keeps close relatives together. Moreover, individuals of many species can sense their degree of relationship to others by chemical or behavioral cues even when they have had no family experience.

Many behaviors have been identified as altruistic. One of the most conspicuous of these is alarm calling. By uttering an alarm in the presence of a predator, the individual warns others of danger but places itself at increased risk by calling attention to its position. Because ground squirrels are social creatures and are exposed to view in their montane meadow habitat (Figure 13.10), predators pose a considerable threat. Reducing this threat with alarm calls warning of danger is an important factor in their lives. Alarm calling in response to terrestrial predators has a cost. When predators approached groups of ground squirrels, at least one of whom gave an alarm call, 13 percent of calling individuals were attacked compared with only 5 percent of silent individuals.

Figure 13.10 Belding's ground squirrel at Tioga Pass, California. Courtesy of P. W. Sherman. From P. W. Sherman, *Science* 197:1246–1253 (1977).

Ground squirrels normally give alarm calls only when close relatives can hear them. Individuals sounding alarms in the presence of terrestrial predators, such as weasels, usually are adult and first-year females; males and yearling females do not exhibit the behavior (Figure 13.11). This makes sense from the standpoint of kin selection because males disperse widely from their place of birth, whereas females remain close to home. Moreover, females called more frequently when they had occupied the same territory for many years (and hence female relatives were likely to be nearby) or when relatives were known to be present in the population.

Parent–offspring conflict

Rather than passively accepting whatever their parents offer, most offspring actively solicit care. Young animals beg for food and solicit brooding. Eggs actively take up yolk from the ovarian tissues or bloodstream of the mother. For the most part, the interests of parent and offspring are compatible; when progeny thrive, so do their parents' genes. But when the selfish accumulation of resources by one offspring reduces the overall fecundity of the parents, parent and offspring can come into conflict.

Each act of parental care benefits the offspring by enhancing its survival but costs the parent by decreasing the number of other contemporary or future offspring. Resources allocated to one child cannot be delivered to others; prolonged care delays the birth of subsequent children; the risks of caring for today's children decrease the probability that parents will survive to rear tomorrow's.

From the standpoint of the parent, its offspring are genetically equivalent, and the parent should not exhibit preferences in the delivery of care to them. From the standpoint of the individual offspring, however, self has twice the value genetically as a sibling; a sibling shares only half of an individual's genes identically by descent. Therefore, when an individual possesses a genetic factor that increases parental care, that trait is favored as long as the cost, in terms of number of siblings, is less than twice the benefit to the individual.

As offspring develop and become more self-sufficient, the ratio of benefit to cost of continuing to care for them decreases (Figure 13.12). The cost of a particular parental act may change little over the age of the offspring, but as the young mature and become better able to care for themselves, the benefits of parental care dwindle. When the benefit : cost ratio drops below 1, the parent should cease to provide care to those offspring in favor of producing additional ones. Suppose, however, that a child has a gene that increases its solicitation of parental care. Because the inclusive fitness of the child discounts by one-half the cost of not delivering care to siblings, it "prefers" that parental care should continue until the benefit : cost ratio is one-half. Thus the period of time between the age at which B/C is 1 and the age at which it is 0.5 is one of conflict between parent and offspring. (Postadolescent humans may encounter this region of conflict if they express a wish to live with their parents after finishing college.)

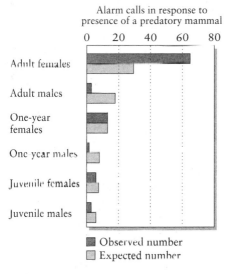

Alarm calls in response to presence of a predatory mammal

■ Observed number
□ Expected number

FIGURE 13.11 Expected and observed frequencies of alarm calling by sex and age classes of Belding's ground squirrels. Expected values were computed by assuming that individuals call in direct proportion to the number of times they are present when a predatory mammal appears. From P. W. Sherman, *Science* 197:1246–1253 (1977).

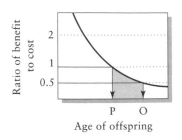

FIGURE 13.12 The ratio of benefit to cost of an act of parental care toward an offspring decreases with the offspring's age as it grows and becomes self-sufficient. Because all their offspring are of equal genetic value to parents, parents should shift care to succeeding offspring when the benefit : cost ratio falls below 1 (age P). Because siblings have a genetic relationship of one-half, however, inclusive-fitness arguments dictate that they should solicit parental care to the point where that benefit : cost ratio equals 0.5 (age O). This difference establishes a region of parent–offspring conflict (shaded). After R. L. Trivers, *Amer. Zool.* 14:249–264 (1972).

Insect societies

The complex societies of the termites, ants, bees, and wasps have presented a formidable challenge to evolutionary ecologists primarily because of the existence of nonreproductive castes. But before considering the evolutionary issues raised by insect societies, we shall discuss a bit of their natural history. There are several grades of sociality, the highest of which is **eusociality**. This grade is characterized by several adults living together in groups; by overlapping generations — that is, parents and offspring living together in the same colony; by cooperation in nest building and brood care; and by reproductive dominance, including the presence of sterile **castes**. Thus defined, eusociality is limited among insects to the termites (Isoptera) and the ants, bees, and wasps (Hymenoptera), although the elements of eusociality are present in one mammal, the naked mole rat.

The complex organization of the insect society is dominated by one or a few egg-laying queens. Nonreproductive progeny of the queen gather food and care for developing brothers and sisters, some of which become sexually mature, leave the colony to mate, and establish new colonies. Most insect societies are huge extended families.

Bee societies are organized simply; females are divided among a sterile worker caste and a reproductive caste that is produced seasonally. Whether an individual will become a sterile worker or a fertile reproductive is controlled by the quality of nutrition given the developing larva. In general, differentiation of sterile castes is stimulated by environmental (usually nutritional) factors. The development of sexual forms can be inhibited by substances produced by the queen and fed to the larvae. In bees, the worker caste represents an arrested stage in the development of the reproductive female, stopped short of sexual maturity.

Ant and termite colonies often have a continuous gradation of worker castes, ranging from very small individuals that are primarily responsible for the nutrition of the colony to larger individuals that are specialized morphologically to defend it against intruders (Figure 13.13). In termite colonies, workers are both male and female; in hymenopteran societies, workers are all females produced from fertilized eggs. Males, which develop from unfertilized eggs, appear in colonies only as reproductives (drones) that leave to seek mates.

From its distribution across taxonomic groups, it is clear that eusociality has evolved independently many times in the bees, wasps, and ants. It is less clear what route was followed. The most widely accepted sequence of evolutionary steps includes a lengthened period of parental care of the developing brood, the parents either guarding the nest or continuously provisioning the larvae in a manner similar to birds feeding their young. If a parent lived and continued to produce eggs after its first progeny emerged as adults, then the offspring would be in a position to help raise the subsequent brood. This overlapping of generations, which is not common among insects, and extensive parental care are necessary ingredients in the recipe for eusociality. Once progeny remain with their mother after they attain adulthood, the way is open to relinquishing their own reproductive function solely to support hers.

Most ideas about the evolution of eusociality have centered on the sterile caste as an alternative strategy of reproduction that maximizes the sterile indi-

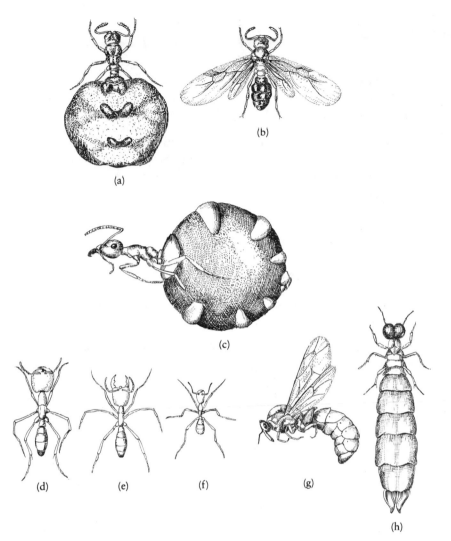

FIGURE 13.13 Castes of several species of ants: a virgin queen (a) and an old egg-laying queen (b) of the workerless social parasite *Anergates atralulus* of Europe; a replete of the honey ant *Myrmecosystus melliger* from Mexico (c); three sizes of blind workers (d, e, and f), a winged male (g), and a queen (h), blind and wingless, of the African visiting ant *Dorylus nigricans*. After W. M. Wheeler, *Social Life Among the Insects*. Harcourt, New York (1923); P. P. Grassé, *Traité de zoologie*. Vol. X, *Insects superieurs et hemipteroides,* Part II. Masson, Paris (1951).

vidual's contribution to the inclusive fitness of its genes. This could occur when close relatives compete for some resource that is essential to reproduction; specialization of function, such as between egg laying and food gathering/brood tending, may then minimize this competition.

Behavioral relationships in the social insects represent one extreme along a continuum of social organization from animals that live alone, except to breed, to those that aggregate in large groups organized by complex behavior. Regardless of its complexity, however, all behavior balances costs and benefits to the individual and to close relatives affected by the behavior. We see this theme over and over in the relationship of organisms to one another and to their environments.

🐟 *SUMMARY*

1. In most species, reproductive function is divided between two sexes. This creates conflicts over the allocation of resources between male and female sexual function, and it affects the interactions between individuals of the same and different sexes.

2. Separation of sexual function between individuals occurs infrequently among plants but commonly among animals. It is favored when sexual function imposes large fixed costs and the two sexes compete strongly.

3. The sex ratio in a population balances the contributions of genes to progeny through male and female function. In general, because the rarer sex is favored, most populations have equal numbers of males and females at evolutionary equilibrium.

4. In some parasitic wasps, males compete with siblings for matings, and the sex ratio is shifted in favor of females.

5. Mating systems may be monogamous (a lasting bond formed between one male and one female), polygamous (more than one mate, usually female, per individual), or promiscuous (mating at large within the population, without lasting pair bonds).

6. Polygyny arises when individuals of one sex can monopolize either resources or mates through intrasexual competition. In polygynous birds, some females gain greater fitness by joining an already mated male that holds a superior territory than by joining an unmated male on an inferior territory.

7. Promiscuity may arise when males contribute little, other than genes, to the number or survival of their offspring; this is the common condition in all plants and most animals.

8. When males compete among themselves for mates, females can choose among them. This leads to sexual selection of traits in males that indicate male fitness. These may appear as "handicaps" that only the more fit males can bear without encumbrance.

9. Selection imposed by behavioral interactions with members of one's family and with unrelated individuals within one's population provides the evolutionary basis for social behavior.

10. Territoriality is the defense of an object or area from intrusion by other individuals. Animals maintain territories when the resources they gain in doing so are rewarding and defensible.

11. Dominance hierarchies order social groups by rank, which is established by direct confrontation. Because rank is generally respected, dominance relationships reduce conflict within the group.

12. Living in large social groups may benefit the individual through better detection of, and defense against, predators and through more efficient location of food. Groups form to the extent that these advantages outweigh the ill effects of competition between group members.

13. Isolated acts of social behavior involve a donor and a recipient. When both benefit, the behavior is termed cooperation or mutualism; when the donor benefits at a cost to the recipient, the behavior is selfish; when the recipient benefits at a cost to the donor, the behavior is altruistic.

14. The presence of altruistic acts in populations has been explained variously in terms of kin selection. Kin selection arises because, when an individual interacts with a relative, it affects the fitness of a portion of its own genotype inherited directly from a common ancestor.

15. Inclusive fitness expresses the benefit (or cost) of a behavior to the donor plus the benefit or cost to the recipient, the latter adjusted by degree of relationship. For interactions between siblings, which are related by one-half, selection will favor any altruistic behavior whose cost to the donor is less than one-half the benefit it confers on the recipient.

16. Conflict may arise between parent and offspring over the optimal level of parental investment. All siblings are genetically equal in the eyes of their parents, but siblings are genetically related by only one-half. Therefore, individual offspring should prefer unequal parental investment in themselves, even when parental fitness is reduced as a result.

17. The social insects (termites, ants, wasps, and bees) represent extended family groups in which most offspring are retained in the colony as sterile workers, increasing their mother's fitness by rearing reproductive siblings.

◈ SUGGESTED READINGS

Alcock, J. 1980. Natural selection and the mating systems of solitary bees. *American Scientist* 68:146–153.

Bell, G. 1982. *The Masterpiece of Nature: The Evolution and Genetics of Sexuality.* Univ. of California Press, Berkeley.

Bradbury, J. W., & M. B. Andersson (eds.). 1987. *Sexual Selection. Testing the Alternatives.* Wiley-Interscience, New York.

Charnov, E. L. 1982. *The Theory of Sex Allocation.* Princeton Univ. Press, Princeton.

Emlen, S. T., and L. W. Oring. 1977. Ecology, sexual selection, and the evolution of mating systems. *Science* 197:215–223.

Honeycutt, R. L. 1992. Naked mole-rats. *American Scientist* 80:43–53.

Rohwer, S. 1982. The evolution of reliable and unreliable badges of fighting ability. *American Zoologist* 22:531–546.

Seeley, T. D. 1989. The honey bee colony as a superorganism. *American Scientist* 77:546–553.

Stevens, G., and R. Bellig (eds.). 1988. *The Evolution of Sex*. Harper & Row, San Francisco.

Trivers, R. L. 1974. Parent–offspring conflict. *American Zoologist* 14:249–264.

Trivers, R. L. 1985. *Social Evolution*. Benjamin/Cummings, Menlo Park, Calif.

Warner, R. R. 1984. Mating behavior and hermaphroditism in coral reef fishes. *American Scientist* 72:128–136.

Werren, J. H. 1987. Labile sex ratios in wasps and bees. *BioScience* 37:498–506.

Wilkinson, G. S. 1990. Food sharing in vampire bats. *Scientific American* 262:76–82.

Populations

14

POPULATION STRUCTURE

A population comprises the individuals of a species within a given area. A population's boundaries may be the natural ones imposed by geographic limits of suitable habitat, or they may be defined arbitrarily at the convenience of the investigator. In either case, the population has a spatial structure, which means that within its geographic boundary, individuals live primarily within patches of suitable habitat, and their abundances may vary with respect to food, predators, nest sites, and other ecological factors within that habitat.

Population structure provides a snapshot at an instant in time. Populations also exhibit dynamic behavior, continuously changing in time because of births, deaths, and movements of individuals. The regulation of these processes depends on the varied interactions of individuals with their environments and with each other. Moreover, evolution by natural selection and the regulation of both community structure and ecosystem function become apparent when cast in terms of population processes. Hence much of ecology focuses on the population level.

The population has temporal continuity because individuals that are alive at one time have descended from others that were alive at an earlier time. The population also has spatial continuity because the individuals in different parts of the distribution have common ancestors; generally, the greater the distance between two individuals, the more remote in time is their common ancestor. As a result, individuals within a population derive their genes from a common pool and thus share a common history of adaptation to the environment.

In this chapter, we consider the distribution of individuals within populations with respect to their spatial position, age, and genotype. The distribution and movements of individuals determine the spatial coherence of the population—the degree to which different parts of the population share the same dynamics or, alternatively, exhibit independence. As populations become fragmented when humans transform the landscape, the issues of population coherence and, particularly, the vulnerability of small subpopulations to extinction or drastic genetic change become important management and conservation concerns. Populations in natural areas may also be subdivided as a result of habitat heterogeneity. Most island populations pass through initial stages of colonization characterized by small numbers of individuals. Because chance events have greater impact on small populations than on large ones, ecologists must pay close attention to population size.

Within populations, individuals may vary with respect to gender, age, experience, social position, genotype, and the accumulated effects of accident and chance. Evolution occurs when genetic differences between individuals result in different birth rates or death rates, and hence different population dynamics. Sex ratio and age structure influence population dynamics because of sex-related and age-related variations in birth and death. These aspects of population structure are therefore an important part of our understanding of how populations change over time.

Habitat and the distribution of populations

The spatial structure of a population has three main properties: distribution, dispersion, and density. The **distribution** of a population describes its geographic and ecological range, which is determined primarily by the presence or absence of suitable habitat. Thus the distribution of the sugar maple in the United States and Canada stops abruptly to the east at the Atlantic Ocean, more gradually to the west with low precipitation, to the north with cold winters, and to the south with hot summers. Undoubtedly, much suitable habitat exists throughout the world, especially in Europe and Asia, but sugar maple evolved in North America and has not had the opportunity to colonize these areas.

The distributional limits imposed by barriers to long-distance dispersal reveal themselves dramatically when introduced species expand successfully into newly colonized regions. For example, in 1890 and 1891, 160 European starlings were released in Central Park, New York City. Within 60 years, the population had expanded to cover more than 3 million square miles and stretched from coast to coast (Figure 14.1). The habitats occupied by starlings in North America (principally city parks, farms, and pastures) resemble preferred habitats in Europe.

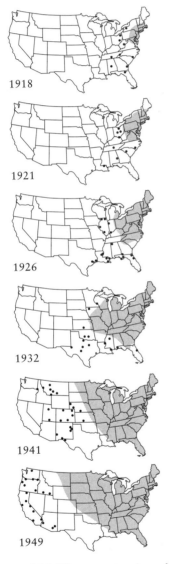

FIGURE 14.1 Western expansion of the range of the European starling (*Sturnus vulgaris*) in the United States. The shaded areas represent the breeding range; dots indicate records of birds in preceding winters. The population now inhabits the entire country. After B. Kessel, *Condor* 55:49–67 (1953).

Not all introductions succeed. Subtle ecological factors, intrinsic qualities of a species that determine whether it is a good or a poor colonizer, and chance determine the outcome of introductions. Even the prolific starling had been introduced at least three times to the United States—in Pennsylvania, Ohio, and Oregon—before it finally took hold in New York.

Within the geographic range of a population, individuals generally live only in suitable habitat. Sugar maples do not grow in marshes, on serpentine barrens, on newly formed sand dunes, in recently burned areas, or in a variety of other habitats that simply lie outside their range of ecological tolerance. Hence the geographic range of the sugar maple appears as a patchwork of occupied and unoccupied areas.

Climate, topography, soil chemistry, and soil texture exert progressively finer influence on the geographic distribution of the perennial shrub *Clematis fremontii* (Figure 14.2). Climate and perhaps interactions with ecologically related species restrict this species of *Clematis* to a small part of the midwestern United States. The variety *riehlii* occurs only in Jefferson County, Missouri.

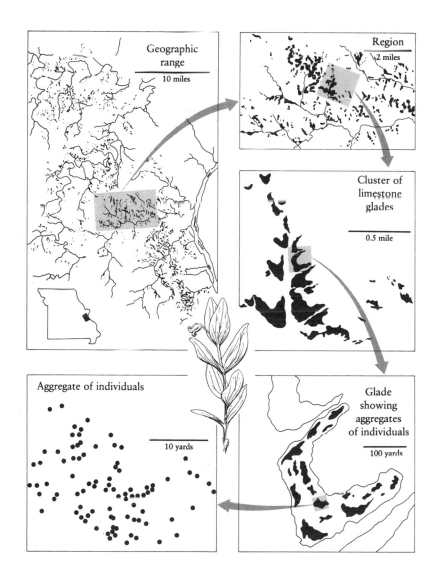

FIGURE 14.2 Hierarchy of patterns of geographic distribution of *Clematis fremontii*, variety *riehlii*, in Missouri. After R. O. Erickson, *Ann. Mo. Bot. Gard.* 32:416–460 (1945).

Within its geographic range, *Clematis fremontii* is restricted to dry, rocky soils on outcroppings of limestone. Small variations in relief and soil quality further confine the distribution of *Clematis* within each limestone glade to sites with suitable conditions of moisture, nutrients, and soil structure. Local aggregations occurring on each of these sites consist of many more or less evenly distributed individuals.

The horned lark (*Eremophila alpestris*), a small songbird of short grasslands, occurs more widely in North America than *Clematis fremontii,* but its local distribution reflects subtle variations in habitat; individuals even use some parts of their territories more than others (Figure 14.3).

The distributions of populations include all the areas they occupy during their life cycle. Thus the geographic range of salmon includes not only the rivers that serve as spawning grounds but also vast areas of the sea where individuals grow to maturity before making the long migration back to their birthplace. Many birds make annual migrations to warm climates during the winter months. Thus the range of the blackburnian warbler lies entirely within the northeastern United States and southeastern Canada during the summer but entirely within Central America and northern South America during the winter (Figure 14.4). The size and year-to-year variations of the population of such migratory species depend on the interactions of individuals with their environments throughout the year. Thus ecological changes in the Amazon Basin or central Africa may affect populations of migratory species that breed in the forests of eastern North America or Europe.

Dispersion

Within the distribution of a population, **dispersion** characterizes the spacing of individuals with respect to one another, forming patterns that vary from tight **aggregation** in discrete clumps to evenly **spaced** distribution (Figure 14.5). Between these extremes one finds **random** dispersion, in which individ-

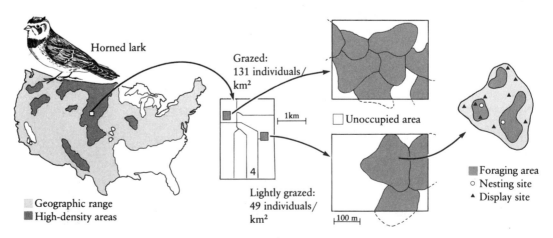

FIGURE 14.3 Hierarchy of patterns of distribution of the horned lark (*Eremophila alpestris*) during the breeding season. After J. A. Wiens, *Ecol. Monogr.* 43:237–270 (1973).

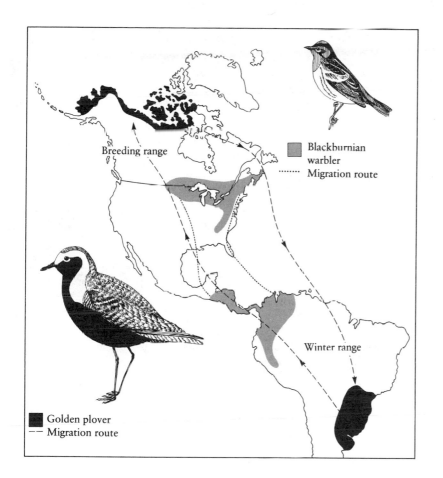

FIGURE 14.4 Breeding and wintering ranges of the golden plover (black) and blackburnian warbler (gray). Migration routes are indicated by dashed lines.

uals are distributed throughout a homogeneous area without regard to the presence of others.

Spaced and clumped distribution patterns derive from different processes. Even spacing—sometimes called **hyperdispersion**—commonly arises from direct interactions between individuals. The maintenance of a minimum distance between each individual and its nearest neighbor results in even spacing; for example, in their crowded colonies, seabirds place their nests just beyond their neighbors' reach (Figure 14.6). Plants situated too close to larger neighbors often suffer the effects of shading and root competition; as close neighbors die, the spacing of individuals becomes more even.

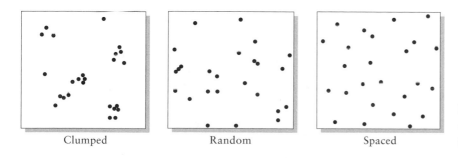

FIGURE 14.5 Diagram of the spatial distributions of individuals in clumped, random, and evenly spaced populations.

FIGURE 14.6 A nesting colony of cape gannets on an island off the coast of South Africa. The densely packed birds space their nests more or less evenly, the distance being determined by behavioral interactions between individuals. Along the entire length of the South African coast, however, seabird populations are clumped during the breeding season on a few offshore islands that offer suitable nesting sites.

Clumping, or aggregation, results from the social tendencies of individuals to form groups, from the clumped distributions of resources, and from the tendency of progeny to remain in the vicinity of their parents. Salamanders that prefer to live under logs exhibit clumped distributions corresponding to the patterns of fallen deadwood. Birds that travel in large flocks often aggregate to find safety in numbers. Finally, in the absence of social antagonism or mutual attraction, individuals may distribute themselves at random, without regard to the positions of other individuals in the population. The three general patterns of dispersion may be distinguished by various statistical tests, one of which is illustrated in Box 14.1.

BOX 14.1	*The poisson distribution*

We may compare patterns of dispersion to the values expected of random spatial distributions. One of the most important of these is the Poisson distribution, which describes the expected number of individuals placed at random within a sample plot.

Suppose we divide a study area into sample plots, or other units of equal area, and count the number of individuals in each plot. Some of the plots will have a single individual, some many, and others none, depending on the density of the population and the size of the plots. Consider, for example, the number of red mites counted on each of 150 apple leaves: 70 of the leaves had no mites, 38 had 1, 17 had 2, and so on, as shown in the following table.

Mites per leaf	Number of mites observed	Poisson distribution	Number of leaves expected
0	70	0.3177	47.65
1	38	0.3643	54.64
2	17	0.2089	31.33
3	10	0.0798	11.98
4	9	0.0229	3.43
5	3	0.0052	0.79
6	2	0.0010	0.15
7	1	0.0002	0.02
8 or more	0	0.0000	0.00
Total	150	1.0000	149.99

Source: R. W. Poole, *An Introduction to Quantitative Ecology.* McGraw-Hill, New York (1974).

How would a random distribution of mites over the leaves appear? Suppose that each mite in the population had an equal probability of finding itself on each of the leaves (a probability of $1/150$, or 0.067, per leaf) irrespective of the number of other mites present. When mites are distributed according to a random process such as this, the expected proportion of leaves (P) with a given number of mites (x) obeys the Poisson distribution

$$P(x) = \frac{M^x c^{-M}}{x!}$$

(*Continued*)

BOX 14.1	*Continued*

where M is the mean number of individuals per sampling unit and $x!$ is the factorial of x (for instance, $5! = 5 \times 4 \times 3 \times 2 \times 1$). In the example, 172 mites were recorded from 150 leaves, so $M = 172/150 = 1.1467$ mites per leaf. To calculate the expected number of leaves with, say, 3 mites, we substitute $M = 1.1467$ and $x = 3$ into the equation for $P(x)$; this gives a value of 0.080, or 8 percent of the 150 leaves (12 leaves). As you can see, fewer leaves than expected had 1 or 2 mites and more leaves than expected had 0 and 4 or more mites, indicating a clumped distribution.

Population density

Biologists define the number of individuals per unit area as the **density** of the population. Density has two important applications in ecology. The first arises from the practical difficulty of estimating the sizes of entire populations distributed over large areas. In such situations, the investigator assumes that the density and trends in density that are observed within a small sampling area mirror those of the population as a whole. The local population size can be measured by directly counting all individuals, by sampling parts of the population, or by using any of several indirect methods, one of which is described in Box 14.2.

The second application arises from the fact that density indicates the ability of the environment to support the particular population under study. In general, individuals are most numerous where resources are most abundant. Thus density provides information about the relationship of a population to the environment, and changes in density reflect changing local conditions. In this respect, density provides a better basis for comparing the ecological relationships of two populations than do their total sizes.

Spatial and temporal variations in population density

Because densities of populations change over time and space, no population has a single structure; one's perception of a population depends on where and when one looks. Long-term records of the population of the chinch bug (*Blissus leucopterus*) in Illinois illustrate this point (Figure 14.7). These records exist because chinch bugs damage cereal crops; the Illinois State Entomologist's Office and later the State Natural History Survey Division determined the importance of monitoring the population, which they estimated from county reports of crop damage attributable to chinch bugs.

Consider the numbers involved. During 1873, when the bugs damaged crops severely over most of the state, ballpark estimates of the population indicated an average density of 1000 chinch bugs per square meter over an area

BOX 14.2	*The Lincoln index to population size*

When it is impossible to count entire populations because of their large size or area of distribution, biologists estimate densities within smaller subareas to serve as indices to population size. When individuals do not move — plants and sessile marine invertebrates, for example — one can estimate their densities by counting individuals within plots of known area. When individuals move between plots faster than the investigator can count the number within a plot, an alternative is the **mark-recapture** method.

Suppose that M specially marked individuals are released into a population of size N. If the marked individuals mix thoroughly with the unmarked ones, a subsequent sample of the population should contain marked individuals in the ratio M to N. Therefore, in a sample of size n from the population, the number of marked individuals recaptured (x) should be

$$x = \frac{nM}{N}$$

Because we know the quantities x, n, and M, we can rearrange this equation to solve for population size:

$$N = \frac{nM}{x}$$

This expression is known as the Lincoln index to population size.

By way of illustration, suppose we capture 25 sunfish in a small pond, mark them by a small, distinctive notch in the tail fin, and then put them back into the pond. Two days later, by which time we suppose the marked fish will have mixed thoroughly with the unmarked ones, we take another sample of 35 fish, of which 14 are marked. The Lincoln index estimate of the population size is $35 \times 25/14 = 62.5$ individuals.

of 300,000 square kilometers, or a total of 3×10^{14} pests (300 trillion, more or less). By contrast, farmers reported hardly any damage in 1870 and 1875. Severe outbreaks were often confined to small portions of the state, sometimes in the north, sometimes in the south. However, spatial and temporal continuity reveals itself in Figure 14.7 in the waxing and waning of infestations.

Temporal variation in population dynamics within a locality may also leaves its mark in the **age structure** of a population — that is, the relative frequencies of individuals of each age. The age composition of samples from the Lake Erie commercial whitefish catch for the years 1945–1951 shows that during 1947, 1948, and 1949, most of the individuals caught belonged to the 1944 year class (Figure 14.8). Biologists estimate the age of fish from growth rings on the scales; their data showed that 1944 was an excellent year for

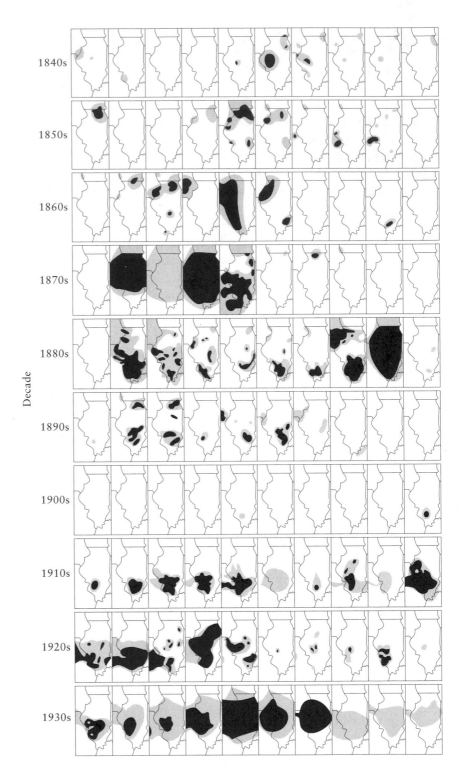

FIGURE 14.7 Distribution of crop damage caused by the chinch bug (*Blissus leucopterus*) in Illinois between 1840 and 1939. From V. E. Shelford and W. P. Flint, *Ecology* 24:435–455 (1943).

spawning and recruitment, particularly compared to the several years that followed.

Variation in the annual recruitment of individuals into a population also appears in the age structure of stands of trees. Age may be estimated from the growth rings in the woody tissue of the trunk of the tree, one ring being added each year under normal circumstances (Figure 14.9). The pattern in a virgin stand of timber surveyed near Hearts Content, Pennsylvania, in 1928 shows that most species established themselves sporadically over the nearly 400-year span of the record (Figure 14.10). Many white pines became established between 1650 and 1710, undoubtedly following a major disturbance, perhaps associated with the serious drought and fire year of 1644. Fire can open a forest enough to permit the establishment of white pine seedlings, which do not tolerate deep shade. In contrast, beech—a species that can become established under the canopy of a closed forest—exhibited a relatively even age distribution.

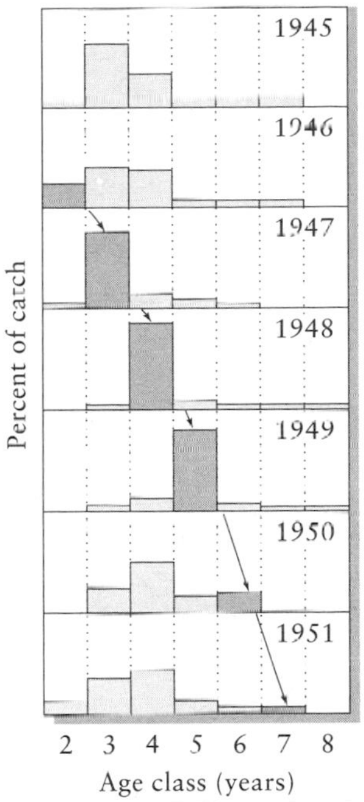

Dispersal and the spatial coherence of populations

The movement of individuals between populations and their movement between local areas within populations influence local population change. Biologists refer to such movements within populations as **dispersal** and to movements between populations as **emigration** (leaving) and **immigration** (entering).

Dispersal, particularly over long distances, is difficult to measure because detecting it requires the recapture of marked individuals. Most attempts to estimate dispersal come from studies of population genetics in which investigators wish to determine the genetic structure of a population and to estimate both the effective population size for evolutionary processes and the amount of gene flow between populations. Indeed, the first attempts to measure dispersal in natural populations involved the movement away from a release point of fruit flies (*Drosophila*) distinguished by a visible mutation.

FIGURE 14.8 Age composition of samples from the commercial whitefish catch from Lake Erie between 1945 and 1951. Fish spawned in 1944 are highlighted by solid bars. From G. H. Lawler, *J. Fish. Res. Bd. Canada* 22:1197–1227 (1965).

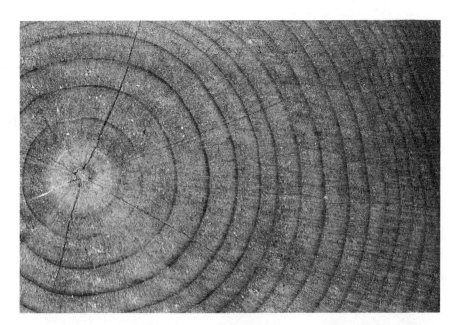

FIGURE 14.9 Cross section of the trunk of a Monterey pine, showing annual rings of winter (dark) and summer (light) growth.

Population biologists describe dispersal by a number of mathematical indices, each of which reflects different assumptions about the structure of the population. The simplest model describes dispersal as random movement through a homogenous environment, analogous to Brownian motion in physics. With respect to a fixed point of origin, some random movements take an individual farther away and some bring it closer, although on average, distance tends to increase with time. The position of any individual at any moment is the sum of many random increments of distance; the probability of finding an individual at a given distance from the release site is described by the normal frequency distribution (Figure 14.11). This is a bell-shaped curve whose peak coincides with the point of origin (that is, distance = 0) and whose breadth is characterized by a single variable, the standard deviation s. The value of s provides a convenient index to dispersal distance; s_t is the dispersal distance estimated over the average life span of an individual, t (Box 14.3). **Neighborhood size** is defined as the number of individuals included within a circle of radius $2s_t$; it indicates the subset of other individuals that a member of a population can potentially interact with over its life span.

Recaptures of small songbirds marked with leg bands as nestlings indicated that dispersal distances (s_t) for 8 species varied between 344 and 1681 meters, densities varied between 16 and 480 individuals per square kilometer, and neighborhood size varied between 151 and 7679 individuals. For three populations of the land snail *Cepaea nemoralis,* dispersal distances (s_1) varied between 5.5 and 10 meters after 1 year, but because individuals have long life spans, neighborhood sizes were quite large: 1800 to 7600 individuals. Note that the neighborhood sizes of these slowly dispersing snails are on the same order as those of small birds. Mark-recapture data on the rusty lizard (*Sceloporus olivaceous*) near Austin, Texas, revealed s_t to be 89 meters and neighborhood size to be between 225 and 270 individuals.

In most studies of dispersal, the principal source of information is the movement of individuals away from a point of release. Other kinds of observations are also pertinent, however, particularly the spread of introduced populations, which can occur only by the movements of individuals. The Euro-

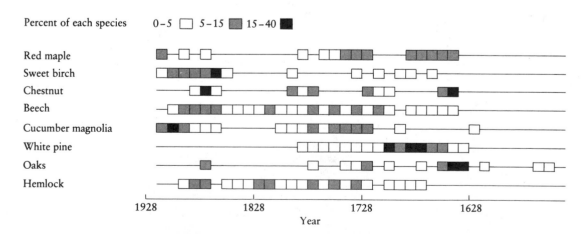

FIGURE 14.10 Age distribution of forest trees near Hearts Content, Pennsylvania, in 1928. After A. F. Hough and R. D. Forbes, *Ecol. Monogr.* 13:299–320 (1943).

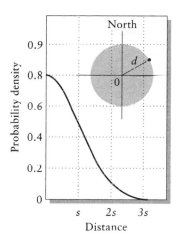 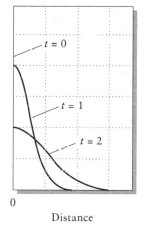

FIGURE 14.11 The probability density of individuals around a release point assumes a normal distribution when movement occurs at random. The distance dispersed may be characterized by the standard deviation of the curve (*s*), which increases in direct proportion to time *t* since release.

BOX 14.3   *Dispersal distance and neighborhood size*

The standard deviation (*s*) of a normal distribution is the square root of its variance (*s*²). The variance is estimated by the expression

$$s^2 = \frac{1}{N} \sum_{i=1}^{N} d_i^2$$

where d_i is distance of the *i*th individual from the point of origin and N is the number of individuals in the sample.

In one study, investigators measured the rate of dispersal of fruit flies (*Drosophila pseudoobscura*) in grasslands in central Colorado. Flies were marked with minute fluorescent dust particles; individuals were caught in traps placed at regular intervals in eight directions of the compass at distances of up to 351 m from the release point. After 1 day of dispersal, the estimated values of *s* were 139 and 171 m at two localities.

A felicitous property of *s* as a measure of dispersal is that variances (*s*²) of the distances add over time. Thus when fruit flies disperse distance s_1 in one day, the variance in distances after 2 days will be $2s_1^2$, and after *t* days it will be ts_1^2. Accordingly, if adult fruit flies live an average of 23 days and the average dispersal parameter (*s*) per day is 151 m, s_{23} is the square root of 23 × 151², or 724 m.

Neighborhood size within a population is the number of individuals within a circle whose radius is twice the dispersal distance (*s*) within the average reproductive life span of the individual. Neighborhood size provides an index to the number of individuals in a population that are potentially coupled by strong interactions. In the case of the fruit flies we just described, densities were about 0.38 flies per 100 m². The area of a circle of radius 2*s* is $4\pi s^2$, which would have included an estimated 26,387 individual flies.

pean starling spread almost 4000 kilometers across the United States in 60 years, at an average rate of about 67 kilometers per year. This figure greatly exceeds the estimates just reported for dispersal distance within populations of small songbirds. Adult starlings tend to nest in the same area year after year, so most dispersal is accomplished by young birds. Furthermore, the westward spread of the starling was characterized by frequent sightings of nonbreeders before breeding populations were established in an area. It is less likely that such long-distance movements of juveniles would be detected within an established population.

Whereas a few long-distance movements may expand the border of a population, they may have less effect on the dynamics of widely separated but established subpopulations. Given the propensity of populations to increase, a small number of colonizing individuals can grow to a large population in a short period. But those same individuals immigrating to an established population may have a negligible impact.

The life table

The accurate projection of change in population size requires knowledge of the number of individuals of each age and their probabilities of survival and rates of fecundity. These statistics, which collectively are known as the **life table,** determine the addition and removal of individuals from a local population (in the absence of immigration and emigration). Because it is hard to ascertain paternity in many species, life tables are usually based entirely on females. For some populations with highly skewed sex ratios or unusual mating systems, this can pose difficulties, but in most cases a female-based life table provides a workable population model.

Age is designated in a life table by the symbol x, and **age-specific** variables are indicated by the subscript x. When reproduction occurs during a brief breeding season each year, each age class is composed of a discrete group of individuals born at approximately the same time. When reproduction is continuous, as it is in the human population, each age class x is designated arbitrarily as comprising individuals between ages $x - \frac{1}{2}$ and $x + \frac{1}{2}$.

The fecundity of females, expressed in terms of female offspring produced per breeding season or age interval, is designated by b_x (think of b for "births"). Life tables portray the statistics of mortality in several ways. The fundamental measure is the probability of **survival** (s_x) between ages x and $x + 1$. The probabilities of survival over many age intervals are summarized by the **survivorship** to age x (l_x), which is the probability that a newborn individual will be alive at age x. Because by definition all newborn individuals are alive at age 0, $l_0 = 1$. The proportion of newborns alive at age 1 is the probability of surviving from age 0 to age 1, hence $l_1 = s_0$. Similarly, $l_2 = s_0 s_1$ and, by extension, $l_x = s_0 s_1 s_2 \cdots s_{x-1}$. An additional measure is sometimes included in the life table: the **expectation of further life** (e_x) of an individual of age x.

The life table variables are summarized in Table 14.1 and illustrated in Table 14.2 by data for the annual meadow grass *Poa annua*. This life table follows the survival and fecundity of a planting of the grass under experimental conditions for 2 years, by which time the last individual had died. Age is tabu-

| TABLE 14.1 | *Summary of life table variables* |

l_x	Survival of newborn individuals to age x
b_x	Fecundity at age x
m_x	Proportion of individuals of age x dying by age $x + 1$
s_x	Proportion of individuals of age x surviving to age $x + 1$
e_x	Expectation of further life of individuals of age x
k_x	$-\log_e s_x$, the exponential mortality rate between age x and $x - 1$

lated in units of 3 months. Because *Poa* is hermaphroditic, sexes are not distinguished. Of 843 plants alive at time 0 (germination), 722, or 85.7 percent, were alive at 3 months ($t = 1$). Hence s_0 and l_1 both equal 0.857. The life table shows that the probability of dying increased with increasing age; hence the expectation of further life decreased with increasing age. Fecundity rose to a peak of 620 seeds per 3-month period at 6 months of age and then declined.

Estimates of survival in natural populations

One can estimate survival with varying degrees of reliability from four kinds of information: (1) the survival of individuals to a particular age (survivorship), (2) the survival of individuals in each age class from one time period to the next,

| TABLE 14.2 | *Life table of the grass* Poa annua |

Age $(x)^*$	Number alive	Survivorship (l_x)	Mortality rate (m_x)	Survival rate (s_x)	Expectation of life (e_x)	Fecundity (b_x)
0	843	1.000	0.143	0.857	2.114	0
1	722	0.857	0.271	0.729	1.467	300
2	527	0.625	0.400	0.600	1.011	620
3	316	0.375	0.544	0.456	0.685	430
4	144	0.171	0.626	0.374	0.503	210
5	54	0.064	0.722	0.278	0.344	60
6	15	0.018	0.800	0.200	0.222	30
7	3	0.004	1.000	0.000	0.000	10
8	0	0.000				

*Number of 3-month periods; in other words, 3 = 9 months.

Source: M. Begon and M. Mortimer, *Population Ecology* (2nd ed.). Sinauer, Sunderland, Mass. (1986). After data of R. Law.

(3) the ages at death within a population, and (4) the age structure of a population. Data of the first kind form the basis of the **cohort life table,** or **dynamic life table,** which follows the fate of a group of individuals born at the same time from birth to the death of the last individual. The life table for *Poa annua* (Table 14.2) is one of this type. This method is readily applied to populations of plants and sessile animals in which marked individuals can be continually resampled over the course of their life spans. Herein lies one of its disadvantages, however: it can take a long time to collect the data (particularly if the subject of the study is redwood trees!). It is also difficult to apply to highly mobile animals.

Another method, employing a **static life table** or **time-specific life table,** sidesteps the time problem by considering the survival of individuals of known age during a single time interval. The investigator estimates each age-specific survival value independently for each age class of the population during the same period. Of course, to apply this technique it is necessary to know the ages of individuals (estimated by growth rings, tooth wear, or other reliable index).

The distribution of ages at death was used to construct a life table for Dall mountain sheep in Mount McKinley (now Denali) National Park, Alaska. The size of the horns, which grow continuously during the lifetime of the sheep (Figure 14.12), provided estimates of the age at death. Of 608 skeletal remains, 121 were judged to have been less than 1 year old at death, 7 between 1 and 2 years, 8 between 2 and 3 years, and so on, as shown in Table 14.3. Now,

FIGURE 14.12 A group of Dall mountain sheep in Alaska. The size of the horns increases with age. Courtesy of the American Museum of Natural History.

TABLE 14.3	Life table for the Dall mountain sheep constructed from the age at death of 608 sheep in Denali National Park		
Age interval (years)	Number dying during age interval	Number surviving at beginning of age interval	Number surviving as a fraction of newborns (l_x)
0–1	121	608	1.000
1–2	7	487	0.801
2–3	8	480	0.789
3–4	7	472	0.776
4–5	18	465	0.764
5–6	28	447	0.734
6–7	29	419	0.688
7–8	42	390	0.640
8–9	80	348	0.571
9–10	114	268	0.439
10–11	95	154	0.252
11–12	55	59	0.096
12–13	2	4	0.006
13–14	2	2	0.003
14–15	0	0	0.000

Source: Based on data of O. Murie, *The Wolves of Mt. McKinley*. U.S. Department of the Interior, National Park Service, Fauna Series No. 5, Washington, D.C. (1944); quoted by E. S. Deevey, Jr., *Quarterly Review of Biology* 22:283–314 (1947).

we reason, all 608 dead sheep must have been alive at birth; all but the 121 that died during the first year must have been alive at the age of 1 year (608 − 121 = 487), all but 128 (the 121 dying during the first year and the 7 dying during the second) must have been alive at the end of the second year (608 − 128 = 480), and so on, until the oldest sheep had died during their fourteenth year. Survival (l_x; the rightmost column in Table 14.3) was calculated by converting the number of sheep alive at the beginning of each interval to a decimal fraction of those alive at birth. Thus, for example, the 390 sheep alive at the beginning of the seventh year represented 64.0 percent (decimal fraction 0.640) of the original newborns in the sample.

The relationship of life tables to the dynamics of populations will be explained in more detail in the following chapters. Population structure emphasizes the complexity of the relationship of individuals to their environments. Heterogeneity of the conditions and resources required for existence and reproduction appear as variations in the distribution and density of individuals within a population. Dispersal evens out this variation to some degree and couples the dynamics of different parts of the population. Thus changes in the density of a population within a small area reflect not only changes in the local conditions, as they affect birth and death rates, but also changes in other localities conveyed by the movement of individuals.

❧ SUMMARY

1. Ecologists define populations either by their geographic limits of distribution or by an arbitrary boundary around a smaller area of intensive study. The primary objectives of population studies are to describe the spatial structure and temporal behavior of populations and to gather data relevant to understanding their dynamics. Populations achieve genetic cohesion through the common ancestry of individuals and achieve spatial cohesion through the movements of individuals.

2. The spatial structure of a population may be defined in terms of its density and its pattern of dispersion. Density is the number of individuals per unit of area; it may be estimated either directly by counts of individuals within plots of known size or indirectly by such methods as the mark-recapture technique.

3. Dispersion describes the relationship of individuals to the position of others in the population. Clumped distributions may result from independent aggregation of individuals in suitable habitats, from spatial proximity of direct descendants, or from tendencies to form social groups. Evenly spaced distributions may result from antagonistic interactions between individuals.

4. The age structure of a population often indicates temporal heterogeneity of recruitment of individuals. For example, certain seedlings tend to become established in forests primarily following such a major disturbance as a fire, drought, or storm. Thus population processes may be sporadic rather than uniform over time.

5. Movement within populations may be characterized by the variance in distances of individuals from their initial position, which may be a point of release in an experimental study. Neighborhood size—the number of individuals within a circle of area 4π times the variance of lifetime dispersal distance—provides an index to the number of potentially interacting individuals within a population. For several species, neighborhood sizes have been estimated as on the order of 10^2 to 10^4 individuals.

6. The life table displays the fecundities (b_x) and probabilities of survival (s_x) of individuals by age class (x). These are the principal variables in models of the dynamics of populations. Survival rates may be estimated from the fates of individuals born at the same time (cohort or dynamic life table), the survival of individuals of known age during a single time period (static or time-specific life table), the age structure of a population at a particular time, or the age distribution of deaths.

Begon, M., and M. Mortimer. 1986. *Population Ecology* (2nd ed.). Sinauer, Sunderland, Massachusetts.

Caughley, G. 1977. *Analysis of Vertebrate Populations*. Wiley, New York and London.

Cook, R. E. 1983. Clonal plant populations. *American Scientist* 71:244–253.

Deevey, E. S., Jr. 1947. Life tables for natural populations of animals. *Quarterly Review of Biology* 22:283–314.

Harper, J. L. 1977. *Population Biology of Plants*. Academic Press, New York and London.

Pielou, E. C. 1977. *Mathematical Ecology*. Wiley, New York and London.

Southwood, T. R. E. 1978. *Ecological Methods* (2nd ed.). Chapman and Hall, London; Wiley, New York.

Varley, G. C., G. R. Gradwell, and M. P. Hassell. 1975. *Insect Population Ecology*. Blackwell Scientific Publications, Oxford.

15

POPULATION GROWTH AND REGULATION

The immense capacity of populations to increase cannot be better illustrated than by the human population, which has grown prolifically, at times doubling each quarter-century. Ever since humankind began to understand the rapid increase in its numbers, population growth has caused concern. This concern led to the development of mathematical techniques to predict the growth of populations — the discipline of **demography** — and to the intensive study of natural and laboratory populations to determine the mechanisms of population regulation. The fruits of this labor have included a general understanding of the effects of crowding on birth and death processes within populations.

In this chapter, we shall explore the nature of population growth and the factors that limit population size, showing how their effect increases with the increasing density of the population in such a way as to bring growth under control.

Exponential growth

A population experiencing **exponential growth** increases in proportion to its size, just as a bank account earns interest more rapidly with a larger principal. Increase in numbers depends on reproduction by individuals in the population. Therefore, a population that grows at a constant exponential rate gains individuals ever faster as the population increases. For example, a 10 percent annual rate of increase adds 10 individuals in 1 year to a population of 100, but 100 new individuals to a population of 1000. Allowed to continue at this rate, the population would rapidly climb toward infinity. Charles Darwin wrote, in *On the Origin of Species*, "There is no exception to the rule that every organic being naturally increases at so high a rate, that, if not destroyed, the earth would soon be covered by the progeny of a single pair." To make his case as forcefully as possible, Darwin offered a conservative example:

> *The elephant is reckoned the slowest breeder of all known animals, and I have taken some pains to estimate its probable minimum rate of natural increase; it will be safest to assume that it begins breeding when thirty years old, and goes on breeding till ninety years old, bringing forth six young in the interval, and surviving till one hundred years old; if this be so, after a period of from 740 to 750 years there would be nearly nineteen million elephants alive, descended from the first pair.*

Because baby elephants grow up, mature, and themselves have babies, the elephant population grows exponentially.

Such a population increases according to

$$N(t) = N(0)\ e^{rt}$$

where $N(t)$ is the number of individuals in the population after t units of time, $N(0)$ is the initial population size ($t = 0$), and r is the exponential growth rate. The constant e is the base of the natural logarithms; it has a value of approximately 2.72. Exponential growth results in a continuously accelerating curve of increase (or decelerating curve of decrease) whose slope varies directly with the size of the population (Figure 15.1).

The rate at which individuals are added to a population undergoing exponential growth is the derivative of the exponential equation; that is,

$$\frac{dN}{dt} = rN$$

This equation expresses two principles. First, the rate of increase (dN/dt) varies in direct proportion to the size of the population (N). Second, the exponential growth rate (r) expresses population increase (or decrease) on an individual basis. In words, this equation could read: (the rate of change in total population size) = (the contribution of each individual to population growth) × (the number of individuals in the population at the present time).

The individual, or **per capita**, contribution to population growth represents the difference between the birth rate (b) and the death rate (d) calculated on a

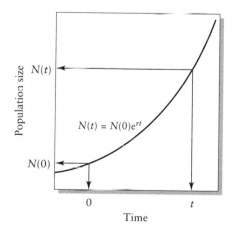

FIGURE 15.1 The curve of exponential growth for a population growing at rate r between time 0 and t. During this period, the number of individuals increases from $N(0)$ to $N(t)$.

per capita basis. That is, $r = b - d$. Rates of birth and death are abstractions having little meaning for the individual. An elephant dies only once, so it can't have a rate of death. Babies are produced in discrete litters separated by intervals required for gestation, not at constant rates, such as 0.05 offspring per day. But when births and deaths are averaged over the population as a whole, they take on meaning as rates of demographic events in the population. If 1000 individuals produced 10,000 progeny in a year, we could reasonably assign a per capita birth rate of 10 per year and assume that a population of 1 million would produce 10 million progeny — still 10 per individual — under the same conditions. If, of the groundhogs alive on their day in one year, only half survived to February 2 of the next year, we would ascribe a death rate of 50 percent per year to the population, even though some of the groundhogs had died "completely" and others hadn't died at all.

Geometric growth

The human population grows continuously because babies are born and added to the population at all seasons of the year. But this situation is unusual in natural populations, most of which restrict reproduction to a particular time of year. Accordingly, the population grows during the breeding season and then declines between one breeding season and the next (Figure 15.2). In the case of the California quail, the number of individuals doubles or triples each summer as adults produce their broods of chicks, but then it dwindles by nearly the same amount during the fall, winter, and spring. Within each year, population growth rate varies tremendously depending on seasonal changes in the balance of birth and death processes. If we were interested in projecting population growth, it would be pointless to compare numbers in August, recently augmented by the chicks born that year, with those in May, after winter has taken its toll. One must count individuals at the same time each year, so that all counts are separated by the same cycle of birth and death processes. Such increase (or decrease) over discrete intervals is referred to as **geometric growth.**

The rate of geometric growth is most conveniently expressed as the ratio of the population in one year to that in the preceding year, or other time interval. Demographers have assigned the symbol λ to this ratio; hence $\lambda = N(t + 1)/N(t)$, where t is an arbitrary time. This definition of λ can be rearranged to provide a formula for projecting the size of the population through a single time interval:

FIGURE 15.2 Growth of populations with discrete (seasonal) reproduction. (a) Rabbits in a subalpine population in New South Wales, Australia. (b) California quail. The curves are shaded differently to show that each represents the population of individuals born (or hatched) in different years. After K. Myers, in P. J. den Boer and G. R. Gradwell (eds.), *Dynamics of Populations.* Centre Agric. Publ. Doc., Wageningen, The Netherlands (1970), pp. 478–506; J. T. Emlen, Jr., *J. Wildl. Mgmt.* 4:2–99 (1940).

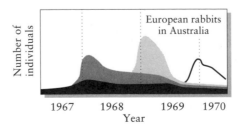

(a)

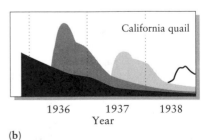

(b)

$$N(t + 1) = N(t)\lambda$$

To project the growth of a population over many time intervals, we multiply the original number by the geometric growth rate λ once for each interval of time passed. Hence $N(1) = N(0)\lambda$, $N(2) = N(0)\lambda^2$, $N(3) = N(0)\lambda^3$, and

$$N(t) = N(0)\lambda^t$$

Note that this equation for geometric growth is identical to that for exponential growth except that λ takes the place of e^r which equals the amount of exponential growth accomplished in one time period. Because of this relationship, curves depicting the two models of growth can be superimposed on each other (Figure 15.3), and there is a direct correspondence between the values of λ and r. When a population's size remains constant, $r = 0$ and $\lambda = 1$ ($r = \log_e \lambda$, and $\log_e 1 = 0$). Decreasing populations have negative exponential growth rates and geometric growth rates less than 1 (but greater than 0; a real population cannot have a negative number of individuals). Increasing populations have positive exponential growth rates and geometric growth rates greater than 1.

Age structure and population growth rate

When birth rates and death rates have the same values for all members of the population, the total population size (N) provides the proper basis for projecting increase. But when birth and death rates vary with respect to age, the contribution of younger and older individuals to population growth must be figured separately; two populations having identical birth and death rates at corresponding ages, but different **age structures** (proportions of individuals in each age class), will grow at different rates, at least for a while. A population composed wholly of prereproductive adolescents and oldsters too feeble to breed cannot increase until young individuals reach reproductive age. This represents an extreme case, but smaller variations in age distribution can also profoundly influence population growth rate.

When age-specific birth and survival rates remain unchanged for a sufficient period, a population will assume a **stable age distribution**. Under such conditions, each age class in the population grows or declines at the same rate and, therefore, so does the total size of the population. A little pencil-and-paper figuring with a hypothetical population will demonstrate this result. Imagine a population of 100 individuals having the characteristics shown in Table 15.1. Because all 3-year-olds die, there are no 4-year-olds in the population; the newborns have a fecundity of zero, as seems biologically reasonable. The population is counted at the end of the reproductive season. Projection of this population into the future is illustrated in Table 15.2.

In this example, the population at first grows very erratically, with λ fluctuating between 1.05 and 1.69. Eventually λ settles down to a constant value of 1.49. At this point, the population has achieved a stable age distribution; the percentage of individuals in each age class after eight time intervals appears in the rightmost column of Table 15.2. Even by the end of the fourth interval, the

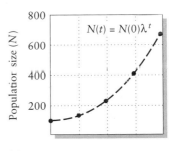

(a)

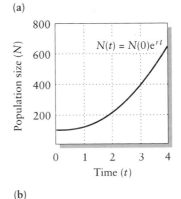

(b)

FIGURE 15.3 Increase in the number of individuals in populations undergoing geometric growth (a) and exponential growth (b) at equivalent rates ($\lambda = 1.6$, $r = 0.47$).

TABLE 15.1 *Life table for a hypothetical population of 100 individuals*

Age (x)	Survival (s_x)	Fecundity (b_x)	Number of individuals (n_x)
0	0.5	0	20
1	0.8	1	10
2	0.5	2	40
3	0.0	2	30

population has approached its stable age distribution, with proportions in each of the age classes of 62.7, 21.8, 10.0, and 5.4 percent. Under stable age conditions, each age class grows at the same rate (Figure 15.4).

The stable age distribution and growth rate depend on birth and survival values. Any change in the life table alters the stable age distribution and results in a new rate of population growth. Consider the example in Table 15.3, in which we reduce the survival and fecundity of our hypothetical population to the point where the population declines. As often happens in a decreasing population, the distribution of individuals shifts toward the older age classes. In this example, the shift to the new stable age distribution involves small changes and occurs quickly, after which the population achieves a growth rate of $\lambda =$

TABLE 15.2 *Projection of population age classes and total size through time, according to survival and fecundity values in Table 15.1**

	0	1	2	3	4	5	6	7	8	Percent
n_0	20	74	69	132	175	274	399	599	889	63.4
n_1	10	10	37	34	61	87	137	199	299	21.3
n_2	40	8	8	30	28	53	70	110	160	11.4
n_3	30	20	4	4	15	14	26	35	55	3.9
N	100	112	118	200	279	428	632	943	1403	100
λ	1.12	1.05	1.69	1.40	1.53	1.48	1.49	1.49		

*The population was projected by multiplying the number of individuals in an age class by the survival to obtain the number in the next older age class in the next time period. Thus $n_x(t) = n_{x-1}(t-1)s_x$. Then the number of individuals in each age class was multiplied by its fecundity to obtain the number of newborns. Thus $n_0(t) = \sum n_x(t)b_x$.

0.82. Early on, however, some age classes briefly increase (n_2 between $t = 0$ and $t = 1$) as the age distribution of the population readjusts to the new life table values. The effect of population growth rate on the age structure stands out in comparisons of human populations with stable and with growing populations (Figure 15.5). Rapid growth leads to a bottom-heavy age structure with large proportions of young individuals.

The intrinsic rate of increase

The **intrinsic rate of increase,** usually indicated by r_m, is the exponential or geometric rate of increase assumed by a population with a stable age distribution. When changing environmental conditions alter life table values, the age structure of the population continuously readjusts to the new schedule of birth and death rates. In practice, then, populations rarely achieve stable age distributions and therefore rarely grow at their intrinsic rates of increase. Because the actual growth performance of a population depends as much on past conditions, which determine its age structure, as on the present life table values, the intrinsic rate of increase provides a more useful index to the effect of environmental conditions or individual attributes on population growth.

Each life table has a single intrinsic rate of increase. Finding the exact value of r_m requires the solution of a complicated equation, but it may be approximated (r_a) by the following expression, which uses the terms of the life table:

$$r_a = \frac{R_0 \log_e R_0}{\sum x l_x b_x}$$

where R_0—the **net reproductive rate**—is the sum of the $l_x b_x$ column. The calculations are illustrated in Table 15.4.

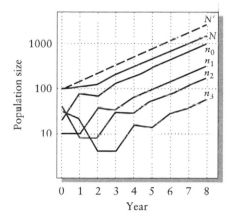

FIGURE 15.4 Growth of age classes as a population achieves its stable age distribution. The data used to make this graph are from Table 15.2.

TABLE 15.3	*Projection of a population with four age classes and a negative exponential growth rate*									
	TIME									
	0	1	2	3	4	5	6	7	8	Percent
n_0	1491	1130	965	776	642	525	431	352	286	57.7
n_1	501	447	339	289	233	183	157	129	106	21.4
n_2	270	301	268	203	174	140	118	94	76	15.3
n_3	90	81	90	81	61	52	42	35	28	5.7
N	2352	1959	1662	1349	1110	910	746	610	496	
λ	0.83	0.85	0.81	0.82	0.82	0.82	0.82	0.81		

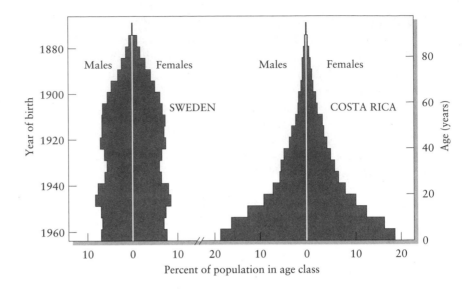

FIGURE 15.5 Population age structure of Sweden in 1965 and of Costa Rica in 1963. Because Sweden's population has grown slowly, its population is distributed toward older ages. Declining birth rates during the Depression and the baby boom that followed World War II were responsible for irregularities in the age structure. Costa Rica's rapid population growth, caused by a high birth rate, has resulted in a bottom-heavy age structure. After data in N. Keyfitz and W. Flieger, *World Population. An Analysis of Vital Data.* Univ. of Chicago Press, Chicago (1968).

	TABLE 15.4	*Estimation of the exponential rate of increase from Table 15.1* *			

x	s_x	l_x	b_x	$l_x b_x$	$x l_x b_x$
0	0.5	1.0	0	0.0	0.0
1	0.8	0.5	1	0.5	0.5
2	0.5	0.4	3	1.2	2.4
3	0.0	0.2	2	0.4	1.2
Net reproductive rate (R_0)				2.1	
Expected number of births weighted by age					4.1

*The sums of the $l_x b_x$ column (net reproductive rate) and the $x l_x b_x$ column are used to estimate r_a according to the equation given in the text. In this case, we calculate r_a to be 0.38; this is equivalent to $\lambda = 1.46$, close to the observed value of about 1.48 after the population achieved a stable age distribution.

The growth potential of populations

We can best appreciate the capacity of a population for growth by following its rapid increase when introduced into a new region with a suitable environment. In 1937, 2 male and 6 female ring-necked pheasants were released on Protection Island, Washington. They increased to 1325 adults within 5 years. The 166-fold increase represents a 180 percent annual rate of increase ($r = 1.02$, $\lambda = 2.78$). When domestic sheep were introduced to Tasmania, a large island off the coast of Australia, the population increased from less than 200,000 in 1820 to more than 2 million in 1850. The tenfold increase in 30 years is equivalent to an annual rate of increase of 8 percent ($r = 0.077$, $\lambda = 1.08$). Even such an unlikely creature as the elephant seal, whose population had been all but obliterated by hunting during the nineteenth century, increased from 20 individuals in 1890 to 30,000 in 1970 ($r = 0.091$, $\lambda = 1.096$). If you are unimpressed, consider that another century of unrestrained growth would find 27 million elephant seals crowding surfers and sunbathers off southern California beaches. Before the end of the *next* century, the shorelines of the Western Hemisphere would give lodging to a trillion of the beasts.

Elephant seal populations do not hold any growth records. Quite the contrary. Life tables of populations maintained under optimal conditions in the laboratory have exhibited potential annual growth rates (λ) as great as 24 for the field vole, 10 billion (10^{10}) for flour beetles, and 10^{30} for the water flea *Daphnia* (Table 15.5).

The intrinsic growth rate of a population varies with environmental conditions and population density. It is strictly determined by the life table values, which express the interaction between the individual and its environment. As a consequence of this interaction, the life table (and hence the intrinsic growth rate) varies with respect to the conditions of the environment.

These conditions vary spatially and temporally, creating differences in population dynamics from place to place and leading to changing population dy-

TABLE 15.5	*The doubling time of the population*			
Species	λ	$\log_e \lambda = r$	t_2 (years)	t_2 (days)*
Elephant seal	1.096	0.091	7.5	—
Ring-necked pheasant	2.78	1.02	0.67	246
Field vole	24	3.18	0.22	79
Flour beetle	10^{10}	23	0.03	11
Water flea	10^{30}	69	0.01	3.6

*Rapid growth rates may be expressed conveniently in terms of the time required for the population to double. The relationship between geometric growth rate (λ) and doubling time (t_2), derived from the equation for geometric growth, is $t_2 = \log_e 2 / \log_e \lambda$; that is, $0.69 / \log_e \lambda$. Hence for the field vole ($\lambda = 24$), $t_2 = 0.69 / \log_e 24$, which is 0.22 year, or 79 days.

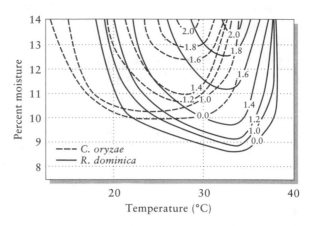

FIGURE 15.6 The influence of temperature and the moisture content of grain on the geometric rate of increase of populations of the grain beetles *Calandra oryzae* and *Rhizopertha dominica* living in wheat. Rates of increase are indicated by contour lines that describe conditions with identical values of λ, which are indicated on each line. After L. C. Birch, *Ecology* 34:698–711 (1953).

namics over time. Such effects of the environment are best revealed in experimental studies in which groups of individuals from the same population are subjected to different conditions. For example, the intrinsic rates of increase (λ) of populations of two species of grain beetles are influenced differently by temperature and moisture (Figure 15.6). Neither species performed well at low temperatures and humidities; moreover, the optimal conditions for growth differed between the species, *Rhizopertha* populations growing most rapidly at somewhat warmer temperatures. Of the two, *Rhizopertha* has the more tropical distribution in nature.

Climate and other environmental conditions have been shown to be important determinants of life table values of populations in their natural settings, as one would expect. For example, the European rabbit, introduced to Australia in the last century and now widespread throughout many habitats, survives longer but produces fewer offspring per year in arid regions than in the more mesic Mediterranean climate regions (Table 15.6).

This particular example raises the important issue of the regulation of population size. Continuous exponential growth leads, with time, to an inconceivable number. As Darwin put it in 1872, "Even slow-breeding man has doubled in twenty-five years, and at this rate, in less than a thousand years, there would literally not be standing-room for his progeny." The growth potential of populations is driven home by the history of the rabbit population in Australia. In 1859, 12 pairs were released on a ranch in Victoria to provide sport for hunters. Within 6 years, the population had increased so rapidly that 20,000 rabbits were killed in a single hunting drive. Even by conservative estimates, the population must have increased by a factor of at least 10,000 in 6 years, an exponential rate (r) of about 1.5 per year (a doubling time of about 5.5 months). Yet the life tables of present-day populations (Table 15.6) suggest growth rates (r_a) of 0.30 and 0.21. Considering the difficulty of estimating survival of the young to reproduction age and the statistical errors involved in such studies, these values probably do not differ from $r = 0$. How can the initial rapid growth be reconciled with the eventual stabilization of the rabbit population? Either birth rates decreased, death rates increased, or both changed when the population became more numerous. When there are more rabbits, there is less food for each; and fewer resources mean that fewer offspring can be nourished and these fewer survive less well.

The regulation of population size

Even the most slowly reproducing species would cover the earth in a short time if its population growth were unrestrained. Nearly two centuries ago, Thomas Malthus understood that this fact "implies a strong and constantly operating check on population from the difficulty of subsistence." In *An Essay on the Principle of Population* (1798), he wrote:

> *Through the animal and vegetable kingdoms, nature has scattered the seeds of life abroad with the most profuse and liberal hand. She has been comparatively sparing in the room*

TABLE 15.6 *Condensed life tables for European rabbits in arid and Mediterranean climates of Australia*

Age (months)	Pivotal age* (months)	Survival (l_x)†	Proportion of female population	Females pregnant at any one time	Litter size (embryos)	b_x	$l_x b_x$	$x l_x b_x$
Arid								
3–6	4.5	0.570	0.140	0.05	3.8	0.2	0.057	0.257
6–12	9.0	0.457	0.265	0.15	4.7	1.8	0.411	3.702
12–18	15.0	0.302	0.217	0.32	4.6	3.1	0.468	7.022
18–24	21.0	0.186	0.192	0.37	4.3	3.1	0.288	6.054
24	37.5	0.061	0.187	0.34	4.5	2.0	0.089	3.317
Total							1.313	20.352
Mediterranean								
3–6	4.5	0.222	0.177	0.23	4.4	1.9	0.198	0.891
6–12	9.0	0.260	0.359	0.43	5.6	8.7	0.696	6.264
12–18	15.0	0.075	0.303	0.53	5.9	9.4	0.357	5.358
18–24	21.0	0.028	0.107	0.45	5.9	2.8	0.039	0.823
24	37.5	0.006	0.053	0.55	6.2	1.8	0.005	0.203
Total							1.295	13.539

*Midpoint of the age interval, frequently used to calculate generation time when reproduction is more or less continuous, as it is in rabbits in Australia.

†Based on rabbits aged 0 to 3 months; hence $l_{1.5} = 1$.

Source: K. Myers, in P. J. der Boer and G. R. Gradwell (eds.), *Dynamics of Populations*. Centre Agric. Publ. Documentation, Wageningen, The Netherlands, pp. 478–506.

and the nourishment necessary to rear them. The germs of existence contained in this spot of earth, with ample food, and ample room to expand in, would fill millions of worlds in the course of a few thousand years. Necessity, that emperious all pervading law of nature, restrains them within the prescribed bounds. The race of plants, and the race of animals shrink under this great restrictive law.

Darwin echoed this view in *On the Origin of Species:*

As more individuals are produced than can possibly survive, there must in every case be a struggle of existence, either one individual with another of the same species, or with the individuals of distinct species, or with the physical conditions of life. It is the doctrine of Malthus applied with manifold force to the whole animal and vegetable kingdoms; for in this case there can be no artificial increase of food, and no prudential restraint from marriage. Although some species may be now increasing, more or less rapidly, in numbers, all cannot do so, for the world would not hold them.

This essentially modern view of the regulation of populations grew out of an awareness of the immense capacity of populations for exponential increase. In a sense, a population's growth potential and the relative constancy of its numbers cannot be logically reconciled otherwise.

The logistic equation

In 1920 Raymond Pearl and L. J. Reed, at the Institute for Biological Research of Johns Hopkins University, published a paper in the *Proceedings of the National Academy of Sciences* entitled "On the Rate of Growth of the Population of the United States Since 1790 and Its Mathematical Representation." Thorough and accurate population data had been gathered even in colonial times. Indeed, the phenomenal population growth of the American colonies had greatly impressed upon Malthus how rapidly humans could multiply; this was not so evident in the more crowded European countries of his time. Pearl and Reed wished to project the future growth of the population, which they supposed must eventually reach a limit. Data for the population to 1910, the latest census then available, had revealed a decline in the exponential rate of growth (Figure 15.7). Pearl and Reed reasoned that if the decline followed a regular pattern that could be described mathematically, it would be possible to predict the future course of the population, as long as the decline in the exponential growth rate continued. They also reasoned that changes in the exponential rate of growth must be related to the size of the population rather than to time, because the time scale is arbitrary with respect to any particular population. And so, in place of a constant value of *r* in the differential equation for unre-

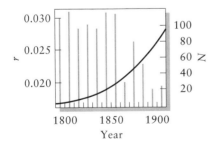

Figure 15.7 Increase in the population of the United States between 1790 and 1910 (curve) and the exponential rate of increase during each 10-year period (vertical bars). From data in R. Pearl and L. J. Reed, *Proc. Natl. Acad. Sci.* 6: 275–288 (1920).

strained population growth ($dN/dt = rN$), Pearl and Reed suggested that r decreases as N increases, according to the relation

$$r = r_0\left(1 - \frac{N}{K}\right)$$

where r_0 represents the intrinsic exponential growth rate of the population when its size is very small (that is, close to 0), and K — the **carrying capacity** of the environment — represents the number of individuals that the environment can support. Accordingly, the differential equation describing restricted population growth became

$$\frac{dN}{dt} = r_0 N\left(1 - \frac{N}{K}\right)$$

This equation, which is called the **logistic equation,** prescribes that the exponential rate of increase decreases as a linear function of the size of the population. Such a decrease reasonably approximated the data for the population of the United States (Figure 15.8).

So long as population size N does not exceed the carrying capacity K — that is, N/K is less than 1 — the population continues to increase, albeit at a slowing rate. When N exceeds the value of K, the ratio N/K exceeds 1, the term in parentheses $(1 - N/K)$ becomes negative, and the population decreases. Because populations below K increase and those above K decrease, K is the eventual equilibrium size of a population growing according to the logistic equation. The relationships among r, dN/dt, and N are illustrated in Figure 15.9. The curve for dN/dt has a maximum at intermediate population size, specifically when $N = K/2$, and falls to 0 as N approaches either 0 or K.

The time course of population growth according to the logistic equation can be found by integrating the differential, which yields

$$N(t) = \frac{K}{1 + be^{-rt}}$$

where b is a constant equal to $[K - N(0)]/N(0)$. The value of b depends arbitrarily on the size of the population at the designated time zero — 1790 in the case of Pearl and Reed's data. This equation describes a sigmoid, or S-shaped, curve (Figure 15.10): the population grows slowly at first, then more rapidly as the number of individuals increases, and finally more slowly, gradually approaching the equilibrium number K. Applied to the growth of the population of United States from 1790 to 1910, this curve is illustrated in Figure 15.11.

Pearl and Reed obtained the best fit of their equation to the population data when the value of K was 197,273,000 and that of r_0 was 0.03134. Thus even though the population in 1910 was only 91,972,000, they extrapolated its future growth to twice the 1910 level on the basis of earlier growth performance. Projections often prove incorrect, however, when circumstances change. The U.S. population reached 197 million between 1960 and 1970, when it was still growing vigorously. Although a leveling off in the mid-200 millions can now be predicted on the basis of a much reduced birth rate in recent years, all this could easily change.

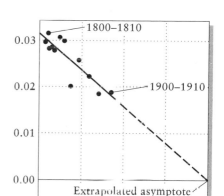

FIGURE 15.8 Exponential rates of population increase in the United States during each decade between 1790 and 1910 plotted as a function of the population size during that decade (the geometric mean of the beginning and ending numbers). Data are from Figure 15.7.

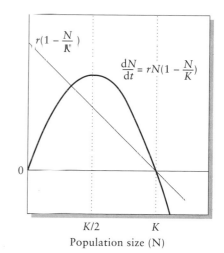

FIGURE 15.9 Various representations of the logistic curve of population growth. (a) The exponential growth rate of the population (dN/Ndt) as a function of population size; r declines to 0 when $N = K$. (b) The absolute rate of growth (dN/dt), which is the product of population size (N) and the exponential growth rate, reaches a maximum when population size is one-half the carrying capacity (K).

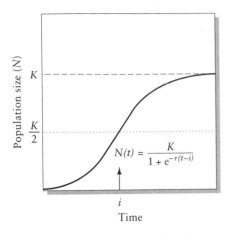

FIGURE 15.10 According to logistic growth, the increase in numbers over time follows an S-shaped curve that is symmetrical about the inflection point (K/2). That is, the accelerating and decelerating phases of population growth have the same shape.

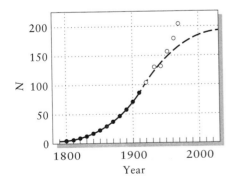

FIGURE 15.11 A logistic curve fitted to the population of the United States between 1790 and 1910 (solid dots). Subsequent censuses (open dots) have departed from the projected population curve.

Density-dependent factors

The logistic equation has been applied successfully to describe the growth of populations in the laboratory and in natural habitats. The equation suggests that factors limiting growth exert stronger effects on mortality and fecundity as the population grows. But what are these factors, and how do they operate? Many things influence the rate of population growth, but only **density-dependent factors,** whose effect increases with crowding, can bring the population under control. Of prime importance among these factors are limitation of food supply and places to live, and predators, parasites, and diseases whose effects are felt more strongly in crowded than in sparse populations. Other factors such as temperature, precipitation, and catastrophic events alter birth and death rates largely without regard to the numbers of individuals in a population. Thus such **density-independent factors** may influence the exponential growth rate of a population, but they do not regulate its size.

Density dependence in animals

Numerous experimental studies have revealed the various mechanisms of density dependence. For example, when fruit flies are confined to a bottle with a fixed supply of food, the descendants of a single pair of flies increase in number rapidly at first but soon reach a limit. When different numbers of pairs of flies are introduced into otherwise identical culture bottles, the number of progeny raised per pair varies inversely with the density of flies in the bottle (Figure 15.12). This effect results from competition among the larvae for food, which causes high mortality in dense cultures. Adult life span also declines at high densities, but well above the levels that affect the survival of larvae. It is often the case that juvenile stages suffer the adverse effects of density-dependent factors more than adults do.

The effects of density on the life tables of grain beetles result from antagonistic behavioral interaction between the larvae under crowded conditions. *Rhizopertha dominica* is a tiny beetle that completes its larval development within a single grain (of wheat, for example), living off the kernel of the seed. Females lay their eggs on the surfaces of seeds. Immediately after hatching, the larva bores its way into the seed, where it commences development. Once inside, the larva enjoys security as long as it is alone; a kernel of wheat cannot support more than one beetle.

What happens when two larvae meet in the same grain? The British ecologist A. C. Crombie described it thus:

> *When two larvae in the first, second, third or fourth instars were put together into a small hole drilled in a wheat grain and watched under a binocular microscope, they were often seen to attack each other with their mandibles, and eventually either one or both left the hole. When a larva entered such a hole it*

always went to the bottom and turned round so as to face outwards. Other larvae trying to enter the hole were fiercely attacked. Sometimes such combats resulted in the body wall of one of the antagonists becoming punctured and its bleeding to death. In their tunnels in wheat grains larvae of all instars were always found curled up with the head facing towards the way they had entered. Furthermore, in all grains dissected during the experiments to be described, whenever two larvae were found in the same tunnel at least one of them was always dead. (J. Exp. Biol. 20:135–151, 1944)

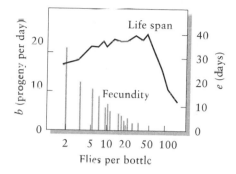

FIGURE 15.12 The continuous line shows the influence of density on life span (*e*, days) and the vertical bars its influence on fecundity (*b*, progeny per day) in laboratory populations of the fruit fly *Drosophila melanogaster*. After R. Pearl, *Q. Rev. Biol.* 2:532–548 (1927).

The experiments Crombie referred to were as follows. He infested a single wheat grain with from one to eight larvae, giving them 6 hours to become established in the grain. He then transferred the infested grain to a dish with five other fresh grains and left them together for 48 hours. At the end of the experiment, he dissected all the grains to determine how many larvae had been killed and how many had migrated to a different grain. The results showed conclusively the strong effects of density on the survival and movement of larvae even within the first 2 days (Figure 15.13).

Crombie's beetles contested each grain of wheat through direct confrontation. After all, sole possession of a single grain is a larva's ticket to success. In this population, density dependence came into play abruptly as the number of larvae approached the number of grains of wheat. Above that level, regardless of the number of larvae relative to wheat grains, the end result was always the same: one grain, one larva.

Water fleas (*Daphnia pulex*) exert a less direct influence on one another through their consumption of resources. Each water flea consumes millions of prey, the single-celled green algae and diatoms of plankton. As prey are consumed, the availability of food to other water fleas decreases gradually, leading to a graded response of birth and death rates to prey density. When water fleas were maintained in small beakers at densities between 1 and 32 individuals per cubic centimeter and were fed identical cultures of green algae, fecundity decreased markedly with increasing population density (Figure 15.14). Somewhat unexpectedly, however, survival increased at densities up to 8 individuals per cubic centimeter before decreasing at higher densities. At densities of 8 individuals per cubic centimeter and above, stunted body growth suggested that depletion of food resources between periodic replenishment of the alga stock culture limited birth rates and survival; the effect was clearly density-dependent.

Geometric rate of population growth (λ) calculated from the water flea life tables decreased linearly with increasing density and fell below 1.0 at a density of about 20 individuals per cubic centimeter (Figure 15.15). Therefore, under the conditions of temperature, light, water quality, and food availability provided in the laboratory, *Daphnia* populations would attain a stable size of about 20 individuals per cubic centimeter, regardless of the initial density of the culture.

Most studies of density dependence have focused on laboratory populations where factors can be controlled experimentally. But the simplicity of

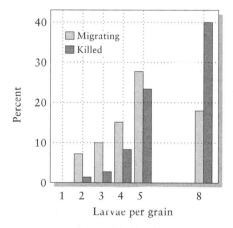

FIGURE 15.13 Effects of density (number of larvae per grain) on migration and on mortality in the grain beetle *Rhizopertha dominica*. From data in A. C. Crombie, *J. Exp. Biol.* 20:135–151 (1944).

such systems also leads us to question the relevance of laboratory findings to populations in more complex natural surroundings, where physical conditions change continuously and food and predation are not controlled by the experimenter. Ideally we would like to conduct the same experiment in nature as in the laboratory — that is, to alter the density of individuals in the population while keeping everything else constant. In practice, this difficult experiment can be accomplished only with populations managed intensively for some other purpose. Game animals are sometimes maintained at altered levels by management practices, and ecologists have taken advantage of such situations to study population processes.

The reproduction and survival of deer depend directly on the quality of the food supply. Deer browse leaves, and they require large quantities of new growth with high nutritional content to maintain high growth rates and normal reproduction. In white-tailed deer in New York State, the proportion of females pregnant and the average number of embryos per pregnant female were found to be directly related to range conditions (Table 15.7). The number of *corpora lutea* in the ovary indicates the number of eggs ovulated and hence the reproductive potential of the female. The difference between the number of *corpora lutea* and the number of embryos shows that poor range condition, resulting in poor nutrition of pregnant females, caused embryo death and resorption. In the central Adirondack area, where the habitat for deer was very poor, even ovulation was greatly reduced.

Range deterioration caused by overgrazing can often be reversed by selective hunting to thin dense populations. When an area of very poor range in the Adirondack Mountains of New York was opened to hunting, the population of white-tailed deer decreased, range quality recovered, and reproduction improved dramatically (Table 15.8).

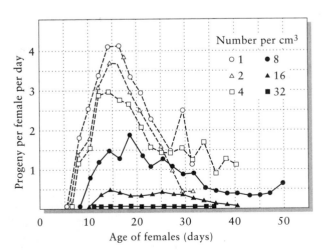

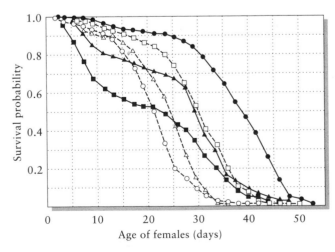

FIGURE 15.14 Fecundity and survival in laboratory populations of *Daphnia pulex* at different densities. From P. W. Frank, C. D. Boll, and R. W. Kelly, *Physiol. Zool.* 30:287–305 (1957).

Region*	Percent of females pregnant	Embryos per female	*Corpora lutea* per ovary
Western (best range)	94	1.71	1.97
Catskill periphery	92	1.48	1.72
Catskill central	87	1.37	1.72
Adirondack periphery	86	1.29	1.71
Adirondack center (worst range)	79	1.06	1.11

TABLE 15.7 *Reproductive parameters of white-tailed deer (Odocoileus virginianus) in five regions of New York State, 1939–1949*

*Arranged by decreasing suitability of range.

Source: E. L. Chaetum and C. W. Severinghaus, Variations in fertility of white-tailed deer related to range conditions. *Trans. North Amer. Wildl. Conf.* 15:170–189 (1950).

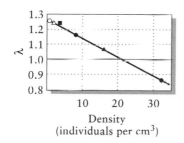

FIGURE 15.15 These values of λ are calculated from the life table data for *Daphnia pulex* portrayed in Figure 15.14. Population growth rate decreases as a function of density. After R. Laughlin, *J. Anim. Ecol.* 34:77–91 (1965).

Density dependence in plants

Like animals, plants experience increased mortality and reduced fecundity at high density. Unlike many animals, however, plants respond to density by slowed growth, along with its consequences for fecundity and, to a lesser ex-

TABLE 15.8 *Reproductive parameters of white-tailed deer (Odocoileus virginianus) in the Adirondack Mountains of New York State before and after hunting*

Region	Percent of females pregnant	Embryos per female	*Corpora lutea* per ovary
DeBar Mountain			
1939–1943 (prehunting)	57	0.71	0.60
1947 (after heavy hunting)	100	1.78	1.86
Moose River			
1939–1943 (prehunting)	91	1.00	0.98
1947 (after heavy hunting)	69	1.00	1.13

Source: E. L. Chaetum and C. W. Severinghaus, Variations in fertility of white-tailed deer related to range conditions. *Trans. North Amer. Wildl. Conf.* 15:170–189 (1950).

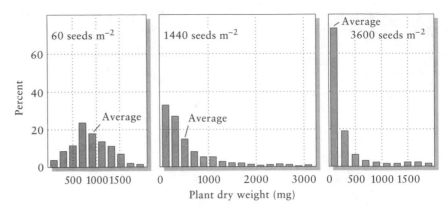

FIGURE 15.16 The distribution of dry weights of individuals in populations of flax plants sown at different densities. After J. L. Harper, *J. Ecol.* 55:247–270 (1967).

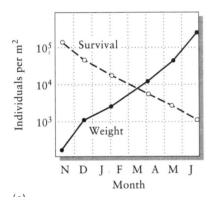

(a)

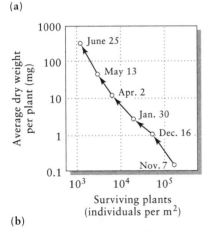

(b)

FIGURE 15.17 (a) Progressive change in plant weight and population density in an experimental planting of horseweed (*Erigeron canadensis*) sown at a density of 100,000 seeds per square meter. (b) The relationship between plant density and plant weight as the season progressed. After J. L. Harper, *J. Ecol.* 55:247–270 (1967).

tent, survival. Starve an animal and most often you will kill it; starve a plant and you will stunt its growth. This difference arises in part from the modular construction of plants. The repeated units of leaves and buds that make up the plant body exist more or less independently of one another; that a branch fails to grow has little effect on the well-being of other branches on the tree. The more complicated arrangement and interdependence of the parts of animals impose greater constraints on size and shape.

Flexibility of plant growth is revealed in the size of flax (*Linum*) plants grown to maturity at different densities (Figure 15.16). When seeds were sown sparsely at a density of 60 seeds per square meter, the modal dry weight of individuals fell between 0.5 and 1 g, and many plants attained weights exceeding 1.5 g. When sown at densities of 1440 and 3600 seeds m⁻², most of the individuals weighed less than 0.5 g, and few grew to large size. The variation in size within a planting results from chance factors early in the seedling stage, particularly the date of germination and the quality of the site in which the seedling grows. Early germination in a favorable spot gives the plant an initial growth advantage over others, which increases as the larger plants grow and crowd their smaller neighbors.

The flexibility of plant growth does not preclude mortality in crowded situations. When horseweed (*Erigeron canadensis*) seed was sown at a density of 100,000 seeds m⁻² (equivalent to about 10 seeds in the area of your thumbnail), the young plants competed vigorously. As the seedlings grew, many died and the density of surviving seedlings decreased (Figure 15.17). At the same time, however, the growth of surviving plants exceeded the decline of the population, and the total weight of the planting increased. Over the entire growing season, a thousandfold increase in the average weight of each plant more than balanced the hundredfold decrease in population density.

When the logarithm of average plant weight is plotted as a function of the logarithm of density, data points recorded during the growing season fall on a line with a slope of approximately −3⁄2 (Figure 15.18). Plant ecologists call this relationship between average plant weight and density a **self-thinning curve.** Such is the regularity of this relationship that many have referred to it as the **−3⁄2 power law.**

Density-dependent factors bring populations under control and regulate their size close to the carrying capacity set by the availability of resources and by the conditions of the environment. Changes in these conditions and resources continually establish new equilibrium values toward which the populations grow or decline. Furthermore, catastrophic changes in the environment brought about by a sudden freeze, a violent storm, or a shift in ocean current, for example, often reduce populations far below their carrying capacity and initiate periods of population recovery. Thus, although density-dependent factors regulate all populations, variations in the environment also cause all populations to vary, to a greater or lesser extent, about their equilibrium sizes. We shall explore population variation in more detail in the next chapter.

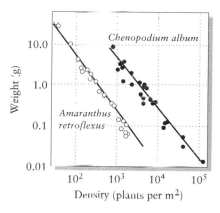

FIGURE 15.18 Change in plant density and mean plant weight with time for plantings of *Amaranthus retroflexus* and *Chenopodium album*. The two variables are related by a slope of $-3/2$ (-1.5), which has been extremely consistent in experiments of this type. After J. L. Harper, *Population Biology of Plants.* Academic Press, New York (1977).

ꙮ SUMMARY

1. Population growth can be described by the exponential rate of increase (r) in the expression $n(t) = N(0)e^{rt}$. The factor by which a population increases in 1 unit of time (e^r) is the geometric growth rate of the population (λ). λ and e^r are interchangeable in population equations.

2. The exponential growth rate is the difference between the birth and death rates averaged over individuals (per capita) in the population (that is, $r = b - d$).

3. The instantaneous rate of increase of an exponentially growing population is $dN/dt = rN$.

4. Populations with discrete breeding seasons increase geometrically by periodic increments according to the relation $N(t + 1) = N(t)\lambda$.

5. When birth rates and death rates vary according to the age of the individual (summarized by the life table), we must also know the proportion of individuals in each age class in order to project population growth.

6. A population with a fixed life table assumes a stable age distribution in which the numbers in each age class, as well as the population as a whole, increase at the same exponential or geometric rate, known as the intrinsic rate of increase of the population.

7. The life table of a population varies with the conditions of the environment and with the density of the population.

8. The contrast between the potential of all populations for rapid growth and the observation of relative constancy of populations over long periods led naturally to the concept of density-dependent effects of environmental factors on population processes. Population growth, early investigators reasoned, could be slowed only if birth rate and survival decreased as populations grew. Dwindling supplies of food and the increasing pressure of predators and disease exert their effects on population processes in such a density-dependent manner.

9. Density-dependent population growth can be described by the logistic equation, which has the differential form

$$\frac{dN}{dt} = rN\left(1 - \frac{N}{K}\right)$$

and the integral form

$$N(t) = \frac{K}{(1 + be^{-rt})}$$

10. Laboratory and field studies of animal and plant populations showed in detail the expression of density-dependent factors in population processes. During the 1930s, experimental laboratory systems were developed with fruit flies, protozoa, flour and grain beetles, and water fleas. Similar studies of plants and of animals in natural habitats, based largely on such economically important species as crop plants, weeds, insect pests, and game species, came later.

❧ SUGGESTED READINGS

Andrewartha, H. G., and L. C. Birch. 1954. *The Distribution and Abundance of Animals.* Univ. of Chicago Press, Chicago.

Andrewartha, H. G., and L. C. Birch. 1984. *The Ecological Web. More on the Distribution and Abundance of Animals.* Univ. of Chicago Press, Chicago.

Clutton-brock, T. H., M. Major, and F. E. Guinness. 1985. Population regulation in male and female red deer. *Journal of Animal Ecology* 54:831–846.

Kingsland, S. E. 1985. *Modeling Nature. Episodes in the History of Population Ecology.* Univ. of Chicago Press, Chicago.

Dack, D. 1954. *The Natural Regulation of Animal Numbers.* Oxford Univ. Press, London.

Pulliam, H. R. 1988. Sources, sinks, and population regulation. *American Naturalist* 132:652–661.

Skogland, T. 1985. The effects of density-dependent resource limitations on the demography of wild reindeer. *Journal of Animal Ecology* 54:359–374.

Weiner, J. 1988. Variation in the performance of individuals in plant populations. In A. J. Davy, M. J. Hutchings, and A. R. Watkinson (eds.). *Plant Population Ecology.* Blackwell Scientific Publications, Oxford, pp. 59–81.

Weller, D. E. 1987. A reevaluation of the $-3/2$ power rule of plant self-thinning. *Ecological Monographs* 57:23–43.

Wilson, E. O., and W. H. Bossert. 1971. *A Primer of Population Biology.* Sinauer, Sunderland, Mass.

Wynne-Edwards, V. C. 1986. *Evolution through Group Selection.* Blackwell Scientific Publications, Oxford and Boston, Mass.

TEMPORAL AND
SPATIAL DYNAMICS

Populations continuously move toward steady-state densities determined by the qualities the organisms exhibit and the conditions that prevail in their environments. But the environment varies, and so do populations. Moreover, patterns of variation derive not only from the changeable nature of the environment but also from the intrinsic dynamics of population responses. Thus the study of variation in population size has its roots in two kinds of phenomena. The first includes the responses of populations to perceptible change in the environment. The second includes regular cycles in numbers that are unrelated to obvious periodic variation in the environment.

Because ecological conditions vary spatially, the dynamics of individual populations may differ from one place to another. When movements couple the dynamics of populations in two localities, they behave as a single population. More often, distance isolates populations that are restricted to different localities, and they behave at least partly independently. Under such circumstances, changes in the total population reflect the sum of changes in all subpopulations. But because the dynamics of small populations differ from those of large populations, subdivided populations have unique properties.

In this chapter, we shall discuss the causes of variation in population size, explore how variation affects small and large populations differently, and examine the consequences of dispersal among subpopulations. The dynamics of small populations have become increasingly relevant as species dwindle to extinction and landscape use by humans fragments habitats into smaller, more isolated parcels.

Fluctuation in natural populations

Variation in the density of a population depends on the magnitude of fluctuation in the environment and on the inherent stability of the population. After sheep became established on Tasmania, their population varied irregularly between 1,230,000 and 2,250,000 — less than a factor of two — over nearly a century (Figure 16.1). Much of the variation was related to changes in grazing practices, markets for wool and meat, and pasture management, which could be considered factors in the environment of the sheep industry, if not in that of the sheep themselves.

In sharp contrast, populations of small, short-lived organisms may fluctuate wildly over many orders of magnitude within short periods. The combined populations of species of green algae and diatoms that make up the phytoplankton exhibit tremendous population increases and crashes over periods of a few weeks (Figure 16.2); short-period fluctuations may overlay changes with longer periods that occur on, say, a seasonal basis. Sheep and algae differ in their degree of sensitivity to environmental change and in the response times of their populations. Because sheep are larger, they have greater capacity for homeostasis and therefore better resist the effects of environmental change. Furthermore, because sheep live for several years, the population at any one time includes individuals born over a long period; this tends to even out the effects on population size of short-term fluctuations in birth rate. The lives of single-celled algae span only a few days or weeks, so populations turn over rapidly and bear the full impact of a capricious environment.

Populations of similar species in the same place may respond to different environmental factors. For example, the densities of four species of moths, whose larvae feed on pine needles, were found to fluctuate more or less independently in a pine forest in Germany (Figure 16.3). The populations varied

FIGURE 16.1 Numbers of sheep on the island of Tasmania since their introduction in the early 1800s. After J. Davidson, *Trans. Roy. Soc. South Australia* 62:342–346 (1938).

over three to five orders of magnitude (1000-fold to 100,000-fold) with irregular periods of a few years. Furthermore, the highs and lows of the populations did not coincide closely, suggesting that even though the species fed on the same resource in the same forest, their populations were governed independently by different factors.

Key-factor analysis

Recognizing that population fluctuations result from the impact of environmental factors on birth and death processes, biologists began to examine the effects of each factor separately to determine which exert the greatest influence on populations and whether they act in a density-dependent fashion. The forest entomologist R. F. Morris explained the approach as follows:

> *A preliminary examination of rather extensive life-table data for the spruce budworm . . . suggested that the factors affecting this species in any one place are of two types — those that cause a relatively constant mortality from year to year and contribute little to population variation, and those that cause a variable, though perhaps much smaller, mortality and appear to be largely responsible for the observed changes in population. . . . A factor of the latter type will here be called a "key factor," meaning simply that changes in population density from generation to generation are closely related to the degree of mortality caused by this factor, which therefore has predictive value.* (Ecology 40:580–588, 1959)

Morris reasoned that the change in a population from time t to time $t + 1$ (λ) depends on the fecundity and survival of individuals alive at time t. Furthermore, survival includes several components representing the survival of individuals through each stage of the life cycle (egg, various larval stages, pupa, and

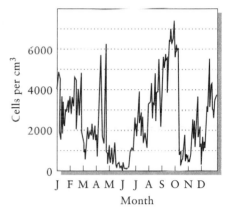

FIGURE 16.2 Variation in the density of phytoplankton in samples of water taken from Lake Erie during 1962. After C. C. Davis, *Limnol. Oceanogr.* 9:275–283 (1964).

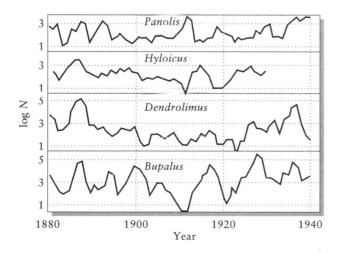

FIGURE 16.3 Fluctuations in the number of pupae of four species of moth (hibernating larvae in the case of *Dendrolimus*) in a managed pine forest in Germany over 60 consecutive midwinter counts. After G. C. Varley, *J. Anim. Ecol.* 18:117–122 (1949).

adult). Each of these stages suffers from its own mortality factors. Accordingly, Morris represented population changes by the simple model

$$\lambda = FS_1S_2S_3 \cdots$$

In logarithmic form,

$$r = \log F + \log S_1 + \log S_2 + \log S_3 + \cdots$$

(remember, $r = \log_e \lambda$). He determined statistically whether variation in r followed upon variation in one or a few of the values of F and S_i. By way of example, Morris applied his analysis to data for the black-headed budworm (Figure 16.4). The budworm belongs to the moth family Tortricidae (leaf-rollers), is native to eastern Canada, and is a major defoliator of fir trees in New Brunswick. One generation of adults appears each year; the population overwinters in the egg stage. Twelve years of observations on the density of budworm populations and the percentage of parasitism of the larvae by wasps and flies revealed a fluctuation in numbers of over two orders of magnitude with a period of about 9 years. The survival of larvae at risk to parasites (S_1) followed a similar course, being high during the population buildup and low during periods of decline. In this example, the exponential growth rate of the population ($r = \log_e \lambda$) strongly correlated with $\log_e S_1$. From this analysis, Morris concluded that larval parasitism is a **key factor** in the population processes of the black-headed budworm.

In other studies of Canadian agricultural and forest insect pests, the critical stage at which key factors acted, as well as the nature of the key factor itself, varied from species to species. In only one, the Colorado potato beetle, was food supply a key factor. Weather was the most important factor in three of the species, and disease, parasitism, and emigration from local populations ranked highly in the rest. In 9 of the 12 studies, the key factor exerted a density-dependent influence; that is, it caused survival to decrease as the initial size of the population increased.

Population cycles and intrinsic demographic processes

Except for factors associated with daily, lunar (tidal), and seasonal cycles, environmental fluctuations tend to be irregular rather than periodic. Historical records reveal that years of abundant rain or drought, extreme heat or cold, and such natural disasters as fires and hurricanes occur irregularly, perhaps even at random. Biological responses to these factors are similarly aperiodic. For example, widths of the growth rings of trees vary in direct proportion to temperature and rainfall; patterns of successive ring widths cannot be distinguished from a random series (Figure 16.5).

Trends in the sizes of many populations do, however, change with periodic frequency (for example, see *Bupalus* in Figure 16.3). For many years, ecologists believed that such **cycles** must be caused by environmental factors that exhibit

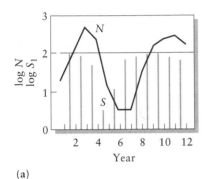

(a)

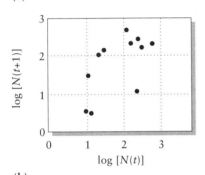

(b)

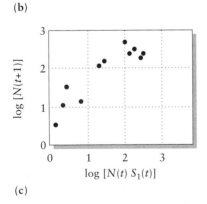

(c)

FIGURE 16.4 (a) Population size (*N*, solid line) and larval survival (*S*, vertical bars) for the black-headed budworm. (b, c) The relationship between exponential growth rate (*r*) and the number of larvae surviving parasitism (S_1). After data in R. F. Morris, *Ecology* 40:580–588 (1959).

similar periodic variation. The regular 11-year cycle in sunspot numbers was frequently mentioned, but the sunspot cycle never matched population cycles well, and no model could be devised to link the two.

With the development of population models in the 1920s and 1930s, it became evident that because of their inherent dynamic properties, populations subjected to even minor, random environmental fluctuations could be caused to oscillate. Such cycling can result from **time delays** in the response of births and deaths to changes in the environment. Just as the momentum imparted to a pendulum by the acceleration of gravity carries it past the equilibrium point and causes it to swing back and forth periodically, the "momentum" imparted to a population by high birth rates at low density or high death rates at high density carries the population past its equilibrium when demographic responses are time-delayed.

Time delays that cause populations to oscillate when displaced from their equilibria are inherent to models based on discrete generations. According to these models, populations respond by discrete increments from one time to the next and therefore cannot continuously readjust growth rate as population size approaches equilibrium. This can cause the population to overshoot the equilibrium, first in one direction and then in the other, as population size N draws closer to carrying capacity K. When the exponential growth rate r is less than 1, the increase in the population between t and $t + 1$ will be less than the difference between the population size and the equilibrium level K. Therefore, the population will approach K directly, as shown in Figure 16.6. When r exceeds 1 but is less than 2, the population will overshoot the equilibrium but will nonetheless end up closer to the equilibrium than before. Thus the population will **oscillate** back and forth across the equilibrium value, getting closer with each generation. This behavior is called **damped oscillation.**

When r exceeds 2, the population ends up further from the equilibrium each generation, and the oscillations increase. With increasing r, these oscillations take on very complex, eventually unpredictable forms referred to as **chaos.** Short of this point, however, populations may settle into stable oscillations called **limit cycles,** in which numbers bounce back and forth between high and low values. In reality, few populations have such a discrete structure

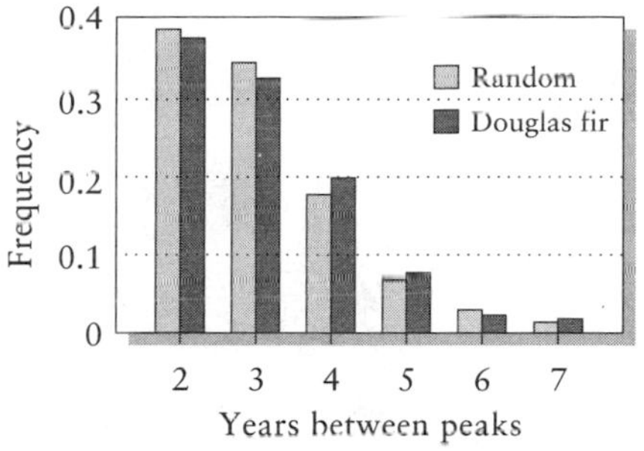

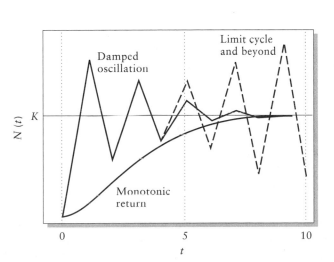

FIGURE 16.5 Frequency distributions of intervals between peaks in the widths of growth rings of Douglas fir and a series of random numbers. After L. C. Cole, *J. Wildl. Mgmt.* 15:233–252 (1951).

FIGURE 16.6 Approach to equilibrium according to a discrete logistic process when $r < 1$ (monotonic return), $1 < r < 2$ (damped oscillation), and $r > 2$ (limit cycle).

as to make these models realistic, and such complicating factors as age structure alter the behavior of the models considerably. Even in continuously breeding populations, however, time lags in population processes caused by developmental periods can create cyclic population behavior.

Time delays and oscillations in continuous-time models

Oscillations are produced in continuous-time models when the response of population growth to density is time-delayed; in other words, they occur when the effect of density dependence reflects the density of the population τ time units in the past (τ is the lowercase Greek letter tau). Modified accordingly, the logistic equation becomes

$$\frac{dN}{dt} = rN(t)\left[1 - \frac{N(t - \tau)}{K}\right]$$

This model produces damped oscillations in N as long as the product $r\tau$ is less than $\pi/2$ (about 1.6). Below $r\tau = e^{-1}$ (0.37), the population increases or decreases **monotonically** — without oscillation — to the equilibrium point. For $r\tau$ greater than $\pi/2$, the oscillations increase to form limit cycles with maximum population size reaching $N/K = e^{r\tau}$. Thus for $r\tau = 2$, oscillations increase in amplitude until the maximum value of N is $e^2 = (7.4)$ times K. Population biologists refer to such stably maintained oscillations as limit cycles. Their periods, from peak to peak, increase from about 4τ to more than 5τ with increasing r.

Cycles have been observed in many laboratory cultures of single species. In one study, populations of the water flea *Daphnia magna* exhibited marked oscillations when cultured at 25°C but showed strong damping at 18°C (Figure 16.7). The period at 25°C appeared to be just over 40 days for two cycles, suggesting a time delay in the density-dependent response of about 10 days; this approximates the average age of water fleas giving birth at 25°C. The time lag arose in the following manner. As population density increased, reproduction decreased, falling nearly to zero when the population exceeded 50 individuals. In contrast to fecundity, survival was less sensitive to density; adults lived at least 10 days, even at the highest densities. As a result, the pulse of deaths in the population lagged about 10 days behind the pulse of births at the beginning of each upswing of the cycle. High density early in the cycle prevented births. Then, when the population fell to densities low enough to permit reproduction, it contained only senescent, nonreproducing individuals. Thus the beginning of a new cycle awaited the buildup of young, fecund individuals. The length of the time delay was approximately the average adult life span at high density.

At the lower temperature, reproductive rate fell quickly with increasing density, and life span increased greatly over that seen at 25°C at all densities. Populations at the colder temperature apparently lacked a time delay because death was more evenly distributed over ages, and some individuals gave birth, even at high population densities; consequently, generations overlapped more

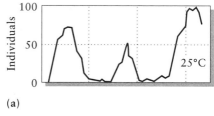

(a)

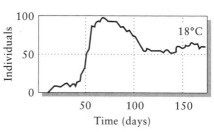

(b)

FIGURE 16.7 Growth of *Daphnia magna* populations at 18°C (a) and 25°C (b), showing the development of population cycles at the warmer temperature. After D. M. Pratt, *Biol. Bull.* 85:116–140 (1944).

broadly. At the higher temperature, the water fleas behaved according to a discrete-generation model with its built-in time delay of one generation. At lower temperature, they behaved according to a continuous-generation model with little or no time delay.

Storage of lipid reserves by some species of water fleas reduces the sensitivity of mortality to density and therefore introduces a time delay into the population processes. *Daphnia galeata,* a large species, stores energy in the form of lipid droplets (Figure 16.8) during periods of abundant food (that is, low density). It can then live on these stored reserves when food supplies dwindle as a result of overgrazing at high population density. Females also pass lipid to each offspring through oil droplets in the eggs, thereby increasing the survival of young, prereproductive water fleas under poor feeding conditions. The smaller *Bosmina longirostris* stores little lipid, so starvation increases directly in response to increases in population density. The consequences for population growth are predictable: *Daphnia* exhibits pronounced limit cycles with a period of 15 to 20 days under the conditions of the experiment, whereas *Bosmina* populations grow quickly to an equilibrium with perhaps a single strongly damped overshoot (Figure 16.9). The r value of *Daphnia* populations was about 0.3 d^{-1}. With a cycle period of 15 to 20 days, τ must have been about 4 to 5 days, and therefore $r\tau$ was about 1.2 to 1.5. Because the value of $r\tau$ was somewhat less than $\pi/2$, the cycles in the *Daphnia* population should have damped out eventually.

Time lags and oscillations in blowfly populations

The behavior of a population with respect to its equilibrium is sensitive to many aspects of life history that govern time delays in responses to density. Slight differences in culture conditions or the intrinsic properties of species can tip the balance between a monotonic approach to equilibrium and a limit cycle. A. J. Nicholson's experimental manipulations of time delay in laboratory cultures of the sheep blowfly *Lucilia cuprina* provide a dramatic demonstration of the relationship of time delays to population cycles. Under one set of culture conditions, Nicholson provided the larvae with 50 g of ground liver per day, while giving the adults unlimited food. The number of adults in the population cycled through a maximum of about 4000 to a minimum of 0 (at which point all the individuals were either eggs or larvae), with a period of between 30 and 40 days (Figure 16.10).

In this experiment, regular fluctuations of the blowfly populations were caused by a time delay in the response of fecundity and mortality to the density of adults in the cages. At high population density, adults laid many eggs, resulting in strong larval competition for the limited food supply. None of the larvae that hatched from eggs laid during adult population peaks survived, primarily because they did not grow large enough to pupate. Therefore, large adult populations gave rise to few adult progeny, and because adults lived less than 4 weeks the population soon began to decline. Eventually, so few eggs were laid on any particular day that most of the larvae survived, and the size of the adult population began to increase again.

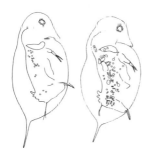

FIGURE 16.8 Distribution of energy stores in the form of oil droplets in individuals of *Daphnia galeata* grown under conditions of low (*left*) and high (*right*) food abundance. From C. E. Goulden and L. L. Hornig, *Proc. Natl. Acad. Sci.* (USA) 77:1716–1720 (1980).

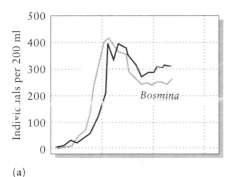

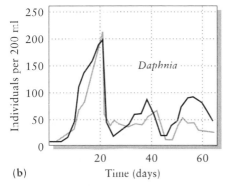

FIGURE 16.9 Densities of two populations of the cladoceran *Bosmina longirostris* (a) and two populations of its larger relative *Daphnia galeata* (b). From C. E. Goulden, L. L. Henry, and A. J. Tessier, *Ecology* 63:1780–1789 (1982).

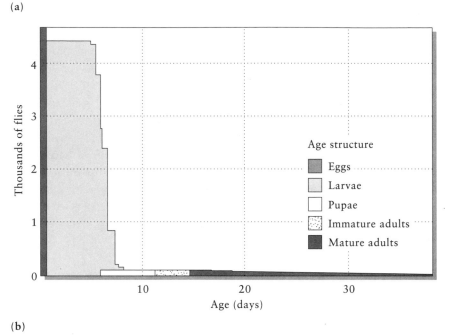

(a)

(b)

FIGURE 16.10 (a) Fluctuations in laboratory populations of the sheep blowfly *Lucilia cuprina*. Larvae were provided with 50 g of liver per day; adults were given unlimited supplies of liver and water. The continuous line represents the number of adult blowflies in the population cage. The vertical lines represent the numbers of adults that eventually emerged from eggs laid on the days indicated by the lines. (b) The average age structure of the population. After A. J. Nicholson, *Cold Spring Harbor Symp. Quant. Biol.* 22:153–173 (1958).

We may interpret Nicholson's result as a time-delayed logistic process, which provides a good fit to the observed oscillations, with a value of $r\tau = 2.1$, in one of Nicholson's experiments. This model predicts the ratio of the maximum to the minimum population to be 84 and the cycle period to be 4.54τ. The experiment clearly reveals that density-dependent factors did not immediately affect the mortality rates of adults as the population increased but were felt a week or so later when the progeny were larvae. Larval mortality did not express itself in the size of the adult population until those larvae emerged as adults about 2 weeks after eggs were laid. The blowfly population resembled the *Daphnia* population at high temperature, in which crowding created discrete, nonoverlapping generations of individuals with an inherent time delay equal to the larval development period, about 10 days.

The hypothesis that time delays cause population cycles could be tested directly by eliminating the time delay in the density-dependent response — that is, by making the deleterious effects of resource depletion at high density felt immediately. Nicholson did this by adjusting the amount of food so that availability of food limited adults as severely as it did larvae. Adults require protein to produce eggs. By restricting the liver available to adults to 1 g per day,

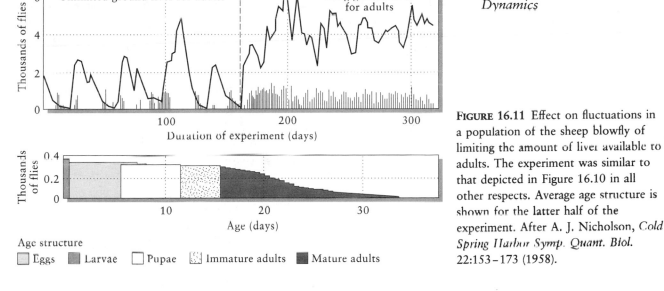

FIGURE 16.11 Effect on fluctuations in a population of the sheep blowfly of limiting the amount of liver available to adults. The experiment was similar to that depicted in Figure 16.10 in all other respects. Average age structure is shown for the latter half of the experiment. After A. J. Nicholson, *Cold Spring Harbor Symp. Quant. Biol.* 22:153–173 (1958).

Nicholson cut egg production to a level determined by the availability of liver rather than by the number of adults in the population. Under these conditions, the recruitment of new individuals into the population was determined at the egg-laying stage by the influence of food supply on per capita fecundity, and most of the larvae survived. As a result, fluctuations in the population all but disappeared (Figure 16.11).

We have seen that responses of populations to density can be delayed by development time and by the storage of nutrients, both of which put off deaths to a later point in the life cycle or to a later time. Density-dependent effects on fecundity can act with little delay when adults produce eggs quickly from resources accumulated over a short period. Populations controlled primarily by such factors should not exhibit marked oscillations.

Regardless of the time delay in the density-dependent response, a population at its equilibrium point will remain there until perturbed by some outside influence, whether a change in the equilibrium level K or a catastrophic change in the population size N. Once displaced from the equilibrium, some populations will move toward stable limit cycles, depending on the nature of the time delay and the response time. Others will return to the equilibrium directly or through damped oscillations. Cycles may be reinforced through interactions with other species — prey, predators, parasites, perhaps even competitors — with similar time constants, as we shall see in later chapters.

Metapopulations

Populations that are divided into discrete subpopulations are called **metapopulations** (Figure 16.12). Two types of process contribute to the dynamics of such populations. The first includes growth and regulation of populations within patches — processes that we have already discussed in detail. The sec-

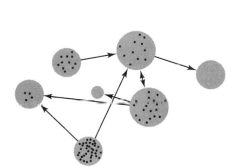

● Boundary of suitable habitat
• Individual
➤ Migration event

FIGURE 16.12 A metapopulation portrayed as a set of discrete subpopulations with partially independent local population dynamics. At any one time, some of the patches of suitable habitat may be occupied while others are not. Each dot represents an individual; arrows depict migration between patches.

ond includes **migration** of individuals between patches, or **colonization** in the case where individuals emigrate to an empty patch. Because subpopulations are typically much smaller than the population as a whole, local catastrophes and chance fluctuations in numbers of individuals have greater impact on population dynamics. Indeed, the smaller the patch, the higher the probability of extinction of the local population during a particular time interval. As a result, the dynamics of the total population depend very critically on the rate of migration between patches, which results in recolonization of empty patches.

When individuals move frequently between subpopulations, local population fluctuations damp out, and changes in local population size mirror those of the larger population. Thus a high rate of migration transforms metapopulation dynamics into the dynamics of a single large population. At the other extreme, when no individuals move between patches, the populations in each patch behave independently. When these subpopulations are small, they have high probabilities of extinction, and the total population gradually goes extinct as individuals die out in each of the patches it occupies. Intermediate levels of migration result in the colonization of some patches left unoccupied by extinction. Under such circumstances, the total population exists as a shifting mosaic of occupied and unoccupied patches. This mosaic has its own dynamics and equilibrium properties, which are illustrated simply in Box 16.1.

Stochastic effects

Small populations respond to chance events, which cause random fluctuations in population size. So far, in the chapters on population growth and regulation, we have considered models that assume large population size and no variation in the average values of birth and death rates due to chance. Such models, whose outcomes we can predict with certainty, are called **deterministic models.** In the real world, however, random variations can influence the course of population growth. Death is a chance event that occurs with a certain probability. But just as in coin flipping, the number of deaths within small populations has a probability distribution. The mean of this distribution is equal to the number of individuals in the population times the probability of death. But the actual number of deaths observed in a particular population will vary above or below this value just by chance. Such processes, which are influenced by chance events, are called **stochastic.**

Coin flipping provides a useful analogy for populations. Suppose you repeatedly flipped sets of 10 pennies. Although the probability of a head turning up on each toss is one-half, any one set of trials might turn up 6 heads and any other set, 3. When the test is repeated frequently enough, the average of the outcomes settles down to 5 heads and 5 tails, but many trials turn up 4 or 6 heads, somewhat fewer yield 3 or 7 heads, and runs with all heads or all tails occur once in 1024 trials, on average.

Turning to population models, let us suppose that adults successfully rear offspring with a probability of 0.5 per year. We would expect a population of 10 individuals to produce 5 offspring on average, but the actual number is

| BOX 16.1 | *A simple model of metapopulation dynamics* |

Consider a population subdivided into discrete patches. Without going into the details of the population dynamics within each patch, we assume that within a given time interval, each subpopulation has a probability of going extinct, which we shall refer to as k. Therefore, if p is the fraction of patches occupied by subpopulations, then subpopulations go extinct at the rate kp. The rate of colonization of empty patches depends on the fraction of patches empty $(1 - p)$ and the fraction of patches sending out potential colonists (p). Thus we may express the rate of colonization as a single constant b times the product $p(1 - p)$. Putting the extinction and colonization terms together yields the metapopulation dynamic

$$\frac{dp}{dt} = bp\,(1 - p) - kp$$

From this equation, it follows that the metapopulation attains an equilibrium ($dp/dt = 0$) when $b(1 - p) = k$. This may be rearranged to give

$$\hat{p} = 1 - \frac{k}{b}$$

the equilibrium proportion of occupied patches. The equilibrium is stable, because when p is below the equilibrium point, colonization exceeds extinction, and vice versa. This simple model shows the critical importance of the relative rates of dispersal and extinction (k/b). When $b = 0$, $\hat{p} = 1$ and all patches are occupied. (This does not mean that the patches cease to have independent dynamics, only that they are large enough or otherwise stable enough not to suffer extinction.) When $k = b$, $\hat{p} = 0$ and the metapopulation heads toward extinction. Intermediate values of k result in a shifting mosaic of occupied and unoccupied patches.

likely to vary from that value. What effect will this have on population growth? Consider a simple birth process (no deaths) in which a population grows exponentially according to $N(t) = N(0)e^{bt}$. Now suppose that the product bt equals 0.5. A deterministic model predicts a population at time t equal to 1.65 times the initial population ($e^{0.5}$). Thus, if the initial population were 500 individuals, the number after interval t would be 824. In very small populations, say $N(0) = 5$ individuals, the realized number would have a mean of 8.24 but would vary from as few as 5 (no births occurred) to as many as 20, just by chance (Figure 16.13).

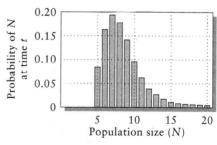

FIGURE 16.13 The probability distribution of the number of individuals N at time t in a population undergoing a pure birth process with initial size $N(0) = 5$ and $bt = 0.5$. After E. C. Pielou, *Mathematical Ecology.* Wiley, New York, 1977.

Stochastic extinction of small populations

Population change follows upon births and deaths. Birth rates and death rates depend on a variety of ecological factors, but whether a particular individual dies or successfully rears one or more progeny during an interval of time depends largely on chance. With an annual probability of death of one-half, for example, some individuals live and, on average, an equal number die. There exists a finite probability, however, that all the individuals in such a population will die, just as 10 out of 10 coin tosses will come up tails with a small but finite probability.

Chance events exert their influence more strongly in small populations than in large ones. This becomes clear when we consider that the probability of obtaining 5 tails in a row with successive tosses of a coin is 1 in 32, compared with the smaller chance of 1 in 1024 of obtaining twice as many tails in a row. When we visualize each individual in the population as a coin, and liken turning up tails to death, it is clear that a population of 5 individuals has a higher probability of extinction than a population of 10.

Theorists have devoted considerable attention to the probability of extinction of populations. They have derived mathematical expressions relating probability of extinction at time t [$p_0(t)$] to birth rate b, death rate d, and population size N. The simplest of these expressions represents the case in which b and d are equal — that is, births balance deaths and the average change in population size is zero — for which

$$p_0(t) = \left[\frac{bt}{1 + bt}\right]^N$$

Accordingly, the probability of extinction decreases with increasing population size and increases with larger b and d, indicating more rapid population turnover.

The relationship of the probability of extinction to population size N within time period t is shown in Table 16.1 for a population in which $b = d = 0.5$. These are reasonable values for adult death and recruitment in a popula-

TABLE 16.1	*Probability of extinction when birth rate = death rate = 0.5 per year, for populations of initial size i within period* t			
Population size (*i*)	TIME (*t*)			
	1	10	100	1000
1	0.33	0.83	0.98	0.998
10	10^{-4}	0.16	0.82	0.980
100	10^{-48}	10^{-7}	0.14	0.819
1000	10^{-99}	10^{-79}	10^{-8}	0.135

tion of terrestrial vertebrates. We see, for example, that for a population with 10 individuals, the probability of extinction is 0.16 within a 10-year period, 0.82 within a 100-year period, and virtually certain (0.98) within 1000 years. Even for an initial population size of 1000, the probability of extinction is more than 10 percent within a millennium and becomes virtually certain (0.999) within a million years.

When populations dwindle to small size, they become more and more susceptible to extinction, particularly on small islands where populations are restricted geographically and are rarely augmented by immigration. In fact, extinction occurs frequently enough that its probability can be determined from historical records. For example, species lists compiled in 1917 and 1968 for birds on the Channel Islands, off the coast of southern California, reveal that during the 51-year interval between censuses, 7 of 10 species disappeared from Santa Barbara Island (3 km² in area), but only 6 out of 36 species disappeared from the larger Santa Cruz Island (249 km²). (Some of the species extinct on each island were replaced by new colonists of different species.) On an annual basis, these figures can be expressed as between 0.10 and 1.7 percent of the avifauna per year, with extinction rate and island size inversely related. Comparable rates have been determined for two tropical islands: 0.2 percent per year on Karkar, an island of 368 km² located 16 km off the coast of New Guinea, and 0.23 percent per year on Mona, 26 km², located between Puerto Rico and Hispaniola in the Greater Antilles.

The disappearance of populations from isolated islands dramatizes the role of extinction of subpopulations in local patches in the dynamics of metapopulations. The extinction rate, which influences the equilibrium number of patches occupied, depends on the number of individuals in the subpopulation —and hence on the size of the patch it occupies. These considerations emphasize the interaction of spatial and temporal dynamics in population processes and the fact that we must understand the spatial structure of populations if we are to manage them intelligently.

✥ SUMMARY

Although ecologists find it easier to think about populations in equilibrium, most populations fluctuate, either because their size reflects variations in the environment or because they express oscillatory properties intrinsic to their dynamics. In this chapter, we have characterized factors that govern changes in populations with the goal of understanding the response of populations to external perturbations.

1. Key-factor analysis identifies factors that cause fluctuations in population size based on the relationship of change in population size (r) to the logarithms of the survival rates (S) of individuals at risk to various factors. Field studies have revealed that population-governing factors are often those that cause high and variable mortality from generation to generation. Many of these factors act in density-dependent fashion.

2. Discrete-time models of populations with density dependence tend to oscillate when perturbed. For r between 0 and 1, population size (N) approaches equilibrium (K) monotonically. For r between 1 and 2, N undergoes damped oscillations and eventually settles down to K. When r exceeds 2, oscillations in N increase in amplitude until either a stable limit cycle is achieved or the population fluctuates irregularly (chaos).

3. Continuous-time models can produce cyclic population change when the density-dependent response is time-delayed. Defining the time delay as τ, we find that such models exhibit monotonic damping when the product $r\tau$ lies between 0 and e^{-1} (0.37), damped oscillations when $r\tau$ lies between e^{-1} and $\pi/2$ (1.6), and limit cycles with a period of 4τ or more when $r\tau$ exceeds $\pi/2$.

4. Many laboratory populations of animals exhibit oscillations that arise from time delays in the response of individuals to density. These time delays are related to the period of development from egg to adult and may be enhanced by the storage of nutrients. In laboratory populations of sheep blowflies, A. J. Nicholson experimentally circumvented the time delay and was able to eliminate cycles in numbers.

5. When populations are subdivided into discrete subpopulations occupying patches of suitable habitat, ecologists refer to them as metapopulations. The dynamics of metapopulations depend on birth and death processes within patches and also on the level of migration of individuals between patches. When extinction of subpopulations is small compared with rates of colonization of unoccupied patches, the metapopulation exists as a changing mosaic of an equilibrium number of occupied patches.

6. The dynamics of small populations, such as those within an individual patch of a metapopulation, depend to a large degree on chance birth and death events. Stochastic models demonstrate that the probability of extinction due to random fluctuation in population size is greater in smaller populations.

⬧ SUGGESTED READINGS

Belovsky, G. 1987. Extinction models and mammalian persistence. In M. Soulé (ed.), *Viable Populations for Conservation.* Cambridge Univ. Press, Cambridge and New York, pp. 35–57.

Elliott, J. K. 1985. Population regulation for different life-stages of migratory trout *Salmo trutta* in a lake district stream, 1966–1983. *Journal of Animal Ecology* 54:617–638.

Finerty, J. P. 1980. *The Population Ecology of Cycles in Small Mammals.* Yale Univ. Press, New Haven, Conn.

Gilpin, M. E., and I. Hanski. 1991. *Metapopulation Dynamics. Empirical and Theoretical Investigations.* Cambridge Univ. Press, Cambridge and New York.

Goulden, C. E., and L. L. Hornig. 1980. Population oscillations and energy reserves in planktonic cladocera and their consequences to competition. *Proceedings of the National Academy of Sciences (USA)* 77:1716–1720.

Hassell, M. P., J. H. Lawton, and R. M. May. 1976. Patterns of dynamical behaviour in single-species populations. *Journal of Animal Ecology* 45:471–486.

Keith, L. B. 1990. Dynamics of snowshoe hare populations. *Current Mammalogy* 2:119–195.

Podoler, H., and D. Rogers. 1975. A new method for the identification of key factors from life-table data. *Journal of Animal Ecology* 44:85–114.

Taylor, A. D. 1990. Metapopulations, dispersal, and predator–prey dynamics: an overview. *Ecology* 71:429–433.

Varley, G. C., and G. R. Gradwell. 1970. Recent advances in insect population dynamics. *Annual Review of Entomology* 15:1–24.

POPULATION GENETICS AND EVOLUTION

Not all individuals within a population are the same. Each is endowed with a unique genetic constitution, made up of a combination of genetic factors from its mother and father. Genetic variation within a population has many important consequences. First, when genetic variation causes differences in fecundity and survival among individuals, evolution results because those individuals whose characteristics enable them to achieve higher rates of reproduction increase in the population at the expense of those that are genetically less well endowed. Second, most genetic variation is harmful to the individual, particularly when an individual inherits copies of a mutant gene from both parents. Because close relatives are more to likely share mutant genes than are individuals picked at random from the population, most species have evolved mechanisms to reduce the chance of close inbreeding (mating among relatives) and its harmful genetic consequences. Finally, the number of individuals and the breeding system can influence the genetic variation of a population and its long-term potential for evolutionary response to a changing environment. In particular, small population size fosters the elimination of rare genetic variants and reduces the genetic diversity of the population.

In this chapter, we shall discuss the origin of genetic variation and its consequences for evolution. We shall also consider some of the mechanisms by which individuals manage the short-term deleterious effects of most genetic variations. Because the interaction between the organism and the environment is subject to genetic influence, ecology and evolution are intimately related. Indeed, we would have difficulty appreciating one without understanding the other.

The origin of genetic variation

All the genetic information in a cell is contained in the sequence of four subunits along the molecule **deoxyribonucleic acid,** or **DNA** for short. The molecular subunits of DNA are adenine, thymine, cytosine, and guanine, referred to by their first letters as A, T, C, and G. The DNA strand forms a template from which the cell manufactures proteins. Proteins themselves consist of chains composed of up to 20 different types of amino acids, each encoded by a sequence of three subunits of the DNA molecule. The coding triplet is referred to as a **codon.** For example, the DNA sequence AAA (adenine-adenine-adenine) specifies the amino acid phenylalanine, GAG specifies leucine, CTT specifies glutamic acid, and so on. Because four different subunits taken three at a time yield 64 (4^3) different sequences, the genetic code contains considerable redundancy. In other words, because DNA must encode only 20 amino acids, and 64 different codons are possible, several codons may specify a single amino acid. For example, at the extreme, leucine is encoded by six different sequences: AAT, AAC, GAA, GAG, GAT, and GAC.

A change (*substitution*) in one of the subunits in a DNA codon may alter the amino acid that it specifies. For example, consider the following sequence of subunits in a DNA molecule and the corresponding amino acids:

Position:	1	4	7	10	13	16
DNA:	GAA	TGG	CGA	GAA	ATA	GGG
Amino acid:	Leucine	Serine	Alanine	Leucine	Tyrosine	Proline

If the guanine subunit occupying the eighth position were changed to thymine, the third codon would be altered from CGA to CTA and would encode the amino acid aspartine instead of alanine. Such changes do occur, and geneticists call them **mutations.** The new protein produced by the mutant gene might have different properties from the original type, and although these properties *might* be beneficial, they are more likely to decrease the survival or the fecundity of the individual in some way. Because of the redundancy in the genetic code, changes in some DNA subunits do not alter the amino acid specified by a codon and therefore are not expressed in the phenotype.

The sequence of DNA subunits is read in groups of three, so the deletion of a single subunit offsets the triplets by one place from their original position and changes the sequences of subunits in all the ensuing codons. For example, if guanine were deleted from the eighth position of the sequence presented above, the third triplet of the code would now be read as the seventh, ninth, and tenth subunits, resulting in the following amino acid sequence:

Position:	1	4	7	10	13	16
DNA:	GAA	TGG	CAG	AAA	TAG	GGT
Amino acid:	Leucine	Serine	Valine	Phenylalanine	Isoleucine	Proline

In general, *deletions* and *additions* of subunits have far more pervasive and negative effects on proteins than do substitutions.

Many proteins have been characterized so well genetically and biochemically that we can demonstrate the connection between genetic material, protein, organism function, and reproductive success. The hemoglobin molecule is one of these. Hemoglobin consists of four protein chains; the predominant type of hemoglobin in adult humans has two kinds, referred to as alpha and beta chains, each determined by a different gene. The disease sickle-cell anemia results from a mutation in a single subunit in the gene that determines the amino acid sequence in the beta chain. When the seventeenth subunit, which is normally thymine, changes to adenine, the amino acid valine (CAT) replaces glutamic acid (CTT) as the sixth amino acid in the sequence. This genetic defect alters the structure of hemoglobin such that when the molecules release oxygen, they become stacked close together in long helices, impairing their function and causing severe and debilitating anemia. The aberrant hemoglobin molecules give red blood cells a peculiar sicklelike shape — hence the name of the disease.

A mutation is a change in the individual DNA subunit. Mutations are caused by random copying error when the genetic material replicates during cell division, by the action of certain chemical agents, or by ionizing radiation. For any particular subunit, the rate of mutation is extremely low, on the order of one in 100 million per generation. However, this low rate multiplied by hundreds or thousands of DNA subunits in a gene, and by a trillion or so in such complex organisms as vertebrates, means that each organism usually has one or more mutations in some part of its genome. Many of these mutations do not express themselves because they occur in portions of the DNA molecule that do not encode proteins, because they create redundant codons that do not change the specified amino acid, or because the amino acid changes caused by the mutation have no physiological effect. Nonetheless, measured rates of expressed mutation average about one in 100,000 to one in a million per generation for each gene.

Most mutations exert a detrimental effect on the organism. The reason for this is simple. Over millions of years of evolution, natural selection weeds out most of the deleterious genes, leaving behind only those genes that suit the organism well to its environment. Any new variant is likely to disrupt the well-tuned interaction between the organism and its surroundings. Mutations may convey an advantage when the environment changes and the organism is no longer so well adapted. Then, small changes in the genetic material inherited from one parent may sometimes shift the structure and functioning of the organism to better match it to its new environment. These changes in the environment come about through variation in climate; the introduction of new organisms into the habitat; or genetic changes in predators, prey, and disease organisms that are already present. Thus the evolution of genetic systems reflects both the generally deleterious nature of mutation and the fact that species sometimes endure because genetic variability has endowed a few members with the capacity to survive a certain kind of change.

Genotype and phenotype

The **genotype** includes all the genetic characteristics that determine the structure and functioning of the organism, which themselves make up the **phenotype.** Thus the genotype is the set of genetic instructions, and the phenotype represents the rendering, or expression, of the genotype in the form of the organism. To put it another way, the genotype is to the phenotype as blueprints are to the structure of a building. Each **gene** codes for a particular protein, which may contribute to the organism's structure or may function as an enzyme or hormone. Different sequences of genetic subunits in a particular gene, caused by mutations, are referred to as **alleles.** In many cases these alleles, or alternative forms of a given gene, have visible effects on the organism's phenotype. For example, blue-eyed and brown-eyed humans have different alleles of a single gene. Many genetic disorders, such as sickle-cell anemia, Tay-Sachs disease, cystic fibrosis, and albinism in humans, also result from mutations in a single gene.

Every organism has two copies of each gene, one inherited from its mother and one from its father (some exceptions include sex-linked genes and organisms that reproduce without the sexual union of gametes). An individual that has two different forms (alleles) of a particular gene is said to be **heterozygous** for that gene. When both copies of the gene are the same, the individual is **homozygous.** When different alleles of a gene reside in an individual, either they may produce an intermediate phenotype, or one may mask the expression of the other. In the latter case, one allele is said to be **dominant** and the other **recessive.** Most deleterious mutations are recessive because the resulting decrease in function of its gene product may be masked by the normal gene product of the dominant allele.

Hardy–Weinberg equilibrium

It is an important principle of genetics that in a large population, with random mating, no selection, and no mutation, the frequencies of homozygotes and heterozygotes would achieve equilibrium proportions, which we may calculate from the proportions of each allele in the population. This is called the **Hardy–Weinberg equilibrium** after the two geneticists who independently described the principle in 1908. Suppose that a particular genetic locus A has two alleles A_1 and A_2 that occur in proportions p and q ($p + q = 1$). At Hardy–Weinberg equilibrium the three genotypes that can result will occur in the following proportions:

Genotype:	A_1A_1	A_1A_2	A_2A_2
Frequency:	p^2	$2pq$	q^2

Thus if one allele (A_1) occurs with a frequency of 0.7, 49 percent ($0.7^2 = 0.49$) of the genotypes that result will be the homogygote of A_1, 42 percent ($2 \times 0.7 \times 0.3 = 0.42$) will be the heterozygote, and 9 percent ($0.3^2 = 0.09$) will be the homozygote of A_2. It is very useful to know this because *deviations* from these genotype frequencies indicate the effects of selection, nonrandom mating, or other factors that influence the genetic makeup of the population.

Variation in quantitative traits

The relationships between organisms and their environments often depend on modifications of continuously varying traits, such as lengths of appendages, body size and shape, thickness of hair or cuticle, and continuous gradations of behavior. These traits are often called **quantitative** characteristics. Variation in these traits depends on the contributions of many genes, and their expression in the phenotype cannot be analyzed in the same manner as genotype frequencies for a single gene. Animal and plant breeders interested in such traits as milk production, the oil content of seeds, and the rate of egg production have developed a mathematical treatment of continuously varying traits, known as **quantitative genetics.** The theory of quantitative genetics assumes that variation within a population results from the additive contributions of many genes with similar effect. Thus the length of an appendage may come under the influence of a dozen genetic loci, each of which may cause a small increase or decrease in length relative to the population average. Individuals with a net excess of length-incrementing alleles at these 12 loci would have appendages longer than the average.

The heart of quantitative genetics is the **variance** of a trait within a population. Variance (V) is a statistical measure of variation; specifically, it is the average of the squared deviations of individuals from the mean of the population (Box 17.1). Continuously variable traits often exhibit a bell-shaped distribution of values in natural populations, with most individuals clustered near the mean and with frequency diminishing at extreme values (Figure 17.1). Each individual's value of a particular trait (the **phenotypic value**) is determined by deviations from the population mean caused by genetic and environmental influences. Because both sources of deviation enter into the calculation of variance for all values in the population, we may speak of phenotypic variance (V_P) as having two components, one attributable to genetic constitution (V_G) and one resulting from environmental factors (V_E). The two components added together equal the total phenotypic variance; that is, $V_P = V_G + V_E$.

The genotypic variance can be subdivided further: V_A is the additive variance determined by the expression of alleles in homozygous form; V_D is the dominance variance determined by the interaction of alleles in heterozygous form; and V_I is the interaction variance, comprising the influences of different genes on the expression of alleles at a particular locus. The principal task of quantitative genetics has been to estimate the magnitude of the several components of phenotypic variance. This is made necessary by the fact that response to selection derives only from the additive genetic component of variance: only V_A reflects the genetic diversity of the population — that is, the different

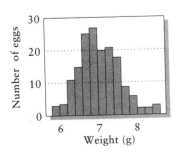

FIGURE 17.1 Frequency distribution of the weights of eggs of the European starling *Sturnus vulgaris* near Philadelphia, Pennsylvania.

Box 17.1	*Calculation of the variance within a small population*

The variance in a trait X (a measurement) in a population of n individuals is

$$V = \frac{1}{n} \sum_{i=1}^{n} (X_i - \overline{X})^2$$

where X_i is the value for each individual i ($i = 1$ to n), and $\overline{X}$ is the mean value of X in the population. The calculation of a variance is shown by example in the following tabulation.

Individual (i)	Height, in inches (X_i)	Deviation from mean ($X_i - \overline{X}$)	Squared deviation ($X_i - \overline{X}$)²
1	75	5	25
2	73	3	9
3	72	2	4
4	72	2	4
5	71	1	1
6	70	0	0
7	68	−2	4
8	67	−3	9
9	67	−3	9
10	65	−5	25
	Total = 700	Sum of squared deviations = 90	
	Mean ($\overline{X}$) = 70	Variance = 9.0	

alleles that replace, and are replaced by, others during evolutionary change. Phenotypic variance can be partitioned into its several components by statistical analyses of the results of breeding programs designed for the purpose. Suffice it to say that these analyses use correlations of phenotypic values between close relatives — usually between parents and their offspring or between siblings.

The proportion of the phenotypic variance that is due to additive genetic factors is often expressed as their ratio and called the **heritability** ($h^2 = V_A/V_P$). Historically, most studies of heritability involve traits of commercial value in livestock, poultry, and crops. Representative values of h^2 for such traits ap-

TABLE 17.1	*Heritabilities of several traits in domesticated and laboratory animals and plants*	
Trait	Organism	Heritability
Size or length		
Plant height	Corn	0.70
Root length	Radish	0.65
Tail length	Mice	0.60
Length of wool	Sheep	0.55
Body length	Pigs	0.50
Weight		
Body weight	Sheep	0.35
Body weight	Pigs	0.30
Body weight	Chickens	0.20
Production		
Butterfat content of milk	Cattle	0.60
Egg weight	Chickens	0.60
Thickness of back fat	Pigs	0.55
Weight of fleece	Sheep	0.40
Milk yield	Cattle	0.30
Fecundity		
Egg production	Chickens	0.30
Yield	Corn	0.25
Litter size	Pigs	0.15
Litter size	Mice	0.15
Conception rate	Cattle	0.05
Life history		
Age at onset of laying	Chickens	0.50
Age at puberty	Rats	0.15
Viability	Chickens	0.10

Source: D. S. Falconer, *Introduction to Quantitative Genetics* (2nd ed.). Ronald Press, New York (1981).

pear in Table 17.1. The data indicate that sizes have higher heritabilities (0.50 to 0.70), and hence are less sensitive to environmental variation, than weights (0.20 to 0.35). Among traits related to production and fecundity, those creating the greater drain on energy and nutrients have the lower heritabilities. Thus the percentage of butterfat in milk is under strong genetic control ($h^2 = 0.60$), whereas total milk production has a low heritability (0.30); and variation in egg size in chickens has a large additive genetic component ($h^2 = 0.60$), whereas rate of egg production has a lower heritability (0.30). Any trait that requires a large commitment of resources is sensitive to environmental variation in those resources. The heritabilities of fecundity and life history characteristics generally are low (0.05 to 0.50). Heritability estimates for traits of individuals in wild populations resemble those of domestic animals and crops.

Fitness and natural selection

Different phenotypes interact with the environment in slightly different ways, leading to variation in fecundity and survival among individuals in a population. The rate of reproduction of a phenotype — the intrinsic rate of increase of the life table for that phenotype — is a measure of its fitness. When differences in fitness among individuals have a genetic basis, those alleles that have the higher fitness reproduce faster and take over the population. The phenotypes with the highest fitness are said to be **selected,** and the change in genotype frequencies resulting from **natural selection** is referred to as **evolution.**

We can now point to many cases of selection producing evolution in natural populations, often in response to environmental changes brought about by human activities. Perhaps the most striking case is that of industrial melanism in the peppered moth in England.

The English have always been avid butterfly and moth collectors, and such enthusiasts look carefully for rare variant forms. Early in the nineteenth century, occasional dark (or **melanistic**) specimens of the common peppered moth (*Biston betularia*) were collected. Over the next 100 years, the dark form, referred to as *carbonaria,* became increasingly common in industrial areas (today it makes up nearly 100 percent of some populations). The phenomenon aroused considerable interest among geneticists, who showed by cross-mating light and dark forms that melanism is an inherited trait determined by a single gene.

In the early 1950s H. B. D. Kettlewell, an English physician who had been practicing medicine for 15 years and was an amateur butterfly and moth collector, changed the course of his life to pursue the study of industrial melanism. Several facts about melanism were known before Kettlewell began his studies. The melanistic trait is an inherited characteristic, so its spread reflected genetic changes (evolution) in the population. The earliest records of the *carbonaria* form were from forests near heavily industrialized regions of England. In the absence of factories and other heavy industry, the light form of the moth still prevailed. It was also known that melanism is not unique to the peppered moth; dark forms have appeared in many other moths and in other insects.

The peppered moth inhabits dense woods and rests on tree trunks during the day. Kettlewell reasoned that where melanistic individuals had become common, the environment must somehow have been altered in such a way as to give the dark form a survival advantage over the light form. Could natural selection have led to the replacement of the typical light form by the *carbonaria* form? To test this hypothesis, Kettlewell had to find some measure of fitness other than the relative evolutionary success of the two forms.

To determine whether the *carbonaria* form had greater fitness than typical peppered moths in areas where melanism occurred, Kettlewell chose the mark-recapture method. He marked adult moths of both forms with a dot of cellulose paint and then released them. The mark was placed on the underside of the wing so that it would not call the attention of predators to a moth resting on a tree trunk. Kettlewell recaptured moths by attracting them to a mercury vapor lamp in the center of the woods and to caged virgin females at the edge of the woods. (Only males could be used in the study because females are attracted neither to lights nor to virgin females.)

In one experiment, Kettlewell released 201 typicals and 601 melanics in a wooded area near industrial Birmingham, where the tree bark was darkened by pollution. The results were as follows:

	Typicals	Melanics
Number of moths released	201	601
Number of moths recaptured	34	205
Percent recaptured	16	34

These figures indicated that more of the dark form survived over the course of the experiment.

Although consistent with Kettlewell's original hypothesis, the results could be interpreted otherwise: as differential attraction of the two forms to the traps or as differential dispersion of the two forms away from the point of release. Variables besides differential mortality had to be accounted for.

To test the hypothesis of natural selection unequivocally, Kettlewell ran a similar experiment in an unpolluted forest near Dorset, where the tree bark was lighter. Of 496 marked typicals released, 62 (12.5 percent) were recaptured; of 473 marked melanics, only 30 (6.3 percent) were recaptured. Thus in the unpolluted forest, light adults had a higher recapture rate than dark adults. If typicals and melanics were differently attracted to light traps or dispersed from a release point at different rates, the level of pollution would not have influenced the results. In fact, only differential survival could account for a reversal in the relative rates of recapture. This confirmed Kettlewell's hypothesis and established natural selection as being responsible for the high frequency of the *carbonaria* form in industrial areas.

The specific agent of selection was easily identified. Kettlewell reasoned that in industrial areas, pollution had darkened the trunks of trees so much that "typical" moths stood out against them and were readily found by predators. Any aberrant dark forms were better camouflaged against the darkened tree trunks, and their coloration conferred survival value (Figure 17.2). Eventually, differential survival of dark and light forms would lead to changes in their relative frequency in a population. To test this idea, Kettlewell placed equal numbers of the light and dark forms on tree trunks in polluted and unpolluted woods and watched them carefully at some distance from behind a blind. (A blind is a tentlike structure intended to conceal observers from their subjects; it is more often called a hide in England.) He quickly discovered that several species of birds regularly searched the tree trunks for moths and other insects and that these birds more readily found a moth that contrasted with its background than one that resembled the bark it clung to. Kettlewell tabulated the following instances of predation:

	INDIVIDUALS TAKEN BY BIRDS	
	Typicals	Melanics
Unpolluted woods	26	164
Polluted woods	43	15

These data were fully consistent with the results of the mark-recapture experiments. Together they clearly demonstrate the operation of natural selection, which over a long period resulted in genetic changes in populations of the peppered moth in polluted areas. Many decades were required for the replacement of one form by the other. The agents of selection were insectivorous

FIGURE 17.2 Typical and melanistic forms of the peppered moth at rest on a lichen-covered tree trunk in an unpolluted countryside (a) and on a soot-covered tree trunk near Birmingham, England (b). From the experiments of H. B. D. Kettlewell.

birds whose ability to find the moths depended on the coloration of the moth with respect to its background. Kettlewell's study shows clearly how the interaction between the organism and its environment determines its fitness.

Population genetics

The evolutionary mechanics of selection and genetic responses are part of the science of **population genetics.** A primary task of population geneticists since the late 1920s has been to develop quantitative predictions of changes in gene frequencies in response to selection. Other issues that have been investigated include the influences of population size and breeding system on the maintenance of genetic variation in populations, as we shall see.

In such simple cases as selection upon a single gene with one allele dominant over the other, the models of population genetics describe mathematically the intuitive result: the allele whose bearers leave the most descendants eventually predominates in the gene pool. The equations also enable us to predict rates of change in gene frequency and, hence, how rapidly a population can respond genetically to a change in the environment.

The time required for a dominant gene to replace its recessive allele depends on its frequency at the beginning and end of the substitution process and on the strength of selection. In the case of the replacement of the typical form of the peppered moth by the *carbonaria* form in polluted woods of England, this substitution is known to have taken about a century. From the results of H. B. D. Kettlewell's experiments on the peppered moth, we can estimate that the fitness of the allele for typical coloration was only 47 percent that of the *carbonaria* allele in polluted woods; hence the fitness differential, or strength of selection, was 0.53. Plugging this value into equations predicting the time required for allele substitution reveals that a change in frequency of the *carbonaria* allele from 0.05 to 0.95 would have required 47 generations, and a change from 0.01 to 0.99 would have required 204 generations. The peppered moth has 1 generation each year, so the population genetics equations appear to be consistent with the observed time course of the gene substitution, assuming an initial frequency before the Industrial Revolution of about 1 percent. Had selection been only a tenth as strong, the same change in frequency would have required 10 times longer.

The same equations can be turned around and used to estimate fitness differentials from changes in gene frequency. For example, since the introduction of pollution control programs in England, the frequency of the *carbonaria* allele has decreased at a rate consistent with a 12 percent selective disadvantage. Records of trapped red and silver foxes kept by the Moravian mission posts in Labrador for a hundred years (1834 to 1933) show that the proportion of silver foxes in the catch declined from about 0.15 to 0.05 during that time. Because silver coat is a recessive phenotype, the frequency of the genotype is equal to q^2, and the frequency of the silver allele (q) must have decreased from 0.39 to 0.22 over a period of 100 years, a change that would have required a fitness differential of 0.035, or 3.5 percent per year. It does not seem unreasonable that a greater demand for silver fox furs might have led trappers to cause an annual mortality of silver foxes 3 percent higher than that of red foxes.

Selection and evolutionary response of quantitative traits

The change in a quantitative trait resulting from a single generation of selection (R, for response) depends on the deviation of selected individuals from the mean value of the population (S, for selection differential) and on the heritability of the trait, according to the relationship

$$R = h^2 S$$

For example, if h^2 were 0.5 and males and females 10 size units larger than the population average were bred together, their progeny would be 5 size units, overall, larger than the average of the unselected population. The greater the heritability of the trait, the more rapidly it can respond to selection.

Values of R and S are conveniently expressed as multiples of the standard deviation of measurements within the population, the standard deviation (SD) being the square root of the variance. Each value expressed in standard deviation units corresponds to a particular percentile rank in the population. Zero SD units represents the population mean; when values are symmetrically distributed about the mean, half the individuals lie above that value and half below. When values have a normal distribution, as in Figure 17.3, 31 percent of the individuals lie above +0.5 SD and 31 percent lie below −0.5 SD from the mean; 16 percent have values more extreme than +1.0 SD or −1.0 SD; 7 percent exceed 1.5 SD; and only 2.3 percent exceed 2.0 SD in each direction from the mean. As a result, the more intense the selection (the larger the value of S), the smaller the number of individuals selected and the smaller the number of resulting progeny for the next generation of selection. When selection is too strong, the population dwindles, eventually to extinction. Even in artificial selection programs, the strength of selection is limited by the size of the stock population and by the reproductive rate of selected individuals, which must be at least as large as the number of individuals eliminated by selection each generation.

The relationship among selection intensity (percentage of individuals selected), phenotypic selection differential (S), and response (R) is illustrated by a program of selection for rate of egg laying in the flour beetle *Tribolium castaneum*. In the stock population, the number of eggs laid by an adult from 7 to 11 days after it emerges from the pupa (the phenotypic trait investigated) had a mean of 19.0, a standard deviation of 11.8, and a heritability of 0.30. The investigators established one unselected line and five lines with different levels of selection ranging from 50 to 5 percent (Table 17.2). Knowing the variability, heritability, and selection differential, we can estimate the initial response of the population to selection. For example, in the C line, a selection intensity of 20 percent corresponds to a selection differential of 1.4 SD, or 16.5 eggs (1.4 × 11.8). With a heritability of 0.30, the response to selection should be about 5 eggs per generation ($R = h^2 S = 0.30 \times 16.5$). The observed response fell somewhat short of this prediction (about 3 eggs per generation), probably because the estimate of heritability included maternal and dominance effects as well as additive genetic variation. But the beetle population did behave as

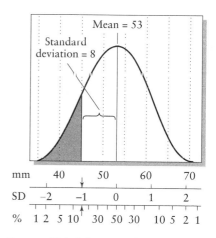

FIGURE 17.3 Schematic diagram of variation in a hypothetical trait showing the relationship among the measurement scale, the standard deviation, and the proportion of the population with phenotypic values more extreme than a particular value. For example, 16 percent of the population have phenotypic values greater than 1 standard deviation below the mean (shaded portion). The same is true of phenotypic values greater than 1 SD above the mean.

	Number of families scored per generation	Number of females scored per family	Total scored	Total selected	Selection intensity (percent removed)	Selection differential (S)*
Line						
A	10	20	200	10	95	2.0
B	20	10	200	20	90	1.8
C	40	5	200	40	80	1.4
D	66	3	198	66	67	1.1
E	100	2	200	100	50	0.8
F	200	1	200	200	0	0.0

TABLE 17.2 *Selection procedure in six lines of* Tribolium castaneum *under selection for fecundity*

*Standard deviation units.

Source: R. G. Ruano, F. Orozco, and C. Lopez-Fanjul, *Genet, Res.* 25:17–27 (1975).

predicted in that the rate of response varied in direct proportion to the intensity of selection (Figure 17.4).

With continued selection pressure on experimental populations, the response to selection eventually stops, as Figure 17.4 shows. The slowed response results from two factors: the erosion of genetic variation by selection and the application of opposing selection by correlated changes in other traits. Selection works only when individuals vary genetically within a population. When all unfit alleles are removed by selection, evolution pauses until new mutations or gene combinations appear.

Laboratory studies show that the large gains predicted by quantitative-genetics models can, in fact, be realized. Because quantitative variation is multigenic, the selection response can be pushed several standard deviations beyond the mean phenotypic value of the unselected population. For example,

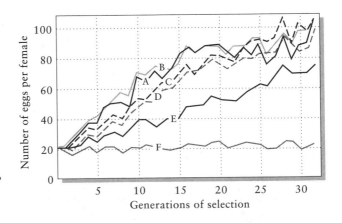

FIGURE 17.4 Change in rate of egg laying in *Tribolium castaneum* lines exposed to different levels of selection, increasing from F to A. From R. G. Ruano, F. Orozco, and C. Lopez-Fanjul, *Genet. Res.* 25:17–27 (1975).

mature body weights of unselected Japanese quail average about 91 g with a standard deviation of 8 g. After 40 generations of selection for high body weight, the population average in one study increased to 200 g, or almost 14 SD units above the mean of the unselected population.

What is often more difficult to explain than the response of a trait to selection is the leveling off of the response, often after only modest progress. This usually does not result from exhaustion of genetic variation for the trait: reverse selection (back toward the mean of the unselected population) typically produces an immediate response, which can happen only when genetic variation remains in the population. Furthermore, when investigators merely relax selection and breed all phenotypes with equal frequency, a selected trait sometimes returns toward the preselection measurement, apparently by itself. The most reasonable explanation of these results is that selection applied to one trait causes changes in other traits that affect the fitness of the organism. Increase in rate of egg laying, for example, may cause physiological or morphological changes that reduce viability and thus oppose the artificial selection regime.

Correlated responses to selection

Evolutionary responses often include traits other than the one selected. Both development and functioning integrate the parts of organisms, bringing about an interdependence of phenotypic traits, particularly those involving size or rate of growth and production. Animal and plant breeders have estimated genetic correlations between traits in many domestic species. For example, in poultry the genetic correlation between body weight and egg weight is 0.50; between body weight and egg production, it is −0.16. Therefore, one cannot apply selection to body weight without also obtaining a relatively rapid increase in egg size and a slower but steady decrease in rate of laying. Simultaneous selection of large egg size and small body size goes against the grain of genetic correlation, and it is usually unsuccessful.

Many traits, such as body weight and egg size, form developmentally, genetically, or functionally related groups that tend to respond to selection in concert. For example, genetic correlations among skeletal measurements of mice reveal four clusters of traits that are highly integrated genetically among themselves but relatively independent of each other: (1) skull length; (2) skull width, body weight, and tail length; (3) skull width (providing a link to group 2), scapula length, and other measurements associated with the pectoral girdle; and (4) limb bones and total body length. In mice, therefore, selection for body weight produces responses in tail length and the proportions of the skull as well as in body weight itself.

Flour beetles (*Tribolium*) selected for fast and for slow larval development exhibit a number of correlated responses that decreased fitness regardless of the direction of selection. Selection for rapid development resulted in decreased size and larval survival and increased incidence of adult abnormalities. Selection for slow development resulted in increased adult weight, increased incidence of adult abnormalities, and decreased fertility of females.

Similar correlated responses in reproductive fitness have appeared in selected lines of chickens. Selection for both increase and decrease in body weight and egg weight caused a decline in reproductive fitness, as indicated by rate of egg production, hatch rate, and survival of chicks to 9 months. Regardless of the character or direction of selection, fitness in the selected lines varied from 54 to 85 percent of that in the nonselected line. These and other experiments emphasize the fact that natural populations are balanced genetically and that their adaptations are both well tuned to the environment and finely adjusted to one another.

Inbreeding and outcrossing

Everyone knows that close inbreeding is bad. **Inbreeding** is simply mating among close relatives. Brother–sister matings and, where possible (especially in plants), self-mating (**selfing**) may result in the expression of deleterious recessive genes. The mechanism is simple. Suppose that an individual is heterozygous for a rare, deleterious, recessive gene (the gene pool is full of them, and most individuals have some). If that individual were to mate with an individual from the general population, which probably would not have the same rare allele, half of their progeny would be heterozygous, like the one parent, and half would be homozygous for the common form of the gene, like the other parent (Figure 17.5). None of the progeny would be disadvantaged by the union. If, however, the individual selfed, one-quarter of its offspring would be homozygous for the deleterious allele and would suffer loss of fitness as a result. Matings between close relatives produce the same result, only less frequently.

FIGURE 17.5 A diagram showing how inbreeding can uncover deleterious recessive genetic variation. With respect to a single gene, each heterozygous parent produces two types of gametes in equal frequency, one having a copy of the gene inherited from the parent's mother and one from the parent's father. Four combinations of these gamete types appear with equal probability in the offspring. With inbreeding, one-quarter of the progeny of two heterozygous parents express a deleterious allele in homozygous form.

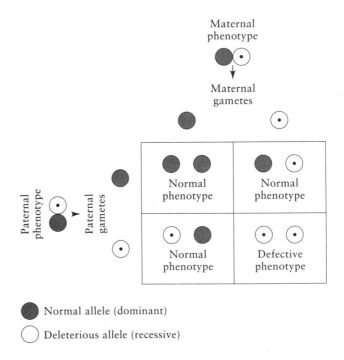

Most species employ mechanisms — including dispersal of progeny, recognition of close relatives, and negative assortative mating — to reduce the occurrence of inbreeding. Hermaphroditic species of plants, in which individuals bear both male and female sexual organs, have additional mechanisms to prevent selfing, including self-incompatibility, temporal separation of male and female function, and elaborate flower structures designed to make self-fertilization difficult.

But although inbreeding generally creates problems, it may confer benefits as well. In particular, selfers can guarantee fertilization of their flowers in habitats that lack suitable pollinators or where individuals are widely spaced. Many weedy species that colonize isolated patches of disturbed habitat (for example, dandelions) are selfers. It is assumed that most deleterious variation was weeded out of such populations as it was exposed in homozygous individuals during the transition between outcrossing and selfing.

Outcrossing at great distance may also reduce fitness when populations of plants include spatially defined variation over small scales of distance, particularly in complex, heterogeneous environments. In such cases, local adaptation to particular habitat patches enhances fitness, and receiving pollen from individuals adapted to different habitat conditions may reduce the fitness of progeny that become established near the female parent.

Several studies have reported an optimal outcrossing distance in populations of plants. Nearby individuals are likely to be close relatives, which creates the problem of inbreeding. Distant individuals are likely to be adapted to different conditions. For example, when flowers of the larkspur *Delphinium nelsoni* in central Colorado were fertilized with pollen obtained from the same individual and from individuals located at distances of 1, 10, 100, and 1000 m, the number of seeds set per flower was greatest when the pollen source came from a distance of 10 m and least for selfed pollen and for that obtained 1000 m distant. Furthermore, when these seeds were planted, survival to 1 and 2 years greatly favored matings across the intermediate distance of 10 m.

Plants may not be able to control the distances that their pollinators travel between flowers, but they may manage genetic variation by establishing competition between pollen grains for the opportunity to fertilize ovules and by selectively aborting developing ovules on the basis of the genotype of the embryo. Most plants produce many more flowers than they can support to maturity as fruits; flowers are relatively cheap, seeds and fruits expensive. Pollen tubes must grow through the style (maternal tissue) to reach the ovule. Biochemical interactions that control the rate of that growth may be sensitive to the genotype of the pollen and the female flower parts. This is the basis of self-incompatibility. A heavy pollen load is advantageous to the female because many more pollen tubes compete to fertilize the ovules, resulting in strong selection among pollen genotypes. Excess fertilized ovules are reduced in part by predation or other extrinsic damage — but also in part by programmed abortion. Abscission of flowers and fruits can be highly selective with respect to pollen loads, number of ovules fertilized per flower, and genotypes of developing embryos.

Genetic changes in small populations

In small populations, the frequencies of alleles and genotypes may change because of random variations in birth and death rates that are due to chance events, even in the absence of selection and mutation. Such changes are referred to as **genetic drift.** When two alleles occur together in a population, the rate of increase or decrease in each of them has a random component. Just by chance, one may increase and the other decrease, resulting in a change of frequency. These changes are not biased in one direction or the other, but the occurrence, by chance, of a long series of changes in one direction may lead to the disappearance of one allele from the population. In this case, we say that the other allele becomes **fixed.** Further variation in allele frequency cannot take place unless mutation introduces new alleles into the population.

The rate of fixation of alleles is inversely related to the size of the population. Thus genetic variation decreases more rapidly over time in small populations than in large ones. Furthermore, a single episode of small population size, such as the colonization of an island or a new habitat by a few individuals out of a large parent population, can reduce genetic variation in the colonizing population. Such episodes are known as **founder events.** When founding populations consist of 10 or fewer individuals, they typically contain a substantially reduced sample of the total genetic variation of the parent population. Continued existence at low population size results in further loss of genetic variation to drift and close inbreeding. The significance of such variation for natural populations is that the fragmentation of populations into small subpopulations may eventually restrict the evolutionary reponsiveness of each subpopulation to the selective pressures of changing environments, thus making a small population more vulnerable to extinction.

Geographic variation in the gene pool

The genotypes of individuals within a population often vary geographically. This may happen because of differences in selective factors in different parts of the population or through random changes (genetic drift, founder effects) in isolated or partially isolated subpopulations.

Botanists have long recognized that individuals of a species that are grown in different habitats may exhibit various forms corresponding to the conditions under which they are grown. In many cases, these differences among habitats result from developmental responses, but experiments on some species have revealed genetic adaptations to local conditions.

About 70 years ago, the Swedish botanist Göte Turesson collected seeds of several species of plants that lived in a variety of habitats and grew them in his garden. He found that even when grown under identical conditions, many of the plants exhibited different forms depending on their habitat of origin. Turesson called these forms **ecotypes,** a name that persists to the present, and suggested that ecotypes represent genetically differentiated strains of a population, each restricted to a specific habitat. Because Turesson grew these plants

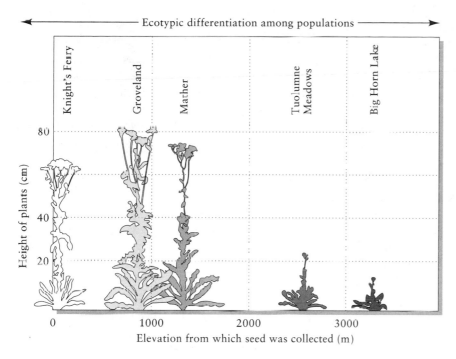

FIGURE 17.6 Ecotypic differentiation in populations of the yarrow, *Achillea millefolium,* demonstrated by raising plants derived from different elevations under identical conditions in the same garden at Stanford, California, which is close to sea level. After J. Clausen, D. D. Keck, and W. M. Hiesey, *Carnegie Inst. Wash. Publ.* 581:1–129 (1948).

under identical conditions, he realized that the differences between the ecotypes must have had a genetic basis and that they must have resulted from evolutionary differentiation within the species according to habitat.

Similar experiments in California on a species of yarrow, *Achillea millefolium,* also revealed ecotypic variation. *Achillea,* a member of the sunflower family, grows in many habitats ranging from sea level to more than 3000 m in elevation. Plants grown at sea level at Stanford, California, raised from seed collected at various points along the altitude gradient, retained the distinctive size and level of seed production typical of the populations from which they came (Figure 17.6). Similar differentiation has been found over distances of only a few meters where contrasting selection pressures are strong enough to overcome the migration of individuals, seeds, or pollen between habitat patches. Such situations frequently arise on soils that develop on mine tailings,

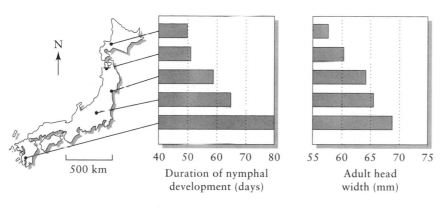

FIGURE 17.7 Clinal geographic variation in duration of nymphal development and the width of the adult head in males from five local populations of the field cricket *Teleogryllus emma* in Japan. The nymphs were raised at 28°C on a cycle of 16 hours of light and 8 hours of dark. Data from S. Masaki, *Evolution* 21:725–741 (1967).

which sometimes exert strong selection pressure for tolerance of toxic metals (for example, copper, lead, zinc, and arsenic).

Frequently, a trait may exhibit a continuously graded response, called a **cline,** to variation in the environment over a geographic gradient. For example, in the Japanese field cricket *Teleogryllus,* body size and the duration of nymphal (larval) development both increase clinally from north to south. We know that these clines have a genetic basis, because individuals from different localities raised under the same temperature and photoperiod retain their regional differences (Figure 17.7).

In contrast, the frequencies of alleles that control the color and banding pattern of the shells of the snail *Bradybaena* bear no relationship to habitat or locality (Figure 17.8). The snail was introduced to Japan, probably many times over the last 200 years, with the widespread cultivation of sugar cane. Variation in the morph frequencies among subpopulations could have occurred if each local population had been established by a small group of colonists that contained a random, but not necessarily representative, sample of the genetic variation of the parent population. A similarly haphazard distribution of genotypes occurs in populations of the small annual plant *Linanthus parryae.* In one small area of southern California, blue flowers (the normal flower color is white) are haphazardly distributed without apparent relation to environmental factors (Figure 17.9). Evidently, gene flow occurs so infrequently between populations isolated by as little as a kilometer that the frequencies of alleles for flower color diverge by genetic drift.

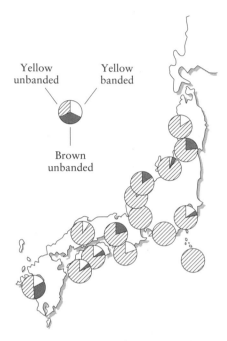

FIGURE 17.8 Geographic variation in the frequencies of morphs of the land snail *Bradybaena similaris* in Japan. Data from T. Komai and S. Emura, *Evolution* 9:400–418 (1955).

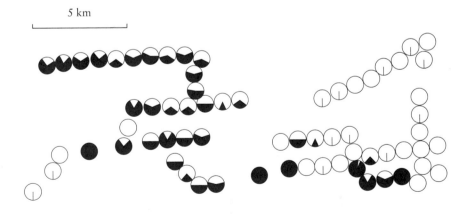

FIGURE 17.9 Frequencies of blue flowers (black portion of circle) in subpopulations of *Linanthus parryae* from a small region of southern California. Genotype frequency varies abruptly over small distances in some cases, suggesting genetic drift or founder effects in small, isolated subpopulations. From C. Epling and T. Dobzhansky, *Genetics* 27:317–332 (1942).

$\mathscr{E}$ SUMMARY

1. Population genetics demonstrates how evolution can proceed by the substitution of alleles according to their relative fitnesses. Genetic changes in populations also occur through mutation, through changes in breeding system, and, in small populations, through random processes.

2. Mutations result from changes in the subunits that make up the DNA molecule. These occur at a very low rate, but they are the ultimate source of all genetic variation. Most mutations are detrimental to the well-being of their carriers.

3. The genotype includes all the genetic factors that determine the structure and functioning (which together constitute the phenotype) of the individual. Many genetic factors have unique, measurable effects on the phenotype. The frequency of homozygous and heterozygous genotypes, in the absence of selection, mutation, and nonrandom mating, can be estimated by the Hardy–Weinberg equilibrium.

4. Many adaptations of ecological interest involve modifications of continuously varying traits. Variation in a trait within a population is described by its variance, which has environmental and genetic components. The science of quantitative genetics has developed statistical analyses to tease apart these components from the results of certain breeding programs.

5. Heritability (h^2) is the ratio of additive genetic variance to phenotypic variance; its value ranges between 0 and 1. Heritabilities of traits in natural populations are on the order of 0.5 to 0.7 for many size traits, but they are often lower for production-related traits.

6. Evolution, by which we mean genetic changes in populations, occurs when genetic factors influence survival and fecundity. Those individuals that achieve the highest reproductive rate are said to be selected, and their proportion increases with time. The example of melanism in the peppered moth illustrates this principle.

7. The response of a trait (R) to selection is equal to the heritability times the selection differential (S). In animal and plant breeding, the stronger the selection, the faster the response, as long as enough offspring are produced to replace the individuals that are selectively removed.

8. Response to selection levels off when genetic variation is exhausted or, more frequently, when correlated responses in other traits reduce the fitness of selected individuals.

9. The response in one trait to selection applied on a second depends on the genetic correlation between them. The phenotype often consists of groups of genetically intercorrelated traits that tend to respond to selection in concert. Such correlations may inhibit the

independent evolutionary response of two traits to opposing selective pressures.

10. Genetic variation maintains the long-term evolutionary potential of a population, and it has been argued that a primary function of sexual reproduction is to increase genetic variability through the recombination of genotypes.

11. Genetic variation may, however, impair the ability of individuals to adapt to local variation in the environment. To some degree, individuals can minimize the effects of genetic variation on the genotypes of their offspring by selective mating to control the level of inbreeding and the outcrossing distance.

12. In small populations, random variations in reproductive success result in changes in gene frequencies and the occasional loss of alleles, causing a decrease in genetic diversity. This process is called genetic drift.

13. Different selection pressures and genetic drift in small populations result in variation in gene frequencies within the geographic range. This variation is accentuated by population subdivision and poor dispersal.

❧ SUGGESTED READINGS

Cook, L. M., C. S. Mani, and M. E. Varley. 1986. Postindustrial melanism in the peppered moth. *Science* 231:611–613.

Falconer, D. S. 1989. *Introduction to Quantitative Genetics* (3rd ed.). Longman, Harlow, England.

Ford, E. B. 1975. *Ecological Genetics* (4th ed.). Chapman and Hall, London; Wiley, New York.

Gould, F. 1991. The evolutionary potential of crop pests. *American Scientist* 79:496–507.

Grant, P. R. 1991. Natural selection and Darwin's finches. *Scientific American* 265:82–87.

Hartl, D. L. 1988. *A Primer of Population Genetics* (2nd ed.). Sinauer, Sunderland, Mass.

Kettlewell, H. B. D. 1959. Darwin's missing evidence. *Scientific American* 200:48–53.

Ralls, K., J. D. Ballou, and A. Templeton. 1988. Estimates of the cost of inbreeding in mammals. *Conservation Biology* 2:185–193.

Schemske, D. W. 1984. Population structure and local selection in *Impatiens pallida* (Balsaminaceae), a selfing annual. *Evolution* 38:817–832.

5

Species Interactions

18

RELATIONSHIPS AMONG SPECIES

Survival and reproduction determine the growth rates of populations and the evolutionary fitness of individuals within populations. And they depend on how well the individual copes with biological as well as physical factors in its environment. To reproduce, individuals must gather sufficient resources to defend territories, attract mates, make eggs, and provision offspring. To survive, they must tolerate the physical stresses of the environment and also avoid detection and capture by predators and infection by disease organisms.

Plants and animals employ a variety of structures and behaviors to obtain food and avoid being eaten or parasitized. Indeed, this variety is one of the most remarkable features of life. Each type of organism has its own unique place in nature. Much of this diversity has resulted from natural selection acting on the ways in which plants and animals procure resources and escape predation. The wing markings that make moths blend into their resting places during the day enable them to escape the notice of most predators. By their insistent colors and fragrances, flowers call attention to themselves and attract the notice of the insects and birds that carry pollen from one flower to the next and effect fertilization.

The agents whose influence has shaped these adaptations are biological, and their effects differ from those of physical factors in two ways. First, biological factors elicit interactive traits: the predator shapes its prey's adaptations for escape, but its own adaptations for pursuit and capture are shaped, just as surely, by the prey. Second, biological factors tend to diversify adaptations rather than promoting similarity. In response to physical stresses of the environment, many kinds of organisms arrive at similar solutions. Thus most desert plants have reduced or finely divided leaves that minimize heat stress and water loss. But in response to biological factors in its environment, each organism specializes, pursuing a different assortment of prey, striving to avoid a different combination of predators and disease organisms, and engaging in cooperative arrangements with a unique set of pollinators, seed dispersers, or gut microorganisms.

Types of species interactions

The relationships among species fall conveniently into three broad categories defined by the effect of the interaction on each of the parties. These are summarized in Table 18.1.

Consumption covers all sorts of interactions, including those between herbivores and plants and those between parasites and their hosts. These are the most fundamental interactions in nature because everything must eat, and most organisms risk being eaten. Predator–prey, herbivore–plant, and parasite–host relationships are all special cases of consumer–resource relationships, which organize biological communities into series of **consumer chains.** We have seen these before in the guise of food chains. In all consumer–resource interactions, while the consumer benefits and its population size may increase, the resource suffers, individual and population alike. Thus, while energy and nutrients move up the consumer chain, population control mechanisms and natural selection exert their influences in both directions. **Detritivore** links in food chains differ from predator–prey links because the detritus eater does not affect the rate of supply of its food from living organisms. For example, earthworms in the soil do not directly influence leaf fall

TABLE 18.1	Categories of relationships among species	
	EFFECT OF INTERACTION ON	
Type of interaction	Species 1	Species 2
Competition	Negative (−)	Negative (−)
Consumer–resource	Positive (+) for consumer	Negative (−) for resource
Detritivore–detritus	Positive (+)	Indifferent (0)
Mutualism	Positive (+)	Positive (+)

from the trees above them, although they may indirectly increase plant production by speeding the return of nutrients to forms in the soil that plants can reuse.

Mutualism and **competition** resemble each other only in that the effects of interacting populations are more or less reciprocal. Mutualists benefit each other. The mycorrhizal fungus extracts from the soil inorganic nutrients that the plant may use, and the plant supplies the fungus with carbohydrates for energy metabolism; the flower provides the bee with a supply of nectar, and the bee carries pollen between plants and effects fertilization. In most cases, each member of the mutualistic association is specialized to perform a complementary function for the other. In the lichen plant, a photosynthetic alga teams up with a fungus that can obtain nutrients from difficult substrates, such as bark and rock surfaces. Such intimate associations, in which the members form a distinctive entity, are referred to as **symbioses**, literally a "living together."

Competition results when many species seek the same resources, and the depressing effect that each has on the availability of their common resource adversely affects the other. This relationship arises whenever two consumer chains join together. Usually, individuals compete indirectly through their mutual effects on shared resources. Less frequently, when consumers can profitably defend resources, competitors may interact directly through various antagonistic behaviors. Hummingbirds chase other hummingbirds, not to mention bees and moths, from flowering bushes. Encrusting sponges use poisonous chemicals to overcome other species of sponges as they expand to fill open space on rock surfaces.

In every case, interactions between species involve adaptations of each party to reduce the negative effects of the interaction and to increase its positive effects. In the case of mutualism, the participants cooperate, and we can expect a fair degree of mutual coordination of their structure and functioning. In the case of predator and prey, each vies to take advantage of the other, and new adaptations continually arise to shift the balance in favor of one or the other. This evolutionary game of back and forth has spawned an array of fascinating mechanisms for eating and for avoiding being eaten. We shall examine some of these in this chapter, wherein we explore the variety of predator – prey, parasite – host, and herbivore – plant interactions and conclude with a look at several examples of mutualism. In subsequent chapters, we shall consider the population consequences of these relationships and their implications for the regulation of populations and for evolution.

A variety of predator and prey adaptations

When we think of predator and prey, we usually think of lynx and rabbit, or bird and beetle — predators that pursue, capture, and eat individual prey. Though smaller than the predator, such prey usually seem large enough to be worth pursuing. But many organisms consume minute prey in vast numbers, and they are also predators. The blue whale weighs many tons but eats small, shrimplike krill, fish fry, and the like. On a smaller scale, clams and mussels

FIGURE 18.1 African lions taking a midday break in Amboseli Park, Kenya. With their powerful legs and jaws, lions can subdue prey somewhat larger than themselves. But because they cannot maintain speed over long distances, successful hunting relies on stealth and surprise.

pump water through tiny filtering devices that trap minute plankton, and many protozoa, sponges, and rotifers filter bacteria and other microorganisms from the water.

As the size of prey increases in relation to that of the predator, prey become more difficult to capture, and predators become specialized for pursuing and subduing their prey (Figure 18.1). Beyond a certain size ratio, however, predators lack sufficient strength and swiftness to capture potential prey items. Lions will attack animals their own size or a little larger, but they are no match for fully grown elephants. A few species, such as wolves, hyenas, and army ants, hunt cooperatively and thus can run down and subdue prey substantially larger than themselves.

At the other end of the relative size spectrum we find the parasites — the myriad viruses, bacteria, protozoa, worms, and others that invade the body of the host and feed on its tissues or blood or on the partially digested food in its intestine. Parasitism differs from predation and filter feeding because the survival of most parasites depends on their hosts' surviving rather than dying; the parasite must not bite the hand that feeds it.

Depending on what parts of a plant they eat, herbivores act as either predators or parasites. Parasites live off the productivity of a host organism without killing it, so a deer, browsing on trees and shrubs, functions as a parasite. A sheep that consumes an entire plant, pulling it up by the roots and macerating it into lifeless shreds, behaves as a predator. The beetle larva that develops within a seed, thereby destroying the embryonic plant organism it contains, is also a predator.

To go about their business, predators need mobility, senses, and the ability to handle prey. For example, the structure of teeth reflects the nature of the diet — that is, the job they must do to secure and process food items (Figure 18.2). Herbivores, especially those that eat grass, have teeth with large grinding surfaces to break down tough, fibrous plant materials. The teeth of predators have cutting and biting surfaces that both immobilize the prey in the mouth and cut it into pieces small enough to swallow. Seemingly simple differences in dentition reflect important ecological differences. The upper and lower incisors of horses, for example, are strongly opposed so that they can cut the fibrous stems of grasses. Other ungulates, such as cows, sheep, and deer, lack upper incisors; their lower teeth press against the upper jaw at an angle for gripping and pulling plant material.

Many predators use their forelegs to help tear their food into small morsels. Among birds, for example, the hawks, eagles, owls, and parrots use their powerful, sharp-clawed feet and hooked beaks for this purpose. Diving birds often eat large fish, but they must swallow them whole because their hind legs are specialized for swimming and diving rather than for grasping and dismantling prey. Some species of snakes compensate for their lack of grasping appendages with distensible jaws that enable them to swallow large prey whole (Figure 18.3).

Quality of diet influences adaptations of the predator's digestive and excretory systems as well as structures directly related to procuring food. Plants contain long, fibrous molecules, such as cellulose and lignin, that form supportive structures in the stems and leaves. Because these components make vegetation more difficult to digest than the high-protein diets of carnivores, the digestive tracts of herbivorous animals are often greatly elongated. In addition, many contain saclike offshoots — **caeca** (the singular is caecum) of rabbits, **rumens** of cows — that, like fermentation vats, house bacteria and protozoa that aid digestion. With a larger volume of intestines, herbivores can keep meals in the digestive tract longer and digest them more thoroughly. However, herbivores must carry quantities of undigested food in their bellies, adding weight and reducing their mobility.

The senses of predators

To locate and capture food, predators have evolved senses commensurate with their habitats, their feeding tactics, and their prey's ability to avoid detection. We ourselves primarily use vision to locate food, particularly as it is now displayed on the shelves of supermarkets. Yet our sight is pitiable compared to that of hawks and falcons, and many insects perceive ultraviolet light, which is

FIGURE 18.2 Skulls of three mammals illustrating adaptations of the jaws and teeth to different diets. (a) Coyote, with daggerlike canine teeth and knifelike premolars for securing and tearing flesh. (b) Fallow deer, with well-developed, flat-surfaced molars and premolars for grinding plant materials; note absence of canines and upper incisors. The lower incisors are used to secure vegetation against the upper jaw; the deer then rips the leaves from the plant. (c) Beaver, with greatly enlarged chisellike incisors that are used to gnaw on wood. After T. A. Vaughan, *Mammalogy* (3rd ed.). Saunders, Philadelphia (1986).

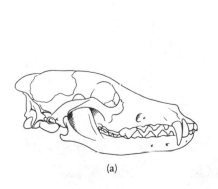

(a)

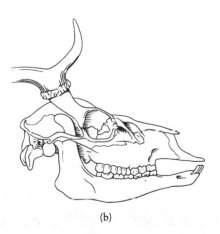

(b)

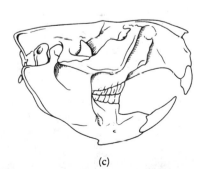

(c)

FIGURE 18.3 Some species of snakes have enlarged their gape as much as 20 percent by shifting the articulation of the jaw with the skull from the quadrate bone (black) to the supratemporal (stippled). After C. Gans, *Biomechanics*. Lippincott, Philadelphia (1974).

invisible to us (Figure 18.4). Insects can also detect rapid movement, such as that of wings beating 300 times per second; we see such movement only as a blur.

Among the more unusual sensory organs of predators are the pit organs of pit vipers, a group that includes the rattlesnake. The pit organs, located on each side of the head in front of the eyes, detect the infrared (heat) radiation given off by the warm bodies of potential prey — a sort of "seeing in the dark" (Figure 18.5). Pit vipers are so sensitive to infrared radiation that they can detect a small rodent several feet away in less than a second. Moreover, because the pits are directionally sensitive, vipers can locate warm objects precisely enough to strike them. The sonar systems of insectivorous and carnivorous bats similarly enable these predators to locate flying prey in the dark. Bats produce very loud, high-pitched pulses of noise — generally beyond the range of our hearing — and sense the echoes that bounce back from the bodies of prey. Owls don't emit sonar sounds, but they have such sensitive and directionally informative hearing that they can locate mice and other prey by the sounds they make as they move through the habitat.

FIGURE 18.4 The appearance of flowers to the human eye (left) and to eyes that are sensitive to ultraviolet light (right). *Above:* marsh marigolds. *Below:* five species of yellow-petaled Compositae from central Florida. Courtesy of T. Eisner. From T. Eisner, R. E. Silberglied, D. Aneshansley, J. E. Carrel, and H. C. Howland, *Science* 166:1172–1174 (1969).

A few aquatic animals have developed the sensory ability to detect electrical fields. Some species of electric fish continuously discharge electricity from specialized muscle organs, creating a weak electric field around them. Nearby objects distort the field, and these changes are picked up by receptors on the surface of the electric fish. Some species use electric signals to communicate between individuals. The specialized electric ray *Torpedo* uses powerful electrical currents (up to 50 volts at several amperes) to defend itself and to kill prey. As one might expect, the production and sensation of electric fields are most highly developed in fish that inhabit murky waters. In other habitats where visibility is poor, bottom-dwelling species such as catfish use elongated fins and barbels around the mouth as sensitive touch and taste receptors.

In contrast to the magnificent senses of many predators, others perceive their surroundings only dimly and rely on chance to bump into prey. (For this tactic to work, their prey must be equally oblivious). But even such "blind" predators adopt searching patterns that increase their chances of encountering prey. For example, the predatory larvae of the ladybird beetle feed on mites and aphids that infest the leaves of certain plants, and they must physically contact their prey to recognize them (Figure 18.6). Their movements on the leaves are not oriented toward the prey, but neither are they random. The veins and rims of leaves make up a small percentage of the leaf surface, yet larvae spend most of their time searching on these areas — which they recognize by touch — where most aphids are distributed.

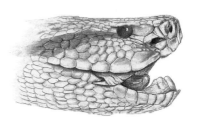

FIGURE 18.5 Head of the western rattlesnake (*Crotalis viridis*), showing the location of the infrared-sensitive pit between, and slightly lower than, the eye and the nostril. After D. Burkhardt, W. Schleidt, and H. Altner, *Signals in the Animal World* (trans. K. Morgan). McGraw-Hill, New York (1967).

Prey escape

The ways in which prey organisms avoid predators are as diverse as the hunting tactics of predators. Organisms that are extremely small compared to their predators — those captured by filter feeders, for example — exhibit few adap-

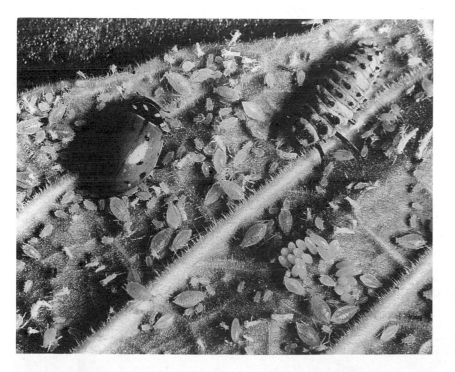

FIGURE 18.6 Adult and larval ladybird beetles (family Coccinellidae) feeding on aphids in a laboratory culture. Note the hairs on the veins of the leaf, which deter the aphids from penetrating the plant and sucking its juices. Courtesy of the U.S. Dept. of Agriculture.

tations to avoid being caught, but larger quarry may hide, fight, or flee. Many circumstances of the predator–prey relationship determine which strategy, or combination of strategies, minimizes the prey's chance of being caught. Grassland offers no hiding places for large ungulates, so escape depends on early detection of predators and on swiftness. Plants cannot flee and may rely on thorns and toxic chemicals to fend off herbivores.

Protective defenses rarely involve physical combat because few prey can match their predators, and predators carefully avoid those that can. Instead, many seemingly defenseless organisms produce foul-smelling or stinging chemical secretions to dissuade predators. Whip scorpions and bombardier beetles direct sprays of noxious liquids at threatening animals. Many plants and animals contain toxic substances, making them inedible or poisonous. Slow-moving animals, such as the porcupine and the armadillo, protect themselves with spines or armored body coverings.

Crypsis and warning coloration

The evolution of camouflaged appearances and positions by which various prey avoid detection by predators fascinates observers of the natural world and testifies to the force and pervasiveness of natural selection.

Many organisms achieve **crypsis** — avoiding predation by blending in with the background — by matching the color and pattern of the bark, twigs, or leaves upon which they rest. Various animals resemble inedible objects: sticks, leaves, flower parts, even bird droppings. The elaborate concealment of their heads, antennae, and legs underscores the importance of these cues to predators. The stick-mimicking phasmids (stick insects) and leaf-mimicking katydids often conceal their legs in the resting position either by folding them back upon themselves or upon the body or by protruding them in a stiff, unnatural fashion. The dead-leaf-mimicking mantis *Acanthops* partially conceals its head under its folded front legs (Figure 18.7). Asymmetry is also a good cover

FIGURE 18.7 This Central American mantis of the genus *Acanthops* resembles a dead, curled-up leaf and thus escapes the notice of most predators.

for animals, but it is difficult to achieve. The leaf-mimicking moth *Hyperchiria nausica* produces the appearance of an asymmetrical midvein by folding one forewing over the other (Figure 18.8). Moths sometimes rest with a leg protruding to one side but not to the other, or with the abdomen twisted to one side to break their symmetry.

When predators discover cryptic organisms, they may be confronted with a variety of second-line defenses, including startle displays and various attack-and-escape mechanisms. The green caterpillar of the hawkmoth *Leucorampha omatus* normally assumes a cryptic position. When disturbed, however, it puffs up its head and thorax, looking for all the world like the head of a small poisonous snake, complete with a false pair of large, shiny eyes; the caterpillar consummates this display by weaving back and forth, while hissing like a serpent (Figure 18.9). The eyespots that many moths and other insects display when disturbed (Figure 18.10) frighten their predators, because they resemble the eyes of large birds of prey.

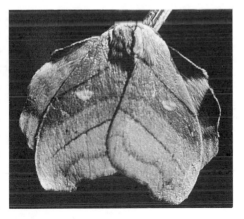

FIGURE 18.8 The moth *Hyperchiria nausica* partly disguises its symmetry by folding one wing over the other.

The appropriate line of defense sometimes depends on features of the environment other than predators. Large-bodied moths at low elevations in the tropics rarely exhibit such protective displays as eyespots or flash coloration. During the day, the air temperature rises to within 6°C of the optimal working temperature of their flight muscles, and moths can escape from predators by flight. At higher elevations, moths require a lengthy preflight warm-up period because the air temperature may be 15°C lower than the working temperature of the flight muscles. The cold precludes escape, so a large proportion of moth species at these elevations do have special protective displays.

Crypsis is a strategy of palatable animals. Others take a bolder approach to predator defense: they produce noxious chemicals or accumulate them from food plants, and they advertise the fact with *conspicuous* color patterns **warning coloration**). Predators learn quickly to avoid such conspicuous markings as the black and orange stripes of the monarch butterfly; its taste guarantees loss of an appetite for monarchs for some time to come. It is not a coincidence that many noxious forms adopt similar patterns: black and either red or yellow stripes characterize such diverse animals as yellow-jacket wasps and coral snakes. Some predators have evolved a generalized aversion to such patterns.

Why aren't all potential prey species noxious or unpalatable? Part of the answer is that chemical defenses can be costly to manufacture and maintain. In many cases, defensive compounds can use up a large fraction of the individual's energy or nutrients that might otherwise be allocated to growth or reproduction. Furthermore, many prey organisms rely on their food plants to supply toxic organic compounds that they cannot manufacture themselves. Of course, the consumer must itself avoid the toxic effects of such chemicals to use them effectively against its potential predators.

FIGURE 18.9 Snake display by the caterpillar of the sphingid moth *Leucorampha omatus*. From M. H. Robinson, *Evol. Biol.* 3:225–259 (1969).

Mimicry

Distasteful animals and plants that display warning coloration often serve as models for the evolution of the mimicking of such color patterns by *palatable* forms. Some potential prey even contrive to resemble their predators (Figure

18.11). These relationships are collectively referred to as **Batesian mimicry,** which was named after its discoverer, the nineteenth-century English naturalist Henry Bates. In his journeys to the Amazon region of South America, Bates found numerous cases of palatable insects that had forsaken the cryptic patterns of close relatives and had come to resemble brightly colored, distasteful species.

Experimental studies have subsequently demonstrated that mimicry does confer advantage on the mimic. For example, toads that were fed live bees thereafter avoided the palatable drone fly, which mimics bees. But when naive toads were fed only dead bees from which the stings had been removed, they relished the drone fly mimics. Similar results were obtained with blue jays as predators; the distasteful monarch butterflies were the models, and their (palatable) viceroy butterfly mimics were the experimental subjects.

In some cases, mimicry relationships may involve several different models. For example, in the African swallow-tailed butterfly *Papilio dardanus,* females are polymorphic, each individual resembling one of a variety of different models. (The males do not mimic other species, presumably because females choose males on the basis of their coloration and prefer the typical swallowtail color pattern. Moreover, because they do not produce or carry eggs or spend time searching for suitable oviposition sites, males may be less vulnerable to predators.) Why do mimicry polymorphisms evolve? When mimics become common relative to models, the predator does not learn to avoid either one so quickly because it often samples the palatable mimics rather than the noxious models. In this case, a *rare* mimetic form has an advantage: predators that are learning to associate warning coloration with unpalatability are not confused by frequent encounters with common mimetic forms. Thus natural selection, by favoring odd types of mimics, acts to diversify the population.

Another type of mimicry, called **Müllerian mimicry** after its discoverer, occurs between species of unpalatable organisms that come to resemble (mimic) each other. Many species form Müllerian mimicry complexes in which each participant is both model and mimic. When a single pattern of warning coloration is adopted by several *unpalatable* species, avoidance learning by predators is made more efficient (no satisfactory predation on prey displaying

FIGURE 18.10 The eyespot display of an automerid moth from Panama. (a) Normal resting attitude. (b) Reaction when touched.

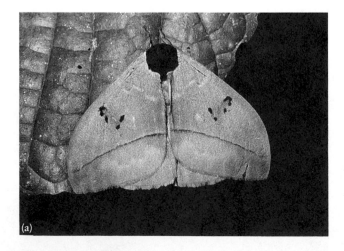

Figure 18.11 The wing markings of the tephritid fly *Rhagoletis pomonella* closely resemble the forelegs and pedipalps of jumping spiders (Salticidae). Courtesy of the U.S. Dept. of Agriculture.

that coloration occurs), and each bad experience benefits all members of the mimicry complex.

Parasites

Parasites are usually much smaller than their prey, or hosts, and live either on their surfaces (**ectoparasites**) or inside their bodies (**endoparasites**). Both types demonstrate characteristic adaptations to their way of life. Because parasites generally live inside of, or in close association with, a larger organism, they expend little effort to maintain their own internal environments. Indeed, parasites living in the gut (tapeworms, for example) bathe in a predigested food supply and retain little more than a highly developed capacity to produce eggs. In addition, however, parasites must disperse through the hostile environment between hosts. Many accomplish this via complicated life cycles, one or more stages of which can cope with the external environment.

Ascaris, an intestinal roundworm that parasitizes humans, has a relatively simple life cycle. A female *Ascaris* may lay tens of thousands of eggs per day, which pass out of the host's body in the feces. Where sanitation is poor or where human excrement is applied to farmland as fertilizer, the eggs may be inadvertently ingested. The egg is the only stage during the life cycle of *Ascaris* that occurs outside the host, and it is well protected by a sturdy, impermeable outer covering. The parasite relies on the host to consume these eggs and complete the life cycle.

Schistosoma, a trematode worm (blood fluke) that commonly infects humans and other mammals in tropical regions, has a more complicated life cycle that involves a freshwater snail as an intermediate host. Male–female pairs of adult worms live in the blood vessels that line the human intestine or the bladder, depending on the species of *Schistosoma.* The eggs pass out of the host body in the feces or urine. When the eggs are deposited in water, they develop into a free-swimming larval form (the miracidium), which locates and burrows into a snail within 24 hours. In the snail, the miracidium produces cells that eventually develop into free-swimming cercariae, which leave the snail and can penetrate the skin of the next host in the cycle. Once inside the body of a human or other host, the cercariae travel a circuitous route through blood vessels until they become lodged in an appropriate place, where they metamorphose into adult worms.

A snail infected with one miracidium may liberate from 500 to 2000 cercariae per day over a period of a month. During their lifetimes, which average between 4 and 5 years in human hosts, adults may produce as many as several hundred eggs per day. Of course, few of these eggs and larvae survive to become sexually mature adults. Nonetheless, *Schistosoma,* which causes the incurable disease schistosomiasis, probably affects hundreds of millions of the human population and countless domestic animals.

The life cycle of the protozoan parasite *Plasmodium,* which causes malaria, resembles that of *Schistosoma* in that it involves two hosts, a mosquito and a human or some other mammal, bird, or reptile. But whereas the sexual phase of the schistosome life cycle occurs in the human host, that phase of the malaria parasite's life cycle takes place within the mosquito (Figure 18.12). When an infected mosquito bites a human, cells called sporozoites are injected into the bloodstream with the mosquito's saliva. The sporozoites enter the red blood cells and feed on hemoglobin. When the malaria cell becomes large enough, it undergoes a series of divisions (asexual reproduction), and the daughter cells break out of the red blood cell. Each daughter cell enters a new red blood cell, grows, and repeats the cycle, which takes about 48 hours. (When the infection has built up to a high level, the emergence of daughter cells corresponds to periods of high fever.) After several of these cycles, some of the cells that enter red blood corpuscles change into sexual forms. If these are swallowed by a mosquito along with a meal of blood, the sexual cells are transformed into eggs and sperm, and fertilization (sexual reproduction) takes place. The fertilized egg then divides and produces sporozoites, which work their way into the salivary glands of the mosquito, from which they may enter a new intermediate host, thereby completing the life cycle.

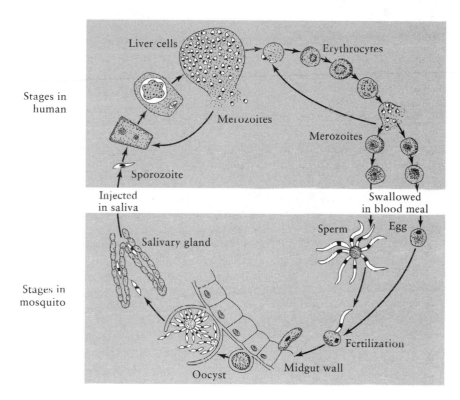

Figure 18.12 Stages in the life cycle of the malaria parasite *Plasmodium*. After R. Buchsbaum, *Animals Without Backbones* (2nd ed.). Univ. of Chicago Press, Chicago (1948); M. Sleigh, *The Biology of Protozoa*. American Elsevier, New York (1973).

Plant defenses

The conflict between herbivore and plant resembles that between parasite and host in that both are waged primarily on biochemical battlegrounds. Plant defenses against herbivores include the inherently low nutritional value of most plant tissues and the toxic properties of so-called **secondary compounds** produced and sequestered for defense. Sessile marine organisms, including both plants and animals, also employ a variety of chemical defenses. Structural defenses, such as spines, hairs, and tough seed coats, are important as well (Figure 18.13).

The nutritional quality and digestibility of plant foods is critical to herbivores. Because young animals have a high protein requirement for growth, the reproductive success of grazing and browsing mammals depends on the protein content of their food. Herbivores usually select plant food according to its nutrient content. Young leaves and flowers are chosen frequently because of their low **cellulose** content; fruits and seeds are particularly nutritious compared to leaves, stems, and buds.

Many plants use chemicals to reduce the availability of their proteins to herbivores. For example, tannins sequestered in vacuoles in the leaves of oaks and other plants combine with leaf proteins and digestive enzymes in a herbivore's gut, thereby inhibiting protein digestion. Thus **tannins** considerably slow the growth of caterpillars and other herbivores, reducing the quality of tannin-laden hosts as food plants. With the buildup of tannins in oak leaves during the summer, fewer and fewer leaves are attacked by herbivores. Insects,

for their part, can reduce the inhibitory effects of tannins by producing deter-
gent-like surfactants in their gut fluids, which tend to disperse tannin–protein
complexes.

Whereas tannins exhibit a generalized reaction with proteins of all types,
many secondary substances of plants (that is, those used not for metabolism
but for other purposes, chiefly defense) interfere with specific metabolic pathways or physiological processes of herbivores. But because the sites of action
of such substances are localized biochemically, herbivores may counter their
toxic effects by modifying their own physiology and biochemistry. Detoxification may involve one or several biochemical steps, including oxidation, re-

FIGURE 18.13 Spines protect the stems and leaves of many plants: (a) a cholla
cactus (*Opuntia*) from Arizona; (b) an agave (century plant) from Baja California;
(c) a *Parkinsonia* (bean family) from the Galapagos Islands.

duction, or hydrolysis of the toxic substance, or its conjugation with another compound.

Several early studies of the chemical give and take between plants and herbivores focused on the larvae of bruchid beetles, many of which infest the seeds of legumes (pea family). Adult bruchids lay their eggs on developing seed pods. The larvae then hatch and burrow into the seeds, which they consume as they grow. To counter this attack, legumes have mounted a variety of defenses, including the evolution of tiny seeds. Each larva feeds on only one seed. To pupate successfully and metamorphose into an adult, the larva must attain a certain size, which is ultimately limited by the amount of food in the seed. The small seeds of some species of legumes contain too little food to support the growth of a single bruchid larva.

Most legumes also contain substances that inhibit proteolytic enzymes produced in the herbivore's digestive organs. Although these toxins provide an effective biochemical defense against most insects, many bruchid beetles have metabolic pathways that either bypass them or are insensitive to them. Among legume species, however, soybeans stand out as being resistant to attack even by most bruchid species. When bruchids lay their eggs on soybeans, the first instar larvae die soon after burrowing beneath the seed coat; chemicals isolated from soybeans have been shown to inhibit the development of bruchid larvae in experimental situations.

Seeds of the tropical leguminous tree *Dioclea megacarpa* contain 13 percent by dry weight of L-canavanine, a nonprotein amino acid that is toxic to most insects because it interferes with the incorporation into proteins of arginine, which it closely resembles. One species of bruchid, *Caryedes brasiliensis,* possesses enzymes that discriminate between L-canavanine and arginine during protein formation and additional enzymes that degrade L-canavanine to forms that can be used as a source of nitrogen. For every defense, a new attack can be devised.

The tobacco hornworm (the larval stage of the moth *Manduca sexta*) can tolerate nicotine concentrations in its food far in excess of levels that kill other insects. Nicotine disrupts normal functioning of the nervous system by preventing the transmission of impulses from nerve to nerve. The hornworm has circumvented this defense by excluding nicotine from the nerve at the cell membrane (in other species of moths, nicotine readily diffuses into nerve cells). Resistance to nicotine enables *M. sexta* to feed on tobacco (*Nicotiana tabacum*), a member of the tomato family (Solanaceae), but some other species of *Nicotiana* produce other alkaloid toxins that the tobacco hornworm cannot tolerate. When tobacco hornworms were grown on 44 species of *Nicotiana* in greenhouse experiments, the larvae grew normally on 25 species but were retarded or stopped completely on the others. In addition, 15 of the species caused moderate to severe mortality.

Most plants produce toxic, defensive compounds. Many of these, like pyrethrin, are important sources of pesticides (which is how plants use them); others, like digitalis, have found use as drugs (some of their pharmacological effects are beneficial in small doses). Secondary plant compounds can be divided into three major classes of chemical structure: nitrogen compounds ultimately derived from amino acids, terpenoids, and phenolics (Table 18.2). Among the nitrogen-based substances are lignin, a highly condensed polymer

TABLE 18.2	*Secondary plant compounds involved in plant–animal interactions*		
Class	Approximate number of structures	Distribution	Physiological activity
NITROGEN COMPOUNDS			
Alkaloids	5500	Widely in angiosperms, especially in root, leaf, and fruit	Many toxic and bitter-tasting
Amines	100	Widely in angiosperms, often in flowers	Many repellent-smelling, some hallucinogenic
Amino acids (nonprotein)	400	Especially in seeds of legumes but relatively widespread	Many toxic
Cyanogenic glycosides	30	Sporadic, especially in fruit and leaf	Poisonous (as HCN)
Glucosinolates	75	Cruciferae and 10 other families	Acrid and bitter
TERPENOIDS			
Monoterpenes	1000	Widely, in essential oils	Pleasant-smelling
Sesquiterpene lactones	600	Mainly in Compositae, but increasingly found in other angiosperms	Some bitter and toxic, also allergenic
Diterpenoids	1000	Widely, especially in latex and plant resins	Some toxic
Saponins	500	In over 70 plant species	Haemolyse blood cells
Limonoids	100	Mainly in Rutaceae, Meliaceae, and Simaroubaceae	Bitter-tasting
Curcurbitacins	50	Mainly in Cucurbitacaeae	Bitter-tasting and toxic
Cardenolides	150	Especially common in Apocynaceae, Asclepiadaceae, and Scrophularaceae	Toxic and bitter
Carotenoids	350	Universal in leaf, often in flower and fruit	Colored
PHENOLICS			
Simple phenols	200	Universal in leaf, often in other tissues as well	Antimicrobial

	Approximate number of structures	Distribution	Physiological activity
Class			
Flavonoids	1000	Universal in angiosperms, gymnosperms, and ferns	Often colored
Quinones	500	Widely, especially in Rhamnaceae	Colored
OTHER			
Polyacetylenes	650	Mainly in Compositae and Umbelliferae	Some toxic

TABLE 18.2 *Continued*

Source: J. B. Harborne, *Introduction to Ecological Biochemistry* (2nd ed.). Academic Press, New York (1982).

that resists digestion; **alkaloids,** such as morphine (derived from poppies) and atropine and nicotine (from various members of the tomato family); nonprotein amino acids, such as L-canavanine; and cyanogenic glycosides, which produce cyanide (HCN). **Terpenoids** include essential oils, latex, and plant resins; among the **phenolics,** many simple phenols have antimicrobial properties.

Where herbivory is more intense, plants have more varied and concentrated toxins. And where plant defenses are strong, adaptations of herbivores to detoxify poisonous substances proliferate. This war between plants and herbivores promotes the biochemical specialization of herbivores to certain restricted groups of plants with similar toxins. Associations of plants and herbivores in groups based on plant chemistry and structure have been referred to as **plant defense guilds.**

Recent investigations have shown that plant defenses may be **induced** by herbivore damage. Alkaloids, phenolics, N-oxidases, and proanthocyanins, all of which are linked to antiherbivore defenses, increased dramatically in many plants following defoliation by herbivores or the clipping of leaves by investigators. Other studies have shown that plant responses to grazing can substantially reduce subsequent herbivory (Figure 18.14). Inducibility suggests that chemical defenses are often too costly to maintain economically under light grazing pressure. Undoubtedly, the offensive biochemical tactics of herbivores are also expensive.

Mutualism

A relationship between two species that benefits both constitutes mutualism. In very general terms, mutualisms fall into three categories: trophic, defensive, and dispersive. Examples of **trophic mutualisms** usually involve partners spe-

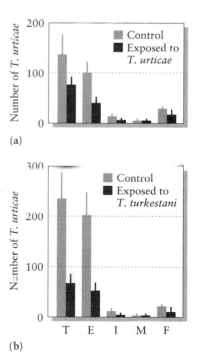

FIGURE 18.14 Mean numbers of the mite *Tetranychus urticae* on cotton plants previously exposed to mites and on controls with no previous exposure. The effect is similar whether the initial exposure is to *T. urticae* (a) or to the related species *T. turkestani* (b). T = total population, E = eggs, I = immatures, M = adult males, F = adult females. From R. Karban and J. R. Carey, *Science* 225:53–54 (1984).

cialized in complementary ways of obtaining energy and nutrients. Hence the term *trophic,* which pertains to feeding relationships. We have seen trophic mutualism in the symbioses of algae and fungi to form lichens, of fungi and plant roots to form mycorrhizae, and of *Rhizobium* bacteria and plant roots to form nitrogen-fixing root nodules. In these cases, each of the partners supplies a limiting nutrient or energy source that the other cannot obtain. *Rhizobium* can assimilate molecular nitrogen (N_2) from the soil atmosphere—a useful feature in nitrogen-poor soils—but requires carbohydrates supplied by the plant for the energy needed to do this. Bacteria in the rumens of cows and other ungulates can digest cellulose in plant fibers, which the cow's own digestive enzymes can't touch. The cow benefits because it assimilates some of the by-products of bacterial digestion and metabolism for its own use (it also digests some of the bacteria themselves). The bacteria benefit by having a steady supply of food in a warm, chemically regulated environment that is optimal for their own growth.

Defensive mutualisms involve a species that receives food or shelter from its mutualistic partner in return for defending that partner against herbivores, predators, or parasites. For example, in marine systems, specialized fish and shrimp clean parasites from the skin and gills of other species of fish. These cleaners benefit from the food value of the parasites they remove, and the groomed fish are unburdened of some of their parasites. Such relationships, often referred to as **cleaning symbioses,** are most highly developed in the clear, warm waters of the tropics, where many cleaners display their striking colors at locations, called cleaning stations, to which other fish come to be groomed. As might be expected, a few species of predatory fish mimic the cleaners: when other fish come and expose their gills to be groomed, they get a bite taken out instead.

Cleaners and the fish they groom do not live in close association, and each can survive without the other. Their mutualism is **facultative,** meaning that the partners can do with it or without it. When two species are inextricably bound through mutual dependence, they form what is known as an **obligate mutualism.** The alga and fungus in a lichen are obligate trophic mutualists. Among defensive mutualisms, the interdependence between certain kinds of ants and swollen-thorn acacias in Central America provides a fine example of obligate mutualism.

The acacia plant provides food and nesting sites for ants in return for the protection that the ants provide from insect pests. The bull's-horn acacia (*Acacia cornigera*) has large hornlike thorns with a tough woody covering and a soft pithy interior (Figure 18.15). To start a colony in the acacia, a queen ant of the species *Pseudomyrmex ferruginea* bores a hole in the base of one of the enlarged thorns and clears out some of the soft material inside to make room for her brood. In addition to housing the ants, the acacias provide food for the ants in nectaries at the bases of their leaves and in the form of nodules, called Beltian bodies, at the tips of some leaves (Figure 18.16). As the colony grows, more and more of the thorns on the plant are filled; in return, the ants protect the plant from insect pests. A colony may grow to more than a thousand workers within a year, and it may eventually include tens of thousands of workers. At any one time, about a quarter of the ants are outside the nest actively gathering food and defending the plant against herbivorous insects. The

FIGURE 18.15 The thorns of *Acacia hindsii* (right), like those of *A. cornigera,* are greatly enlarged and have a soft pith that the ants excavate for nests. Thorns of non-ant acacia are shown at left for comparison. Courtesy of D. H. Janzen, from D. H. Janzen, *Evolution* 20:249–275 (1966).

relationship between *Pseudomyrmex* and *Acacia* is obligatory: neither the ant nor the acacia can survive without the other. Other ant–acacia associations are facultative; that is, the ant and the acacia *can* co-occur to mutual benefit, but both can exist independently as well. Species of acacia that altogether lack the protection of ants frequently produce toxic compounds to defend their leaves against herbivores.

To test the influence of ants on the growth and survival of acacia plants, one can keep ants off new acacia shoots and compare their growth to that of shoots that have ants. After 10 months of one such experiment, shoots lacking ants weighed less than one-tenth those with intact ant colonies, and they produced fewer than half the number of leaves and a third the number of swollen thorns.

The mutualism between ants and acacias has been accompanied by adaptations of both species to increase the effectiveness of the association. For example, *Pseudomyrmex* is active both night and day, an unusual trait for ants, and thereby provides protection for the acacia at all times. In a similar adaptive gesture, the acacia retains its leaves throughout the year, and thereby provides a year-round source of food for the ants. Most related species lose their leaves during the dry season.

FIGURE 18.16 The leaves of *Acacia collinsii*, like those of *A. cornigera*, provide ants with food in the form of Beltian bodies at the tips of leaflets (a) and nectaries at the leaf base (b). Courtesy of D. H. Janzen. From D. H. Janzen, *Evolution* 20:249–275 (1966).

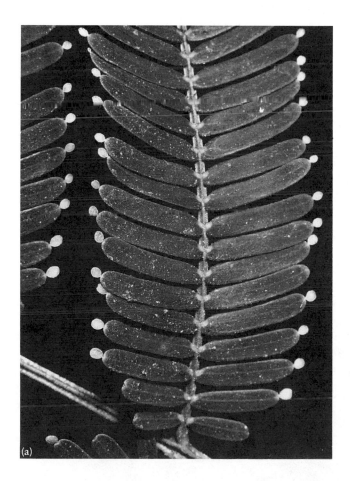

Pollination and seed dispersal

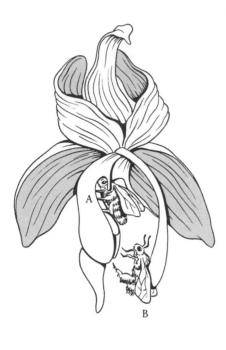

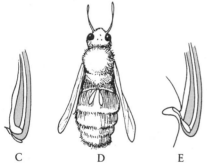

FIGURE 18.17 Pollination of the orchid *Stanhopea grandiflora* by *Eulaema meriana*. The bee enters from the side and brushes at the base of the orchid lip (A). If it slips (B), the bee may fall against the pollinarium, which is placed on the end of a column (C), and the pollinarium becomes stuck to the hind end of the thorax (D). If a bee with an attached pollinarium later falls out of another flower, the pollinarium may catch in the stigma (E), which is so placed on the column that the flower cannot be self-fertilized. After R. L. Dressler, *Evolution* 22:202–210 (1968).

Dispersive mutualisms generally involve animals that transport pollen between flowers for the reward of nectar or that disperse seeds to suitable habitats in return for nutritional fruits that contain the seeds. Plant–pollinator relationships are highly developed in the orchid family, with its variety of flower shapes, colors, and smells. The intricate tie between flower and pollinator is exemplified by the orchid *Stanhopea grandiflora* and the tropical bee *Eulaema meriana*. The attraction of these bees to the particular orchids they visit is unusual in that the flowers produce no nectar and only male bees visit them. The flowers are extremely fragrant, and each species of *Stanhopea* orchid has its own unique combination of odors so that bees may find them without confusion. Each type of orchid tends to attract a single type of bee.

When a male *Eulaema* bee visits an orchid, it brushes parts of the flowers with specially modified forelegs and then transfers collected substances to the tibia of the hind leg, which is enlarged and has a storage cavity. This behavior provides bees with a perfume that the males use to attract females.

For *Eulaema* to pollinate *Stanhopea*, the bee enters the flower from the side and brushes at a saclike modification on the lip of the orchid (Figure 18.17). The surface of the lip is very smooth, and the bee often slips when it withdraws from the flower. (The orchid fragrances may also intoxicate the bees and cause them to lose their footing on the lip of the orchid.) If the bee slips, it may brush against the column of the orchid flower, where the pollinaria (saclike structures filled with pollen) are precisely placed to stick to the hindmost part of the thorax of the bee. If a bee with an attached pollinarium slips and falls out of another flower, the pollinarium catches on the stigma and pollinates the flower. Thus flower structure and bee behavior are mutually adapted to increase the efficiency of pollen transfer.

Plants also confront the problem of seed dispersal. Often, seeds that germinate close to the parent fare poorly because they must compete with other established individuals. Seedlings of many species, even those that eventually form the canopies of dense forests, become established most readily in disturbed habitats where sunlight abounds and mature plants do not compete with them. Of course, seed dispersal requires a mechanism to disperse the seeds. Many plants rely on wind to carry their seeds away, but such seeds must be tiny to be carried effectively on currents of air, and they land more or less at random.

Many other plants, particularly in the tropics, entice various animals— usually birds and mammals—to disperse their seeds for them. In return, the plants offer a reward in the form of a fruit or other edible structure attached to the seed. The species of the disperser depends on the size, color, and structure of the fruit and on its position on the tree. Fruits suspended on the ends of branches may be accessible only to birds that can pick them off while hovering. When a bird removes a fruit from a plant, it may consume the entire structure, digest and assimilate the fruit pulp, and pass the undigested seeds in the feces. Birds often perch over suitable germination sites, such as at the edges of clearings in forests. The seed dispersed in this manner may find itself in a good habitat with a modest supply of highly nutritive manure. Indeed, the seeds of some species of plants will germinate only after passing through the digestive tract of an animal.

Because mammals have teeth, they can often gnaw through tough seed coats and get at the seed itself. This does not make for effective dispersal. But many mammals, and some birds, bury or otherwise cache excess seeds that they cannot eat at a single sitting, putting them aside as insurance against poorer times ahead. (Blue jays are notorious hoarders, as anyone can attest who has tried to keep a bird feeder east of the Rockies stocked with sunflower seed.) Many of these seeds are never recovered; they may germinate and grow, the squirrel or chipmunk having effectively planted them. Plants can take advantage of this **hoarding** behavior by producing large crops of seeds all at once, overwhelming the capacity of seed predators to consume them. As a result, many are buried, never to be retrieved.

🙟 SUMMARY

The environments of organisms include biological factors — food, predators, and diseases — that interact with the physical environment in molding adaptations. Unlike physical factors, biological factors themselves evolve and diversify. The relationship between predators and their prey illustrates the mutual influence between organisms and biological factors in the environment.

1. Depending on the relative sizes of predators and their prey, predatory behavior may consist of active pursuit of individual prey or the filtering of tiny organisms from large quantities of water.

2. Predators are well adapted to pursue, capture, and eat particular types of prey. Carnivorous and herbivorous mammals, for example, differ in the conformation of their teeth and in the size and design of their digestive systems.

3. Organisms escape predation by avoiding detection; by means of chemical, structural, and behavioral defenses; and by escape. Crypsis and warning coloration are examples of the defenses of edible and unpalatable organisms, respectively.

4. Host–parasite relations are specialized interactions characterized by complex life cycles of the parasite, which the difficulties of locating and infecting new hosts make necessary. Parasite and host often evolve a delicate balance when the parasite's well-being depends on the survival of its host.

5. Plants have evolved numerous structural and chemical defenses to deter herbivores. These include factors that influence the nutritional quality and digestibility of plant parts, and specialized chemicals — secondary compounds — that have toxic effects on animals and microorganisms. Most of these chemicals are nitrogen-containing compounds, terpenoids, or phenolics derived by modification and elaboration of normal metabolic pathways.

6. Herbivores themselves have evolved ways to detoxify many secondary chemicals of plants, enabling them to specialize on plant hosts that are poisonous to most other species.

7. Mutualisms are relationships that benefit both parties. They may be classified as trophic, defensive, or dispersive, depending on the nature of the relationship. In trophic mutualisms, each partner is specialized to provide a different limiting nutrient. In defensive mutualisms, one partner provides protection, usually in return for food.

8. Dispersive mutualisms are plant–animal interactions in which the animal disperses pollen or seeds in the course of harvesting or processing food that the plant supplies. The structures of flowers and fruits limit the variety of animals that perform this function for a particular species of plant, thereby increasing the efficiency of pollen transfer and the likelihood that seeds will reach suitable sites for germination and growth.

❧ SUGGESTED READINGS

Allen, M. F. 1991. *The Ecology of Mycorrhizae.* Cambridge Univ. Press, New York.

Bryant, J. P. 1981. Phytochemical deterrence of snowshoe hare browsing by adventitious shoots of four Alaskan trees. *Science* 213:889–890.

Cheng, T. C. 1970. *Symbiosis. Organisms Living Together.* Pegasus, New York.

Griffin, D. R. 1986. *Listening in the Dark. The Acoustic Orientation of Bats and Man.* Comstock, Ithaca, New York.

May, M. 1991. Aerial defense tactics of flying insects. *American Scientist* 79:316–329.

Real, L. (ed.). 1983. *Pollination Biology.* Academic Press, Orlando, Fla.

Rice, E. L. 1984. *Allelopathy* (2nd ed.). Academic Press, Orlando, Fla.

Robinson, M. H. 1969. Defenses against visually hunting predators. *Evolutionary Biology* 3:225–259.

Trager, W. 1986. *Living Together. The Biology of Animal Parasitism.* Plenum, New York.

Whittaker, R. H., and P. P. Feeny. 1971. Allelochemics: chemical interactions between species. *Science* 171:757–770.

Wickler, W. 1968. *Mimicry in Plants and Animals.* World University Library, London.

19

COMPETITION

Competition is the use or defense of a resource by one individual that reduces the availability of that resource to other individuals. Thus competition is one of the most important mechanisms by which the activities of individuals affect the well-being of others, whether they belong to the same species (**intraspecific** competition) or to different species (**interspecific** competition). As we saw earlier, competition within populations of the same species causes density-dependent decreases in levels of resources and thereby affects fecundity and survival. The more crowded the population, the stronger the competition between individuals. Thus intraspecific competition underlies the regulation of population size. Furthermore, when genetic factors cause individuals to differ in the efficiency with which they exploit resources, the more efficient individuals may produce more offspring, and their genes will increase in the population. In this way, intraspecific competition is intimately related to evolutionary change.

Competition between individuals of different species causes a mutually depressing effect on the populations of both; each species contributes to regulating the population of the other as well as its own. Under some conditions, particularly when interspecific competition is intense, this may lead to the elimination of one species by the other. Thus competition is an important factor in determining which species can coexist within a habitat.

The outcome of competition between two populations depends on the relative efficiencies with which they exploit their shared resources. Every population consumes resources. When these are scarce relative to the demand for them, each act of consumption by one individual makes the resource less available to others, as well as to itself. As consumption continues, resources decline to a level that no longer supports the growth of the consuming population, and the population may reach an equilibrium size. When one population can continue to grow at a resource level that curtails the growth of a second population, the first will eventually replace the second. Thus competition and its various outcomes depend on the relationship of consumers to their resources.

In this chapter, we shall consider some of the general principles of competition between species, illustrate the potential effects of competition by examining the results of laboratory experiments, and demonstrate the importance of competition in natural systems.

Resources

Ecologist David Tilman of the University of Minnesota defines a **resource** as any substance or factor that is consumed by an organism and that can lead to increased population growth rates as its availability in the environment is increased. Two qualifications are key to this definition. First, a resource is consumed and its amount is thus reduced. Second, a resource is used by the consumer for its own maintenance and growth. Thus food is always a resource, and water is a resource for terrestrial plants and animals. Water is consumed and it is critical to maintenance and growth; furthermore, when its availability is reduced, biological processes are affected in such a way as to reduce population growth.

Consumption connotes more than the act of eating. For sessile animals, space (open sites) should be considered a resource. Among barnacles growing on rocks within the intertidal zone, individuals require space to grow, and larvae require space to settle and take up adult life (Figure 19.1). Hence crowding increases adult mortality and reduces fecundity by limiting the growth of adults and the recruitment (settling) of larvae. Open space fosters reproduction and recruitment, and individuals "consume" open sites as they colonize and grow on them. Hiding places and other safe sites constitute another kind of resource. Each area of habitat has a limited number of holes, crevices, or patches of dense cover in which an organism may escape predation or seek refuge from adverse weather. As some individuals occupy the best sites, others must settle for less favorable places; they may suffer higher mortality as a consequence.

What factors are not resources? Temperature is not a resource. Higher temperature may stimulate increased reproductive rate, but individuals do not consume temperature. Of course, temperature and other nonconsumable physical and biological factors are important, but they must be considered differently from resources. Temperature, humidity, salinity, hydrogen ion concentration (pH), buoyancy, and viscosity are all **conditions** that influence the rates of processes and therefore the individual's ability to consume resources, but they are not themselves used and thereby transformed by the activities of organisms.

Resources can be classified according to how their consumers affect them. **Nonrenewable resources,** such as space, are not altered by use. Once occupied, space becomes unavailable; it is "replenished" only when the consumer leaves. In contrast, **renewable resources** are constantly regenerated, or renewed. Births in a population of prey continually supply food items for predators. The continuous decomposing of organic detritus in the soil provides a fresh supply of nitrate-nitrogen to plant roots.

Among renewable resources, we recognize three types. The first type includes resources that have a source that is external to the system, beyond the influence of consumers. Sunlight strikes the surface of the earth regardless of whether plants "consume" it; local precipitation is largely independent of the consumption of water by plants; for all practical purposes, detritus rains down from the sunlit surface of the sea to the abyssal depths uninfluenced by the consumers groping there in everlasting darkness.

Renewable resources of the second type are generated within the ecosystem and are directly affected by the activities of consumers. Most predator–prey, plant–herbivore, and parasite–host interactions involve resources of this type.

Renewable resources of the third kind issue from within a system, but resource and consumer are linked indirectly either through other resource–consumer steps or through abiotic processes. For example, in the nitrogen cycle of a forest, plants assimilate nitrate from the soil. Herbivores and detritivores consume plant biomass, returning large quantities of organic nitrogen compounds to the soil. These are attacked by microorganisms, which release the nitrogen in a form the plants can use. Consumption thus follows the cycle:

FIGURE 19.1 Competition for space among barnacles on the Maine coast. Above their optimal range in the intertidal zone, the barnacles are sparse, and young can settle in the bare patches (a). Lower in the intertidal zone, dense crowding of barnacles precludes further population growth (b); young barnacles can settle only on older individuals. Courtesy of the American Museum of Natural History.

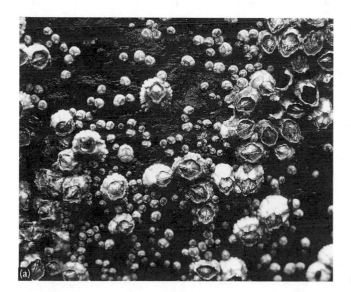

Soil → Plant → Detritivore → Microorganism → Mineral soil

Uptake of nitrate by plants has little direct effect on its release by detritivores. Similarly, consumption of detritus cannot immediately influence plant production. Clearly, however, detritivores and microorganisms do influence plant production indirectly through the rate at which they release nutrients into the soil.

Limiting resources

Consumption reduces the level of a resource. What is used by one organism cannot be used by another. By diminishing their resources, consumers limit their own population increase. As a population grows, its overall resource requirement grows as well, eventually to be balanced by a decreasing supply of resources available to fulfill the need. But whereas all resources, by definition, are reduced by their consumers, not all resources limit consumer populations. All animals require oxygen, for example, but they do not depress its level in the atmosphere even noticeably before some other resource, such as food supply, limits population growth.

The potential of a resource to limit population growth depends on its availability relative to demand. At one time, ecologists believed that populations were limited by the single resource that had the greatest relative scarcity. This principle has been called **Leibig's law of the minimum,** after Justus Leibig, who set forth the idea in 1840. According to this law, each population increases until the supply of some resource (the **limiting resource**) no longer satisfies the population's requirement for it. The growth of a population under a given set of conditions responds uniquely to the level of each of its resources. For the diatom *Cyclotella meneghiniana* grown under silicate and phosphate limitation in a laboratory culture, population growth and resource depletion ceased when phosphate levels were reduced to 0.2 micromolar (μM) or silicate levels were reduced to 0.6 μM. According to Leibig's law of the minimum, whichever of these resources is reduced to this limiting value first regulates the growth of the *Cyclotella* population.

Leibig's law applies, however, only to resources having independent influence on the consumer. In many cases, two or more resources *interact* to determine the growth rate of a consumer population. That is, the growth rate of a consumer at a particular level of one resource depends on the level of one or more other resources.

When amounts of two resources enhance the growth of a consumer population more than the sum of both individually, the resources are said to be **synergistic** (from the classical roots *syn,* "together," and *ergon,* "work"). This principle can be illustrated by a study of the small herbaceous plant *Impatiens parviflora* common in woodlands of England. In one experiment, fertilized (with nitrate and phosphate) and nonfertilized *Impatiens* were exposed to different levels of light from the time of seed germination until the end of the experiment at 5 weeks. Added light enhanced growth of the fertilized plants more than that of the controls (Figure 19.2); hence the ability of *Impatiens* to

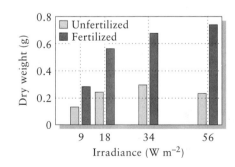

FIGURE 19.2 Joint influence of light levels and fertilizer on the growth of *Impatiens.* After W. J. H. Peace and P. J. Grubb, *New Phytol.* 90:127–150 (1982).

use light depends on the presence of other resources. Plant growth requires both the carbon assimilated by photosynthesis, as a source of energy and for structural carbohydrates, and nitrogen and phosphorus for the synthesis of proteins and amino acids. At the highest light intensities used in the experiment, nitrogen and phosphorus also were shown to interact in their effect on plant growth (Figure 19.3), demonstrating that both are required for normal growth.

The experimental demonstration of competition

When ecologists began to consider the dynamics of population growth in the late 1920s, many investigators initiated experiments to determine the effects of one species on the population growth of another. These experiments were similar in design: two species were grown separately under controlled conditions and resources to determine their carrying capacities, and then they were grown together under the same conditions to determine the effect of each on the other. The difference in the population growth of one species in the presence and in the absence of the other was a measure of the competition between them. The experiments of the Russian biologist G. F. Gause on protozoans were among the first and most influential on subsequent work in population biology. As shown in Figure 19.4, when the protozoans *Paramecium aurelia* and *P. caudatum* were established separately on the same type of nutritive medium, both populations grew rapidly to limits imposed by resources. When grown together, however, only *P. aurelia* persisted. Similar experiments with fruit flies, mice, flour beetles, and annual plants have almost always produced the same result: one species persists and the other dies out, usually after 30 to 70 generations.

The outcome of competition depends on the conditions of the environment. When the flour beetles *Tribolium castaneum* and *T. confusum* were grown together in vials of wheat flour under cool, dry conditions, *T. castan-*

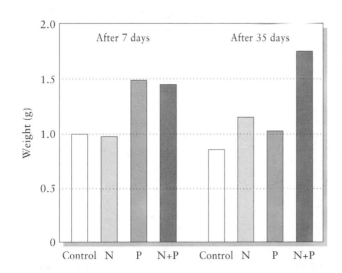

FIGURE 19.3 Joint influence of nitrogen (N) and phosphorus (P) fertilization on growth of *Impatiens*. Note that in the first week of growth, adding nitrogen alone had little effect (there was so much in the soil that it was not a limiting factor for young plants). After 35 days, however, addition of *neither* nutrient alone enhanced growth as much as the addition of both. Clearly, nitrogen and phosphorus were synergistic in promoting plant growth. After W. J. H. Peace and P. J. Grubb, *New Phytol.* 90:127–150 (1982).

eum usually excluded *T. confusum*. Under warm, moist conditions, however, it was *T. confusum* that persisted (Figure 19.5). The experiments showed not only that physical conditions were critical to the outcome of competition, but also that only the relative performance of the two species mattered. *T. castaneum* consistently achieved higher population densities under moist conditions than under dry conditions when grown alone, but just as consistently it was excluded under moist conditions by *T. confusum,* whose populations when grown alone were yet higher.

The competitive exclusion principle

The accumulating results of laboratory experiments on competition eventually appeared so general as to warrant their elevation to the status of a general rule, the **competitive exclusion principle.** The principle can be summarized as follows: two species cannot coexist on the same limiting resource. The qualification *limiting* is required in the definition of the principle because competition expresses itself only when consumption depresses resources and thereby limits population growth.

The competitive exclusion principle states that species having identical resource requirements cannot coexist. Similar species do, of course, coexist in nature. But as we shall see in later chapters, detailed observations always reveal ecological differences between such species, often based on subtle differences in habitat or diet preference. These observations prompt us to ask how much ecological segregation is sufficient to allow coexistence. Although this question has been very difficult to answer, theoretical analyses of competition have suggested some of the general conditions under which species may coexist.

The theory of competition and coexistence

Most competition theory springs from the formulations of A. J. Lotka and G. F. Gause, who used the logistic equation for population growth as their starting point. Remember that according to the logistic equation, the rate of increase of population *i* is expressed by

FIGURE 19.4 Increase in populations of two species of *Paramecium* when grown in separate cultures (a) and when grown together (b). Although both species thrive when grown separately, *P. caudatum* cannot survive together with *P. aurelia*. After G. F. Gause, *The Struggle for Existence.* Williams & Wilkins, Baltimore (1934).

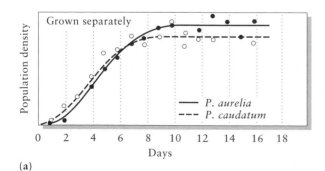

(a)

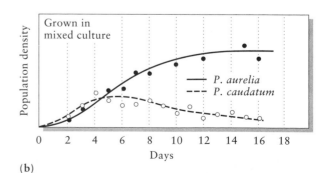

(b)

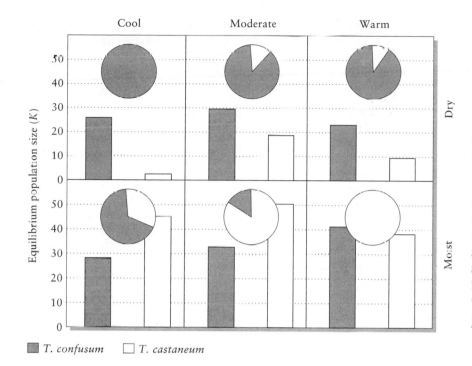

FIGURE 19.5 The outcome of competition between two species of flour beetles of the genus *Tribolium* at different temperatures and relative humidities. The vertical bars illustrate the equilibrium populations (total number of adults, larvae, and pupae per gram of flour) when each species is grown alone; the pie charts present the percentage of 20 to 30 contests won by each species. After data in T. Park, *Physiol. Zool.* 27:177–238 (1954); T. Park, *Science* 138:1369–1375 (1962).

$$\frac{dN_i}{dt} = r_i N_i \left(1 - \frac{N_i}{K_i}\right)$$

where r is the exponential rate of increase in the absence of competition and K is the number of individuals that the environment can support. Intraspecific competition appears as the term N/K; as N approaches K, N_i/K approaches 1 and the quantity $(1 - N/K)$ approaches 0. As we have seen before, a stable equilibrium is reached when $N = K$.

Volterra incorporated interspecific competition in the logistic equation by adding the term $-a_{ij}N_j/K_i$ to the quantity within the parentheses. Hence

$$\frac{dN_i}{dt} = r_i N_i \left(1 - \frac{N_i}{K_i} - \frac{a_{ij}N_j}{K_i}\right)$$

where N_j is the number of individuals of a second species (j), and a_{ij} is the coefficient of competition — that is, the effect of an individual of species j on the exponential growth rate of the population of species i. The term a_{ij} is a dimensionless constant whose value is a fraction of the effect of an individual of species i on its own population growth rate. We may think of the competition coefficient a_{ij} as the degree to which individuals of species j use the resources of individuals of species i. That is why $a_{ij}N_j$ is divided by K_i, the carrying capacity of species i. It is the degree to which individuals of species j usurp the resources of species i — expressed by the value K_i — that determines the effect of j on i's rate of population growth.

Strictly speaking, the term N_i/K_i should have coefficient a_{ii}, but this is assumed to be 1 and is left out. Although it need not be, the value of a_{ij} is usually less than 1; individuals of the same species are likely to compete more intensely than individuals of different species. Because each species of a pair exerts an effect on the other, the mutual relationship between them requires two equations: one, presented above, for the effect of species j on species i and a second, similar equation for the effect of species i on species j. If two species are to coexist, the populations of both must reach a stable size greater than 0. That is, both dN_i/dt and dN_j/dt must equal 0 at some combination of positive values of N_i and N_j. From the previous equation, we see that $dN_i/dt = 0$ when

$$\hat{N}_i = K_i - a_{ij}N_j$$

The little "hat" (^) over the N indicates that it is an equilibrium value. In the absence of interspecific competition ($a_{ij} = 0$), the equilibrium population size $\hat{N}_i$ is equal to K_i, a measure of the resources available to species i. Interspecific competition reduces the effective carrying capacity of the environment for species i by the amount $a_{ij}N_j$—that is, in proportion to the population size and coefficient of competition of the second species.

By making some algebraic substitutions, we can express the equilibrium population size of species i in terms of just the carrying capacities and competition coefficients, specifically

$$\hat{N}_i = \frac{(K_i - a_{ij}K_j)}{(1 - a_{ij}a_{ji})}$$

Because competition coefficients generally are less than 1, the denominators of these equations normally assume positive values. Therefore, the equilibrium value N_i will be positive only when the numerator of the equation is positive and hence when a_{ij} is less than the ratio K_i/K_j. In words, coexistence is more likely when coefficients of competition are relatively weak and the carrying capacities of both competitors are similar. Any number of species may coexist as long as these criteria are met for all pairs of them. In the most general terms, the coexistence of two species requires that

$$a_{ij}a_{ji} < 1$$

Competition in nature

Competitive exclusion is a transient phenomenon. The evidence of exclusion having taken place is lost when the poorer competitor disappears. We can observe the process in the laboratory when we mix populations according to our whim and follow the course of their interaction. The closest natural analogy to the laboratory experiment is the accidental or intended introduction of species by humans. For example, when many species of parasites are introduced simultaneously to control a weed or insect pest, the control species are brought together in the same locality to exploit the same resource. It is not surprising that competitive exclusion has occurred under these conditions.

Between 1947 and 1952, the Hawaii Agriculture Department released 32 potential parasites to combat several species of fruit pests, including the Oriental fruit fly. Of these species, 13 became established, but only 3 kinds of braconid wasps proved to be important parasites of fruit flies. Populations of these wasps, all closely related members of the genus *Opius*, successively replaced each other from early 1949 to 1951, after which only *Opius oophilus* was commonly found to parasitize fruit flies (Figure 19.6). As each parasite population was replaced by a more successful species, the level of parasitism of fruit flies by wasps also increased, suggesting superior competitive ability.

A similar pattern of replacement, in this case involving wasps that parasitize scale insects, has been more thoroughly documented in southern California. Scale insects are pests of citrus groves and can cause extensive damage to the trees. As the evolution of resistance by pests reduced the effectiveness of chemical pesticides, agricultural biologists turned to the importation of insect parasites and predators. Yellow scales have infested California citrus groves since oranges and lemons were first planted there. In the late 1800s, the red scale was accidentally introduced and replaced yellow scale throughout most of its range, perhaps itself a case of competitive exclusion. Of the many species introduced in an effort to control citrus scale, tiny parasitic wasps of the genus *Aphytis* (from the Greek *aphyo*, "to suck") have been most successful. One species, *A. chrysomphali*, was accidentally introduced from the Mediterranean region and became established by 1900.

Despite its tremendous population growth potential, *A. chrysomphali* did not effectively control scale insects, particularly not in the dry interior valleys. In 1948 a close relative from southern China, *A. lingnanensis*, was introduced as a control agent. This species increased rapidly and widely replaced *A. chrysomphali* within a decade. When both species were grown in the laboratory, *A. lingnanensis* was found to have the higher net reproductive rate, whether the two species were placed separately or together in population cages.

Although *A. lingnanensis* had excluded *A. chrysomphali* throughout most of southern California, it still did not provide effective biological control of scale insects in the interior valleys because cold winter temperatures greatly reduced parasite populations. Wasp larval development slows to a standstill at temperatures below 16°C (60°F), and adults cannot tolerate temperatures below 10°C (50°F).

In 1957 a third species of wasp, *A. melinus*, was introduced from areas in northern India and Pakistan where temperatures range from below freezing in winter to above 40°C in summer. As was hoped, *A. melinus* spread rapidly

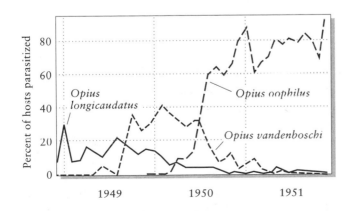

FIGURE 19.6 Successive change in the predominance of three species of wasps of the genus *Opius* parasitic on the Oriental fruit fly. After H. A. Bess, R. van den Bosch, and F. A. Haramoto, *Proc. Hawaiian Entomol. Soc.* 17:367–378 (1961).

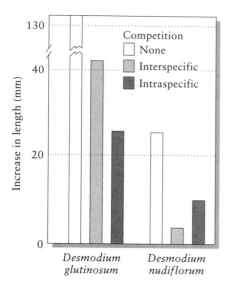

FIGURE 19.7 The growth responses of two species of *Desmodium* when planted near individuals of the same species, near individuals of the other species, and at a distance from individuals of either species. After W. G. Smith, *Amer. Midl. Nat.* 94:99–107 (1975).

throughout the interior valleys of southern California, where temperatures resemble the wasp's native habitat, but it did not become established in the milder coastal areas.

The depressing effect of interspecific competition on the growth of plants has been demonstrated in several experimental studies, such as the following on two species of *Desmodium,* small herbaceous legumes (members of the pea family) common in oak woodlands in the midwestern United States. Small individuals of each species, *D. glutinosum* and *D. nudiflorum,* were planted 10 cm from a large individual of the same species (intraspecific test), 10 cm from a large individual of the other species (interspecific test), or at least 3 m from any *Desmodium* plant (control). The total increase in length of all leaves, both old and new, served as an index to subsequent growth.

The results of the experiment (Figure 19.7) showed that both species grew best in the absence of individuals of either species (though surrounded by unrelated plants that occur in the habitat). It was also clear, however, that (1) the growth of *D. nudiflorum* was depressed more by interspecific than by intraspecific competition, and (2) interspecific competition was asymmetrical, with *D. glutinosum* exerting the stronger effect and *D. nudiflorum* the weaker.

Experimental studies of competition among species of animals

Plants occupy space. In dense plantings, roots and leaves crowd together so closely that individuals constantly vie for sunlight, water, and soil nutrients. The animals that most closely resemble plant systems in this regard are the space-occupying invertebrates of rocky shores. Among the most prominent of these are barnacles, which may form dense, continuous populations. Just as plants rely on light as a source of energy, barnacles gather food in the form of plankton from the water that washes over them.

One of the first experimental demonstrations of competition in nature resulted from the work of Joseph Connell on two species of barnacles within the intertidal zone of the rocky coast of Scotland. Adults of *Chthamalus stellatus* normally occur higher in the intertidal zone than those of *Balanus balanoides,* the more northerly of the two species. Although the vertical distributions of newly settled larvae of the two species overlap broadly within the tide zone, the line between the vertical distributions of adults is sharply drawn.

Connell demonstrated that adult *Chthamalus* live only in that portion of the intertidal zone above *Balanus* not because of physiological tolerance limits but because of interspecific competition. When Connell removed *Balanus* from rock surfaces, *Chthamalus* thrived in the lower portions of the intertidal zone where they normally did not occur.

The two species compete directly for space. *Balanus* have heavier shells and grow more rapidly than *Chthamalus;* as individuals expand, the shells of *Balanus* edge underneath those of *Chthamalus* and literally pry them off the rock! *Chthamalus* can occur in the upper parts of the intertidal zone because they are more resistant to desiccation than *Balanus;* even when surfaces in the upper levels are kept free of *Chthamalus, Balanus* do not invade.

Competition between barnacles results from physical **interference** rather than from differential **exploitation** of food or other resources. Mobile animals may exhibit similar interference competition through occasional aggressive encounters. For example, two species of voles (small mouselike rodents of the genus *Microtus*) co-occur in some areas of the Rocky Mountain states. In western Montana, the meadow vole normally inhabits wet habitats surrounding ponds and water courses, whereas the montane vole is restricted to dry habitats. When meadow voles were trapped and removed from an area of wet habitat, montane voles began to move in from surrounding dry habitats. After montane voles were trapped and removed from a dry habitat, which they occupied exclusively, meadow voles began to be caught, presumably after having moved in from moister surrounding habitats. Each species excludes the other from one habitat by aggressive behavior.

Although territorial defense and social aggression occur frequently within species, they are not common between species, where competition occurs more regularly through exploitation of resources. Because exploitative competition expresses its effects indirectly, through differential survival and reproduction of individuals of different species, it may be difficult to detect.

In Big Bend National Park, Texas, the canyon lizard *Sceloporus merriami* and the tree lizard *Urosaurus ornatus* search for insect prey on exposed surfaces of large rocks. Where *Sceloporus* were removed from experimental areas, the numbers of *Urosaurus* increased over those of controls during 2 years of the study (Figure 19.8). In contrast, removal of *Urosaurus* did not result in an increase in *Sceloporus* populations. Fluctuations in control and experimental populations of *Sceloporus* were closely related to rainfall (1975 and 1977 were dry years), suggesting that physical factors may limit *Sceloporus* more severely whereas competition influences *Urosaurus* more strongly, a situation recalling the relative ecological positions of *Balanus* and *Chthamalus* in the intertidal zone.

Mechanisms of competition

We have distinguished **exploitation** competition, in which individuals, by using resources, deprive others of the benefits of those resources, and **interference** competition, in which individuals directly affect others by physical means

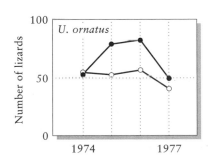
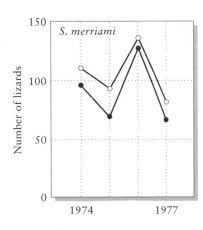

FIGURE 19.8 Populations of two species of lizards (individuals per hectare) on plots from which one or the other of the species was removed and on control plots where both populations remained. From A. E. Dunham, *Ecol. Monogr.* 50:309–330 (1980).

(chasing and fighting, for example) and chemical means (toxins). Thomas Schoener, an ecologist at the University of California at Davis, further subdivided competition into six categories, according to its mechanisms:

1. Consumptive competition, based on the use of some renewable resource

2. Preemptive competition, based on the occupation of open space

3. Overgrowth competition, which occurs when one individual grows upon or over another, thereby depriving the second of light, nutrient-laden water, or some other resource

4. Chemical competition, by production of a toxin that acts at a distance after diffusing through the environment

5. Territorial competition—that is, the defense of space

6. Encounter competition, which involves transient interaction over a resource that may result in physical harm, loss of time or energy, and theft of food

The occurrence of each of these mechanisms of competition depends on the capabilities of organisms and the habitats in which they occur (Figure 19.9). Preemptive and overgrowth competition appear among sessile space users, primarily terrestrial plants and marine plants and animals living on hard substrates. Territorial and encounter competition appear among actively moving animals; chemical competition, among terrestrial plants (toxins are diluted too readily in aquatic systems). By Schoener's count, consumptive competition is the commonest, especially in terrestrial environments. Preemptive and overgrowth competition predominate in marine habitats, in part because most studies have involved sessile organisms living on hard substrates.

Of Schoener's categories, one that we have not talked about and that merits some discussion is chemical competition, or **allelopathy**. Although the caus-

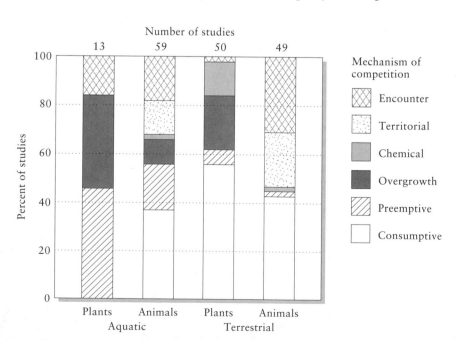

FIGURE 19.9 Frequencies of various mechanisms of competition among plants and animals in aquatic and terrestrial habitats, as revealed by experimental field studies. From data in T. W. Schoener, *Amer. Nat.* 103:277–313 (1983).

ing of injury (-*pathy*) to other individuals (*allelo-*) by chemical means has been reported most frequently in terrestrial plants, such interactions may take on a variety of forms. Thus some parasites appear to exclude other, potentially competing parasites from a host by stimulating the host's immune system against them. It has also been suggested that the abundant oils in the eucalyptus trees of Australia promote frequent fires in the leaf litter, which kill the seedlings of competitors. More frequently, it is the direct effect of a toxic substance that does the damage.

In shrub habitats in southern California, several species of sage of the genus *Salvia* use chemicals to inhibit the growth of other vegetation. Clumps of *Salvia* are usually surrounded by bare areas separating the sage from neighboring grassy areas (Figure 19.10). Observed over long periods, *Salvia* may be seen to expand into the grassy areas. But because sage roots extend only to the edge of the bare strip and not beyond, it is unlikely that a toxin is extruded into the soil directly by the roots. The leaves of *Salvia* produce volatile terpenes (a class of organic compounds that includes camphor and gives foods spiced with sage part of their distinctive taste), which apparently affect nearby plants directly through the atmosphere.

Asymmetry of competition

Field experiments have revealed many cases in which one member of a species pair responded to the addition or removal of individuals of the other species, but not vice versa. Among 98 reciprocal tests of competition between species reported in the literature, no interaction was reported for 44 pairs, reciprocal negative competition for 21 pairs, and response by only one species for 33 pairs. The high percentage of cases in the last category suggests that asymmetry is more the rule than the exception in competitive interactions.

Asymmetry in competition must derive from asymmetry in ecology. Moreover, the superior competitor is almost always more strongly limited by some factor — environmental tolerances or predators — that is external to the com-

FIGURE 19.10 (a) Bare patch at the edge of a clump of sage includes a 2-m-wide strip with no plants (A to B) and a wider area of inhibited grassland (B to C) lacking wild oat and bromegrass, which are found with other species to the right of C in unaffected grassland. (b) Aerial view shows sage and California sagebrush invading annual grassland in the Santa Inez Valley of California. Courtesy of C. H. Muller. From C. H. Muller, *Bull. Torrey Bot. Club* 93:332–351 (1966).

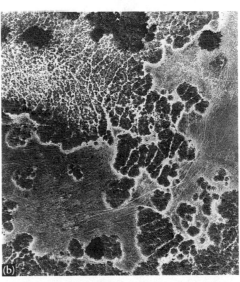

petitive interaction. In the case of the barnacles mentioned earlier, *Balanus* could effectively exclude *Chthamalus* from the lower levels of the intertidal zone, but it lacked the physiological tolerance that allowed *Chthamalus* to exist higher up. Therefore, when *Balanus* was removed from the lower portion of the tidal zone, where it dominated, *Chthamalus* could become established, showing a strong effect of *Balanus* on *Chthamalus*. However, when *Chthamalus* was removed from rocks at high tide levels, where it dominated, *Balanus* failed to colonize.

Predators often balance strong asymmetry among competitors by their preference for the competitively dominant species. On marine hard substrates (such as rocky shores and reef rubble), rapidly growing, leafy species of algae often outcompete slowly growing, encrusting forms, which they overgrow and shade out of existence. But such predators as sea urchins and herbivorous fish readily graze the leafy forms of algae while leaving the less accessible, encrusting ones untouched. In this situation, competition appears to be asymmetrical: removal of leafy algae results in the establishment of encrusting forms (an experiment possible only in the absence of herbivores); removal of encrusting forms (whose presence depends on those herbivores) does not lead to the establishment of leafy species, because these are eaten as soon as they colonize the rock surface.

Along the shores of the Caribbean, patch reefs present three ecological zones: the reef flat, the topmost part of the reef that is covered by shallow water and is often exposed at low tide; the reef slope, the outer edge of the reef as it descends to deep water; and the sand plain, the area of sandy bottom between patches of coral reef. Because few fish can inhabit the shallow water on the reef flat, herbivory there is slight. On the sand plain, predatory fish drive away herbivorous species, which cannot find hiding places in the exposed habitat. Therefore, only the structurally complex reef slope experiences strong herbivory by fish. And as a result, species of algae typical of the reef slope must effectively resist herbivory.

This was discovered by transplanting species of algae from each of the three habitats to locations within each habitat. As we might expect, algae from the reef slope were virtually untouched in all the habitats, but species from the reef flat and sand plain were heavily grazed (40 to 90 percent of their biomass was consumed in 48 hours) when transplanted to the reef slope (Figure 19.11). In another experiment, algae were transplanted to the reef slope, where some individuals were completely enclosed in cages and others were exposed to her-

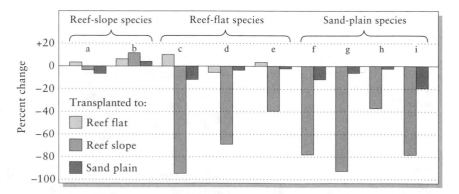

Figure 19.11 Percent change in biomass of species of algae growing in three coral reef habitats after transplanting to each of the other habitats. Reef-slope species resist grazing by fish; other species are heavily grazed when transplanted to the reef slope. After M. E. Hay, T. Colburn, and D. Downing, *Oecologia* 65:591–598 (1983).

bivory in partial cages. The sand-plain species were grazed heavily in the partially exposed cages but grew vigorously when protected from herbivores; reef-flat species transplanted to the same conditions did not grow well when protected, possibly because of low levels of light.

Sand-plain species of algae grow more rapidly than those from the slope and flat; they would probably exclude species from other habitats through overgrowth competition were it not for their vulnerability to herbivory on the reef slope and their inability to tolerate the physically stressful conditions of desiccation and temperature fluctuation on the reef flat. Reef-flat species form a dense turf of algae, resistant to both herbivory and desiccation, but the price of this adaptation is slow growth. These examples illustrate that asymmetric competition results when species that share one resource (space or food) are differently specialized with respect to other aspects of the environment (physical conditions or consumers).

Competition between distantly related species

Charles Darwin emphasized that competition should be most intense between closely related species or organisms. In *On the Origin of Species,* he remarked, "As species of the same genus have usually, though by no means invariably, some similarity in habits and constitution, and always in structure, the struggle will generally be more severe between species of the same genus, when they come into competition with each other, than between species of distinct genera." Darwin reasoned that similar structure indicates similar ecology, especially resource requirements. But many resources are jointly used by distantly related organisms. Barnacles and mussels, as well as algae, occupy space in the intertidal zone. Both fish and aquatic birds prey on aquatic invertebrates. Krill (*Euphausia superba*), shrimplike crustacea that abound in subantarctic waters, are fed upon by virtually every other type of large animal, including fish, squid, diving birds, seals, and whales. Recent increases in seal and penguin populations in the southern oceans have been related to decreased competition from whales, whose populations had been decimated by commercial exploitation.

In terrestrial habitats, invertebrates in forest litter are consumed by spiders, ground beetles, salamanders, and birds. Birds also compete with lizards for many of the same prey in other habitats. Seeds are a major resource for ants, rodents, and birds in desert ecosystems. Because many of the ants and rodents in deserts eat little else but seeds, the opportunity for interspecific (interphylum!) competition certainly presents itself.

Predation and the outcome of competition

Darwin noted that grazing can maintain a high diversity of plants in grasslands. In the absence of grazers, dominant competitors grow rapidly and exclude others. Similar results have been obtained from experiments on marine algal communities under the pressure of grazing by limpets, snails, and urchins. These studies indicate that predation has a strong hand in shaping the structure of biological communities by influencing the outcome of competitive interactions between prey species.

University of Washington ecologist Robert Paine was one of the first to demonstrate this point experimentally. On the exposed rocky coast of the state of Washington, the intertidal zone harbors several species of barnacles, gooseneck barnacles, mussels, limpets, and chitons (a kind of grazing mollusk); these are preyed upon by the starfish *Pisaster* (Figure 19.12). In one experiment, Paine removed starfish from a study area 8 m in length and 2 m in vertical

FIGURE 19.12 (a) Congregation of starfish (*Pisaster*) at low tide on the coast of the Olympic Peninsula, Washington. The starfish (b) is an important predator on mussels (c).

extent; an adjacent area of similar size was left undisturbed. Following the removal of the starfish, the number of prey species in the experimental plot decreased rapidly, from 15 at the beginning of the study to 8 at the end. Diversity declined in the experimental areas when populations of barnacles and mussels increased and crowded out many of the other species. Paine concluded that starfish were a major factor in maintaining the diversity of the area.

Studies conducted in artificial ponds have shown that predatory salamanders can reverse the outcome of competition among frog and toad tadpoles. In one experiment, ponds were supplied with 200 hatchlings of the spadefoot toad (*Scaphiopus holbrooki*), 300 of the spring peeper (*Hyla crucifer*), and 300 of the southern toad (*Bufo terrestris*). Each of the ponds, which were identical in all respects, received either 0, 2, 4, or 8 of the predatory broken-striped newt (*Notophthalmus viridescens*). In the absence of newt predation, *Scaphiopus* tadpoles grew rapidly, survived well, and dominated the ponds along with smaller numbers of *Bufo. Hyla* tadpoles were all but eliminated (Figure 19.13). *Notophthalmus* apparently prefer toad tadpoles, and at higher numbers of predators in the tanks, the survival of both *Scaphiopus* and *Bufo* decreased markedly. With fewer toads per pond, levels of food increased, and the survival and growth of *Hyla* tadpoles improved immensely, as did the growth of surviving *Scaphiopus* and *Bufo* tadpoles.

⊱ SUMMARY

1. Competition is the use or contesting of a resource by more than one individual consumer. When the individuals belong to the same species, their interaction is called intraspecific competition; when they belong to different species, it is called interspecific competition. Intraspecific competition is expressed demographically as density dependence and may lead to the regulation of population size. Interspecific competition is expressed as a reduction in the carrying capacities of competing populations and, in its extreme, may lead to the exclusion of a species.

FIGURE 19.13 Effect of predators on percent survival and on weight at metamorphosis in three species of anurans (frogs and toads) raised in large tanks. From P. J. Morin, *Science* 212:1284–1286 (1981).

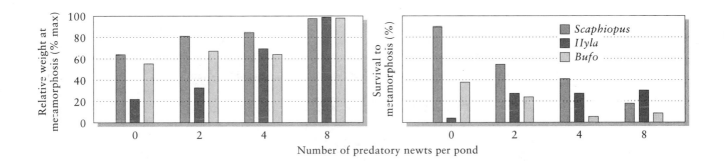

2. A resource may be defined as any factor that is consumed and whose increase promotes population growth. Thus light, food, water, mineral nutrients, and space are resources. Temperature, salinity, and other such conditions are not.

3. Resources may by classified as nonrenewable (space) or renewable (light and food), and the latter may be further distingushed according to the influence of the consumer on the provisioning of the resource; no influence, direct influence, or indirect influence through other consumers.

4. Of all the resources consumed, only one or a few limit the population growth of the consumer. These are normally those whose supply relative to demand is least. This principle is known as Leibig's law of the minimum. Many pairs of resources, however, exert a synergistic influence on population growth; that is, their combined effect is greater than the sum of their individual effects.

5. Competition may be demonstrated in the laboratory and in the field by change in the population size of one species following the addition or removal of another. When two species compete strongly, the population of the first is sensitive to changes in numbers of the second, and vice versa.

6. The outcome of competition depends on the conditions of the environment for any given pair of species.

7. Theoretical investigations and laboratory studies led to the generalization that no two species of competitors could coexist on the same limiting resource. This has come to be known as the competitive exclusion principle.

8. Some mathematical treatments of competition are based on the logistic equation of population growth. Competition between two species is incorporated into the equation for the population growth rate of the first species. A term called the coefficient of competition expresses the effect of individuals of the second species on the population growth rate of the first species.

9. The joint equilibrium of two species can be described by an equation including the carrying capacities and competition coefficients of the two species. In the most general terms, coexistence requires that the product of the competition coefficients of the first and second species is less than 1.

10. Laboratory experiments present clear evidence of competition among species, but natural populations are limited by physical conditions and consumers as well as by shared resources. Since the early 1960s, ecologists have conducted numerous field studies designed to reveal the influence of competition on the sizes of natural populations.

11. In pest control programs, the disappearance of one parasite following the introduction of a second, similar species has provided indirect evidence of competition.

12. Transplant experiments with plants, creating varying conditions of intraspecific and interspecific competition, often show striking effects of interspecific competition.

13. Removal experiments involving intertidal invertebrates have demonstrated strong competition among such space-filling animals as barnacles, mussels, and encrusting sponges. Competitive exclusion is accomplished by direct physical interaction.

14. Rapid invasion of a habitat by a species of small mammal following the removal of another, related species illustrates direct interference through aggressive behavior.

15. Exploitation competition is most convincingly demonstrated in studies that show appropriate changes in resource levels accompanying the demographic response of one species after removal of a competitor.

16. Field studies have revealed many mechanisms of competition: consumptive, preemptive, overgrowth, chemical, territorial, and encounter. The first is usually referred to as exploitation competition, and the last three as interference competition. Preemptive and overgrowth competition are based on the use of space and renewable resources, respectively, but involve close contact of competing individuals.

17. Experiments have demonstrated the prevalence of asymmetry in competition, in which the effect of one species is clearly greater than that of the other. Coexistence is possible, however, when the asymmetry is balanced by physical conditions or by consumers whose actions favor the poorer competitor.

18. Although competition is likely to be strongest between similar species, distantly related organisms may also share resources. Ants, rodents, and birds consume the seeds of desert annuals; barnacles and algae occupy the same rock surfaces in intertidal zones.

❦ SUGGESTED READINGS

Connell, J. H. 1961. The influence of interspecific competition and other factors on the distribution of the barnacle *Chthamalus stellatus*. *Ecology* 42:710–723.

Connor, E. F., and D. Simberloff. 1986. Competition, scientific method, and null models in ecology. *American Scientist* 74:155–162.

Grace, J. B., and D. Tilman (eds.). 1990. *Perspectives on Plant Competition*. Academic Press, San Diego, Calif.

Hardin, G. 1960. The competitive exclusion principle. *Science* 131:1292–1297.

Hay, M. E., T. Colburn, and D. Downing. 1983. Spatial and temporal patterns in herbivory on a Caribbean fringing reef: the effects on plant distribution. *Oecologia* 58:299–308.

Morin, P. J. 1981. Predatory salamanders reverse the outcome of competition among three species of anuran tadpoles. *Science* 212:1284–1286.

Paine, R. T. 1974. Intertidal community structure: experimental studies on the relationship between a dominant competitor and its principal predator. *Oecologia* 15:93–120.

Peace, W. J. H., and P. J. Grubb. 1982. Interaction of light and mineral nutrient supply in the growth of *Impatiens parviflora*. *New Phytologist* 90:127–150.

Schoener, T. W. 1982. The controversy over interspecific competition. *American Scientist* 70:586–595.

Schoener, T. W. 1983. Field experiments on interspecific competition. *American Naturalist* 122:240–285.

Tilman, D. 1982. *Resource Competition and Community Structure*. Princeton Univ. Press, Princeton.

20

PREDATION

The basic question of population biology is this: What factors influence the size and stability of populations? In previous chapters, we saw how density-dependent factors and time delays influence the responses of birth and death rates to population density. As we expanded our perspective on populations to include interactions among species, the basic question of population biology was elaborated: How do species influence each other's populations? Exploitation competition and interference competition from other species depresses the growth rate of a population. But because most species are both consumers and resources for other consumers, it also becomes necessary to ask whether populations are limited primarily by what they eat or by what eats them.

In this chapter, we shall explore the impact of consumers—predators, parasites, herbivores, and others—on resource populations. Consumer-resource interactions have been investigated theoretically and experimentally for both laboratory and natural populations. The initial motivation for these studies came from concerns about managing populations of game species and controlling populations of agricultural pests. The insights and practical knowledge gained through these studies have also contributed importantly to our understanding of the structure and dynamics of natural systems.

A few distinctions

Consumers go by many names, the most familiar of which are predators, parasites, parasitoids, herbivores, and detritivores. From the standpoint of population interactions, some of these distinctions are useful, whereas others are confusing. Let's start with **predators**. The images of an owl eating a mouse and a spider eating a fly capture the essentials of predation. Predators catch individuals and consume them, thereby removing them from the prey population. In contrast, a **parasite** consumes a living host. Although it may increase the probability of a host's dying from other causes or reduce its fecundity, a parasite does not by itself remove an individual from the resource population.

The term **parasitoid** is applied to species of wasps and flies whose larvae consume the tissues of living hosts, usually the eggs, larvae, and pupae of other insects, inevitably leading to the death of the host (Figure 20.1). Parasitoids are like parasites in consuming living tissue; in most other ways, however, they resemble predators.

Herbivores eat whole plants or parts of plants. From the standpoint of consumer–resource relations, however, herbivores may function either as predators, consuming whole plants, or as parasites, consuming living tissues but not killing their victims. When a sparrow eats a seed, it acts as a predator because it kills the entire living embryo of a plant contained in that seed. A deer that browses on shrubs and trees acts as a parasite, like a mosquito (or a vampire bat) taking its blood meal. **Detritivores** consume dead organic material — leaf litter, feces, carcasses — and thus have no direct effect on the populations that produce detritus.

The limitation of prey populations by predation

The cyclamen mite is a pest of strawberry crops in California. Populations of the mites are usually kept under control by a species of predatory mite of the genus *Typhlodromus*. Cyclamen mites typically invade a strawberry crop shortly after it is planted, but their populations do not reach damaging levels until the second year. Predatory mites usually invade fields during the second year, rapidly subdue the cyclamen mite populations, and keep them from reaching damaging levels a second time.

Greenhouse experiments have demonstrated the role of predation in keeping the cyclamen mites in check. One group of strawberry plants was stocked with both predator and prey mites; a second group was kept predator-free by regular application of parathion, an insecticide that kills the predatory species but does not affect the cyclamen mite. Throughout the study, populations of cyclamen mites remained low in plots shared with *Typhlodromus,* but their

FIGURE 20.1 The parasitoid wasps, *Apanteles rubecula* (a) and *Ptermalus puparum* (b), laying eggs in larvae of the cabbage worm moth at different stages of development. Courtesy of the U.S. Department of Agriculture.

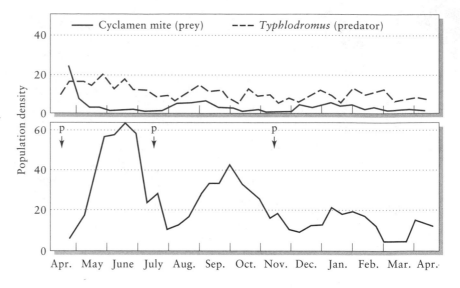

FIGURE 20.2 Infestation of strawberry plots by cyclamen mites (*Tarsonemus pallidus*) in the presence of the predatory mite *Typhlodromus* (*above*) and in its absence (*below*). Prey populations are expressed as numbers of mites per leaf; predator levels are the numbers of leaflets in 36 that had 1 or more *Typhlodromus*. Parathion treatments are indicated by p. After C. B. Huffaker and C. E. Kennett, *Hilgardia* 26:191–222 (1956).

infestation attained damaging proportions on predator-free plants (Figure 20.2). In field plantings of strawberries, the cyclamen mites also reached damaging levels where predators were eliminated by parathion, but they were effectively controlled in untreated plots (a good example of an insecticide having an undesired effect). When cyclamen mite populations began to increase in an untreated planting, the predator populations quickly burgeoned and reduced the outbreak. On average, cyclamen mites were about 25 times more abundant in the absence of predators than in their presence.

Typhlodromus owes its effectiveness as a predator to several factors in addition to its voracious appetite. Its population can increase as rapidly as that of its prey. Both species reproduce parthenogenetically; female cyclamen mites lay 3 eggs per day over the 4 or 5 days of their reproductive life span; female *Typhlodromus* lay 2 or 3 eggs per day for 8 to 10 days. Seasonal synchrony of *Typhlodromus* reproduction with the growth of prey populations, the ability to survive at low prey densities, and strong dispersal powers also contribute to the predatory efficiency of *Typhlodromus*. During the winter, when cyclamen mite populations dwindle to a few individuals hidden in the crevices and folds of leaves in the crown of the strawberry plants, the predatory mites subsist on the honeydew produced by aphids and white flies. They do not reproduce except when they feed on other mites. Whenever predators appear to control prey populations, the predators usually exhibit a high reproductive capacity compared with that of the prey, combined with strong dispersal powers and the ability to switch to alternative food resources when primary prey are unavailable.

Inadequate dispersal is perhaps the only factor that keeps the cactus moth from exterminating its principal food source, the prickly pear cactus. When prickly pear (*Opuntia*) was introduced to Australia, it spread rapidly over the island continent, covering thousands of acres of valuable pasture and range land. After several unsuccessful attempts to eradicate the plant, the cactus moth (*Cactoblastis cactorum*) was introduced from South America. The caterpillar of the moth feeds on the growing shoots of the prickly pear and quickly destroys the plant—literally by nipping it in the bud. After it became established in Australia, the moth population exerted such effective control that within a few years, the prickly pear became a pest of the past (Figure 20.3).

The cactus moth has not eradicated the prickly pear, however, because the cactus manages to disperse to predator-free areas, thereby keeping one jump ahead of the moth and maintaining a low-level equilibrium in a continually shifting mosaic of isolated patches. Indeed, one would probably not guess that the cactus moth keeps the prickly pear at its present low population levels; the moths are scarce in the remaining stands of cactus in Australia today. (The same moth probably controls prickly pear populations in some areas of Central and South America, but its decisive role might have gone unnoticed if the appropriate experiment had not been performed in Australia.)

Experiments on the effect of sea urchins on populations of algae have demonstrated predator control in some biological communities of rocky shores. The simplest experiments consist of removing sea urchins, which feed on attached algae, and following the subsequent growth of their algal prey. When urchins are kept out of tide pools and subtidal rock surfaces, the biomass of algae quickly increases, indicating that predation reduces algal populations below the level that the environment can support. Different kinds of algae also appear after predator removal. Large brown algae flourish and begin to replace both coralline algae (whose hard, shell-like structure deters grazers) and small green algae (whose short life cycles and high reproductive rates enable algal population growth and reestablishment to keep ahead of grazing pressure by sea urchins). In subtidal plots kept free of predators, brown kelps became established in thick stands that shaded out most small species.

Parasite – host systems

Parasitism resembles predation in that a living resource is consumed, but it differs in that the resource is not killed, at least not immediately. Parasites and disease organisms nonetheless have profound physiological and behavioral

FIGURE 20.3 Photographs of a pasture in Queensland, Australia, 2 months before (a) and 3 years after (b) introduction of the cactus moth to control the prickly pear cactus. From A. P. Dodd, in A. Keast, R. L. Crocker, and C. S. Christian (eds.). *Biogeography and Ecology in Australia.* W. Junk, The Hague (1959). Courtesy of W. H. Haseler, Department of Lands, Queensland, Australia.

consequences for their hosts, adversely affect reproduction in many instances, and may limit geographic distribution. Parasite–host interactions also differ from predator–prey interactions in that many parasites pass through complex life cycles alternating between different hosts and free-living stages (see Figure 18.12) and in that hosts can mount immune responses to prevent, control, or eliminate parasite infections.

The complex life histories of parasites involve a variety of interactions with hosts, and different sets of factors affect each stage, both parasitic and free-living, of the life cycle. Problems of locating hosts (contagion) are crucial in parasite and, particularly, disease populations. In the case of endoparasites — those living within the bodies of their hosts — population growth may occur within a single host, resulting in a subdivision of the parasite population and fostering population interactions among the descendants of infecting individuals.

The balance between parasite and host populations is influenced by the **immune response** and other defenses of the host. Immunity depends in part on the production of **antibodies** that recognize and bind to foreign proteins (antigens), such as the protein molecules on the outer surfaces of parasites and disease organisms. The immune response takes time to develop, which gives the parasite a chance to develop and multiply within the host. However, antibodies persist long after an infection has been controlled, thereby reducing the probability of subsequent infection.

Parasites have their own means of circumventing the immune mechanism. Some microscopic and submicroscopic disease organisms produce chemical factors that suppress the immune system of the host; this is the most troublesome feature of the AIDS virus. Others have surface proteins that mimic host antigens and thus escape notice by the immune system. Some schistosomes are known to excite an immune response when they enter the host but do not succumb to antibody attack because they coat themselves with proteins of the host before antibodies become numerous. As a consequence, parasites that subsequently infect a host face a barrage of antibodies stimulated by the earlier entrance of now-entrenched parasite individuals. When this response affects closely related species of parasites, it is known as **cross-resistance**. For example, most of the predominantly human forms of schistosomiasis are extremely virulent. But when a person has been infected previously by other schistosome organisms, some of which have little effect on humans, the impact of the parasite is moderated considerably.

On a population level, a serious outbreak of a viral or bacterial disease is often followed by a period during which most of the individuals in a population have achieved some degree of immunity to reinfection. Until this immunity is lost, or until susceptible individuals are recruited into the population, the disease organisms may be unable to spread.

Herbivores and plant populations

We have seen the role the cactus moth plays in controlling the population of prickly pear cactus in Australia. Herbivorous insects have been employed in many other situations to control imported weeds. Consider the example of

Klamath weed, a European species toxic to livestock (and the source of the drug hypericin, its toxic ingredient), which accidently became established in northern California in the early 1900s. By 1944 the weed had spread over 2 million acres of range land in 30 counties. Biological-control specialists borrowed a herbivorous beetle of the genus *Chrysolina* from an Australian control program. In a success of similar proportions to that of the war against prickly pear, within 10 years after the first beetles were released, Klamath weed was all but obliterated as a range pest. Its abundance is estimated to have been reduced by more than 99 percent.

The proportion of net primary production consumed by herbivores is least in forests, intermediate in grasslands, and greatest in aquatic environments. In the littoral zones of aquatic communities, herbivores consume most of the plant production; in terrestrial habitats, the bulk of plant production moves up the food chain through the detritus pathway. The difference between the functional roles of herbivores in these two habitats is related to the greater digestibility of aquatic versus terrestrial vegetation.

In grasslands, herbivores (mostly insects and grazing mammals) consume 30 to 60 percent of the aboveground vegetation. Their influence on plant production is revealed by exclosure experiments. In one study in California, wire fences were constructed to keep voles out of small areas of grassland. At the end of the 2-year study, food plants of the voles (mostly annual grasses) had grown more and produced more seed within the fenced plots than outside the exclosures where voles continued to graze. Perennial grasses and herbs *not* included in the voles' diet were not directly affected by the exclosure (Figure 20.4).

Although herbivores rarely consume more than 10 percent of forest vegetation, occasional outbreaks of tent caterpillars, gypsy moths, and other insects can completely defoliate or otherwise eradicate entire forests. Long-term studies of the growth and survival of trees after defoliation by insects demonstrate that there may be a considerable lag between an infestation and the expression of its effects.

Grazing resembles infection by parasites and pathogens in that, like the infected host, the individual grazed-upon plant is usually not killed. Thus plants

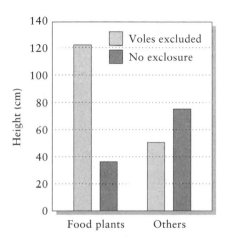

(a) Vegetative growth

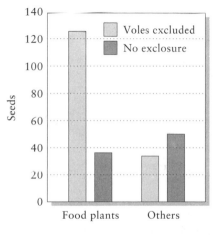

(b) Seed production

FIGURE 20.4 Relative biomass (summed height per 100 cm²) (a) and seed production (seeds per 100 cm²) (b) of food plants and nonfood plants in grassland plots fenced to exclude voles and in unfenced control plots. The bar graphs present results of the experiment after 2 years. Food plants are mostly annual grasses; nonfood plants include perennial grasses and herbs. After G. O. Batzli and F. A. Pitelka, *Ecology* 51:1027–1039 (1970).

have time to respond to grazing and browsing by producing defensive chemical compounds in a manner recalling the immune mechanisms of animals. Depending on how long the plant takes to produce these defenses, time lags may be incorporated into the grazer–plant system, thereby promoting oscillations.

Plants have many inducible defenses against herbivory, most of them chemical. In response to wounding, toxic, noxious, or nutrition-reducing compounds may be produced—either in the area of the wound or systemically throughout the plant—that reduce subsequent herbivory. In some cases, these responses may take only minutes or hours; in others, they require a new season of growth.

When shoots of aspen, poplar, birch, and alder are heavily browsed by snowshoe hares, shoots produced during the following year have exceptionally high concentrations of terpene and phenolic resins, which are extremely unpalatable to the hares. When resins were applied in varying concentrations to shoots of unbrowsed trees, hares were found to avoid shoots containing 80 mg or more of resin per gram dry weight, regardless of the amount of other food available.

Predator–prey cycles

Population cycles have contributed to the lore of population ecology since Charles Elton's paper "Periodic Fluctuations in the Numbers of Animals: Their Causes and Effects" was published in the *British Journal of Experimental Biology* in 1924. Most of Elton's data concerned fur-bearing mammals in the Canadian boreal forests and tundra, where the Hudson's Bay Company had kept detailed records of the numbers of furs brought in by trappers each year. Data for the lynx and its principal prey, the snowshoe hare, revealed regular fluctuations of great magnitude (Figure 20.5). Each cycle lasted approximately 10 years, and cycles of the two species were highly synchronized, with peaks in lynx abundance tending to trail those in hare abundance by a year or two.

FIGURE 20.5 Population cycles of the lynx and the snowshoe hare (see photograph) in the Hudson's Bay region of Canada, as indicated by fur returns to the Hudson's Bay Company. Photograph courtesy of the U.S. Fish and Wildlife Service. After D. A. MacLulich, University of Toronto Studies, Biol. Ser. No. 43 (1937).

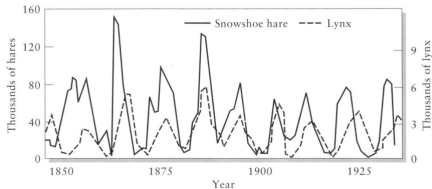

The periods of population cycles vary from species to species, and even within a species. In Canada, most cycles have periods either of 9 to 10 years or of 4 years. Although colored foxes have a 10-year cycle over most of their range, they exhibit a pronounced 4-year cycle in Labrador and on the Ungava Peninsula, where they prey mostly on lemmings, which also have 4-year cycles. In general, small herbivores such as voles and lemmings have 4-year cycles; large herbivores — snowshoe hares, muskrat, ruffed grouse, and ptarmigan — have 9- to 10-year cycles. Predators that feed on short-period herbivores (arctic fox, rough-legged hawk, snowy owl) themselves have short population cycles. Predators of larger herbivores (red fox, lynx, marten, mink, goshawk, horned owl) have longer cycles. The length of the cycle also appears to be related to habitat; longer periods are observed in forest-dwelling species and shorter ones in tundra-dwelling species.

The closely linked population cycles of some predators and prey suggest that the oscillations could result from the way in which predator and prey interact with each other. We saw in an earlier chapter that delays in the responses of birth rates and death rates to changes in the environment can cause population cycles. Most predator–prey interactions have response lags because of the time required to produce offspring or to mount an immune response. Population cycles appear to represent a stable interaction between predator and prey. One of the earliest goals of population biologists was to establish such cycles in experimental populations, for which one could work out the dynamics of the relationship and the cause of the cycles.

When azuki bean weevils (*Callosobruchus chinensis*) are maintained in cultures with parasitoid braconid wasps (*Heterospilus*), the populations of predator and prey fluctuate out of phase with each other in regular cycles of population change (Figure 20.6). Introduced to a population of weevils that is ultimately limited by a constant ration of seeds, the wasps rapidly increase in number. As the wasp population grows, parasitism becomes a major source of mortality for the weevils. When parasitism exceeds the reproductive capacity of the weevil population, the number of weevils begins to decline, but because *Heterospilus* is an efficient parasitoid, it continues to prey heavily even as the population is reduced. Eventually the weevils are nearly exterminated, and *then* the predator population, lacking adequate food, decreases rapidly. The wasp is not so efficient that all weevil larvae are attacked; hence a small but persistent reserve of weevils always remains to initiate a new cycle of prey population growth after the predators have become scarce.

Extremely efficient predators often eat their prey populations to extinction and then become extinct themselves. This hopeless situation can be stabilized, however, if some of the prey can find refuges in which to escape the predators.

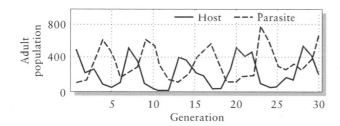

FIGURE 20.6 Population fluctuations of the azuki bean weevil (host) and its braconid wasp parasitoid. After S. Utida, *Cold Spring Harbor Symp. Quant. Biol.* 22:139–151 (1957).

G. F. Gause demonstrated this principle in some of the earliest experimental studies on predator–prey systems. Gause employed *Paramecium* as prey and another ciliated protozoan, *Didinium*, as predator. In one experiment, predator and prey individuals were introduced to a nutritive medium in a plain test tube. By creating so simple an environment, Gause had "stacked the deck" against the prey; the predators readily found all of them, and when the last *Paramecium* had been consumed, the predators starved. In a second experiment, Gause added some structure to the environment by placing glass wool, in which the *Paramecium* could escape predation, at the bottom of the test tube. The tables thus having been turned, the *Didinium* population starved after consuming all readily available prey, but the *Paramecium* population was restored by individuals concealed in the glass wool.

Gause finally achieved recurring oscillations in the predator and prey populations by periodically adding small numbers of predators—restocking the pond, so to speak. The repeated addition of individuals to the culture corresponds, in natural predator–prey interactions, to repopulation by colonists from other areas of a locality in which extinction of either predator or prey has occurred. This is reminiscent of the interaction between the cactus moth and the prickly pear, in which the cactus escapes complete annihilation by dispersing to predator-free areas.

C. B. Huffaker, a University of California biologist who pioneered the biological control of crop pests, attempted to produce a mosaic environment in the laboratory that would allow predator and prey to persist without restocking of either population. The six-spotted mite *Eotetranychus sexmaculatus* was prey; another mite, *Typhlodromus occidentalis*, was predator; oranges provided the prey's food. Huffaker established experimental populations on trays within which he could vary the number, exposed surface area, and dispersion of the oranges (Figure 20.7).

Each tray had 40 positions arranged in 4 rows of 10 each; where Huffaker did not place oranges, he substituted rubber balls of about the same size. The exposed surface area of the oranges was varied by covering them with different amounts of paper, the edges of which were sealed in wax to keep the mites from crawling underneath. In most experiments, Huffaker first established the prey population with 20 females per tray, then introduced 2 female predators 11 days later. Both species reproduce asexually; males are not required.

When six-spotted mites were introduced to the trays alone, their populations leveled off at between 5500 and 8000 mites per orange area. When predators were added, their numbers increased rapidly and they soon wiped out the prey population. Their own extinction followed shortly.

Although predators always eliminated the six-spotted mites, the position of the exposed areas of the oranges influenced the course of extinction. When the areas were in adjacent positions, minimizing dispersal distance between food sources, the prey reached maximum populations of only 113 to 650 individuals and were driven to extinction within 23 to 32 days after the beginning of the experiment. The same area of exposed oranges dispersed randomly throughout the 40-position tray supported prey populations that reached maxima of 2000 to 4000 individuals and persisted for 36 days. Thus survival of the prey population could be prolonged by providing remote areas of suitable habitat to which predators could disperse slowly.

Huffaker reasoned that if predator dispersal could be further retarded, the two species might coexist. To accomplish this, he increased the complexity of the environment and introduced barriers to dispersal. The number of possible food positions was increased to 120, and a feeding area equivalent to 6 oranges was dispersed over all 120 positions. A mazelike pattern of Vaseline barriers was placed among the food positions to slow the dispersal of the predators. *Typhlodromus* must walk to get where it is going, but the six-spotted mite spins a silk line that it can use like a parachute to float on wind currents. To take advantage of this behavior, Huffaker placed vertical wooden pegs throughout the trays, which the mites used as jumping-off points in their wanderings. This arrangement finally produced a series of three population cycles over the 8 months of the experiment (Figure 20.8). The distribution of the predators and the prey throughout the trays continually shifted as the prey, on the way to extermination in one feeding area, recolonized the next a jump ahead of their predators.

Despite the tenuousness of the predator–prey cycle that was achieved, this experiment illustrates that a spatial mosaic of suitable habitats enables predator and prey populations to coexist stably. But, as we saw in Gause's experiments with protozoa, predator and prey may also coexist locally if some prey can take refuge in hiding places. And when the environment is so complex that predators cannot easily find scarce prey, stability again can be achieved, as we shall see.

FIGURE 20.7 One of Huffaker's experimental trays with 4 oranges, half exposed, distributed at random among the 40 positions in the tray. Other positions are occupied by rubber balls. Each orange was wrapped with paper and its edges sealed with wax. The exposed area was divided into numbered sections to facilitate counting the mites. Courtesy of C. B. Huffaker. From C. B. Huffaker, *Hilgardia* 27:343–383 (1958).

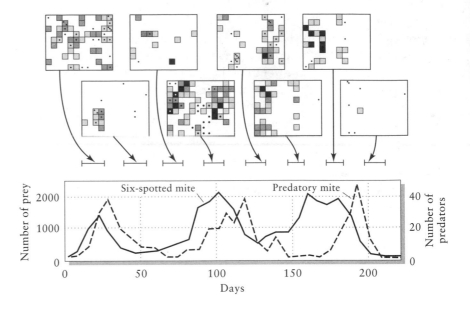

FIGURE 20.8 *Below:* Population cycles of the six-spotted mite and the predatory mite *Typhlodromus* in a laboratory situation. *Above:* The boxes show the relative density and positions of the mites in the trays at the eight times indicated. Shading records the relative density of six-spotted mites; dots indicate the presence of predatory mites. After C. B. Huffaker, *Hilgardia* 27:343–383 (1958).

A simple predator–prey model

In an attempt to understand the origin of population cycles, Alfred J. Lotka and the Italian biologist Vito Volterra independently provided the first mathematical descriptions of predator–prey interactions during the 1920s. As we shall see, the Lotka–Volterra model predicts oscillations in the abundances of predator and prey populations, with predator numbers lagging behind those of their prey. In discussing the model, we shall follow the common practice of designating the number of predator individuals by P and the number of prey by H (think of H for *herbivore*).

The growth rate of the prey population has two components: (1) the unrestricted exponential growth of the prey population in the absence of predators, rH, where r is the exponential growth rate (the difference between the per capita birth and death rates); and (2) the removal of prey by predators, over and above other causes of death. Lotka and Volterra assumed that predation varies in direct proportion to the product of the prey and predator populations, HP, and hence in proportion to the probability of a random encounter between predator and prey. Accordingly, the rate of increase of the prey population is given by

$$\frac{dH}{dt} = rH - pHP$$

where p is a coefficient expressing the efficiency of predation.

The growth rate of the predator population balances its birth rate, which depends on the number of prey captured, against a constant death rate imposed from outside the system (for example, by weather):

$$\frac{dP}{dt} = apHP - mP$$

The temperate zone

Temperate forests superficially resemble tropical rain forests in their structure but differ in being less diverse and less complex and in having a cold season of dormancy. The canopy may reach 40 meters in this red oak forest in Michigan. Most of the production of a forest enters detritus food chains in which fungi play a special role of attacking lignin in wood. Although wind plays a more prominent role in pollination and seed dispersal in temperate zones, many species have showy flowers (dogwood) and conspicuous fruits (cherry) to attract animals.

Photo by R. E. Ricklefs.

Photo by Marcos A. Guerra, courtesy of Smithsonian Tropical Research Institute.

Photo by R. E. Ricklefs.

Photo by R. E. Ricklefs.

A cliff in the Red Sea off Israel reveals the abundant and often colorful life that abounds in tropical oceans.

Photo by Eric Hanauer.

The coral reefs of the Caribbean Sea off the northern coast of Panama reveal a seemingly endless variety of animals. The fanlike gills and feeding appendages of polychaete worms contrast with the coral in which they burrow. The green color of the coral comes from the symbiotic algae that live within the coral organism. Encrusting sponges cover bits of dead coral. A nudibranch—a snail without a shell—grazes algae and small animals on the surface of coral rubble. Anemones, which are sessile relatives of jellyfish, trap small prey from the water with their stinging tentacles; parrotfish chew up the living coral with their powerful jaws. The abundant food and complex structure of the coral reef provide habitat for hundreds of species of fish, many of which travel in schools to reduce the risk of predation.

Photo by Carl C. Hansen, courtesy of Smithsonian Tropical Research Institute.

Photo by Carl C. Hansen, courtesy of Smithsonian Tropical Research Institute.

Photo by Carl C. Hansen, courtesy of Smithsonian Tropical Research Institute.

Photo by Carl C. Hansen, courtesy of Smithsonian Tropical Research Institute.

Photo by Carl C. Hansen, courtesy of Smithsonian Tropical Research Institute.

Satellite image of the California coast on June 15, 1981, showing concentrations of phytoplankton (red and yellow) near the coast and in large eddy currents that extend out from the coast. This image emphasizes the spatial heterogeneity of the marine system resulting from water currents. Satellite image of the North Atlantic Ocean during the first week of June, 1984, in which warm water is indicated by red and cold water by green or blue. The image shows the Gulf Stream along the coast of Florida breaking up into large eddies as it crosses the Atlantic. The Gulf Stream not only transports considerable heat to the vicinity of northern Europe but also conveys a tropical community of marine organisms far to the north, where the warm-water environment becomes increasingly heterogeneous in time and space.

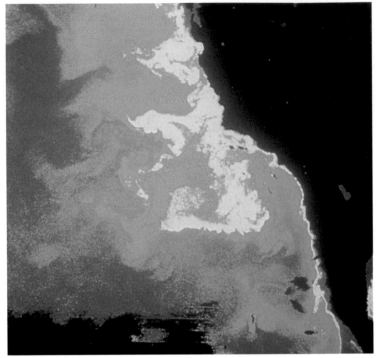

Coastal Zone Color Scanner image courtesy of J. A. McGowan, from J. Paláez and J. A.McGowan, *Limnol. Oceanogr.* 31:927-950 (1986).

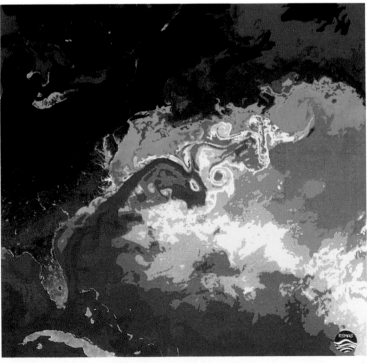

Image courtesy of Otis Brown, Robert Evans, and Mark Carle, University of Miami Rosentiel School of Marine and Atmospheric Science.

The birth term is the number of prey captured (pHP) times a coefficient (a) for the efficiency with which food is converted to population growth. The death rate is a constant (m, for *mortality*) times the number of predator individuals.

When both predator and prey populations achieve equilibrium ($dH/dt = 0$ and $dP/dt = 0$), $rH = pHP$ and $apHP = mP$. We can rearrange these equations to give

$$\hat{P} = \frac{r}{p} \qquad \text{and} \qquad \hat{H} = \frac{m}{ap}$$

where $\hat{P}$ and $\hat{H}$ are the equilibrium sizes of the predator and prey populations, respectively. Note that both $\hat{P}$ and $\hat{H}$ are constant values, each being independent of the abundance of the other population.

According to the Lotka–Volterra model, when the populations stray from their joint equilibrium, rather than returning to the equilibrium point, the populations oscillate around it in a continuous cycle. The period of the oscillation (T) is approximately $2\pi/\sqrt{rm}$. Hence the higher the population growth potential of the prey or the death rate of the predator — that is, the higher the rate of population turnover — the faster the system oscillates.

The relationship between predator and prey can be portrayed on a graph with axes representing the sizes of the populations (Figure 20.9). By convention, predator numbers increase along the vertical axis and prey numbers along the horizontal axis. The equilibrium population values of the predator ($\hat{P}$) and of the prey ($\hat{H}$) partition the graph into four regions. The horizontal line $\hat{P} = r/p$, representing the condition $dH/dt = 0$, is called the **equilibrium isocline** of the prey. For any combination of predator and prey numbers that lies in the region below this line, the prey increase because there are few predators to eat them. In the region above the prey isocline, prey populations decrease because of overwhelming predator pressure. For the predators, population increases are limited to the region to the right of the predator equilibrium isocline ($\hat{H} = m/ap$), where prey are abundant enough to sustain population growth. To the left of the line, predator populations decrease as a result of insufficient prey.

The change in both populations together follows a path, shown in each of the four sections of the graph, that combines the individual changes of the predator and prey populations. In the lower right, for example, both predator and prey increase, and the joint population trajectory moves up and to the right. The trajectories in the four regions together define a counterclockwise cycling of the predator and prey populations one-quarter cycle out of phase, with the prey population increasing and decreasing just ahead of its predator population (Figure 20.10).

The equilibrium isocline for the predator ($dP/dt = 0$) defines the minimum level of prey ($\hat{H} = m/ap$) that can sustain the growth of the predator population. That of the prey ($dH/dt = 0$) defines the greatest number of predators ($\hat{P} = r/p$) that the prey population can sustain. If the reproductive rate of the prey (r) increased, or the hunting efficiency of the predators (p) decreased, or both, the prey isocline (r/p) would increase — that is, the prey population would be able to bear the burden of a larger predator population. If the death rate of the predators (m) increased and either the predation efficiency (p) or the

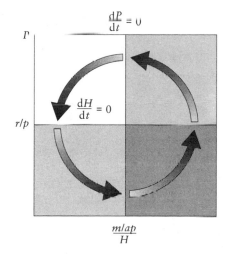

FIGURE 20.9 Representation of the Lotka–Volterra predator–prey model on a population graph. The trajectories of the populations show that the predator and prey will continually oscillate out of phase with each other.

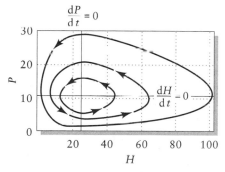

FIGURE 20.10 Simulated population trajectories of predator and prey according to a Lotka–Volterra model. The degree of oscillation about the equilibrium point reflects the initial population sizes in the simulation. From G. D. Elseth and K. D. Baumgardner, *Population Biology*. Van Nostrand, New York (1981).

reproductive efficiency of the predators (*a*) decreased, the predator isocline (*m/ap*) would move to the right, and more prey would be required to support the predator population. Increased predator hunting efficiency (*p*) would simultaneously reduce both isoclines; fewer prey would be needed to sustain a given capture rate (the predator isocline would decrease), and the prey population would be less able to support the more efficient predators (the prey isocline would decrease).

The functional response

According to the Lotka–Volterra model, when either the predator or the prey population is displaced from its equilibrium, the system will oscillate in a closed cycle. Any further perturbation of the system will give the population fluctuations a new amplitude and duration until some other outside influence acts. This equilibrium is said to be **neutral** because no internal forces act to restore the populations to the intersection of the predator and prey isoclines. Therefore, random perturbations will eventually increase the fluctuations to the point where the trajectory strikes one of the axes of the predator–prey graph, and one or both populations die out. This property in itself suggests that the Lotka–Volterra equations greatly oversimplify nature.

Lotka pointed out that adding terms representing density-dependent limitation of predators and prey would tend to create an inward spiraling of the population trajectory toward the joint equilibrium. For example, the expression $dH/dt = rH - pHP - cH^2$ — a logistic equation ($dH/dt = rH - cH^2$) with an added predation term ($-pHP$) — leads to a damped oscillation and a stable equilibrium. Density dependence in the predator population has the same effect.

Other concerns about the adequacy of the Lotka–Volterra model focus on the predation term ($-pHP$). In particular, for a given density of predators, the rate of consumption increases in direct proportion to the density of prey (H). Accordingly, predators cannot be satiated. They just keep on eating, no matter how many prey they can catch. Clearly this aspect of the model is unrealistic. How would adding a bit more reality here affect the behavior of the model?

The relationship of an individual predator's rate of food consumption to prey density has been labeled the **functional response** by Canadian entomologist C. S. Holling, now at the University of Florida. Three general types of functional response curves are shown in Figure 20.11. Type I is the linear rela-

FIGURE 20.11 Three types of functional responses of predators to increasing prey density. Type I: Predator consumes a constant proportion of the prey population regardless of its density. Type II: Predation rate decreases as predator satiation sets an upper limit to food consumption. Type III: Predator response lags at low prey density because of low hunting efficiency or the absence of search image. (a) The functional response in terms of the number of prey consumed. (b) The functional response in terms of the proportion of prey consumed.

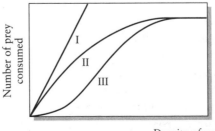

(a)

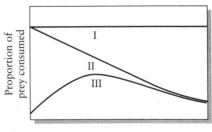

Density of prey population

(b)

tionship of the Lotka–Volterra model. Type II describes a situation in which the number of prey consumed per predator initially rises quickly as the density of prey increases but then levels off with further increases in prey density. Type III resembles type II in having an upper limit to prey consumption, but it differs in that the response of predators to prey is depressed at low prey density.

Two factors dictate that the functional response should reach a plateau. First, predators may become satiated — constantly full — at which point their rate of feeding is limited by the rate at which they can digest and assimilate food. Second, as the predator captures more prey, the time spent handling and eating the prey cuts into searching time. Eventually, the two reach a balance, and prey capture rate levels off (Box 20.1).

At high prey densities, type II and type III response curves differ little; they are both inversely density-dependent. That is, as the density of prey increases, the proportion of prey consumed decreases. Over the lower range of prey density, however, type III responses differ from those of type II in that the proportion of prey consumed increases with density of prey. Several factors may cause the predator response to decrease at lower prey density, thereby re-

---

BOX 20.1   *Holling's disc equation*

C. S. Holling described the leveling off of consumption rates by a predator as prey abundance increased by a simple expression reflecting the fact that the time required to handle prey items reduces hunting time as prey abundance increases. This expression is known as the **disc equation** because of its application to experiments in which blindfolded human subjects were required to discover and pick up small discs of paper on a flat surface. Any such task, including the capture and eating of prey, requires a certain handling time, T_h. The total handling time is therefore the handling time per item times the number of encounters E. The time left over for searching (T_s) is the total time minus the total handling time; that is, $T_s = T - T_h E$. The number of encounters can itself be defined as the product of search time, prey density (P), and a constant (a) for the efficiency of searching: $E = a(T - T_h E)P$. Combining the expressions for T_s and E and solving for E yields

$$E = \frac{aPT}{1 + aPT_h}$$

When prey are scarce, the denominator term aPT_h is small compared to 1, and the number of prey encountered approaches aPT. Hence encounters are directly proportional to prey density. When prey are dense, aPT_h is large compared to 1, and E approaches the ratio T/T_h, which is a constant value defining the maximum number of prey that can be captured in time T. Thus at high prey density, search time drops to near zero, and the number of prey captured is limited only by how long the predator requires to handle each one; the shorter the handling time, the more prey can be captured.

sulting in a type III functional response: (1) a heterogeneous habitat may afford a limited number of safe hiding places, which protect a larger proportion of the prey at lower densities than at higher densities; (2) lack of reinforcement of learned searching behavior owing to low rate of prey encounter may reduce hunting efficiency at low prey density; and (3) switching to alternative sources of food when prey are scarce may reduce hunting pressure. The relationship of prey consumption to prey density depends on change in average prey vulnerability in the case of factor 1, on search and capture efficiency in the case of factor 2, and on motivation to hunt or searching time in the case of factor 3.

Learning influences the searching behavior of many predators, especially vertebrates with complex nervous systems. An object is always easier to find if one has a preconception of what it looks like and where it might be. Such **search images** are acquired by experience. The denser a prey population, the more frequently predators encounter prey; search efficiency increases and handling time decreases as a result of experience. At low prey density, neither experience nor efficiency is so well developed, and the capture rate is therefore depressed.

Regardless of the mechanisms underlying the type III response, predators often switch to a second prey as the density of the first is reduced. For example, when the predatory water bug *Notonecta glauca* was presented with two types of prey—isopods and mayfly larvae—the predators consumed the more abundant prey species in greater proportion than its percentage of occurrence (Figure 20.12). This **switching** depended to some degree on variation in the success of attacks on isopods as a function of their relative density. When water bugs encountered them infrequently, fewer than 10 percent of attacks were successful. At higher densities, and therefore higher encounter rates, attack success rose to almost 30 percent.

The numerical response

Individual predators can increase their consumption of prey only to the point of satiation. Predator response to increasing prey density above the level that results in satiation can be achieved only through increase in the number of

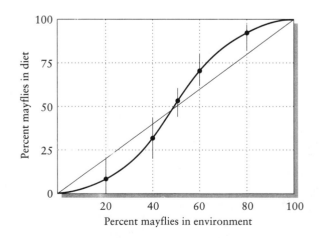

FIGURE 20.12 The percentage of mayfly larvae in the diet of the water bug *Notonecta* as a function of the mayfly's relative abundance among available prey. The straight line indicates no preference; data are presented as means and ranges of separate trials. From M. Begon and M. Mortimer, *Population Ecology* (2nd ed.). Sinauer, Sunderland, Mass. (1981); after J. H. Lawton, J. R. Beddington, and R. Bonser, in M. B. Usher and M. H. Williamson (eds.). *Ecological Stability*. Chapman & Hall, London (1974), pp. 141–158.

predators, by either immigration or population growth, which together constitute the **numerical response**. Populations of most predators grow slowly, especially when the reproductive potential of the predator is much less than that of the prey and the predator's life span is longer. Immigration from surrounding areas is an important contribution to the numerical response of mobile predators, which may opportunistically congregate where resources become abundant. The bay-breasted warbler, a small insectivorous bird of eastern North America, exhibits such behavior during periodic outbreaks of the spruce budworm. During years of outbreak in a particular area, the density of warblers may reach 120 breeding pairs per 100 acres, compared with about 10 pairs per 100 acres during nonoutbreak years. This population behavior clearly shows how a predator may take advantage of a shifting mosaic of prey abundance.

Consider the different responses of three predatory birds, the pomarine jaeger, the snowy owl, and the short-eared owl, to varying densities of lemmings on the arctic tundra (Table 20.1). Lemming populations exhibit great fluctuations; high and low points in a population cycle may differ by a factor of 100. At Barrow, Alaska, during the summer of 1951 when lemmings were scarce, none of the predatory birds bred; short-eared owls did not even appear in the area. During the following summer, one of moderate lemming density, both the jaeger and the snowy owl bred, but short-eared owls again were absent. In 1953, a peak year for lemmings, all three species of avian predators bred. Jaegers were four times more abundant in 1953 than in 1952, showing a strong numerical response. In contrast, the density of snowy owls did not increase, but each pair of birds reared more young. Whereas most snowy owl nests contained 2 to 4 eggs during the year of moderate lemming abundance, clutches of up to 12 eggs were laid during the peak year of 1953.

TABLE 20.1	*Response of predatory birds to different densities of the brown lemming near Barrow, Alaska*		
	1951	1952	1953
Brown lemming (individuals per acre)	1 to 5	15 to 20	70 to 80
Pomarine jaeger	Uncommon, no breeding	4 breeding pairs per square mile	18 breeding pairs per square mile
Snowy owl	Scarce, no breeding	0.2 to 0.5 breeding pairs per square mile; many nonbreeders	0.2 to 0.5 breeding pairs per square mile; few nonbreeders
Short-eared owl	Absent	One record	3 to 4 breeding pairs per square mile

Source: F. A. Pitelka, P. O. Tomich, and G. W. Treichel, *Ecol. Monogr.* 25:85–117 (1955).

Stability in predator–prey systems

Studies of the functional and numerical responses of predators, as well as of the population dynamics of prey populations, have revealed the biological limitations of simple predator–prey models, such as the Lotka–Volterra equations. Many complications of biological reality have been incorporated into models and their contributions to the stability of predator–prey interactions studied. The findings can be summarized by a number of general rules of predator–prey stability. Five factors tend to damp predator–prey cycles and thus promote the long-term persistence of relationship: (1) predator inefficiency (or enhanced prey escape or defense), (2) density-dependent limitation of the population of either the predator or the prey by factors external to their relationship, (3) alternative food sources for the predator, (4) refuges from predation at low prey densities, and (5) reduced time lags in predator population response to changes in prey abundance.

Predator inefficiency (reduced p in the Lotka–Volterra model) results in increased equilibrium levels for both prey and predator populations (more predators can be supported by the larger prey populations) and in lower turnover rates for both at equilibrium. Both these consequences would seem to enhance the stability of the system. Alternative food sources stabilize predator populations because individuals may switch between food types in response to changing prey abundance. Similarly, refuges safe from predation allow prey populations to maintain themselves at higher levels in the face of intense predation, thereby facilitating the recovery phase of the population cycle. Indeed, so many factors tend to stabilize predator–prey relationships that the cyclic behavior observed in many of their populations requires additional explanation. Most probably, population cycles reflect a balancing of the stabilizing effects of density dependence, alternative prey, and refuges against the destabilizing effects of time lags in predator–prey interactions. Time lags are ubiquitous in nature, arising from the development periods of animals and plants, the time required for numerical responses of predator populations, and the time course of immune responses and induced defenses in plants.

Multiple stable states in predator–prey systems

The size of a population is influenced by the abundance of its resources and by its own consumers. Extremely efficient predators may depress a prey population to levels far below the capacity of the environment to support them in the absence of predators. Alternatively, prey may be limited primarily by their own food supplies while predators remove an inconsequential number of prey. In most situations, equilibrium population size reflects a balance between the limiting influences of food supply and predators. Under some circumstances, however, a population may have two stable equilibrium points, only one of which may be occupied at a given time. This situation is referred to as having **multiple stable states.** How do multiple stable states arise? What factors determine which of the stable points is occupied?

A simple model based on a type III functional response curve illustrates how a multiple stable state can develop (Figure 20.13). This model of a hypothetical system describes changes in recruitment and predation in prey populations as a consequence of increasing prey density. The recruitment curve represents the net contribution of births and deaths in the absence of predators. Hence the per capita recruitment is high when the prey population is small and decreases to zero as that population approaches its carrying capacity. The predation curve is the sum of the functional and numerical responses of predator populations. Predation rate may be low at low prey densities because of switching to alternative prey or difficulty in locating scarce prey; it also tails off at high prey densities because of predator satiation and extrinsic limits to predator populations. The predation curve in the graph thus represents a type III functional response.

The recruitment and predation curves shown in Figure 20.13 produce three equilibrium points. The highest and lowest of these three points represent stable equilibria around which populations are regulated; the middle equilibrium is unstable. The lower (in terms of prey density) equilibrium point (A) corresponds to the situation in which predators regulate a prey population substantially below its carrying capacity (*K*). The upper equilibrium (C) corresponds to the situation in which a prey population is regulated close to *K* primarily by availability of food and other resources; here predation exerts a minor depressing influence on population size.

The implications of Figure 20.13 for such practical concerns as the control of crop pests are clear. Predators maintain a shaky hold on prey populations at point A. When a heavy frost or an introduced disease reduces the predator population long enough to allow the prey population to slip above point B, the prey may continue to increase to the higher (in terms of prey density) stable equilibrium point (C), regardless of how quickly the predator population recovers. To the farmer, this means that a crop pest that is normally controlled at harmless levels by predators and parasites suddenly becomes a menacing epidemic. After such an outbreak, predators exert little control over the pest population until some quirk of the environment brings its numbers below point B, back within the realm of predator control.

Using the predation–recruitment diagram in Figure 20.13, we can examine the consequences of different levels of predation for control of the prey population (Figure 20.14). Inefficient predators cannot regulate prey populations at low densities; they depress prey numbers slightly, but the prey population remains near the equilibrium level set by resources (Fig. 20.14a, point C). Increased predation efficiency at low prey density, however, can result in predator control at point A (Fig. 20.14b). When functional and numerical responses are sufficient to maintain high densities of predators, predation may effectively limit prey growth under all circumstances, and equilibrium point C disappears (Fig. 20.14c). Finally, predation may be so intense at all prey densities that the prey are eaten to extinction (Fig. 20.14d, no equilibrium point).

We might expect this situation only in simple laboratory systems or when predator populations are maintained at high densities by the availability of some alternative but less preferred prey (hence, there is no switching). Indeed, many ecologists have advocated providing parasites and predators of the pest with innocuous alternative prey to enhance biological control. At the very

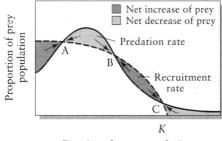

FIGURE 20.13 Predation and recruitment rates in a hypothetical predator–prey system. When predation exceeds recruitment, prey populations decrease, and vice versa (as shown by the arrows). Points A and C are stable equilibria for the prey population; the lower point (A)—lower in terms of prey density—represents population control by predators; the higher point (C) represents population control by food and other resources.

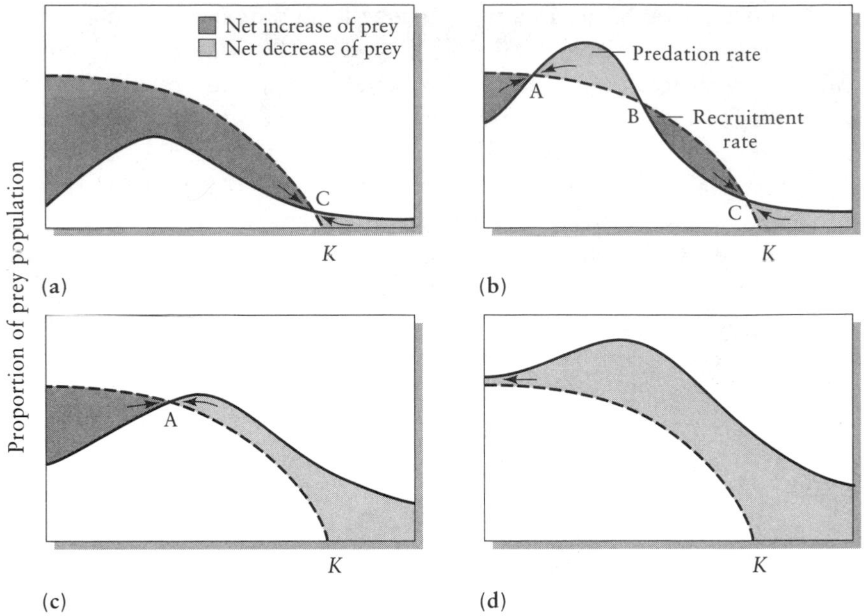

FIGURE 20.14 Predation and recruitment curves at different intensities of predation, showing the effect on the number of equilibria.

least, the curves suggest that the position of a predator–prey equilibrium, whether it is at very low levels of prey or close to their carrying capacity, may shift between extremes with small changes in closely matched predation and recruitment curves. Such considerations would appear to make equilibria at *intermediate* prey densities very unlikely.

Predator-prey population ratios and maximum sustainable yield

The ability of a prey population to support predators varies with its density. A small prey population can support few predators because, although each prey individual's reproductive potential may be high, the total recruitment rate of a small population is low. Prey populations near their carrying capacities are also unproductive because, though members are numerous, each individual's reproductive potential is severely limited by the effects of crowding.

At some intermediate density, the overall recruitment rate of the prey population reaches a maximum. Because predators can remove a number of individual prey equal to the recruitment rate without reducing the size of the prey population, the prey population density that yields the maximum recruitment generally will support the greatest number of predators. The rate of recruitment at this point is known as the **maximum sustainable yield** because it is the maximum rate at which predators can remove individuals without depressing the prey population.

Ranchers and game managers strive to maintain populations of beef cattle, deer, and geese at their most productive levels to maximize the harvest of these species without reducing their populations. Do predators also prudently manage their prey populations in such a way as to maximize the productivity of their own populations? And if so, how can such behavior evolve?

Territorial animals, which exclude competitors from their feeding areas, could indeed space themselves with respect to their prey to achieve maximum yields. When the feeding areas of predators overlap, however, intraspecific competition dictates that each predator must maximize its immediate harvest at the expense of long-term yields. Our own species behaves no differently. Intelligently managed ranches, with fences to exclude competing livestock, can achieve maximum sustainable yields. Alas, in highly competitive situations— fishing in international waters, to name one—we have shown ourselves to be pathetically shortsighted and imprudent, and stocks of many species of fish and other seafoods have declined dramatically.

In most cases, predators exploit prey populations to the degree that their ability to capture prey exceeds the ability of the prey to avoid being captured. Both skills are evolved characteristics. Regardless of whether predators act prudently to manage prey populations for maximum sustainable yield, they often achieve a characteristic equilibrium with their prey populations. The relationship between wolves and various prey populations in several areas demonstrates this equilibrium particularly well (Table 20.2). Population ratios and, particularly, biomass ratios are relatively constant (1 pound of predator for each 150 to 300 pounds of prey) despite the fact that the species and density of the principal prey of the wolf vary considerably with locality.

Different predator–prey systems may achieve different equilibria. The population ratio of mountain lions to deer in California is 1:500 to 1:600, which is equivalent to a biomass ratio of about 1:900, and the exploitation rate is only 6 percent, compared to values of 18 and 37 percent for wolves in

TABLE 20.2 — The relationship between populations of predators and their prey in several localities

Locality	Predator	Principal prey	Density of predators (individuals per 100 square miles)	RATIO BETWEEN PREDATOR AND PREY POPULATIONS Numbers	Biomass
Jasper National Park	Wolf	Elk, mule deer	1	1:100	1:250
Wisconsin	Wolf	White-tailed deer	3	1:300	1:300
Isle Royale	Wolf	Moose	10	1:30	1:175
Algonquin Park	Wolf	White-tailed deer	10	1:150	1:150
Canadian arctic	Wolf	Caribou	1.7	1:84	1:186
Utah	Coyote	Jackrabbits	28	1:1000	1:100
Idaho (primitive area)	Mountain lion	Elk, mule deer	7.5	1:116	1:524
Ngorongoro Crater, Tanzania	Hyena	Ungulates	110	1:135	1:46
Nairobi Park, Kenya	Felids	Ungulates	96	1:97	1:140
Alaska	Pomarine jaeger	Lemmings		1:1263	1:90

Source: R. E. Ricklefs, *Ecology* (2nd ed.). W. H. Freeman, New York (1990).

two areas. A mountain lion–elk–mule deer system in Idaho has a biomass ratio of 1 : 524 and an exploitation rate of 5 percent for the elk population and 3 percent for the mule deer population. Evidently, wolves exploit their prey more efficiently than mountain lions, perhaps because of their social hunting habits.

Where predators feed on more abundant populations of prey, as in savanna, grassland, and tundra habitats, predators not only are more numerous but also achieve higher biomass ratios (1:50 to 1:150, Table 20.2). Large cats—lions and cheetahs—remove 16 percent of prey biomass in Nairobi Park, Kenya, where their biomass ratio is 1 : 140.

❧ SUMMARY

The study of predator–prey interactions attempts to answer two major questions. First, do predators reduce the size of their prey populations well below the carrying capacity? Second, do the dynamics of the predator–prey interaction cause the populations to oscillate? The first question is of great practical concern to those interested in the management of crop pests, game populations, and endangered species. It also has far-reaching implications for understanding the interactions among species that share resources and, therefore, for understanding the regulation of biological communities. The second question was motivated by observations of predator–prey cycles in nature and directly addresses the issue of stability in natural systems. Ecologists have approached these problems with a combination of observation, theory, and experiment.

1. Ecologists distinguish among three basic kinds of consumers: predators, which remove each prey individual from the prey population as they consume it; parasitoids (mostly small flies and wasps), which kill their hosts, but only after the parasitoid larvae have pupated; and parasites and grazers, both of which consume portions of the living animal or plant organism. From the standpoint of population theory, each of these has been treated differently.

2. Experimental studies of a number of pest species and their natural predators have demonstrated that, in many cases, consumers could reduce resource populations far below their carrying capacities.

3. Experimental studies have demonstrated that predator and prey populations can be made to oscillate in the laboratory. Maintenance of population cycles usually requires a complex environment in which prey are able to establish themselves in refuges.

4. Alfred J. Lotka and Vito Volterra, in the 1920s, devised simple models of predator and prey dynamics that predicted population cycles. The models used differential equations in which the rate of prey removal was directly proportional to the product of the predator and prey populations.

5. The functional response describes the asymptotic relationship between prey removal rate per predator and the density of the prey. Whereas the Lotka–Volterra models were inherently unstable, one form of the functional response curve (type III) resulted in stable regulation of the prey population at low density.

6. The numerical response describes the response of a predator population to increasing prey density by population growth and immigration.

7. Stability in predator–prey interactions is promoted by density dependence of either predator or prey, by refuges or hiding places in which the prey can escape predation, by reduced predator efficiency, and under some circumstances by availability of alternative prey. Stable population cycles in nature apparently express the balance of these stabilizing factors and the destabilizing influence of time delays in population responses.

8. One source of time delay in parasite and grazing systems is the acquisition of immunity or other resistance. Plants may exhibit chemical responses to browsing and grazing that discourage further consumption. Time delays in these responses may be responsible for the cycles observed in disease outbreaks and in populations of herbivores (and those of the carnivores that prey on them!).

9. Models of consumers suggest that systems can have two stably regulated points (multiple stable states) between which populations may shift, depending on environmental conditions. The lower equilibrium is determined by the strong depressing influence of predators on prey populations; the upper equilibrium lies close to the carrying capacity of the prey in the absence of predation. Changes in climate may shift a system from one to the other of these points, resulting in successions of endemic and epidemic conditions.

10. Each prey population has a density at which harvesting of that population without causing population decrease is greatest. This is called the point of maximum sustainable yield. When a single predator can completely control its prey, as humans can do with many domestic and game species, maximum sustainable yields can be achieved. When predators compete for the same resources, maximization of short-term yields generally precludes the achievement of a maximum sustainable yield.

☙ SUGGESTED READINGS

Anderson, R. M. (ed.). 1982. *Population Dynamics of Infectious Diseases.* Chapman & Hall, London and New York.

Anderson, R. M., and R. M. May. 1980. Infectious diseases and population cycles of forest insects. *Science* 210:658–661.

Brooks, J. L., and S. I. Dodson. 1965. Predation, body size and composition of the plankton. *Science* 150:28–35.

Crawley, M. J. 1983. *Hervibory. The Dynamics of Animal–Plant Interactions.* Univ. of California Press, Berkeley.

DeBach, P., and D. Rosen. 1991. *Biological Control by Natural Enemies* (2nd ed.). Cambridge Univ. Press, New York.

Duffy, J. E., and M. E. Hay. 1960. Seaweed adaptations to herbivory. *BioScience* 40:368–375.

Errington, P. L. 1963. The phenomenon of predation. *American Scientist* 51:180–192.

Hassell, M. P. 1978. *The Dynamics of Arthropod Predator–Prey Systems.* Princeton Univ. Press, Princeton.

Lindroth, R. L., and G. O. Batzli. 1986. Inducible plant chemical defences: a cause of vole population cycles? *Journal of Animal Ecology* 55:431–449.

May, R. M. 1983. Parasite infections as regulators of animal populations. *American Scientist* 71:36–45.

May, R. M., J. R. Beddington, C. W. Clark, S. J. Holt, and R. M. Laws. 1979. Management of multispecies fisheries. *Science* 205:267–277.

McNaughton, S. J. 1985. Ecology of a grazing ecosystem: the Serengeti. *Ecological Monographs* 55:259–294.

Price, P. W. 1980. *Evolutionary Biology of Parasites.* Princeton Univ. Press, Princeton.

21

EVOLUTIONARY
RESPONSES AND
COEVOLUTION

The outcomes of population interactions depend on carrying capacities, competition coefficients, and predation efficiencies. These in turn are determined by the adaptations of individuals to their environments and therefore are subject to selection and evolutionary change. Populations of predators, prey, and competitors fashion the environment of every species. Each selects traits in the others that tend to alter their interactions. For example, by capturing those prey that it finds most easily, a predator leaves behind more cryptic prey. These prey are selected, in terms of evolution — they reproduce and pass on to future generations their protective coloring or other effective defenses against predators. As a result, the efficiency with which the population of predators exploits that prey population as a whole tends to diminish over evolutionary time, all other things being equal.

When two populations interact, both respond by undergoing evolutionary change. This is referred to as **coevolution.** When their relationship is antagonistic, as it is between predator and prey, the species can become locked into an evolutionary battle to increase their own fitness, each at the other's expense. Such a struggle can lead to an evolutionary stalemate in which both antagonists continuously evolve in response to each other or run out of the genetic variation needed to fuel further evolutionary change. In either event, the net outcome of their interaction may be an equilibrium. Coevolution between mutualists can lead to a stable arrangement of complementary adaptations that promote their interaction. In the most highly symbiotic mutualisms, such coevolved complexes of adaptations make further change very difficult, because both members of the mutualistic pair must evolve in parallel.

In this chapter, we shall explore some of the consequences of evolutionary responses for the interactions between predators and their prey, between competitors, and within mutualistic associations. The following example provides a particularly vivid case study of evolution in action within a pathogen–host system.

The myxoma virus and the rabbit

Shortly after the release of a few pairs on a ranch in Victoria in 1859, the European rabbit became a major pest in Australia. The hundreds of millions of rabbits distributed throughout most of the continent destroyed range and pasture lands and threatened wool production. The Australian government tried poisons, predators, and other potential controls, all without success. After much investigation, the answer to the rabbit problem seemed to be a myxoma virus (a relative of smallpox) discovered in populations of a related South American rabbit. Myxoma produced a small, localized fibroma (a fibrous cancer of the skin). Its effect on South American rabbits was not severe, but European rabbits infected by the virus died quickly of myxomatosis.

In 1950, myxoma virus was introduced locally in Victoria. An epidemic of myxomatosis broke out and spread rapidly. The virus was transmitted primarily by mosquitoes, which bite infected areas of the skin and carry the virus on their snouts. The first epidemic killed 99.8 percent of the infected rabbits, reducing their populations to very low levels. But during the following myxomatosis season (which coincided with the mosquito season), only 90 percent of the remaining population was killed. During the third outbreak only 40 to 60 percent of infected rabbits succumbed, and their population began to grow again.

The decline in the lethality of myxomatosis in the Australian rabbits resulted from evolutionary responses in both the rabbit and the virus populations. Before the introduction of myxoma, some rabbits had genetic factors that conferred resistance to the disease. Though nothing had spurred an increase in these factors before, they were strongly selected by the myxoma infection until most of the surviving rabbit population consisted of their resistant progeny (Figure 21.1). At the same time, virus strains with less virulence predominated because reduced virulence lengthened the survival time of infected rabbits and thus increased the mosquito-borne dispersal of the virus (mosquitoes bite only living rabbits). A virus organism that kills its host quickly has less chance of being carried by mosquitoes to other hosts.

Left on its own, the Australian rabbit–virus system would probably evolve to an equilibrial state of benign, endemic disease, as it had in the population of South American rabbits from which the myxoma was isolated. Currently, pest

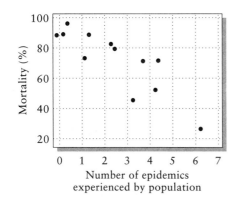

FIGURE 21.1 Decrease in susceptibility of wild rabbits to myxoma virus with grade III virulence (that is, causing 90 percent mortality in genetically unselected wild rabbits). From F. Fenner and F. N. Ratcliffe, *Myxomatosis*. Cambridge Univ. Press, London (1981).

management specialists keep the system out of equilibrium and maintain the effectiveness of myxoma as a control agent by finding new strains of the virus to which the rabbits have yet to evolve immunity.

Evolution in predator–prey systems

Low virulence of myxoma virus results from the fact that mosquitoes disperse the virus to new hosts more efficiently when the infected host has a long life span. Thus the less virulent strains of virus have a higher rate of growth in the population of hosts as a whole, if not within the individual host organism. Highly contagious diseases that are spread directly through the atmosphere or water have no such constraint on dispersal and often exhibit high levels of virulence with debilitating or even fatal consequences for the host. Similarly, most predators do not rely on a third party to find prey, and rather than evolve toward a benign equilibrium of restraint and tolerance, predator and prey tend to become locked in an evolutionary battle of persistent intensity. The outcome of the battle depends on which population gets the evolutionary upper hand.

David Pimentel and his colleagues at Cornell University explored the evolution of host–parasitoid relationships with the housefly and a wasp parasitoid of the fly pupae, *Nasonia vitripennis* (Figure 21.2). In one population cage, *Nasonia* was allowed to parasitize a fly population that was kept at a constant level by replenishment from a stock that had not been exposed to the wasp. None of the flies that escaped wasp parasitism were returned to the population cage, so the wasps were provided only with evolutionarily "naive" hosts.

In a second population cage, the fly hosts were kept at the same constant number, but because emerging flies were allowed to remain, the population could evolve resistance to the wasps. The population cages were maintained for about 3 years, long enough for evolutionary change to occur. Over the course of this experiment, the reproductive rate of the wasps in the cage that permitted evolution dropped from 135 to 39 progeny per female, and their longevity decreased from 7 to 4 days. The average level of the parasite population also decreased (1900 adult wasps versus 3700 in the nonevolving system), and the population size was more constant. These results suggest that flies evolved additional defenses when subjected to intense parasitism.

Experiments were then established in 30-compartment population cages in which the numbers of flies were allowed to vary freely. One such experiment was started with flies and wasps that had had no previous contact with each other, and a second was established with animals from the evolving population discussed above. In the first cage, wasps were efficient parasites, and the system underwent severe oscillations. In the second cage, however, the wasp population remained low and the flies attained a high and relatively constant population (Figure 21.3). This result strongly reinforces the conclusion, drawn from earlier experiments, that the flies had evolved resistance to the wasp parasites.

The genetics of parasites and diseases of economic crops and the genetic basis of crop resistance have been scrutinized closely by agronomists. Control of such pathogens as wheat rust (a fungus) is accomplished primarily by breed-

FIGURE 21.2 The wasp *Nasonia* parasitizing a pupa of the housefly. Courtesy of D. Pimentel. From D. Pimentel, *Science* 159:1432–1437 (1968).

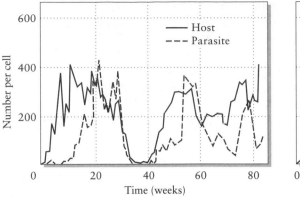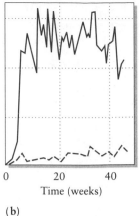

FIGURE 21.3 Populations of houseflies and a wasp parasitoid, *Nasonia vitripennis*, in 30-cell laboratory cages. (a) Control; flies had no previous experience with the wasp. (b) Experimental; flies had been exposed to wasp parasitism for more than 1000 days. After D. Pimentel, *Science* 159: 1432–1437 (1968).

ing strains of wheat with genes that confer resistance to the pathogen. But the defenses provided by resistance genes have proved to be easily circumvented by rapid evolutionary change in the pathogen, probably a single gene replacement in many cases. Agricultural geneticists keep track of such changes in plant disease organisms by routinely exposing different genetic strains of the crop plant to a variety of races of the pathogen and recording the virulence. A surprising result of a Canadian survey of wheat rust was that for a given race, the virulence on strains of wheat with different resistance genes appeared and disappeared sporadically.

Altered virulence in this system apparently results from changes in single genes. For example, in 1969, race 15 B-1L of the wheat rust was virulent on strains of wheat with resistance genes (Sr) 8, 10, and 11, and it was avirulent on strains with genes 15 and 17. In 1970 a subrace of the rust was isolated that had become avirulent on Sr 11. In 1971 a new subrace appeared that was virulent on Sr 15. That same subrace lost its virulence on Sr 8 the following year; another lost its virulence on Sr 11. In 1973 a new subrace virulent on Sr 17 appeared.

The significance of virulence changes for the rust is not clear. The virulence of different resistance genes was determined only in experimental plantings, so the changes in virulence did not appear in the population of rust as a whole. What the rust story does reveal, however, is that natural populations of predators and parasites continuously generate genetic variants that challenge the resistance of their prey and hosts, which presumably respond in kind. Of course, changes in many aspects of predatory behavior and prey defense do not have such simple genetic bases as virulence and resistance in the wheat–rust system. Nevertheless, it is clear that some consumers and their resource populations are locked into an endless evolutionary struggle.

Evolutionary equilibria of predator and prey

The evolutionary responses of predator and prey populations can be depicted by a simple graphical model that relates the rates of evolution of the two to the efficiency of predation (Figure 21.4). For the prey, the rate at which new adaptations useful in the escape or avoidance of predators are selected should vary

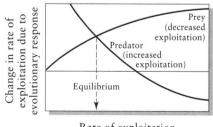

FIGURE 21.4 A graphical model of the evolutionary equilibrium between predator and prey adaptations that determine the level of exploitation in the system. When exploitation rates are low, there is little selection on the prey, but predators are strongly selected to increase rate of prey consumption. The reverse is true at high exploitation rates.

in direct proportion to the predation rate. In the absence of predation, there can be no selection of adaptations for predator avoidance. But as predation increases, so do selection and evolutionary response, at least up to limits set by the availability of genetic variation.

Selection of new adaptations useful to the predator in exploiting the prey should vary in the opposite fashion. When a particular prey species is not heavily exploited, adaptations of predators to use this resource will be selected, and predation on that prey population will increase. As exploitation of the prey increases, however, intraspecific competition among predators reduces the selective value of further increases in predation. Very high rates of predation conceivably could select individuals that shifted their diets toward *other* prey species. Hence evolution by a predator population could result in *decreased* efficiency in the use of a particular prey species, as indicated in that portion of Figure 21.4 where the "predator" curve sinks below zero on the vertical axis.

In this simple model, the balancing influences of predator and prey adaptation achieve a stable evolutionary steady state where the two curves cross. When predator adaptations are relatively effective and the prey are exploited at a high rate, selection on the prey population tends to improve its escape mechanisms faster than selection on the predator population improves its ability to exploit the prey. Conversely, when the exploitation rate is low, the prey evolve more slowly than the predators. This steady state between the adaptations of predators and prey should result in a relatively constant rate of exploitation regardless of the specific predator and prey adaptions.

Pimentel's experiments on host–parasite interactions, which we have discussed, illustrate the dynamics of the predator–prey equilibrium. The housefly (host) and the parasitic wasp *Nasonia* undoubtedly had achieved an evolutionary equilibrium in their natural habitats. When brought into a simple laboratory habitat, *Nasonia* wasps were able to exploit housefly populations at a greatly increased rate because they required little time to search out hosts. (Setting up these experimental conditions was equivalent to shifting the exploitation rate of *Nasonia* on houseflies far above the equilibrium level in Figure 21.4.) This shift increased the selective pressure on the housefly to escape parasitism much more than the selective pressure on the predator to further increase its exploitation rate. As a result, the ability of the housefly to escape parasitism increased and the level of exploitation by *Nasonia* decreased toward a new steady state.

Genetic variation in competitive availability

Evolution proceeds by the replacement of genotypes within a population — an expression of intraspecific competition between individuals having different genes. When two species are closely matched in competition, the different competitive abilities of genotypes within a population may result in a different outcome when a population is dominated by one genotype than when it is dominated by another. Sometimes such genes express themselves in the phenotype so subtly as to avoid detection by the direct examination of individuals. Instead, they must be inferred from the outcome of competition. In this sense, competitive ability summarizes the interaction of a phenotype with its environment. The experiments that follow demonstrate genetic variation in competitive ability. Clearly, such genetic variation also makes it possible for competitive ability to evolve.

Competition experiments involving the flour beetles *Tribolium confusum* and *T. castaneum* sometimes have uncertain outcomes. That is, at certain combinations of temperature and humidity, the two species are so closely matched that small differences in the individuals used in the experiments might tip the balance of competition in favor of one species or the other. At 29°C and 70 percent relative humidity, for example, *T. castaneum* "won" in about five-sixths of the experiments and *T. confusum* in one-sixth. These experiments were begun with 2 pairs of individuals of each species. Each individual carries a small sample of the genetic diversity of the population, and the genetic makeup of any 2 males and 2 females is likely to vary widely between samples. When the experiments were repeated under identical conditions but were begun with 10 pairs of each species, *T. castaneum* won 20 out of 20 times. These results suggest that in the first set of experiments, genetic variation in the parent individuals might have altered the outcome of competition. The larger parental populations used in the second set of experiments probably resembled the average genetic makeup of the species more closely, thereby producing more consistent results.

Several strains of each species of *Tribolium* were inbred to reduce their genetic variation, and then the strains were tested against each other in competition experiments. When populations were inbred for more than 12 generations, certain strains of *T. confusum* consistently outcompeted certain strains of *T. castaneum*—a reversal of their normal competitive relationship. These experiments indicated that changes, even minor ones, in the genetic constitution of a population can greatly influence its competitive ability.

When laboratory populations of the fruit flies *Drosophila serrata* and *D. nebulosa* were grown together, the competitive ability of *D. serrata* appeared to increase over time. The two species were established in population cages at 19°C, and they quickly achieved a pattern of stable coexistence with 20 to 30 percent *D. serrata* and 70 to 80 percent *D. nebulosa*. In one such case, however, the frequency of *D. serrata* began to increase after the 20th week and attained about 80 percent by the 30th week, a reversal of the initial predominance of *D. nebulosa*. When individuals of both species were removed from these populations after the 30th week and tested against stocks maintained in single-species cultures, the competitive ability of each species was found to have increased after exposure to the other in the competition experiment. In one of the population cages with both species, the competitive ability of *D. serrata* evidently had evolved much more rapidly than that of *D. nebulosa,* and their equilibrium frequencies were greatly altered. The difference between these replicates also appeared when the competitive ability of *D. serrata* was tested against that of the unselected stocks of *D. nebulosa.*

It is quite clear now that competitive ability has a genetic basis and can evolve in laboratory populations. The particular adaptations responsible for changes in competitive ability have not been determined in these experiments; they could conceivably include any increase in the efficiency of using the food resource, in the rate of producing offspring per unit of food consumed, and in survival at several stages of the life cycle. One of the few generalizations to come from this body of work is that sparse populations may evolve interspecific competitive ability more rapidly than dense populations. Why? Because if different, somewhat conflicting adaptations determine the outcomes of intraspecific and interspecific competition, then selection for increased interspecific competitive ability will be stronger in the rarer of two competitors (in which intraspecific competition presumably is less important).

We return to the work of David Pimentel for laboratory evidence that a poor competitor can evolve a competitive advantage (judged by relative population density) over a formerly superior adversary. Pimentel and his colleagues conducted laboratory experiments with flies to determine whether two species could coexist on one food resource by frequency-dependent evolutionary changes in their competitive ability. That is, can one species, as it is being excluded by the second and becoming rare, evolve increased interspecific competitive ability rapidly enough to regain the upper hand?

The housefly (*Musca domestica*) and the blowfly (*Phaenicia sericata*), which have similar ecological requirements and comparable life cycles (about 2 weeks), were chosen for the experiments. Both species feed on dung and carrion in nature, and they are often found together on the same food resources. The flies were raised in small population cages at 27°C, with a mixture of agar and liver provided as food for the larvae and sugar provided for the adults. The outcomes of an initial series of four competition experiments between individuals from wild populations of the housefly and the blowfly were split, with each winning twice. The mean extinction time for the blowfly, when the housefly won, was 92 days; it was 86 days for the housefly when the blowfly won. Thus the two species were close competitors, but the small cages used did not allow enough time for evolutionary change before one of the populations was excluded.

To prolong the housefly–blowfly interaction, Pimentel started a population in a 16-cell cage, which consisted of single cages in 4 rows of 4 cages with connections between them (Figure 21.5). Under these conditions, populations of houseflies and blowflies coexisted for almost 70 weeks and showed a strik-

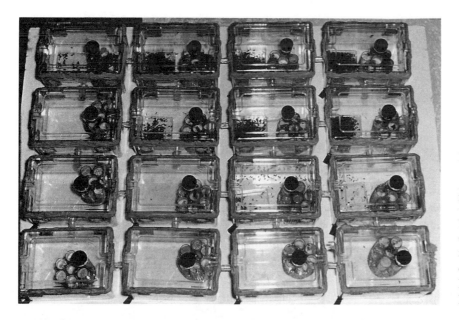

FIGURE 21.5 The 16-cell cage used by Pimentel to study competition between populations of flies. Note the vials with larval food in each cage and the passageways connecting the cells. The dark objects concentrated in the upper right-hand cells are fly pupae. Courtesy of D. Pimentel. From D. Pimentel, E. H. Feinberg, P. W. Wood, and J. T. Hayes, *Amer. Nat.* 99:97–109 (1965).

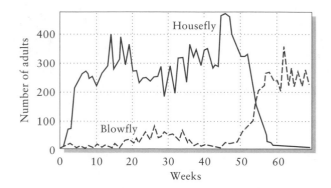

FIGURE 21.6 Changes in competing populations of houseflies and blowflies in a 16-cell cage. After D. Pimentel, E. H. Feinberg, P. W. Wood, and J. T. Hayes, *Amer. Nat.* 99:97–109 (1965).

ing reversal of numbers between the two species (Figure 21.6). After 38 weeks, when the blowfly population was still low, and just a few weeks prior to its sudden increase, individuals of both species were removed from the population cage and tested in competition with each other and with wild strains of the housefly and blowfly. Captured wild blowflies turned out to be inferior competitors to wild and experimental strains of the housefly. But blowflies that had been removed from the population cage at 38 weeks consistently outcompeted both wild and experimental populations of the housefly. Apparently, the experimental blowfly population had evolved superior competitive ability while it was rare and on the verge of extermination.

Character displacement and evolutionary divergence of competitors

Theory suggests that if resources are sufficiently varied, competitors should diverge and specialize. Laboratory experiments have shown that competitive ability may have a genetic component and therefore be under the influence of selection, although particular adaptations have not been identified. But if competition exerts a potent evolutionary force in nature, we should find evidence that competitors have partly molded each other's adaptations.

Although species differ in their ecological requirements, these differences between species have not necessarily evolved as a result of their interaction. An alternative explanation for such differences is that the species adapted to different resources in different places and when their populations subsequently overlapped as a result of range extensions, these ecological differences remained.

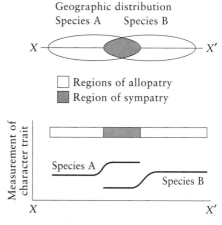

FIGURE 21.7 The phenomenon of character displacement, in which character traits of two closely related species are more similar in allopatric portions of their geographic distributions than where they occur together (sympatry).

We may get around this objection by comparing the ecology of a species in an area in which it co-occurs with a competitor to its ecology in another area where the competitor is absent. If the areas are otherwise similar, any differences between the species in the two areas can be uniquely explained by the presence or absence of a competing species. Specifically, we may infer that competition has caused divergence when two ecologically similar species differ more where they are found together (**sympatry**) than they do in nonoverlapping parts of their ranges (**allopatry**; Figure 21.7). This phenomenon is called **character displacement,** but ecologists disagree on its prevalence in nature.

A number of examples seem to fit the pattern of character displacement, one of which involves ground finches (*Geospiza*) of the Galapagos Islands. On islands with more than one species, the finches usually have beaks of different sizes, indicating different ranges of preferred food size. For example, on Marchena Island and Pinta Island, the ranges in beak size of the three resident species of ground finches do not overlap (Figure 21.8). On Floreana and San Cristobal, the two species *G. fuliginosa* and *G. fortis* have beaks of different sizes. On Daphne Island, however, where *G. fortis* occurs in the absence of *G. fuliginosa,* its beak is intermediate in size between those of the two species on Floreana and San Cristobal. On Los Hermanos Island, *G. fuliginosa* occurs in the absence of *G. fortis,* and its beak is intermediate in size.

The Galapagos ground finches illustrate the diversifying influence of competition because of the chance distribution of species on small islands within the archipelago: some islands have two or three species and some only one. In many other cases, it is difficult to know whether differences between two species arose because of competition between them or because they evolved in response to selection by other environmental factors in different places and retained their differences when their populations reestablished contact. In most cases, certainly, the genetic differences that led to speciation occurred in allopatry. So why not the differences that allow two species to avoid strong competition? In either case, however, coexistence depends on some degree of ecological difference between species, whether it is achieved in allopatry or as an evolutionary consequence of competition in sympatry.

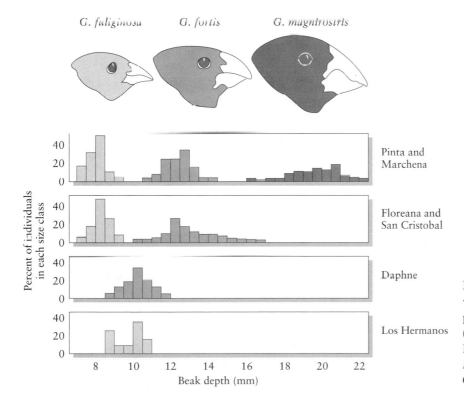

FIGURE 21.8 Proportions of individuals with beaks of different sizes in populations of ground finches (*Geospiza*) on several of the Galapagos Islands. After D. Lack, *Darwin's Finches.* Cambridge Univ. Press, Cambridge, England (1947).

Adaptive radiation

Diversifying evolution is easily appreciated in faunas or floras derived from a single ancestor. Many island groups have been colonized by a small number of species, some of which have undergone extensive proliferation of species. Characteristically, the ecological diversity of the species group expands as the number of species increases, a process referred to as **adaptive radiation**. Because all the species descended from the same initial colonist, the ecological diversity could have arisen only by local evolutionary change. And because colonists that do not form new species change much less than those that do, competitive relationships between members of the species group are implicated in diversification. The avifauna of the Hawaiian Islands provides a striking example of adaptive radiation. The numerous species of honeycreepers (many of them now extinct) descended from a single finchlike colonist from North America some millions of years ago. Over the course of time, different descendant species have evolved differences in body size and beak shape that enable them to exploit a wide variety of food resources (Figure 21.9).

Coevolution

Reciprocal evolutionary responses between populations are referred to as **coevolution**. Coevolution raises several questions. First, do pairs of populations undergo reciprocal evolution, or do "coevolved" traits arise from the response of populations to selective pressures exerted by a variety of environmental factors, followed by an ecological sorting out of species with compatible features? Second, are species organized into interacting sets based on their evolved adaptations, whether "coevolved" or not? And third, do these adaptations enhance such system properties as productivity of the biological community and its resistance to perturbation?

Coevolution has been attributed to complementary adaptations of pairs or small groups of species, without evidence having been presented for the evolutionary history of the relationship itself. But matching adaptations do not sufficiently prove coevolution. The adaptations of organisms clearly match many physical aspects of the environment, yet the physical environment does not respond adaptatively to its living inhabitants. Thus the match between organism and environment cannot be called coevolution.

This logical difficulty is illustrated by an apparent case of coevolution involving ants that protect aphids and leafhoppers from predators and, in return, harvest the nutritious honeydew that they excrete. In one such system, aphids, which are small, sedentary insects, form dense colonies on the inflorescences of ironweed (*Veronia*) in New York State. Also occurring on ironweed is the larger membracid bug (leafhopper) *Publilia,* which sucks plant juices from the leaves. These insects are tended by three species of ants. One, in the genus *Tapinoma,* is tiny (2 to 3 mm) but abundant. The other two (which both represent *Myrmica*) are larger (4 to 6 mm) and more aggressive but less common. The two genera of ants rarely co-occur on the same plant.

The presence of *Tapinoma* greatly enhances the survival of aphid colonies but has less effect on the survival of leafhoppers. The larger *Myrmica* offers

Figure 21.9 Heads of representative species of honeycreepers from the Hawaiian Islands illustrate the variety of beak shapes that have accompanied the adaptive radiation of the group to exploit different food resources. Paintings by H. D. Platt. From J. M. Scott, C. B. Kepler, C. van Riper III, and S. I. Fefer, *BioScience* 38:238 (1984).

substantial protection to leafhoppers but is less effective in warding off predators of aphids. Where both species of ants were excluded, predators were more numerous (recalling experiments on acacias in Central America, discussed in Chapter 18); when the predatory larvae of ladybird beetles were added to the system, *Myrmica* ants and, to a lesser extent, *Tapinoma* ants effectively reduced their predation on leafhoppers.

The ant–aphid–leafhopper system has all the elements expected of coevolution, but how can we be sure that the adaptations of the ant and homopteran participants evolved in response to each other? Most insects that suck plant juices produce large volumes of excreta from which they either do not or cannot extract all the nutrients. Thus honeydew production may reflect diet rather than having evolved to encourage protection by ants. For their part, ants are voracious generalists that are likely to attack any insect they encounter; they may need no special adaptation to confer benefit on the aphids and leafhoppers upon whose excreta they also feed. The fact that the different genera of ants more effectively protect different honeydew sources may simply reflect their different sizes and levels of aggression, which probably evolved in response to unrelated environmental factors.

Why don't ants eat the aphids and leafhoppers they tend? Perhaps this restraint is an evolved trait of ants that facilitates the ant–homopteran mutualism. It may even have arisen as an extension of the common ant behavior of defending plant structures that produce nectar—flowers or specialized nectaries (see, for example, Figure 18.16).

An analogous situation in the Cape region of South Africa highlights the importance of particular adaptations of ants, whether they are coevolved or not, to maintaining an ant–plant mutualism. There, many species of plants in the family Proteaceae have seeds with fleshy, edible, attached structures called elaiosomes (illustrated for some Australian plants in Figure 21.10). Foraging ants pick up seeds and transport them to their underground nests, where the elaiosomes are eaten. The seeds themselves, which the ants cannot eat, are then discarded either in underground chambers or in refuse heaps on the surface. In many regions these disposal sites are suitable for seed germination and seedling establishment. But in the fynbos (brushy, chaparral-like habitat), germination of many species of plants occurs only after fires have swept the habitat, and the only seeds that germinate are those stored by ants in underground nest chambers.

Recently, the Argentine ant *Iridomyrmex humilis* has invaded areas of fynbos shrublands and displaced many of the less aggressive native ants. The Argentine ant differs from the native ants in that it does not store seeds within its nests; the elaiosomes are removed on the surface, and the seeds are dropped there. As a result, germination of one species (*Mimetes*) following fire was drastically reduced in areas invaded by *Iridomyrmex* (Figure 21.11). With the continued persistence of the Argentine ant, it is likely that much of the native Cape flora will disappear as underground seed reserves are depleted. In a similar case, the virtual extinction of the tree *Calvaria major* on the island of Mauritius followed upon the extinction more than 300 years ago of the dodo bird, the only native species capable of effectively dispersing *Calvaria*.

Clearly, particular adaptations of ants and dodos influence their effectiveness as seed dispersers. But these adaptations may be evolutionarily independent of the plants themselves. The South African ants that disperse the seeds of *Mimetes* are diet generalists, and their food-caching behavior probably evolved for reasons unrelated to their role as seed dispersers. The plants may have evolved merely to take advantage of this fortuitous element in their environment without any reciprocal evolution on the part of the ant. Whereas some mutualisms, such as that between ant and acacia, clearly involve specialized adaptations of both parties and seem to represent cases of coevolution, many mutualisms may involve more serendipitous arrangements.

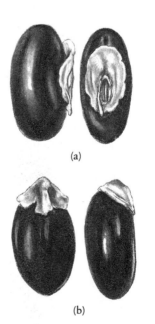

(a)

(b)

FIGURE 21.10 Ant-dispersed seeds of two Australian plants, *Kennedia rubicunda* (a) and *Beyeria viscosa* (b), showing the edible, light-colored appendage (elaiosome) that attracts ants. After R. Y. Berg, *Aust. J. Bot.* 23:475–508 (1975).

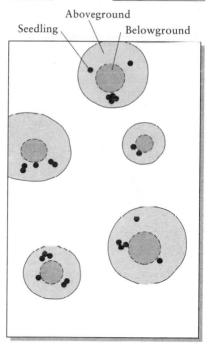

Extent of
Mimetes

1 meter

Seedling
Aboveground
Belowground

(a) *Iridomyrmex* present

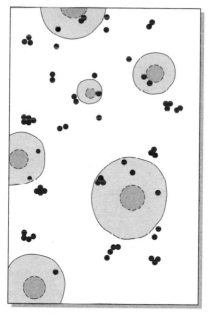

(b) *Iridomyrmex* absent

FIGURE 21.11 Seedling dispersion in *Mimetes cucullatus* populations after a burn in the presence of *Iridomyrmex* (a) and in its absence (b). The extent of aboveground and belowground parts of mature plants are indicated by shading. In the absence of *Iridomyrmex,* native ants distribute the seeds of *Mimetes* widely throughout the habitat. From W. Bond, and P. Slingsby, *Ecology* 65:1031–1037 (1984).

Coevolution in plant – pathogen systems

Plant geneticists have developed strains of domestic crops, such as flax and wheat, that are resistant to particular genetic strains of various pathogens, such as rusts (teliomycetid fungi). The crop strains differ from one another by a few (perhaps even by single) genetic changes that make them either susceptible or resistant to infection by particular strains of rust. Over the course of crop improvement programs, when new strains of rust have appeared either by mutation or by immigration from other areas, crop geneticists select new resistant strains of the crop by exposing experimental populations to the pathogen. New genotypes of the pathogen appear in the area either by migration or by mutation, creating continual evolutionary flux in such a system. A survey of genetic strains of wheat rust (*Puccinia graminis*) in Canada revealed that new virulence genes appear from time to time and sweep through the population (Figure 21.12). Genetic races of wheat rust are distinguished both by physiological characteristics and by virulence when tested on lines of wheat contain-

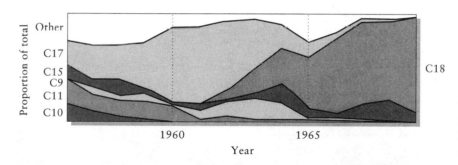

FIGURE 21.12 Relative proportions of virulence genes in the rust *Puccinia graminis* infecting Canadian wheat. From G. J. Green, *Canad. J. Bot.* 53:1377–1386 (1975).

ing different resistance alleles. As shown in Table 21.1, most of the virulence strains within a single physiological race of the rust differ by only one gene, and it is possible to outline a plausible evolutionary relationship between the races. For example, strains 1 and 29 differ only in their virulence on wheat with resistance gene 9b; strains 43 and 20, although in different physiological races, differ only in their virulence with respect to resistance gene 6.

The rust–wheat system contains the essential element of coevolution: an interaction between the fitnesses of genotypes of the host and those of the pathogen. The system is kept in flux by the introduction of new virulence genes in the rust — and perhaps by new resistance genes in the wheat, although the latter are pretty much controlled by plant geneticists nowadays.

Genotype–genotype interactions have been found in several natural systems and may turn out to be the rule in populations of plants and herbivores or of hosts and pathogens. Variation *between trees* in the defenses of ponderosa pines are paralleled by variation in the genotypes of scale insects that infest them (Figure 21.13). The scales are extremely sedentary, exhibiting so little migration from tree to tree that local populations on individual trees have evolved independently of those on other trees. Transferring scales both be-

TABLE 21.1 *Physiological races of wheat rust* Puccinea graminis *and virulence formulas based on the ability of strains to infect wheat with certain resistance genes*

Physiological race	Virulence formula	RESISTANCE GENE							
		5	6	7a	8	9a	9b	10	11a
11	C12	−	+	+	−	+	+	+	+
	C56	−	+	−	+	+	+	−	+
	C20	−	−	+	+	−	−	−	+
32	C21	−	−	−	−	+	−	−	+
	C22	−	−	−	−	+	−	−	−
	C32	−	−	−	−	+	+	−	+
	C34	−	+	+	−	+	+	−	+
	C43	−	+	+	+	−	−	−	+
17	C1	+	+	+	−	+	+	+	+
	C24	+	−	+	−	+	+	+	−
	C29	+	+	+	−	+	−	+	+
29	C3	+	+	−	−	+	−	−	+
	C5	+	−	−	−	+	+	−	+
	C6	+	−	−	−	+	+	−	+
	C30	−	−	−	−	+	+	−	−

Note: + indicates lack of virulence and hence effective host genes against a particular strain of rust.

Source: G. J. Green, *Canad. J. Bot.* 53:1377–1386 (1975).

tween trees and between branches on the same tree reveals this local adaptation. The survival of scales after being transplanted between trees is greatly reduced compared to that of controls transferred within the same tree. It is reasonable to assume that the differences between trees and between local populations of scales are genetic, so this represents a case of genotype–genotype interaction.

The genetics of chemical defenses in plants

Differences in the defensive chemicals of plants can be related to genetic changes, particularly when the pathways of biochemical synthesis and the enzymes responsible are known. University of Illinois biologist May Berenbaum has placed elements of the relationship between certain butterflies and their umbelliferous host plants in the context of coevolution. Umbellifers (parsley family) produce many noxious chemicals, among the most prominent of which are the furanocoumarins. The biosynthetic pathway leads from *para*-coumaric acid (which, being a precursor of lignin, is found in virtually all plants), to hydroxycoumarins such as umbelliferone, and then to furanocoumarins. These last include linear and angular forms, which are produced directly from hydroxycoumarins by different enzyme reactions.

As one proceeds down the biosynthetic pathway from *para*-coumaric acid to hydroxycoumarins and to linear or angular furanocoumarins, toxicity increases and occurrence among plant families decreases. Hydroxycoumarins have some biocidal properties; linear furanocoumarins interfere with DNA replication in the presence of ultraviolet light; angular furanocoumarins interfere with growth and reproduction quite generally.

Para-coumaric acid is widespread among plants, occurring in at least 100 families; only 31 families possess hydroxycoumarins. Linear furanocoumarins (LFCs) are restricted to 8 plant families and are widely distributed in only 2 — Umbelliferae and Rutaceae (citrus family). Angular furanocoumarins (AFCs) are known only from 2 genera of Leguminosae (pea family) and 10 genera of Umbelliferae.

Among species of herbaceous umbellifers in New York, some (especially those growing in woodland sites with low levels of ultraviolet light) lack furanocoumarins, others have linear furanocoumarins only, and some have both linear and angular furanocoumarins. Surveys of the herbivorous insects collected from these species show that host plants containing both angular and linear furanocoumarins were attacked by more species of insects than were plants with only linear furanocoumarins or with none; that the herbivores on AFC/LFC plants tended to be extreme diet specialists, most having been found on no more than 3 genera of plants; and that these specialists tended to be abundant compared to the numbers of the few generalists found on AFC/LFC plants and compared to levels of any herbivores either on LFC plants or on umbellifers lacking furanocoumarins.

Although linear and (especially) angular furanocoumarins are extremely effective deterrents to most species of herbivorous insects, some genera that have evolved to tolerate these chemicals have become successful specialists.

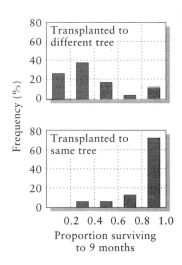

FIGURE 21.13 Black pineleaf scale on needles of ponderosa pine, illustrating the damage caused by feeding. The survival of individual scale insects, which depends on adaptation of localized populations to the genotypes of particular trees, decreases markedly when scales are transplanted to different trees. Courtesy of D. N. Alstad and G. F. Edmunds, Jr. From G. F. Edmunds and D. N. Alstad, *Science* 199:941–945 (1978).

One can make a strong case for coevolution here. The taxonomic distribution of hydroxycoumarins, linear furanocoumarins, and angular furanocoumarins across host plants suggests that plants containing LFCs are a subset of those containing hydroxycoumarins and that those containing AFCs are an even smaller subset of those containing LFCs. This is consistent with an evolutionary sequence of plant defenses progressing from hydroxycoumarins to LFCs and AFCs. Furthermore, insects specialized on plants containing LFCs belong to groups that characteristically feed on plants containing hydroxycoumarins, and those specialized on AFCs have close relatives that feed on plants containing LFCs. These patterns of taxonomic distribution of insects across host plants are consistent with coevolution within the system.

Coevolution between the yucca moth and the yucca

The curious pollination relationship that occurs between species of yucca plants (*Yucca*, in the lily family) and moths of the genus *Tegeticula* (Figure 21.14) was first described nearly a century ago. The moth enters the yucca

FIGURE 21.14 The mohave yucca (*Yucca shidigera*) and the yucca moth (*Tegeticula maculata*) that pollinates it. After J. A. Powell and R. A. Mackie, *Univ. Calif. Publ. Entomol.* 42:1–59 (1966).

flower and deposits 1 to 5 eggs on the ovary. Later, when the eggs hatch, the larvae burrow into the ovary, where they feed on the developing seeds. But after the moth has laid her eggs, she scrapes pollen off the anthers in the flower and rolls it into a small ball, which she grasps with specially modified mouthparts. She then flies to another plant, enters a flower, and proceeds to place the pollen ball onto the stigma of the flower before laying another batch of eggs.

The relationship between the moth and the yucca is obligatory. *Tegeticula* can grow nowhere else; *Yucca* has no other pollinator. In return for pollinating its flowers, the yucca seemingly tolerates the moth larvae feeding on its seeds, but the extent of this loss of potential reproduction is small, rarely exceeding 30 percent. *Yucca* and *Tegeticula* are specialized with respect to each other. The moth has a highly idiosyncratic pollination behavior. The yucca is specially adapted to make the moth's behavior effective: the pollen is sticky and can easily be formed into a ball, and the stigma is specially modified as a receptacle.

A puzzling aspect of the relationship is the restraint the moth exercises in laying a small number of eggs in each flower. Over the short term, it would seem that moths laying larger numbers of eggs per flower might have higher individual fitnesses, even though such behavior over the long term might lead to extinction of the yucca. This system poses the same difficulty as altruistic behaviors within populations, without the option of falling back on kin selection. One possibility is that yuccas can effectively regulate the relative fitness of moths laying different numbers of eggs per flower by selectively aborting developing fruits that are too highly infested by moth larvae to produce seed. Selective abortion of insect-damaged fruits is well known, and yuccas are known to possess mechanisms for fruit abortion. The particular evolutionary steps of moth behavior leading to the present system are lost in time, but the advantage to the moth of ensuring that the flower within which it lays its eggs is properly pollinated should be obvious. Most pollination relationships are neither as intricate nor as specialized as that of the yucca and moth. What "preadaptations" of *Yucca* and *Tegeticula* ancestors, or what accidents of history, started them along their coevolutionary pathway are considerations for the future.

❧ SUMMARY

Populations of predators, prey, and competitors respond to one another by evolutionary change in the characteristics that determine predation efficiency and coefficients of competition. Hence their interactions have evolutionary as well as population dynamics.

1. Evidence of such evolutionary changes in consumer–resource systems has been obtained in laboratory studies on host–parasite interactions. After periods of co-occurrence, rates of parasitism decreased and host populations increased, apparently following the selection of improved defenses against parasites. Further evidence of this change was obtained when stocks of hosts from these experiments were pitted against fresh stocks of parasites.

2. Studies on pathogens of plant crops — wheat rust, for example — have revealed a simple genetic basis for the outcome of the interaction.

3. Because selection for defenses of prey increases in proportion to predation rate, and selection for predatory efficiency decreases as predation rate increases, predator and prey reach an evolutionary steady state at some intermediate level of predation.

4. The outcome of competition in many experimental systems has been shown to depend on the genotypes of the competitors, which indicates that the coefficient of competition, the carrying capacity, or both have a genetic basis.

5. Experiments with competition between species of flies have revealed reversals of competitive ability over the course of tens of generations. By testing populations against unselected controls, investigators confirmed genetic changes in competing populations.

6. A test of whether competition can result in evolutionary divergence in nature is to compare ecological (or related morphological) traits of a population in the presence and absence of a competitor. When the two differ, the pattern is referred to as character displacement, but ecologists have found little solid evidence for such patterns.

7. Coevolution is the interdependent evolution of species that interact ecologically. The interactions may be antagonistic (consumer – resource) or cooperative (mutualism). Because each species in the coevolved pair is an important component of the environment of the other, changes in one select adaptive responses in the other, and vice versa.

8. The most convincing demonstrations of coevolution involve cases of mutualism, such as the obligate interdependence of *Pseudomyrmex* ants and *Acacia*. The ant keeps the plant free of herbivores while the plant provides the ant with food and housing; both have adaptations of structure and behavior (or phenology) that promote the relationship.

9. Plant – pathogen systems (such as fungal rusts on crop plants) have provided genetic models of the coevolutionary process. Genes that increase the virulence of the rust select resistance genes in the plants, leading to a continuous cycle of coevolutionary change.

10. Analysis of metabolic pathways shows the genetic (enzyme) steps in the development of toxic chemicals in plants. When variations in these pathways (and in the abilities of insects to detoxify the chemicals) are overlaid upon taxonomic relationships within each group, we can infer the evolutionary history of the plant – insect interaction.

11. The interaction between yucca moth and yucca is an obligate mutualism in which the moth pollinates the plant but its larvae consume developing seeds. Determining how the level of seed predation is controlled in this coevolved system presents a challenging, yet-unsolved problem.

Boucher, D. H. (ed.) 1985. *The Biology of Mutualism*. Croom Helm, London.

Brower, L. P. 1969. Ecological chemistry. *Scientific American* 220:22–29.

Davies, N. B., and M. Brooke. 1991. Coevolution of the cuckoo and its hosts. *Scientific American* 264:92–98.

Ewald, P. W. 1983. Host–parasite relations, vectors, and the evolution of disease severity. *Annual Review of Ecology and Systematics* 14:465–485.

Futuyma, D. J., and M. Slatkin (eds.). 1983. *Coevolution*. Sinauer, Sunderland, Mass.

Gilberg, L. E., and P. H. Raven (eds.). 1975. *Coevolution of Animals and Plants*. Univ. of Texas Press, Austin.

Handel, S. N., and A. J. Beattie. 1990. Seed dispersal by ants. *Scientific American* 263:76–83.

Janzen, D. H. 1966. Coevolution of mutualism between ants and acacias in Central America. *Evolution* 20:249–275.

Janzen, D. H. 1985. The natural history of mutualisms. In D. H. Boucher (ed.), *The Biology of Mutualism*. Croom Helm, London, pp. 40–99.

Nitecki, M. H. (ed.). 1983. *Coevolution*. Univ. of Chicago Press, Chicago.

Price, P. W. 1977. General concepts on the evolutionary biology of parasites. *Evolution* 31:405–420.

Communities

22

COMMUNITY STRUCTURE

Every place on earth — each meadow, each pond, each rock at the edge of the sea — is shared by many coexisting organisms. These plants, animals, and microorganisms are linked to one another by their feeding relationships and other interactions, forming a complex whole often referred to as the **biological community.** Interrelationships within the community govern the flow of energy and the cycling of elements within the ecosystem. These interrelationships also influence population processes, thereby determining the relative abundances of organisms. Finally, interrelationships within the community select among genotypes and therefore influence the evolution of coexisting species.

We have discussed these three types of interactions at length in earlier chapters. Observations and experiments have clarified the mechanisms underlying these interactions, which are now understood in broad outline. But several related issues have perplexed and polarized ecologists for decades and have not yet been clearly resolved. The common element underlying these issues is a century-old debate over the unity of the community.

One extreme—the **holistic** concept—champions the view that the community is a superorganism whose functioning and organization we can appreciate only when we consider its place in nature as a whole entity. Common sense tells us that we cannot ponder the significance of a kidney's functioning apart from the organism to which it belongs. Many ecologists have argued that it is equally pointless to consider soil bacteria without reference to the detritus they feed on, their own predators, and the plants nourished by their wastes. Accordingly, we can understand each species only in terms of its contribution to the dynamics of the whole system.

The other extreme viewpoint—the **individualistic** concept—holds that community structure and functioning simply express interactions of the individual species that make up the local association and do not reflect any organization, purposeful or otherwise, above the species level. Because natural selection tends to maximize the reproductive output of the individual, each population in the community fosters only its own self-interest. The flow of energy and the cycling of nutrients within the ecosystem result from the predatory endeavors of the individuals that make up the community.

Independently of the debate over the subtle nature of the community, ecologists have devoted considerable effort to describing community-level properties. The simplest measure of a community's structure is the number of species it includes, which is often referred to as **species richness** or **diversity**. Early naturalists knew that more species live in tropical localities than in temperate and boreal zones. Barro Colorado Island, a 16-km² island in Gatun Lake, Panama, supports 211 species of trees that grow to be taller than 10 m, more species than are found in all of Canada. Plots of 0.1 ha in some regions of Amazonian Peru contain that many species. High tropical diversity characterizes most taxanomic groups of organisms except those especially adapted to conditions that are unique to higher latitudes.

Ecologists have made additional comparisons of communities on the basis of numbers of species at each trophic level (that is, primary producers, herbivores, carnivores) and, within trophic levels, among different **guilds** distinguished by method or location of foraging (for example, herbivores comprise leaf-eaters, stem-borers, root-chewers, nectar-sippers, and bud-nippers). Another community-level property is relative abundance. In most associations, a few species are abundant and many more are rare.

The presence of regular patterns of community structure does not argue for or against a holistic (superorganismic) interpretation of the community. Organization can arise either according to some design imposed on the entire system or by the independent activities and interactions of the system's components. In the latter case, the structure of the community is a collective property of its individual components as each endeavors to function in its own right within the community rather than to enhance the functioning of the community as a whole.

Definition of the community

Ecologists have given **community** a variety of meanings. Usually, the term applies to a group of populations that occur together, but there any similarity among definitions ends. Throughout the development of ecology as a science, the term has often denoted **associations** of plants and animals occurring in a particular locality and dominated by one or more prominent species or by a physical characteristic. We speak of an oak community, a sagebrush community, and a pond community, meaning all the plants and animals found in the particular place dominated by its namesake. Used in this way, the term is unambiguous; a community is spatially defined and includes all the populations within its boundaries.

Ecologists also define communities on the basis of interactions among associated populations. This implies a functional rather than a descriptive use of the term. Ecologists sometimes use *association* for groups of populations that occur in the same area without regard to their interactions and reserve *community* to denote an association of interacting populations.

Communities defy delineation when interactions among populations extend beyond arbitrary spatial boundaries. The migrations of birds between temperate and tropical regions link the communities in each area; within some tropical localities, as many as half the birds present during the northern winter are migrants. Salamanders, which complete their larval development in streams and ponds but pursue their adult existence in the surrounding woods, tie together aquatic and terrestrial communities, just as trees do when they shed their leaves into streams and thereby support aquatic, detritus-based food chains.

Community structure and functioning blend a complex array of interactions, directly or indirectly tying together all members of a community into an intricate web. The influence of each population extends to ecologically distant parts of the community. Insectivorous birds do not eat trees, but they do prey on many of the insects that feed on foliage or pollinate flowers. By eating pollinators, birds may indirectly affect the number of fruits produced, the amount of food available to animals that feed on fruits and seedlings, and the predators and parasites of those animals. The ecological and evolutionary impact of a population extends in all directions throughout the trophic structure of the community by way of its influence on predators, competitors, and prey, but this influence dissipates as it passes through each successive link in the chain of interaction. Its impact similarly spreads through space and forward through time by way of the movements of individuals and the momentum of population processes. Because of this momentum, the present-day community also bears the imprint of the past.

The community as a natural unit of ecological organization

Ecologists describe functional relationships among an association of species just as physiologists connect the functions of the various parts of the body. The analogy between community and organism is obvious, as we have seen. Victor E. Shelford, in a 1931 paper entitled "Some Concepts of Bioecology," wrote:

> It is an old practice to liken organisms to cosmic systems, and cosmic systems to organisms. Again in this case, it is

convenient to liken the biome (plant–animal formation) to an amoeboid organism, a unit of parts, growing, moving, and manifesting internal processes which may be likened to metabolism, locomotion, etc., in an organism.

(Ecology *12:455–467*)

Shelford even drew parallels between certain processes of community change and wound healing.

Certainly the most influential advocate of the organismic viewpoint was the American plant ecologist Frederick E. Clements, who, early in this century, perceived communities as discrete units with sharp boundaries and a unique organization. The conspicuousness of many dominant vegetation types reinforced Clements's view. A forest of ponderosa pine, for example, is distinct from the fir forests that grow in moister habitats and from the shrubs and grasses typical of drier sites. The boundaries between these community types are often so sharp that we may cross them within a few meters along a gradient of climate conditions. Some community boundaries, such as that between deciduous forest and prairie in the midwestern United States and that between broad-leaved forest and needle-leaved forest in southern Canada, are respected by most species of plants and animals.

An opposite view of community organization was held at about the same time by H. A. Gleason, who suggested that the community, far from being a distinct unit like an organism, is merely a fortuitous association of organisms whose adaptations enable them to live together under the particular physical and biological conditions that characterize a particular place. A plant association, he said, is "not an organism, scarcely even a vegetational unit, but merely a coincidence."

Clements's and Gleason's concepts of community organization predict different patterns in the distribution of species over ecological and geographic gradients. On the one hand, Clements believed that the species belonging to a community are closely associated with one another; the ecological limits of distribution of each species coincide with the distribution of the community as a whole. Ecologists call this type of community organization a **closed community.** On the other hand, Gleason believed that each species is distributed independently of others that co-occur in a particular association. Such an **open community** has no natural boundaries; therefore, its limits are arbitrary with respect to the geographic and ecological distributions of its member species, which may extend their ranges independently into other associations.

Ecotones

The structure of closed and open communities is shown schematically in Figure 22.1. In the diagram depicting closed communities, the distributions of species in each community coincide closely along a gradient of environmental conditions — for example, from dry to moist. Closed communities are natural ecological units with distinct boundaries. The edges of such communities,

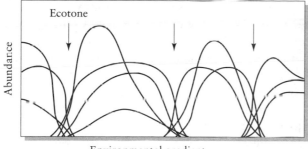

(a) Closed communities

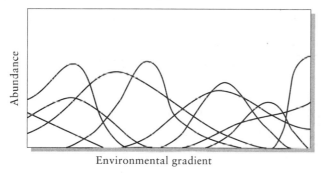

(b) Open communities

FIGURE 22.1 Hypothetical distributions of species organized into distinct assemblages (closed communities, a) or distributed at random along a gradient of environmental conditions (open communities, b). Ecotones between closed communities are indicated by arrows.

called **ecotones,** are regions of rapid replacement of species along the gradient. In the diagram depicting open communities, species are distributed at random with respect to one another, giving an open structure. We may arbitrarily delimit a "community" at some point, perhaps a dry forest community near the left-hand end of the moisture gradient, while recognizing that some of the species included are more characteristic of drier portions of the gradient and that others reach their greatest abundance in wetter sites.

The concepts of open and closed communities both have some validity in nature. We observe distinct ecotones between associations under two circumstances: first, when the physical environment changes abruptly — for example, at the transition between aquatic and terrestrial communities, between distinct soil types, and between north-facing and south-facing slopes of mountains; second, when one species or life form so dominates the environment that the edge of its range signals the distributional limits of many other species.

A change in soil acidity often accompanies the transition between broad-leaved and coniferous forest. The decomposition of needles produces organic acids more abundantly than does the breakdown of leaves; furthermore, because needles tend to decompose slowly, a thick layer of organic duff accumulates at the soil surface. At the boundary between grassland and shrubland or between grassland and forest, sharp changes in surface temperature, soil moisture, light intensity, and burning frequency result in many species replacements. The boundaries between grasslands and shrublands are often sharp, because when one or the other vegetation type holds a slight competitive edge, it dominates the community. Grasses prevent the growth of shrub seedlings by reducing the moisture content of the surface layers of the soil; shrubs depress the growth of grass seedlings by shading them. Fire evidently maintains a sharp edge between prairies and forest in the midwestern United States. Perennial grasses resist fire damage that kills tree seedlings outright, but fires do not penetrate deeply into the moister forest habitats.

Sharp physical boundaries create well-defined ecotones. These occur at the interface between most terrestrial and aquatic (especially marine) communities (Figure 22.2) and where underlying geologic formations cause the mineral content of soil to change abruptly. The ecotone between plant associations on serpentine-derived soils and on nonserpentine soils in southwestern Oregon is examined in more detail in Figure 22.3. Levels of nickel, chromium, iron, and

FIGURE 22.2 A sharp community boundary (ecotone) in the Bay of Fundy, New Brunswick, associated with an abrupt change in the physical properties of adjacent habitats. Seaweeds extend only to the high-tide mark. Between the high-tide mark and the spruce forest, waves wash soil from the rocks, and salt spray kills pioneering land plants, leaving the area devoid of vegetation.

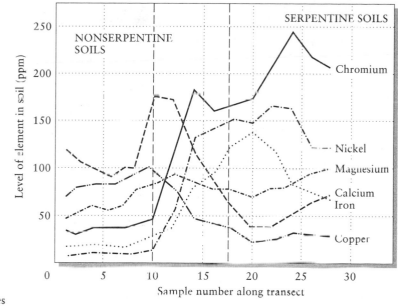

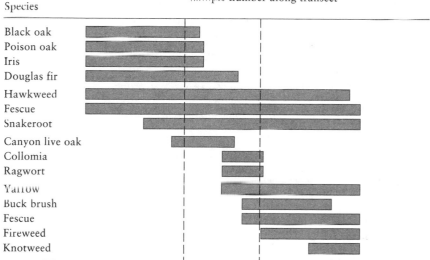

FIGURE 22.3 Changes in the concentration of elements in soil (a) and replacement of plant species (b) across the boundary between nonserpentine soils (samples 1 to 10) and serpentine soils (samples 18 to 28) in southwestern Oregon. The transect diagrammed here is somewhat atypical in that magnesium does not increase so abruptly as usual across the serpentine ecotone. After C. D. White, *Vegetation–Soil Chemistry Correlations in Serpentine Ecosystems.* Ph.D. dissertation, University of Oregon, Eugene (1971).

magnesium increase across the boundary into serpentine soils; the copper and calcium contents of the soil drop off. The edge of the serpentine soil marks the boundaries of many species that are either excluded from, or restricted to, serpentine outcrops. A few species exist only within the narrow zone of transition; others, seemingly unresponsive to variations in soil minerals, extend across the ecotone.

Plant ecologists have long recognized the influence of climate on plant associations. For example, Forrest Shreve described the chaparral–desert transition in Baja California in relation to moisture and freezing temperatures. Desert species, particularly the cacti, do not tolerate prolonged frost and dwindle north of the frost line; chaparral species drop out to the south within the transition zone as a consequence of water stress. As Shreve concluded:

> *Plants of the desert are more sharply confined to their own formation than are the species and genera of the chaparral and*

other northern types of vegetation. This appears to be due to the fact that the only requirement for the long southward extension of a chaparral plant is the occurrence in the desert region of relatively moist habitats, however restricted in area, while the northward extension of a desert plant requires a well-drained soil, a high percentage of sunshine and freedom from freezing temperatures of more than a few hours' duration. These more exacting requirements are met only in close proximity to the edge of the desert or else in light soils or on steep south slopes near the sea.

(Madroño 3:257–264, 1936)

The continuum concept

The deciduous forests of eastern North America are bounded to the north by cold-tolerant needle-leaf forests, to the west by drought- and fire-resistant grasslands, and to the southeast by fire-resistant pine forests. Within the region of their distribution, deciduous forests present a physiognomically uniform appearance. As a result of early botanical explorations, ecologists were aware that different species of trees and other plants occur in different areas within the forest biome. According to Clements's closed-community viewpoint, the distinctive vegetation of each area represented a distinct community separated by sharp vegetational transitions from other communities. But as ecologists described plant distributions in more detail, they found that plant associations fit less and less well the closed-community concept; classifications of plant communities became more and more finely split until absurd levels of distinction were reached.

Out of this mounting chaos there arose a new concept of community organization, referred to as the **continuum.** Within broadly defined habitats, such as forest, grassland, and estuary, populations of plants and animals gradually replace one another along gradients of physical conditions. The environments of the eastern United States form a continuum, with a north–south temperature gradient and an east–west rainfall gradient. Species of trees found in any one region (for example, those native to eastern Kentucky) have different geographic ranges, suggesting a variety of evolutionary backgrounds and ecological relationships. Some species reach their northern limits in Kentucky, some their southern limits. Because few species have broadly overlapping geographic ranges, associations of plant species found in eastern Kentucky do not represent closed communities. Each species has a unique evolutionary history and present-day ecological position, with a variable degree of association with other species in the local community.

A more detailed view of Kentucky forests would reveal that many of the tree species segregate along local gradients of conditions. Some dwell along ridge tops; others along moist river bottoms; some on poorly developed, rocky soils; others on rich, organic soils. The species represented in each of these more narrowly defined associations might exhibit correspondingly closer ecological distributions, but the open-community concept would still better describe these associations.

Gradient analysis

A first step toward resolving the problem of open versus closed communities was to devise methods of portraying the distributions of species along ecological gradients. One way of doing this involves plotting the abundances of species along some continuous gradient of ecological conditions. In such **gradient analysis,** closed-community organization would reveal itself by the presence of sharp ecotones in species distributions, as indicated in Figure 22.1. The gradient itself might embrace any number of physical variables, such as moisture, temperature, salinity, exposure, or light level. It is usually constructed by measuring both the abundances of species and the physical conditions at a number of localities and then plotting the abundances of each species as a function of the value of the physical condition. Ecologists might pick sampling localities at regular intervals along a known physical gradient, such as that of temperature as it decreases up an elevation gradient.

The principal proponent of gradient analysis was Cornell University ecologist Robert Whittaker, and his work was influential in putting to rest the extreme Clementsian view of the closed community. Whittaker conducted most of his work in mountainous areas where moisture and temperature vary over short distances according to elevation, slope, and exposure, which in turn determine light, temperature, and moisture levels at a particular site. When Whittaker plotted the abundances of each species at sites at the same elevation distributed along a continuum of soil moisture, he found that the species occupied unique ranges with peaks of abundance scattered along the environmental gradient (Figure 22.4). Compared with communities in the mountains of southeastern Arizona, fewer species of plants reside in the mountains of Oregon, but each species has a wider ecological distribution, on average.

In the Great Smoky Mountains of Tennessee, dominant species of trees occur widely outside the plant associations that bear their names (Figure 22.5). For example, red oak occurs most abundantly in relatively dry sites at high elevations, but its distribution extends into forests dominated by beech, white

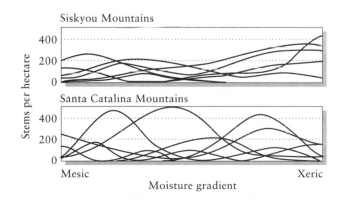

FIGURE 22.4 Distribution of species along moisture gradients at 460–470 m elevation in the Siskyou Mountains of Oregon and at 1830–2140 m elevation in the Santa Catalina Mountains of southeastern Arizona. Species in the more diverse Arizona flora occupy narrower ecological ranges. Thus, despite the greater total number of species in the flora of the Santa Catalina Mountains, they and the Siskyou Mountains have similar numbers of species at each sampling locality. After R. H. Whittaker, *Ecol. Monogr.* 30:279–338 (1960); R. H. Whittaker and W. A. Niering, *Ecology* 46:429–452 (1965).

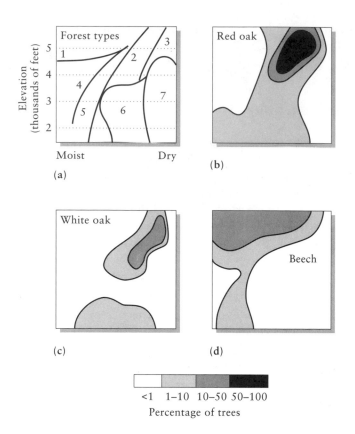

FIGURE 22.5 Distribution of (b) red oak, (c) white oak, and (d) beech with respect to altitude and soil moisture in the Great Smoky Mountains of Tennessee. The approximate boundaries of the major forest associations are shown in (a). Forest types are 1, beech; 2, red oak–chestnut; 3, white oak–chestnut; 4, cove; 5, hemlock; 6, chestnut oak–chestnut; 7, pine. Relative abundance, represented by degree of shading, corresponds to percentage of tree stems more than 1 cm in diameter in samples of approximately 1000 stems. After R. H. Whittaker, *Ecol. Monogr.* 26:1–80 (1956).

oak, chestnut, and even hemlock (an evergreen, coniferous species) and reaches throughout the entire range of elevation in the Smoky Mountains. Beech prefers moister situations than red oak, and white oak reaches its greatest abundance in drier situations, but all three species occur together in many areas. To cite another example of open community structure, the distributions of species of birds along an elevation gradient in Peru similarly failed to reveal evidence of distinct ecotones between associations of species. Even Forrest Shreve, who emphasized the distinct boundary between desert and chaparral plant associations, recognized that

> *As is true of the meeting ground between any two great plant formations, the dominant plants of each formation are found to vary in the distance to which they extend into the other. This indicates that their habitat requirements are not so nearly identical as their close association in the midst of their respective formations would suggest.*

> *(Madroño 3:257–264, 1936)*

Trophic structure and food webs

An ecosystem perspective on the community places species in functional groups whose members occupy similar trophic positions. Thus plants are lumped together as producers, all herbivores (from ant to zebra) share the herbivore label, and so on. Because describing trophic structure in this way conforms to certain thermodynamic criteria, functional descriptions of communities tend to emphasize similarities. Placing species together in functional categories obscures the distinctiveness of communities that arises from the differences in numbers of species or in their evolutionary histories. A food web perspective, on the other hand, although based on functional relationships, emphasizes the connections between populations and recognizes, for example, that not all herbivores consume all producers. Because food web analysis includes species-level information about the community, it has greater power than ecosystem analysis to differentiate structure.

Food web analysis has progressed through two phases: descriptive and analytical. The descriptive phase began early in this century. Initially, ecologists described food webs by drawing diagrams in which arrows connected the species in the community according to their feeding relationships. These diagrams were often complex and did not lend themselves to recording statistics suitable for comparison between communities. To some ecologists, food web diagrams merely emphasized the overwhelming complexity of natural systems and the need to simplify structure into trophic groups.

The analytical phase of the study of food webs, which asks whether the structure of a food web influences the dynamics of its constituent populations, began in the mid-1950s. It followed upon the suggestion that the more complex the community, the greater its stability. The reasoning was simple. When predators have alternative prey, their own numbers depend less on fluctuations in the numbers of a particular prey species. Where energy can take many routes through a system, disruption of one pathway merely shunts more energy through another, and the overall flow continues uninterrupted. This idea linked community stability directly to species diversity and food web complexity, and it stimulated a flurry of theoretical, comparative, and experimental work. These studies have yet to produce a consensus, partly because both structure and stability elude definition and measurement, and partly because different theories lead to different predictions about stability. For example, an alternative to the idea that diversity generates stability is that as communities become more diverse, the species exert greater influence on one another through various interactions; these biological links in turn may create pervasive time lags in population processes and therefore destabilize diverse systems.

Among the most influential of experimental studies was the research of Robert T. Paine, of the University of Washington, on the role of consumers in determining the structure of rocky-shore, intertidal communities. Paine emphasized the role of predators. He compared food webs in the Gulf of California and on the coast of Washington, both of which are dominated by imposing predators, the sea stars *Pisaster* and *Heliaster* (Figure 22.6). By removing sea stars from experimental areas on the coast of Washington, Paine demonstrated the crucial role of the predators in maintaining the structure of the community. Released from predation by this manipulation, mussels (*Mytilus*) spread very rapidly, crowding other organisms out of the study areas and reducing the diversity and complexity of the local food web. Removal of the urchin *Strongylocentrotus,* a herbivore, similarly allowed a small number of competitively superior algae to dominate the system, crowding out many ephemeral or grazing-resistant species. Paine showed that predators and her-

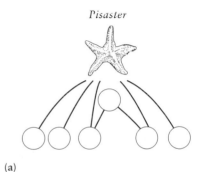

(a)

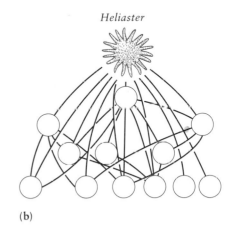

(b)

FIGURE 22.6 Intertidal food webs on the coast of Washington, dominated by the sea star *Pisaster* (a), and in the northern Gulf of California, dominated by *Heliaster* (b). The lowest trophic levels of the food webs illustrated include such herbivores as chitons, limpets, herbivorous gastropods, and barnacles. After R. T. Paine, *Amer. Nat.* 100:65–75 (1966).

bivores can manipulate competitive relationships among species at lower trophic levels and thereby control the structure of the community. Such species as *Pisaster* are called **keystone predators** because when they are removed, the entire edifice of the community tumbles.

Whereas many ecologists emphasized interactions among competitors on the same trophic level, Paine and others who followed his example stressed the additional importance of consumer–resource relationships as a key to understanding community organization. Paine also pointed out the different ways of conceptualizing food webs. His connectedness webs, energy flow webs, and functional webs describe different ways in which populations influence one another within the community. **Connectedness webs** emphasize the feeding relationships among organisms, portrayed as links in a food web. **Energy flow webs** represent an ecosystem viewpoint, in which the connections between species are quantified by the flux of energy between a resource and its consumer. In **functional webs,** the importance of each population in maintaining the integrity of the community reflects its influence on the growth rate of other populations. This controlling role, which only experiments can reveal, need not correspond to the functional importance of a feeding link in an intact community, as shown dramatically for an intertidal-zone food web in Figure 22.7.

As Paine and others employed food web diagrams to portray the structure of biological communities, a few ecologists questioned whether differences in the structure of webs could affect the dynamics, stability, and persistence of communities. The issue is a crucial one in ecology, being a part of the fundamental contrast between holistic and individualistic philosophies. We raise the issue by posing such questions as the following: Is a particular arrangement of

FIGURE 22.7 Three approaches to depicting trophic relationships, illustrated for the same set of species in a rocky intertidal habitat. From R. T. Paine, *J. Anim. Ecol.* 49:667–685 (1980).

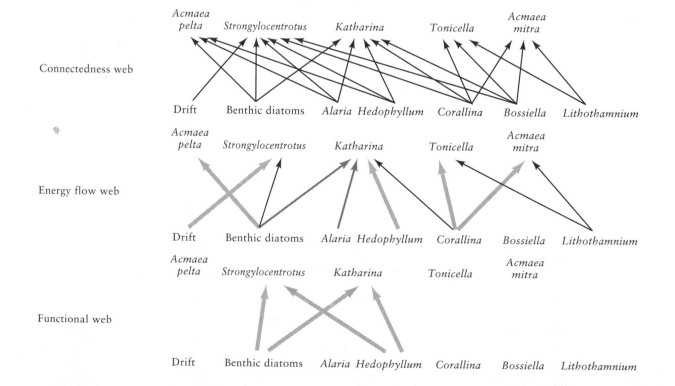

feeding relationships among species intrinsically more stable than a different arrangement among the same number of species? How important is the consideration of food web stability to the structure of natural communities?

In the two food webs illustrated in Figure 22.8, similar numbers of species are organized in strikingly different structures. The mudflat community is relatively simple, having 7 links among 7 species, and with only 1 species preying on more than one trophic level. By contrast, the plant–insect–parasitoid system is complex; it exhibits 12 links among 8 species, several cases of feeding on more than one trophic level, and one case in which 2 species feed on each other. Theoretical work indicates that, all other things being equal, increasing the degree of feeding on more than one trophic level (omnivory) reduces a food web's stability — that is, its resistance to perturbation and its ability to return to equilibrium.

With respect to the food web structure of the community, theory and observation clearly are divided by a great chasm. On the one hand, the attributes of food web design are consistent with and depend on qualities that enhance the intrinsic dynamic stability of the food web. On the other hand, we observe radically different web designs in nature, such as those shown in Figure 22.8. Does this variation mean that the rules of food web stability vary depending on the organisms and ecological circumstances involved, or that feeding relationships depend on the attributes of species and that the design of a system has little influence on its stability?

These are tough questions. An important first step toward answering them must be to characterize community organization in ways that match theory. Five attributes of communities seem important: diversity, **connectance** (basically, complexity), food chain length, omnivory, and **compartmentalization** (subdivision of the food web). The last attribute once again raises the issue of the discreteness of the community. A highly compartmentalized community is one in which subsets of species interact regularly among themselves but only infrequently with species in other subsets. At the extreme, each compartment could be considered a separate community. Compartments might coincide with patches of habitats or distinctive species within habitats. For example, in southern Canada, broad-leaved trees, pines, firs, and hemlocks all have distinctive associations of lepidopteran herbivores; few species or genera of moths and sawflies feed on plants in more than one of the groups. In mixed forests, therefore, one might consider the faunas of broad-leaved and needle-leaved trees as separate communities. In contrast, moths feed widely among species of broad-leaved trees, and although most specialize to a greater or lesser degree, they seem not to distinguish discrete subsets. Indeed, the analysis of many food webs turns up little evidence for compartmentalization despite considerable feeding specialization within associations of species occupying more or less uniform habitats.

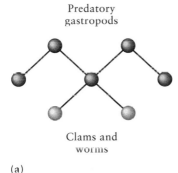

Clams and worms

(a)

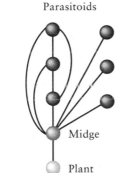

Parasitoids

Midge

Plant

(b)

FIGURE 22.8 Examples of food webs with little (a) and frequent (b) omnivory. (a) is based on intertidal gastropods, bivalves, and their prey; (b) on the plant *Bacharis*, its insect herbivores, and their parasitoids. After S. L. Pimm, *Food Webs*. Chapman & Hall, London and New York (1982).

Relative abundance

The Danish botanist Christen Raunkiaer noted early in this century that the abundances of populations within local assemblages assume regular distributions. When he plotted the numbers of species in each of several abundance classes, the points followed a reversed J-shape, such as that shown in Figure 22.9. This pattern suggested that within a particular community, a few species attain high abundance — they are the **dominants** in the community — whereas most of the other species are represented by relatively few individuals. Raunkiaer did not use a mathematical expression to describe his "law" of frequency, but its generality and pervasiveness among different assemblages have tempted others to mathematical description ever since.

Mathematics can serve two purposes here. We can describe many data (species abundances, in this case) with a simple equation and use the variables therein to make comparisons among different samples of species. Or we can use the logic of the mathematical model to infer processes that produce the distributions observed. Without going into the mathematical details of models of relative abundance, let us simply say that the models have served better as descriptive devices than they have in elucidating the processes that regulate relative abundance. We may fairly say that within every community, some species are common and others are rare. Species abundance appears to reflect the variety and abundance of resources available to each population, as well as the influences of competitors, predators, and diseases. Determining why some species fare well in this arena and others do not continues to pose a major challenge to ecologists.

The lognormal distribution

The abundance of a particular species reflects the balance between a large number of factors and processes, variations in each of which result in small increments or decrements in abundance. Statisticians have shown that the sums of many independent factors with small effects tend to assume a **normal distribution,** the familiar bell-shaped curve. Because factors that affect population size tend to exert a multiplicative influence, one might expect a sample of population sizes to assume a **lognormal distribution:** there being many species with intermediate levels of abundance and relatively few rare or common species.

In 1948 Frank Preston published a seminal paper entitled "The Commonness, and Rarity, of Species," in which he characterized the distribution of species abundances by a lognormal curve. Preston assigned species to classes of abundance on the basis of a logarithmic scale of numbers of individuals per species: 1–2 individuals, 2–4 individuals, 4–8 individuals, 8–16 individuals, and so on. Preston called these classes **octaves** because each is twice as large as the preceding class. (In the musical scale, the vibration frequency of each note is twice that of the note one octave lower.) In a large sample of individuals, species often distribute themselves normally over the logarithmic abundance categories, as shown in Figure 22.10.

The normal distribution is described by the equation

$$n_R = n_0 e^{-(1/2)(R/\sigma)^2}$$

in which n_R represents the number of species whose abundance is R octaves greater or less than the modal abundance of species within the community, n_0

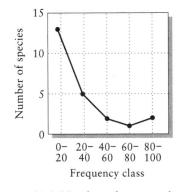

FIGURE 22.9 Number of species of plants in a peat bog near Kalamazoo, Michigan, in each of five frequency classes based on percentages of twenty-five 0.1-m² sampling areas occupied. From data in L. A. Kenoyer, *Ecology* 8:341–349 (1927).

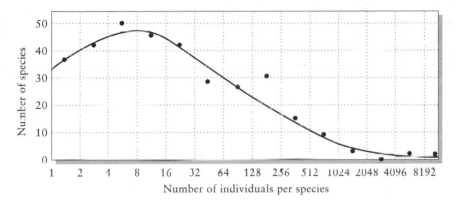

FIGURE 22.10 Relative abundances of species of moths attracted to light traps near Orono, Maine. Size classes (octaves), which increase by a factor of 2 from one class to the next, are on the horizontal axis; the number of species in each size class is plotted on the vertical axis. The distribution of abundances is hump-shaped with a mode of 48 species in the "4–8 individuals" size class. After F. W. Preston, *Ecology* 29:254–283 (1948).

represents the modal number of species (that is, the number in the most abundant class), and σ (the standard deviation) measures dispersion (the breadth of the normal curve). We can see these relationships in a graph of the lognormal distribution (Figure 22.11). The dispersion of the curve — whether it is narrow or broad — is proportional to the constant σ.

In theory at least, the entire lognormal distribution of species abundances in a community can never be fully sampled. Some species are too rare to be represented by one or more individuals in a sample of any size. These species fall below the **veil line** of the distribution, and their presence can be revealed only by increasing the total number of individuals examined. When the size of a sample doubles, the modal abundance of species moves one octave to the right (the abundance of all species doubles, on average), and additional species, each represented by one individual, appear in the distribution at the veil line.

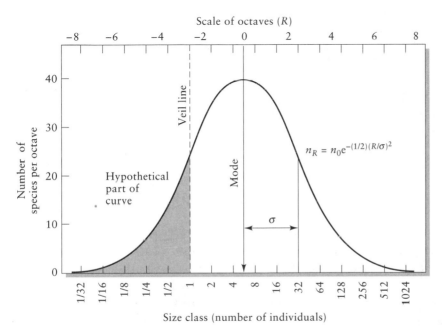

$$n_R = n_0 e^{-(1/2)(R/\sigma)^2}$$

FIGURE 22.11 Lognormal distribution of species abundances. The part of the curve to the left of the veil line (shaded), which corresponds to species with less than one individual in the sample, and thus not represented, is hypothetical. The scale of octaves (R) begins at the modal octave (0). One standard deviation (σ) on both sides of the mode includes about two-thirds of all the species in the sample. After F. W. Preston, *Ecology* 29:254–283 (1948).

A useful feature of Preston's lognormal curve is that it takes sample sizes into account. We can predict the total number of species (N) in a community, including those not represented in the sample, if we know only the number of species in the modal abundance class (n_0) and the dispersion of the lognormal distribution (σ). The appropriate equation is

$$N = n_0 \sqrt{2\pi\sigma^2} = 2.5\sigma n_0$$

For the sample of moths graphed in Figure 22.10, n_0 is 48 and σ is 3.4 octaves; therefore, $N = 2.5 \times 3.4 \times 48 = 408$ species. The actual sample of over 50,000 specimens contained only 349 species, 86 percent of the number that would theoretically be present in the sample area if abundances were distributed lognormally.

The dispersion (σ) of lognormal curves fitted to large samples of associations is remarkably similar for various groups of organisms. Preston obtained values of 2.3 for birds and 3.1 to 4.7 for moths; for diatoms, values ranged between 2.8 and 4.7. Censuses of forest birds revealed σ values of 0.98 for lowland tropical localities, 1.36 for temperate localities, and 1.97 for islands. These data indicate that greater discrepancies exist between abundances of species on islands than in temperate areas or, especially, in lowland tropical areas.

Diversity indices

Differences in the abundances of species in communities pose two practical problems for ecologists. First, the total number of species included varies with sample size, because as more individuals are sampled, the probability of encountering rare species increases. Thus we cannot compare diversity among areas sampled with different intensities merely by counting species. Second, not all species should contribute equally to our estimate of total diversity, because their functional roles in the community vary, to some degree, in proportion to their overall abundance.

Ecologists have tackled the second problem by formulating **diversity indices** in which the contribution of each species is weighted by its relative abundance. Two such indices are widely used in ecology: Simpson's index and the Shannon–Weaver index. In both cases, we calculate the indices from the proportions (p_i) of the species (i) in the total sample of individuals. Simpson's index is

$$D = \frac{1}{\Sigma p_i^2}$$

For any particular number of species in a sample (S), the value of D can vary from 1 to S, depending on the **evenness** of species abundances. When five species have equal abundance, each p_i is 0.20. Therefore, each $p_i^2 = 0.04$, the sum of the p_i^2s is 0.20, and the reciprocal of the sum is 5, the number of species in the sample. Similar calculations for some hypothetical communities are pre-

TABLE 22.1	*Comparison of diversity indices D, H, and e^H for hypothetical communities of five species having different relative abundances*

PROPORTION OF SAMPLE REPRESENTED BY SPECIES					DIVERSITY INDEX		
A	B	C	D	E	D	H	e^H
0.25	0.25	0.25	0.25	0.00	4.00	1.386	4.00
0.20	0.20	0.20	0.20	0.20	5.00	1.609	5.00
0.24	0.24	0.24	0.24	0.04	4.31	1.499	4.48
0.25	0.25	0.25	0.25	0.001	4.02	1.393	4.03
0.50	0.30	0.10	0.07	0.03	2.81	1.229	3.42

sented in Table 22.1, where it is apparent that rarer species contribute less to the value of the diversity index than do common species.

The Shannon–Weaver index, developed from information theory, is calculated by the equation

$$H = -\Sigma\, p_i \log_e p_i$$

and, like Simpson's index, gives less weight to rare species than to common ones. Because *H* is roughly proportional to the logarithm of the number of species, it is sometimes preferable to express the index as e^H, which is proportional to the number of species. Table 22.1 presents e^H, which we may compare directly to Simpson's index.

Sample size and species richness

A serious problem in estimating the number of species in an association arises from the fact, readily apparent from the lognormal distribution, that the number of species increases in direct proportion to the number of individuals sampled. If we wish to standardize measurements of diversity for comparison, we must base them on comparable samples. When samples include different numbers of individuals, comparability can be achieved via a statistical procedure known as **rarefaction,** in which equal-size subsamples of individuals are drawn at random from the total. The effect of rarefaction on number of species, which can be thought of as yielding the relationship between number of species and sample size, was portrayed by Howard Sanders of the Woods Hole Institute of Oceanography for samples of benthic marine organisms dredged from soft sediments in various localities. The total number of specimens in each sample varied because of the different densities of organisms in the substrate and unavoidable variation in sampling procedures. These samples re-

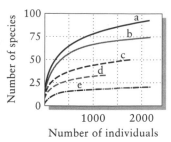

FIGURE 22.12 Number of species of bivalves and polychaete worms as a function of sample size for different marine environments. Tropical and deep-sea faunas tend to be more diverse than faunas in less constant environments. The habitats are a, tropical shallow water; b, slope (deep sea); c, outer continental shelf; d, tropical shallow water; e, boreal shallow water. After H. L. Sanders, *Brookhaven Symp. Biol.* 22:71–81 (1969).

vealed a general relationship between sample size and diversity, but it was not possible to tell whether that relationship appeared as an artifact of sampling or reflected consistent differences between localities, unless one rarefied the samples to make them comparable (Figure 22.12). As the computer randomly removed specimens from the samples, diversity in all the samples decreased. But the rarefaction curves clearly distinguished localities, showing that for comparable samples, diversity varied considerably.

Species and area

As a rule, more species occur within large areas than within small areas. The botanist Olaf Arrhenius first formalized this species–area relationship in 1921. Since then, common practice has portrayed the relationship between species (S) and area (A) with a power function of the form

$$S = cA^z$$

where c and z are constants fitted to data. Graphical portrayals of species–area relationships plot the logarithm of species number against the logarithm of area, as shown in Figure 22.13. After log-transformation, the species–area relationship becomes

$$\log S = \log c + z \log A$$

which is the equation for a straight line.

Analysis of species–area relationships among many groups of organisms revealed that most values of z fall within the range 0.20–0.35. The consistency of z suggested two possibilities. First, observed z values might be a simple statistical consequence of the lognormal distribution of species abundances. As we increase the area of a sample, we usually increase the number of individuals included within it, and as sample size increases, the veil line moves to the left, exposing more and more species. Empirical studies have demonstrated variation in z associated with differences in biological attributes of samples. For example, z-values obtained for continental areas of different size tend to be lower than those obtained for series of islands within a comparable size range.

A second possibility also exists. Where a flora or fauna is perfectly known (that is, all species have been sampled), the sampling properties of the lognormal distribution cannot be held accountable for any relationship between species and area; no species hide behind the veil line. For example, we possess near-perfect knowledge of the land-bird fauna of the West Indies, among which species increase with area with a slope of about $z = 0.24$ (Figure 22.14). Here, differences in diversity between large and small islands must signify differences in their intrinsic qualities. Likely candidates include habitat heterogeneity, which undoubtedly increases with the size (and resulting topographic heterogeneity) of the island, and size per se. Larger islands make better targets for potential immigrants from mainland sources of colonization. In addition, the larger populations on larger islands probably persist longer, being en-

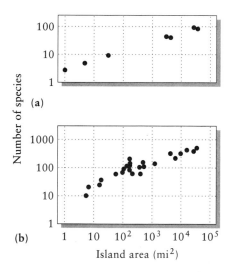

FIGURE 22.13 Species–area curves for (a) amphibians and reptiles in the West Indies and (b) birds in the Sunda Islands, Malaysia. From R. H. MacArthur and E. O. Wilson, *Evolution* 17:373–387 (1963), and *The Theory of Island Biogeography.* Princeton Univ. Press, Princeton (1967).

dowed with greater genetic diversity, broader distributions over area and habitat, and numbers large enough to prevent chance extinction.

❦ SUMMARY

1. The concept of community encompasses associations of interacting populations. Questions about communities address the regulation of species diversity, the relationship between community organization and stability, and the evolutionary origin of community properties.

2. Generally speaking, communities do not form discrete units separated by abrupt transitions in species composition. Species tend to distribute themselves over ecological gradients of conditions independently of the distributions of other species. Ecologists refer to this pattern as open-community structure.

3. Discontinuities between associations of plants and animals, called ecotones, sometimes occur at sharp physical boundaries or accompany change in what growth forms dominate the habitat. The aquatic-terrestrial transition provides an example of the first kind of ecotone, the prairie-forest transition an example of the second.

4. To analyze the distributions of species with respect to environmental conditions and one another, ecologists have devised various types of gradient analysis, in which they position sample localities with respect to gradients of physical conditions. The distributions of species along these environmental gradients emphasize the open structure of communities.

5. Within local areas, ecologists have characterized communities in terms of the number of species present, their relative abundances, and their feeding and other ecological relationships.

6. The lognormal distribution of species abundances has proved to be a useful empirical device, characterizing frequency distributions by a modal abundance class (n_0) and dispersion of abundances about the mode (σ). The term n_0 depends on the size of the sample, but the value of σ is an intrinsic property of the community. Species with predicted abundances of less than one individual—those beyond the veil line—are not detected unless the size of the sample is increased. The lognormal concept stresses the dependence of diversity estimates on sample size.

7. Various indices of diversity, most notably Simpson's index and the Shannon-Weaver (information) index, have been devised to enable us to take into account variations in abundance when making comparisons between samples. The number of species also increases as sample size increases, as predicted by the lognormal distribution of abundances. Accordingly, ecologists have devised rarefaction procedures and other statistical techniques to make samples of species (communities) comparable.

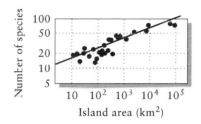

FIGURE 22.14 Species-area curve for land birds of the West Indies, including both the Greater and the Lesser Antilles. After R. E. Ricklefs and G. W. Cox, *Amer. Nat.* 106:195-219 (1972).

8. The number of species increases in direct proportion to the area sampled. This results in part from larger areas giving rise to larger total samples. But studies of well-known faunas and floras also indicate that larger areas are more heterogeneous ecologically, providing opportunities to sample more kinds of habitats, and that larger islands have more species because they are better targets for colonization and because larger populations better resist extinction.

❧ SUGGESTED READINGS

Connor, E. F., and E. D. McCoy. 1979. The statistics and biology of the species–area relationship. *American Naturalist* 113:791–833.

Davis, M. B. 1976. Pleistocene biogeography of temperate deciduous forests. *Geoscience and Man* 13:13–26.

Diamond, J., and T. J. Case (eds.). 1986. *Community Ecology.* Harper & Row, New York.

Gee, J. H. R., and P. S. Giller. 1987. *Organization of Communities. Past and Present.* Blackwell Scientific Publications, Oxford.

Gleason, H. A. 1926. The individualistic concept of the plant association. *Torrey Botanical Club Bulletin* 53:7–26.

MacArthur, R. H., and E. O. Wilson. 1967. *The Theory of Island Biogeography.* Princeton Univ. Press, Princeton.

Magurran, A. E. 1988. *Ecological Diversity and Its Measurement.* Princeton Univ. Press, Princeton.

Paine, R. T. 1980. Food webs: linkage, interaction strength and community infrastructure. *Journal of Animal Ecology* 49:667–685.

Pimm, S. L. 1982. *Food Webs.* Chapman & Hall, London and New York.

Sebens, K. P. 1985. The ecology of the rocky subtidal zone. *American Scientist* 73:548–557.

Terborgh, J. 1985. The role of ecotones in the distribution of Andean birds. *Ecology* 66:1237–1246.

Whittaker, R. H. 1953. A consideration of climax theory: the climax as a population and pattern. *Ecological Monographs* 23:41–78.

Whittaker, R. H. 1967. Gradient analysis of vegetation. *Biological Reviews* 42:207–264.

23

COMMUNITY DEVELOPMENT

Communities exist in a state of continuous flux. Organisms die and others are born to take their places; energy and nutrients pass through the community. Yet the appearance and composition of most communities do not change over time. Oaks replace oaks, squirrels replace squirrels, and so on, in continual self-perpetuation. But when a habitat is disturbed — a forest cleared, a prairie burned, a coral reef obliterated by a hurricane — the community slowly rebuilds. Pioneering species that are adapted to the disturbed habitat are successively replaced by others until the community attains its former structure and composition.

The sequence of changes initiated by disturbance is called **succession,** and the ultimate association of species achieved is called a **climax.** These terms describe natural processes that caught the attention of early ecologists, including Frederic Clements. By 1916 Clements had outlined the basic features of succession, supporting his conclusions with detailed studies of change in plant communities in a variety of environments. Since then, the study of community development has grown to include the processes that underlie successional change, the adaptations of organisms to the different conditions of early and late succession, and the interactions between colonists and the species that replace them. Ecologists have come to realize that succession and community perpetuation differently express the same processes and, further, that so-called climax communities consist of patchwork quilts of successional stages following upon localized disturbances.

Succession and the sere

The creation of any new habitat — a plowed field, a sand dune at the edge of a lake, an elephant's dung, a temporary pond left by a heavy rain — invites a host of species particularly adapted as good invaders. These first colonists are followed by others that are slower to take advantage of the new habitat but are eventually more successful than the pioneering species. In this way, the character of the community changes with time. Successional species themselves change the environment. For example, plants shade the earth's surface, contribute detritus to the soil, and alter soil moisture. These changes often inhibit the continued success of the species that cause them and make the environment more suitable for other species, which then exclude those responsible for the change.

The opportunity to observe succession presents itself conveniently in abandoned fields of various ages (Figure 23.1). On the piedmont of North Carolina, bare fields are quickly covered by a variety of annual plants. Within a few years, most of the annuals are replaced by herbaceous perennials and shrubs. The shrubs are followed by pines, which eventually crowd out the earlier successional species; pine forests are in turn invaded and then replaced by a variety of hardwood species that constitute the last stage of the successional sequence. Change comes rapidly at first. Crabgrass quickly enters an abandoned field, hardly allowing time for the plow's furrows to smooth over. Horseweed and ragweed dominate the field in the first summer after abandonment, aster in the second, and broomsedge in the third. The pace of succession falls off as slower growing plants appear: the transition to pine forest requires 25 years, but another century must pass before the developing hardwood forest begins to resemble the natural climax vegetation of the area.

The transition from abandoned field to mature forest is only one of several successional sequences leading to the same climax. In the eastern United States and Canada, forests are the end point of several different successional series, or **seres,** each of which has a different beginning. The sequence of species on newly formed sand dunes at the southern end of Lake Michigan differs from the sere that develops on abandoned fields a few miles away. The sand dunes are first invaded by marram and bluestem grasses. Plants of these species established in soils at the edge of a dune send out rhizomes (runners) under the surface of the sand, from which new shoots sprout. These grasses stabilize the dune surface and add organic detritus to the sand. Numerous annuals follow the perennial grasses onto the dunes, further enriching and stabilizing them and gradually creating conditions suitable for the establishment of shrub species. Sand cherry, dune willow, bearberry, and juniper form shrub layers be-

FIGURE 23.1 Stages of secondary succession in the oak–hornbeam forest in southern Poland. From (a) to (f), the time since clear-cutting progresses from immediately after clearing (a) to 7, 15, 30, 95, and 150 years. Photographs by Z. Glowacinski, courtesy of O. Jarvinen. From Z. Glowacinski and O. Jarvinen, *Ornis Scand.* 6:33–40 (1975).

fore pines become established. As in the abandoned fields in North Carolina, pines persist for only one or two generations, with little reseeding after initial establishment, giving way in the end to the forest of beech, oak, maple, and hemlock characteristic of the region.

Succession follows a similar course on Atlantic coastal dunes, where the beach grass that initially stabilizes the dune surface is followed by bayberry, beach plum, and other shrubs. Shrubs act like the snow fencing often used to

FIGURE 23.2 Initial stages of plant succession on sand dunes along the coast of Maryland. (a) Beach grass on the frontal side of a dune. This grass is used widely to stabilize dune surfaces. (b) Invasion of back dune areas by bayberry and beach plum. Courtesy of the U.S. Soil Conservation Service.

keep dunes from blowing out; they are called dunebuilders because they intercept blowing sand and cause it to pile up around their bases (Figure 23.2). Succession in estuaries leading to the establishment of terrestrial communities begins with salt-tolerating plants and progresses as sediments and detritus build the soil surface above the water line.

Primary succession

Beginning with Clements, ecologists have classified seres into two groups according to their origin. The establishment and development of plant communities in newly formed habitats that were previously without plants — sand dunes, lava flows, rock bared by erosion or exposed by a receding glacier — is called **primary succession**. The return of an area to its natural vegetation following a major disturbance is called **secondary succession**. The distinction between the two blurs because disturbances vary in the degree to which they destroy the fabric of the community and its physical support systems. A tornado that levels a large area of forest usually leaves intact the soil's bank of nutrients, seeds, and sproutable roots. In contrast, a severe fire may burn through the organic layers of the soil, destroying hundreds or thousands of years of biologically mediated development.

Species colonizing the thin deposits of clay left by receding glaciers in the Glacier Bay region of southern Alaska (Figure 23.3) must cope with deficien-

Figure 23.3 A valley exposed by a receding glacier, visible at top center, in North Tongass National Forest, Alaska. The recently bared rock surfaces at the bottom of the valley just below the glacier have not yet been recolonized by shrubby thickets. Courtesy of the U.S. Forest Service.

cies of nutrients, particularly nitrogen, and with stressful wind and cold. Here the sere begins with mat-forming mosses and sedges and then progresses through prostrate willows, shrubby willows, alder thicket, and sitka spruce to spruce–hemlock forest. Succession moves ahead rapidly, reaching the alder thicket stage within 10 to 20 years and achieving tall spruce forest within 100 years.

Succession provides a means by which dry land is reclaimed from certain aquatic habitats, such as the **bogs** that form in kettleholes or beaver ponds in cool north temperate and subarctic regions. Bog succession begins when rooted aquatic plants become established at the edge of the pond (Figure 23.4). Some species of sedges (rushlike plants) form mats on the water surface extending out from the shoreline. Occasionally these mats grow completely over the pond before it fills in with sediments, producing a more or less firm layer of vegetation over the water surface, a so-called quaking bog. The detritus produced by the sedge mat accumulates in layers of organic sediments on the bottom of the pond, where the stagnant water contains little or no oxygen to sustain microbial decomposition. Eventually these sediments become peat, which is used by humans as a soil conditioner and sometimes as a fuel for heating (Figure 23.5).

As the bog accumulates sediments and detritus, sphagnum moss and shrubs, such as Labrador tea and cranberry, become established along the edges, themselves adding to the development of a soil with progressively more terrestrial qualities. The shrubs are followed by black spruce and larch, which eventually give way to climax species of forest trees, including birch, maple, and fir, depending on the locality.

Disturbance and the sere

Breaks in the canopy of a forest tend to close over as surrounding individuals take advantage of new opportunities. A small gap, such as that left by a falling limb, is quickly filled by the growth of branches from surrounding trees. A big

FIGURE 23.4 Stages of bog succession illustrated by a bog formed behind a beaver dam in Algonquin Park, Ontario. The open water in the center is stagnant, poor in minerals, and low in oxygen. These conditions result in the accumulation of detritus from the vegetation at the edge and lead to a gradual filling-in of the bog, passing through stages dominated by shrubs and, later, black spruce.

Figure 23.5 A 1-m vertical section through a peat bed in a filled-in bog in Quebec, Canada. The layers represent the accumulation of organic detritus from plants that successively colonized the bog as it was filled in. The peat beds are probably several meters thick. Vegetation on the surface of the bog consists largely of sphagnum, blueberry, and Labrador tea.

gap left by a fallen tree may provide saplings in the understory with a chance to reach the canopy and claim a permanent place in the sun. A large area cleared by fire may have to be colonized anew by seed blown or carried in from the surrounding intact forest. Even when reseeding initiates the successional sequence, the size and type of disturbance influence which species become established first. Some require abundant sunlight for germination and establishment, and their seedlings are intolerant of competition from other species. These usually have strong powers of dispersal; they often have small seeds that are easily blown about and can reach the centers of large disturbances inaccessible to members of the climax community.

The influence of gap size on succession has been investigated in several marine habitats, where disturbance and recovery frequently follow upon each other. In southern Australia, Michael Keough investigated the colonization of artificially created patches, ranging in size from 25 to 2500 cm² (5 to 50 cm on a side), by various encrusting invertebrates. The major epifaunal taxa varied considerably in colonizing ability and competitive ability, which are generally inversely related (Table 23.1). When patches of different sizes were created within larger areas of uniform habitat, the exposed areas were quickly filled by growth of tunicates and sponges from the surrounding intact areas. In this case, patch size had little influence on community development, because the distance from the edge to the center of the patches (less than 25 cm) was easily spanned by growth. The many bryozoan and polychaete larvae that settled the patches were quickly overgrown.

Among isolated patches — hard substrates were placed in sand to mimic the shells of *Pinna* clams — size was found to be very important. Just by chance, a few of the small patches were colonized by tunicates and sponges, which produce relatively few propagules, thereby allowing bryozoans and polychaetes to obtain a foothold. Because they make bigger targets, many of the large patches were settled by a few larvae of tunicates and sponges, which then spread rapidly and eliminated other species that had colonized along with them. As a result, tunicates and sponges predominated the larger isolated

TABLE 23.1	*Summary of life history attributes of the major epifaunal (surface-growing) taxa at Edithburgh, southern Australia*			
Taxon	Form	Colonizing ability	Competitive ability	Capacity for vegetative growth
Tunicates	Colonial	Poor	Very good	Very extensive, up to 1 m²
Sponges	Colonial	Very poor	Good	Very extensive, up to 1 m²
Bryozoans	Colonial	Good	Poor	Poor, up to 50 cm²
Serpulid polychaetes	Solitary	Very good	Very poor	Very poor, up to 0.1 cm²

Source: M. J. Keough, *Ecology* 65:423–437 (1984).

patches, but bryozoans and polychaetes — which, once established, can deter the colonization of tunicate and sponge larvae — dominated many of the smaller patches. In this system, bryozoans and polychaetes are disturbance-adapted species — what botanists call **weeds.** They get into open patches quickly, mature and produce offspring at an early age, and then are often eliminated by more slowly colonizing but superior competitors. Such weedy species require frequent disturbances to stay in the system.

Predators and herbivores may be influenced by patch size and in turn influence the sere, either because the behavior of consumers is sensitive to patch size or because consumers require the cover of intact habitat, from whose edges they venture to feed in a newly exposed area. Rabbits rarely feed far from the cover of brush or trees, to avoid being discovered by predators far from safety. Limpets (grazing mollusks) similarly do not venture far from the safety of mussel beds to feed on algae. In an intertidal, rocky shore habitat in central California, Wayne Sousa cleared patches of either 625 cm² or 2500 cm² in mussel beds and excluded limpets from half the patches in each of these sets by applying a barrier of copper paint along their edges. He then monitored the colonization of the cleared patches by several species of algae during the following 3 years (Figure 23.6). Limpets live in the crevices between mussels when they are not feeding; doing so enables them to avoid predation. Because this behavior limits their foraging range, the densities of limpets in small patches (surrounded by much edge compared to area) exceeded those in larger patches. Also, and not surprisingly, through the course of the experiment algae grew more densely in the larger patches.

Where limpet grazing was prevented, total cover by all species of algae was high and did not differ between patches of different size. As we would expect, limpet grazing depressed the establishment and growth of most species of

FIGURE 23.6 A natural cleared patch in a bed of mussels (*Mytilus californianus*) on the central coast of California. The patch is about 1 m across and has been colonized by a heavy growth of the green alga *Ulva*. Note the distinct browse zone around the perimeter of the patch. It is created by limpets, which feed only short distances away from refuge in the mussel bed. Courtesy of W. P. Sousa. From W. P. Sousa, *Ecology* 65:1918–1935 (1984).

algae but favored three species of rare, presumably inferior, competitors: the brown alga *Analipus*, the green *Cladophora*, and the red *Endocladia*. All these algae have a low-lying, crustose growth form that makes them vulnerable to shading and overgrowth by other species. In addition, establishment of *Endocladia* was sensitive to patch size, generally being more common in larger patches regardless of limpet grazing. Where limpets grazed freely, colonizing mussels, which eventually crowd out all other space-occupying species, reached greater abundance in large patches than in small ones. This resulted from the interaction of patch size and limpet grazing, rather than from patch size per se, because mussels colonized areas protected by grazing independently of patch size.

The climax

Ecologists traditionally view succession as leading inexorably to an ultimate expression of community development, the climax community. Early studies of succession demonstrated that the many seres found within a region, each developing under a particular set of local environmental circumstances, progress toward the same climax. These observations led to the concept of the mature community as a natural unit — even as a closed system — which was clearly stated by Frederic Clements in 1916:

> *The developmental study of vegetation necessarily rests upon the assumption that the unit or climax formation is an organic entity. As an organism the formation arises, grows, matures, and dies. Its response to the habitat is shown in processes or functions and in structures which are the record as well as the result of these functions. Furthermore, each climax formation is able to reproduce itself, repeating with essential fidelity the*

stages of its development. The life history of a formation is a complex but definite process, comparable in its chief features with the life history of an individual plant.

(Carnegie Inst. Wash. Publ. *242:1 – 512*)

Clements recognized 14 climaxes in the terrestrial vegetation of North America, including 2 types of grassland (prairie and tundra), 3 types of scrub (sagebrush, desert scrub, and chaparral), and 9 types of forest ranging from pine – juniper woodland to beech – oak forest. He believed that climate alone determines the nature of the local climax. Aberrations in community composition caused by soils, topography, fire, or animals (especially grazing) represent interrupted stages in the transition toward the local climax — immature communities.

In recent years, the concept of the climax as an organism or a discrete unit has been greatly modified, to the point of outright rejection by many ecologists, because it has become clear that communities are open systems whose composition varies continuously over environmental gradients. Whereas in 1930 plant ecologists described the climax vegetation of much of Wisconsin as a sugar maple – basswood forest, by 1950 ecologists placed this forest type on an open continuum of climax communities extending over both broad, climatically defined regions and local, topographically defined areas. To the south, beech increased in prominence; to the north, birch, spruce, and hemlock were added to the climax community; in drier regions bordering prairies to the west, oaks became prominent. Locally, quaking aspen, black oak, and shagbark hickory, long recognized as successional species on moist, well-drained soils, came to be accepted as climax species on drier upland sites.

Mature stands of forest in Wisconsin, representing the end points of local seres, were ordered by J. T. Curtis and R. P. McIntosh along a continuum index ranging from dry sites dominated by oak and aspen to moist sites dominated by sugar maple, ironwood, and basswood. A continuum index for Wisconsin forests was calculated from the species composition of each forest type, and its value varied between arbitrary extremes of 300 for a pure stand of bur oak to 3000 for a pure stand of sugar maple. Although increasing values of the index correspond to seral stages leading to the sugar maple climax, they may also represent local climax communities determined by topographic or soil conditions. Thus the so-called climax vegetation of southern Wisconsin actually represents a continuum of forest (and, in some areas, prairie) types (Figure 23.7).

The causes of succession

Two factors determine the position of a species in a sere: the rate at which it invades a newly formed or disturbed habitat, and changes that occur in the environment over the course of succession. Some species disperse slowly, or grow slowly once established, and therefore become dominant late in the sequence of associations in a sere. Rapidly growing plants that produce many

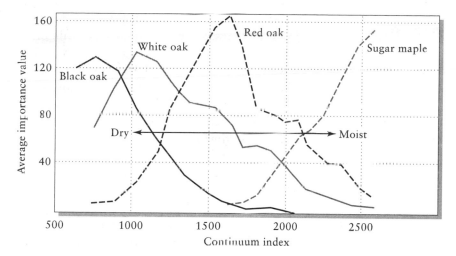

FIGURE 23.7 Relative importance of several species of trees in forest communities of southwestern Wisconsin arranged along a continuum index. Soil moisture, exchangeable calcium, and pH increase to the right on the continuum index. After J. T. Curtis and R. P. McIntosh, *Ecology* 32:476–496 (1951).

small seeds, carried long distances by the wind or by animals, have an initial advantage over species that disperse slowly. They dominate early stages of the sere. Where a habitat burns frequently, many species have fire-resistant seeds or root crowns that germinate or sprout soon after a fire and quickly reestablish their populations.

Early successional species sometimes modify the environment in such a way as to allow later stage species to become established. The growth of herbs on a cleared field shades the soil surface and helps the soil to retain moisture, providing conditions more congenial to the establishment of less tolerant plants. Conversely, colonization by some species may inhibit the entrance of others into the sere either by superior competition for limiting resources or by direct interference.

This diverse array of processes governing the course of succession was summarized by Joseph Connell and R. O. Slatyer under three classes of mechanisms—facilitation, inhibition, and tolerance—which relate the consequences of early stages of succession for the development of later stages. Facilitation, inhibition, and tolerance describe the effect of the presence of one species on the probability of establishment of a second, whether that effect is positive, negative, or neutral.

Facilitation embodies Clements's view of succession as a developmental sequence in which each stage paves the way for the next, just as structure follows structure during an organism's development—and during the building of a house. Colonizing plants enable climax species to invade, just as wooden forms are essential to the pouring of a concrete wall but have no place in the finished building. As we have noted, early stages facilitate the development of later stages by contributing to the nutrient and water levels of soil and by modifying the microenvironment of the soil surface. Alder trees (*Alnus*), which harbor nitrogen-fixing bacteria in their roots, provide an important source of nitrogen to soils developing on sand bars in rivers and in areas exposed by retreating glaciers. Black locust plays the same role in early succession in the southern Appalachian region of the United States.

Soils do not develop in marine systems, but facilitation often occurs when one species enhances the quality of settling and establishment sites for an-

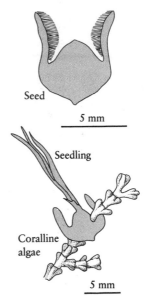

Seed

5 mm

Seedling

Coralline
algae

5 mm

FIGURE 23.8 Seeds of the surfgrass *Phyllospadix* have barbs that enable them to become attached to certain types of erect algae. After T. Turner, *Amer. Nat.* 121:729–738 (1983).

other. Working with experimental panels placed subtidally in Delaware Bay, T. A. Dean and L. E. Hurd found that for some species combinations, the presence of one inhibited the establishment of a second, but that hydroids enhanced the settlement of tunicates and both facilitated the settlement of mussels. In southern California, early-arriving, fast-growing algal stands provided dense protective cover for the reestablishment of kelp plants following their removal by winter storms. In areas kept clear of early successional species of algae, grazing fish quickly removed settling kelp sporophytes. The establishment of the surfgrass *Phyllospadix scouleri* in rocky intertidal communities depended on the presence of certain successional algae to which its seeds cling before germinating (Figure 23.8). In the absence of these algae, the seagrass could not invade the community.

Inhibition of one species by the presence of another is a common phenomenon that we have discussed in detail in the parts of this book dealing with competition and predation. One species may inhibit another by eating it, by reducing resources to a level the second species can barely subsist on, or by confronting it with noxious chemicals or antisocial behavior. With respect to succession, climax species by definition inhibit species characteristic of earlier stages: the latter cannot invade the climax community except following disturbance.

Because inhibition is so intimately related to species replacement, it forms an integral part of the orderly succession from early stages of the sere through the climax. Inhibition can give rise to an interesting situation when the outcome of an interaction between two species depends on which becomes established first. Colonizing propagules are often the most sensitive stage of the life history, and sometimes neither species of a pair can become established in the presence of the competitively superior adults of the other. In this case, the course of succession depends on precedence. Precedence, in turn, may be strictly random, depending on which species reaches a disturbed site first, or it may depend on certain properties of the disturbed site—its size, location, season, and so on. We have seen such a case in subtidal habitats of southern Australia, where bryozoans, when they become established first, can prevent the establishment of tunicates and sponges. Because of their stronger powers of dispersal, this is more likely to happen on small, isolated substrates than elsewhere.

According to Connell and Slatyer's inhibition model, succession follows upon the establishment of one species or another only through the death and replacement of established individuals. Thus, just by chance, successional change moves toward predominance of the longer lived species.

The **tolerance** model of Connell and Slatyer holds that "succession leads to a community composed of those species most efficient in exploiting resources, presumably each specialized on different kinds or proportions of resources." According to this model, species can invade newly exposed habitat and become established with equal ease. The ensuing sere is then determined by the life spans and competitive abilities of the colonists. Early stages will be dominated by poor competitors that have short life cycles but become established quickly; superior competitors will constitute climax species, but they may grow more slowly and may not express their dominance in the sere until others have grown up and reproduced.

Old-field succession on the piedmont of North Carolina

Clearly, all three of Connell and Slatyer's mechanisms — facilitation of establishment, inhibition of establishment, and competitive exclusion (replacement of established populations) — together with the life history characteristics of successional species, are important factors in every sere; none operates exclusively of the others. Early stages of plant succession on old fields in the piedmont region of North Carolina (Figure 23.9) demonstrate how these factors combine in a particular sere. The first 3 to 4 years of old-field succession are dominated by a small number of species that replace one another in rapid sequence: crabgrass, horseweed, ragweed, aster, and broomsedge. The life history cycle of each species partly determines its place in the succession (Figure 23.10). Crabgrass, a rapidly growing annual, is usually the most conspicuous plant in a cleared field during the year in which the field is abandoned. Horseweed is a winter annual whose seeds germinate in the fall. Through the winter, the plant exists as a small rosette of leaves; it blooms by the following midsummer. Because horseweed disperses well and develops rapidly, it usually dominates 1-year-old fields. But because seedings require full sunlight, horseweed is quickly replaced by shade-tolerant species.

Ragweed is a summer annual; seeds germinate early in the spring, and the plants flower by late summer. Ragweed dominates the first summer of succession in fields that are plowed under in the late fall, after horseweed normally germinates. Aster and broomsedge are biennials that germinate in the spring and early summer, exist through the winter as small plants, and bloom for the first time in their second autumn. Broomsedge persists and flowers during the following autumn as well.

Horseweed and ragweed both disperse their seeds efficiently and, as young plants, tolerate desiccation. These abilities enable them to invade cleared fields rapidly and produce seed before competitors become established. De-

FIGURE 23.9 An old field on the piedmont of North Carolina. Such habitats develop after the abandonment of agricultural land.

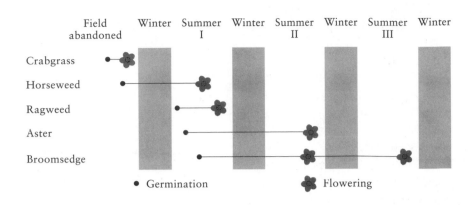

FIGURE 23.10 Schematic summary of the life histories of five early successional species of plants that colonize abandoned fields in North Carolina.

caying horseweed roots stunt the growth of horseweed seedlings; this self-inhibiting effect, whose function and origin are not understood, cuts short the life of horseweed in the sere. Such growth inhibitors presumably are the by-products of other adaptations that increase the fitness of horseweed during the first year of succession. Or perhaps, because horseweed plants have little chance of persisting during the second year as a result of invasion of the sere by superior competitors, self-inhibition has little negative selection value. At any rate, the phenomenon is fairly common in early stages of succession.

Aster successfully colonizes recently cleared fields, but it grows slowly and does not dominate the habitat until the second year. The first aster plants to colonize a field thrive in the full sunlight; the seedlings, however, are not shade-tolerant, and adult plants shade their progeny out of existence. Furthermore, asters do not compete effectively with broomsedge for soil moisture. Catherine Keever observed this when she cleared a circular area, 1 m in radius, around several broomsedge plants and planted aster seedlings at various distances (Figure 23.11).

Approaching the climax

Succession continues until the addition of new species to the sere and the exclusion of established species no longer change the environment of the developing community. The progression of different growth forms modifies conditions of light, temperature, moisture, and (for primary seres) soil nutrients. The replacement of grasses by shrubs and then by trees on abandoned fields brings a corresponding modification of the physical environment. Conditions change more slowly, however, when the vegetation reaches the tallest growth form that the environment can support. The final biomass dimensions of the climax community are limited by climate independently of events during succession.

Once forest vegetation establishes itself, patterns of light intensity and soil moisture do not change, except in the smallest details, with the introduction of new species of trees. For example, beech and maple replace oak and hickory in northern hardwood forests because their seedlings are better competitors in the shade of the forest floor environment, but beech and maple seedlings probably develop as well under their own parents as they do under the oak and hickory trees they replace. At this point, succession reaches a climax; the community has come into equilibrium with its physical environment.

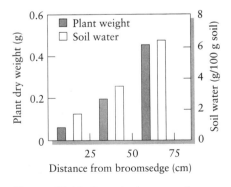

FIGURE 23.11 Growth response of asters (dry weight) and soil water as a function of distance from broomsedge plants in an old field. From C. Keever, *Ecol. Monogr.* 20:230–250 (1950).

To be sure, subtle changes in species composition usually follow the attainment of the climax growth form of a sere. For example, a site near Washington, D.C., that was left undisturbed for nearly 70 years developed a tall forest community dominated by oak and beech. The community had not reached an equilibrium at the time it was studied, because the youngest individuals — the saplings in the forest understory that eventually replace the existing trees — included neither white nor black oak. In another century, this forest will probably be dominated by species with the most vigorous reproduction: red maple, sugar maple, and beech (Figure 23.12).

The time required for succession to proceed from a cleared habitat to a climax community varies with the nature of the climax and the initial quality of the soil. Obviously, succession is slower to gain momentum when starting on bare rock than when starting on a recently cleared field. A mature oak–hickory forest climax will develop within 150 years on cleared fields in North Carolina. Climax stages of western grasslands are reached in 20 to 40 years of secondary succession. Radiocarbon dating methods suggest that primary succession to a beech–maple climax forest on Michigan sand dunes requires up to 1000 years. In the humid tropics, forest communities regain most of their climax elements within 100 years after clear-cutting, provided that the soil is not abused by farming or prolonged exposure to sun and rain. But the development of a truly mature tropical forest devoid of any remnants of successional species requires many centuries.

The character of the climax

Clements's idea that a region has only one true climax (the monoclimax theory) forced botanists to recognize a hierarchy of interrupted or modified seres by attaching such names as subclimax, preclimax, and postclimax. This terminology naturally gave way before the polyclimax viewpoint, which recognized the validity of many different types of vegetation as climaxes, depending on the habitat. More recently, the development of the continuum index and gradient analysis fostered the broader **pattern-climax theory** of Robert Whittaker, which recognizes a regional pattern of open climax communities whose composition at any one locality depends on the particular environmental conditions at that point.

Many factors determine the climax community, among them soil nutrients, moisture, slope, and exposure. Fire is an important feature of many climax communities, favoring fire-resistant species and excluding species that otherwise would dominate. The vast southern pine forests in the Gulf Coast and southern Atlantic Coast states are maintained by periodic fires. The pines have become adapted to withstand scorching under conditions that destroy oaks and other broad-leaved species (Figure 23.13). Some species of pines do not even shed their seeds unless triggered by the heat of a fire passing through the understory below. After a fire, pine seedlings grow rapidly in the absence of competition from other understory species.

Any habitat that is occasionally dry enough to create a fire hazard but is normally wet enough to produce and accumulate a thick layer of plant detritus is likely to be subject to the influence of fire. Chaparral vegetation in season-

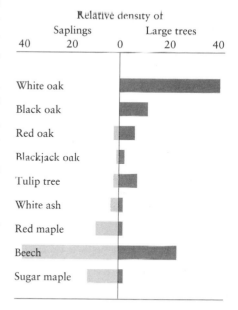

FIGURE 23.12 Composition of a forest undisturbed for 67 years near Washington, D.C. The relative predominance of beech and maple saplings in the understory foretells a gradual successional change in the community beyond the present oak–beech stage. After R. L. Dix, *Ecology* 30:663–665 (1957).

FIGURE 23.13 (a) A stand of longleaf pine in North Carolina shortly after a fire. Although the seedlings are badly burned (b), the growing shoot is protected by the dense, long needles (shown on an unburned individual, c) and often survives. In addition, the slow-growing seedlings have extensive roots that store nutrients that support the plant following fire damage.

ally dry habitats in California is a fire-maintained climax that gives way to oak woodland in many areas when fire is prevented. The forest–prairie edge in the midwestern United States separates "climatic climax" and "fire climax" communities. Frequent burning eliminates seedlings of hardwood trees, but the perennial grasses sprout from their roots after a fire. The forest–prairie edge occasionally shifts back and forth across the countryside, depending on the intensity of recent drought and on the extent of recent fires. After prolonged wet periods, the forest edge advances out onto the prairie as tree seedlings grow up and begin to shade out the grasses. Prolonged drought followed by intense fire can destroy tall forest and permit rapidly spreading prairie grasses to gain a foothold. Once prairie vegetation establishes itself, fires become more frequent because of the rapid buildup of flammable litter. Reinvasion by forest species then becomes more difficult. By the same token, mature forests resist fire and are rarely damaged enough to allow the encroachment of prairie grasses. Hence the stability of the forest–prairie boundary.

Grazing pressure also can modify the climax. Grassland can be turned into shrubland by intense grazing. Herbivores kill or severely damage perennial grasses and allow shrubs and cacti that are unsuitable for forage to establish themselves. Most herbivores graze selectively, suppressing favored species of plants and bolstering competitors that are less desirable as food. On the African plains, grazing ungulates follow a regular succession of species through an area, each using different types of forage. When wildebeest, the first of the successional species, were excluded from large fenced-off areas, the subsequent wave of Thompson's gazelles preferred to feed in areas previously used by wildebeest or other large herbivores (Figure 23.14). Apparently, heavy grazing by wildebeest stimulates growth of the food plants that gazelles prefer and reduces cover within which predators of the smaller gazelles could conceal themselves.

Transient and cyclic climaxes

We view succession as a series of changes leading to a climax, whose character is determined by, and that exists in equilibrium with, the local environment. Once established, the beech–maple forest perpetuates itself, and its general appearance does not change despite the constant replacement of individuals within the community. Yet not all climaxes persist. Simple cases of **transient climaxes** include the development of animal and plant communities in seasonal ponds — small bodies of water that either dry up in the summer or freeze solid in the winter and thereby regularly destroy the communities that become established each year during the growing season. Each spring the ponds are restocked either from larger, permanent bodies of water or from resting stages

FIGURE 23.14 Zebras and Thompson's gazelles feed side by side in the Serengeti ecosystem of East Africa but eat different plants.

that were left by plants, animals, and microorganisms before the habitat disappeared the previous year.

Succession recurs whenever a new environmental opportunity appears. For example, excreta and dead organisms provide resources for a variety of scavengers and detritus feeders. On African savannas, carcasses of large mammals are devoured by a succession of vultures (Figure 23.15), beginning with large, aggressive species that engorge the largest masses of flesh, followed by smaller species that glean smaller bits of meat from the bones, and ending with a kind of vulture that cracks open bones to feed on the marrow. Scavenging mammals, maggots, and microorganisms enter the sere at different points and ensure that nothing edible remains. This succession has no climax because all the scavengers disperse when the feast concludes. We may, however, consider all the scavengers a part of a climax: the entire savanna community.

In simple communities, particular life history characteristics in a few dominant species can create a **cyclic climax**. Suppose, for example, that species A can germinate only under species B, B only under C, and C only under A. This situation creates a regular cycle of species dominance in the order A, C, B, A, C, B, A, . . . , the length of each stage being determined by the life span of the dominant species.

Stable cyclic climaxes usually follow the scheme just outlined, often with one of the stages being bare substrate. Wind or frost heaving sometimes drives the cycle. When heaths and other vegetation forms suffer extreme wind damage, shredded foliage and broken twigs create an opening for further damage, and the process becomes self-accelerating. Soon a wide swath is opened in the vegetation; regeneration occurs on the protected side of the damaged area while wind damage further encroaches upon the exposed vegetation. As a result, waves of damage and regeneration move through the community in the direction of the wind (Figure 23.16). If we watched the sequence of events at any one point, we would witness a healthy heath being reduced to bare earth by wind damage and then regenerating in repeated cycles (Figure 23.17). Simi-

FIGURE 23.15 Vultures feeding on a wildebeest carcass in Masai Mara Park, Kenya.

lar cycles occur where hummocks of earth form in windy regions around the bases of clumps of grasses. As the hummocks grow, the soil becomes more exposed and better drained. With these changes in soil quality, shrubby lichens take over the hummock and exclude the grasses around which the hummock formed. Shrubby lichens wear down with wind erosion, eventually giving way to prostrate lichens, which resist wind erosion but, lacking roots, cannot hold the soil. Eventually the hummocks wear away completely, and grasses once more become established and renew the cycle.

Mosaic patterns of vegetation types typify any climax community where the death of individuals alters the environment. Tree falls open the forest canopy and create patches of habitat that are dry, hot, and sunlit compared to the forest floor under unbroken canopy. These openings are often invaded by early seral forms, which persist until the canopy closes. Thus tree falls create a mosaic of successional stages within an otherwise uniform community. Indeed, adaptation by some species to growing in particular conditions created by different-sized openings in the canopy could enhance the overall diversity of the climax community. Similar ideas have developed about intertidal regions of rocky coasts, where wave damage and intense predation continuously open new patches of habitat.

The concept of the community climax must include cyclic patterns of changes and mosaic patterns of distribution. The climax is a dynamic state,

FIGURE 23.16 Waves of regeneration and wind damage in balsam fir forests on the slopes of Mt. Katahdin, Maine. Courtesy of D. G. Sprugel. From D. G. Sprugel and F. H. Bormann, *Science* 211:390–393 (1981).

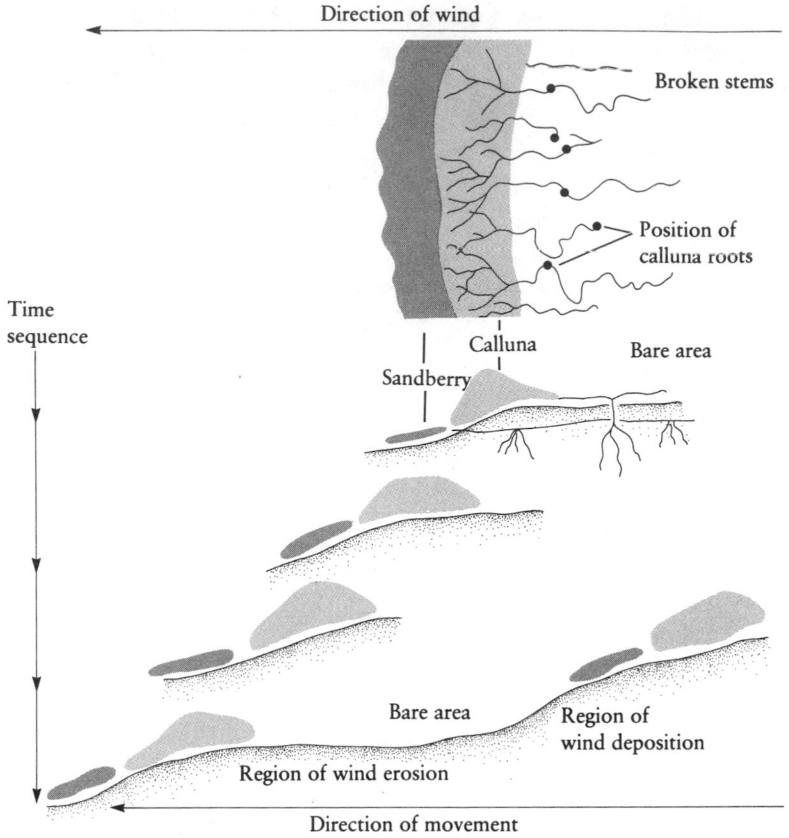

FIGURE 23.17 Sequence of wind damage and regeneration in the dwarf heaths of northern Scotland. At top is a view of the heath from above, showing a band of sandberry growing in the protected lee of the calluna heath and the wind damage to the heath on its upwind side (to the right). Below is a side view of the band of sandberry and heath as it appears to "migrate" downwind over time. After A. S. Watt, *J. Ecol.* 35:1–22 (1947).

self-perpetuating in composition, even if by regular cycles of change. Persistence is the key to the climax, and a persistent *cycle* defines a climax as well as an unchanging steady state does.

Species characteristics through the sere

Succession in terrestrial habitats entails a regular progression of plant forms. Plants characteristic of early stages of succession and those typical of late stages employ different strategies of growth and reproduction. Early-stage species capitalize on their high dispersal ability to rapidly colonize newly created or disturbed habitats. Climax species disperse and grow more slowly, but shade tolerance as seedlings and large size as mature plants give them a competitive edge over early successional species. Plants of climax communities are adapted to grow and prosper in the environment they create, whereas early successional species are adapted to colonize unexploited environments.

Some characteristics of early and late successional plants are compared in Table 23.2. To enhance their colonizing ability, early seral species produce many small seeds that are usually wind-dispersed (dandelion and milkweed, for example). Their seeds can remain dormant in soils of forest and shrub habitats for years until fires or tree falls create the bare-soil conditions required for germination and growth. The seeds of most climax species, being relatively

TABLE 23.2	General characteristics of plants during early and late stages of succession	
Characteristic	Early stage	Late stage
Number of seeds	Many	Few
Seed size	Small	Large
Dispersal	Wind, stuck to animals	Gravity, eaten by animals
Seed viability	Long, latent in soil	Short
Root:shoot ratio	Low	High
Growth rate	Rapid	Slow
Mature size	Small	Large
Shade tolerance	Low	High

large, provide their seedlings with the ample nutrients they need to get started in the highly competitive environment of the forest floor.

The survival of seedlings in shade is directly related to seed weight (Figure 23.18). The ability of seedlings to survive the shade conditions of climax habitats is inversely related to their growth rate in the direct sunlight of early successional habitats. When placed in full sunlight, early successional herbaceous species grew 10 times more rapidly than shade-tolerant trees. Shade-intolerant trees, such as birch and red maple, had intermediate growth rates. Thus plants must balance shade tolerance and growth rate against each other; each species must reach a compromise between those adaptations that best suit it for survival in the sere.

The rapid growth of early successional species results in part from the relatively large proportion of seedling biomass that is allocated to stems and leaves. Leaves sustain photosynthesis, and their productivity determines the net accumulation of plant tissue during growth. Hence the allocation of tissue to the root and the aboveground parts (shoot) influences growth rate. In the seedlings of annual herbaceous plants, the shoot typically accounts for 80 to 90 percent of the entire plant; in biennials, 70 to 80 percent; in herbaceous perennials, 60 to 70 percent; and in woody perennials, 20 to 60 percent.

The allocation of a large proportion of production to shoot biomass in early successional plants leads to rapid growth and the production of large crops of seeds. Because annual plants must produce seeds quickly and copiously, they never attain large size. Climax species allocate a larger proportion of their production to root and stem tissue to increase their competitive ability; thus they grow more slowly. The progression of successional species is therefore accompanied by a shift in the balance between adaptations promoting dispersal and adaptations enhancing competitive ability.

The biological properties of a developing community change as species enter and leave the sere. As a community matures, the ratio of biomass to productivity increases. The maintenance requirements of the community also in-

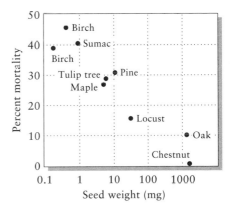

FIGURE 23.18 Relationship between seed weight and mortality of seedlings after 3 months under shaded conditions. After J. P. Grime and D. W. Jeffrey, *J. Ecol.* **53**:621–642 (1965).

crease until production can no longer meet the demand, at which point the net accumulation of biomass in the community stops. The end of biomass accumulation does not necessarily signal the attainment of climax; species may continue to invade the community and replace others whether the biomass of the community increases or not. The attainment of a steady-state biomass does mark the end of major structural change in the community, but further changes are limited to the adjustment of details.

As plant size increases with succession, a greater proportion of the nutrients available to the community come to reside in organic materials. Furthermore, because the vegetation of mature communities has more supportive tissue, which is less readily digestible than photosynthetic tissue, a larger proportion of their productivity enters the detritus food chain rather than the consumer food chain. Other aspects of the community change as well. Forest soils hold nutrients more tightly because tree roots protect soils from erosion. The well-developed root systems of trees take up minerals more rapidly and store them to a greater degree than do the root systems of early successional plants. The forest canopy protects the environment near the ground from extremes of heat and humidity. Conditions in the litter are more favorable to detritus-feeding organisms.

Ecologists generally agree that communities become more diverse and complex as succession progresses, although in some seres, intermediate stages of succession may be more diverse because they contain elements of early seral stages as well as elements of the climax community. We do not know whether this increase in the diversity of a community during its early stages of succession is related to increased production, greater constancy of physical characteristics of the environment, or greater structural heterogeneity of the habitat. Furthermore, we have no reason to suspect that gradients of diversity along a successional continuum respond to the same factors that determine diversity along a structurally analogous gradient of mature communities.

Succession emphasizes the dynamic nature of biological communities. By upsetting their natural balance, disturbance reveals to us forces that determine the presence or absence of species within a community and processes responsible for the regulation of community structure. Succession also emphasizes that the structure of the community comprises a patchwork mosaic of successional stages and that community studies must consider disturbance cycles on many scales of time and space.

❦ *SUMMARY*

1. Succession describes community change at a locality following either habitat disturbance or the colonization of newly exposed substrate. The particular sequence of communities at a given point is referred to as a sere, and the ultimate stable association of plants and animals that is achieved is called the climax.

2. Succession on newly formed substrates, such as sand dunes, landslides, and lava flows — referred to as primary succession — involves substantial modification of the environment by early

colonists. Moderate disturbances, which leave much of the physical structure of the ecosystem intact, are followed by secondary succession.

3. The initial stages of the sere depend on the intensity and extent of the disturbance, but its end point reflects climate and topography. That is, within a region, seres tend to converge on a single climax.

4. Especially for secondary succession, the entrance and persistence of a species in a sere depend on its colonizing and competitive abilities. Members of early stages tend to disperse well and grow rapidly; those of later stages tend to tolerate low resource levels or to dominate direct interactions with other species.

5. Joseph Connell and R. O. Slatyer categorized the processes that govern succession as facilitation, inhibition, and tolerance. All three involve the effect of one established species on the probability of colonization by a second, potential invader. Facilitation predominates in the early stages of primary succession. Inhibition is a more common feature of secondary succession. It may be expressed in precedence effects, conferring competitive dominance on, for example, the first arrival.

6. Succession continues until the community is dominated by species capable of becoming established in their own and one another's presence. At this point the community becomes self-perpetuating.

7. The character of the climax may be influenced profoundly by local conditions, such as fire and grazing, that alter interactions among seral species.

8. Transient climaxes develop on ephemeral resources and habitats, such as vernal pools and the carcasses of individual animals. In such cases, we may think of the regional climax as including transient seres.

9. Cyclic local climaxes may develop where each species can become established only in association with some other one. Cyclic climaxes are often driven by harsh physical conditions, such as frost and strong winds. The regional climax includes any local cyclic seres.

10. The characteristics of species vary according to their place in the sere, so the overall structure and function of the community change accordingly. In general, biomass increases with time, whereas net production and diversity tend to be greatest in middle stages.

❦ SUGGESTED READINGS

Christensen, N. L., and R. K. Peet. 1984. Convergence during secondary forest succession. *Journal of Ecology* 72:25–36.

Connell, J. H., and R. O. Slatyer. 1977. Mechanisms of succession in natural communities and their role in community stability and organization. *American Naturalist* 111:1119–1144.

Grubb, P. J. 1977. The maintenance of species diversity in plant communities: the importance of the regeneration niche. *Biological Reviews* 52:107–145.

Keever, C. 1950. Causes of succession on old fields of the piedmont, North Carolina. *Ecological Monographs* 20:230–250.

Keough, M. J. 1984. Effects of patch size on the abundance of sessile marine invertebrates. *Ecology* 65:423–437.

McIntosh, R. P. 1985. *The Background of Ecology. Concept and Theory.* Cambridge Univ. Press, Cambridge and New York.

Parrish, J. A. D., and F. A. Bazzaz. 1976. Underground niche separation in successional plants. *Ecology* 57:1281–1288.

Parrish, J. A. D., and F. A. Bazzaz. 1982. Niche responses of early and late successional tree seedlings on three resource gradients. *Bulletin of the Torrey Botanical Club* 109:451–456.

Pickett, S. T. A., and P. S. White (eds.). 1985. *The Ecology of Natural Disturbance and Patch Dynamics.* Academic Press, Orlando, Fla.

Riggan, P. J., S. Goode, P. M. Jacks, and R. N. Lockwood. 1988. Interaction of fire and community development in chaparral of southern California. *Ecological Monographs* 58:155–176.

Sousa, W. P. 1984. Intertidal mosaics: patch size, propagule availability, and spatially variable patterns of succession. *Ecology* 65:1918–1935.

Watt, A. S. 1947. Pattern and process in the plant community. *Journal of Ecology* 35:1–22.

West, D. C., H. H. Shugart, and D. B. Botkin (eds.). 1981. *Forest Succession. Concepts and Application.* Springer-Verlag, New York.

24

Biodiversity

Comparisons of communities of plants and animals have revealed certain patterns that suggest the regulation of community properties. One example of such a pattern is the trophic organization of the community, for which the laws of thermodynamics dictate that energy flux decreases at each higher level in the food chain. This organizing principle produces certain regularities in the distribution of numbers of individuals and biomass among trophic levels within the community.

Ecologists have also noted patterns in communities that appear to be indifferent to energetic constraints. The most important of these include certain regularities in the numbers of species within communities (species diversity). As we noted earlier, larger islands tend to support more species than smaller islands, suggesting that diversity is somehow regulated with respect to area or to some ecological factor correlated with area. To cite another example, biologists have found more kinds of organisms in the tropics than at higher latitudes.

The great naturalist explorers of the last century — Charles Darwin, H. W. Bates, Alfred Russel Wallace, and others — recognized that the tropics held a great store of undescribed species, many having bizarre forms and habits. This remains true to the present. Taxonomists have so far catalogued fewer than 2 million species. But by extrapolating the rate of discovery of new insects and other life forms, some biologists have estimated that as many as 30 million kinds of animals and plants may inhabit the earth.

Why are there so many species of organisms in the tropics (and why are there so few toward the poles)? These questions and, more generally, the factors that regulate the diversity of natural communities are the subject of this chapter. Biologists hold two views. One maintains that diversity increases without limit over time; tropical habitats, being much older than temperate and arctic habitats, have had time to accumulate more species. The second view holds that diversity reaches an equilibrium at which those factors that remove species from a system balance those that add species. Accordingly, factors that add species weigh more heavily in the balance as one moves toward the tropics.

Throughout the first half of the twentieth century, the first viewpoint enjoyed the broadest favor. Tropical habitats were thought to have persisted since the beginning of life, whereas the vicissitudes of climate (particularly during the last ice age) occasionally destroyed most temperate and arctic habitats, resetting the diversity clock, so to speak. More recently, however, with the integration of population ecology into community theory, ecologists have come to consider diversity as an equilibrium between opposing diversity-dependent processes, just as equilibrium population size represents the balance between opposing density-dependent birth and death processes. This viewpoint challenges ecologists to identify the processes responsible for adding species to, and removing species from, communities and to understand why the balance between these processes differs systematically from place to place.

Geographic patterns in species diversity

Within most large taxonomic groups of organisms — plant, animal, and perhaps microbial — the number of species increases markedly, with a few exceptions, toward the equator (Figure 24.1). For example, within a small region at 60° north latitude, we might find 10 species of ants; at 40°, between 50 and 100 species; and in a similar sampling area within 20° of the equator, between 100 and 200 species. Greenland is home to 56 species of breeding birds, New York to 105, Guatemala to 469, and Colombia to 1395. Diversity in marine environments follows a similar trend: arctic waters harbor 100 species of tunicates, but more than 400 species are known from temperate regions and more than 600 from tropical seas.

Within a given belt of latitude around the globe, the number of species may vary widely among habitats according to productivity, degree of structural heterogeneity, and suitability of physical conditions. For example, censuses of birds in small areas (usually 5 to 20 hectares) of relatively uniform habitat reveal about 6 species of breeding birds in grasslands, 14 in shrublands, and 24 in floodplain deciduous forests (Table 24.1). Habitat structure apparently over-

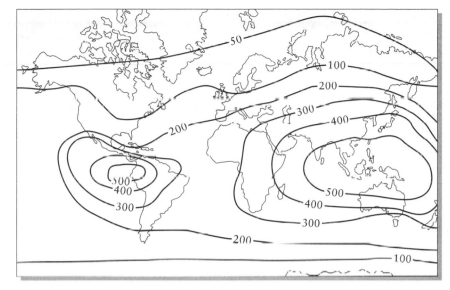

(a)

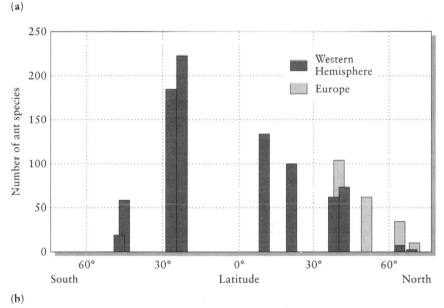

(b)

FIGURE 24.1 Two examples of global patterns of diversity. (a) The contour lines on the map indicate the number of species of near-shore and continental-shelf bivalves (clams, mussels, scallops, and their relatives) that may be found at locations within the contour intervals. The maximum diversity occurs in the tropics, particularly within Australasia and the eastern Pacific Ocean. From F. G. Stehli, A. L. McAlester, and C. E. Helsley, *Geol. Soc. Amer. Bull.* 78:455 (1967). (b) The number of species of ants within small sampling areas as a function of latitude. European localities support greater diversity than is found at similar latitudes in North America, because of their generally warmer temperatures. After data in N. Kusnezov, *Evolution* 11:298 (1957).

rides productivity in determining species diversity. Marshes are productive but structurally uniform, and they have relatively few species. Desert vegetation is less productive, but its greater variety of structure apparently makes room for more kinds of inhabitants (Figure 24.2). The relationship between structure and diversity is apparent to bird watchers and other naturalists, but Robert MacArthur and John MacArthur were the first to set it forth in a plain, graphical form by plotting the diversity of birds observed in different habitats according to diversity in foliage height, a measure of the structural complexity of vegetation (Figure 24.3). Web-building spiders exhibit a similar trend: the diversity of species varies in direct relation to heterogeneity in the heights of the tips of vegetation to which such spiders attach their webs. Lizard species diversity closely parallels the total volume of vegetation per unit area in desert habitats of the southwestern United States.

TABLE 24.1	*Plant productivity and the number of species of birds in representative temperate-zone habitats*	
Habitat	Approximate productivity (g m^{-2} yr^{-1})	Average number of bird species
Marsh	2000	6
Grassland	500	6
Shrubland	600	14
Desert	70	14
Coniferous forest	800	17
Upland deciduous forest	1000	21
Floodplain deciduous forest	2000	24

Source: E. J. Tramer, *Ecology* 50:927–929; productivity data from R. H. Whittaker, *Communities and Ecosystems* (2nd ed.). Macmillan, New York (1975).

FIGURE 24.2 Desert and marsh vegetation illustrate extremes of production with an inverse relationship between production and diversity. The desert is the Sonoran Desert of Baja California; the marsh is the Malheur Refuge, Oregon. Courtesy of U. S. Department of Interior.

On a regional basis, the number of species varies according to the suitability of physical conditions, the heterogeneity of habitats, and isolation from centers of dispersal. In North America, the number of species in most groups of animals and plants increases from north to south, but the influence of geographic heterogeneity and the isolation of peninsulas is apparent. Within the area of North America to the Isthmus of Panama, the number of mammal spe-

cies occurring in square blocks of 150 miles on a side increases from 15 in northern Canada to more than 150 in Central America (Figure 24.4). Across the same latitude in the middle of the United States, more species of mammals live in the topographically heterogeneous western mountains (90 to 120 species per block) than in the more uniform environments of the east (50 to 75 species per block). The number of species of breeding land birds follows a similar pattern, but reptile and amphibian faunas do not. Reptiles are more diverse in the eastern half of the United States than in the mountainous western regions; amphibians are strikingly underrepresented in the deserts of the southwest, because most species require abundant water.

Diversity and niche relationships

Ecologists use the term **niche** to express the relationship of the individual or population to all aspects of its environment—and hence the ecological role of the species within the community. The niche describes the range of conditions and resource qualities within which the individual or species functions. Thus the boundaries of a particular species's niche might extend between temperatures of 10 and 30°C, prey sizes of 4 and 12 mm, and the hours between dawn and dusk. Of course, the species niche includes many more variables than these three, and ecologists often cite the multidimensional nature of the niche to acknowledge the complexity of species–environment relationships. The degree to which the niches of two species overlap determines how strongly

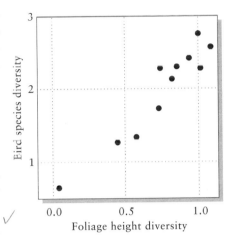

FIGURE 24.3 Relationship between bird species diversity and foliage height diversity, determined for areas of deciduous forest in eastern North America. From R. H. MacArthur and J. MacArthur, *Ecology* 42:594–598 (1961).

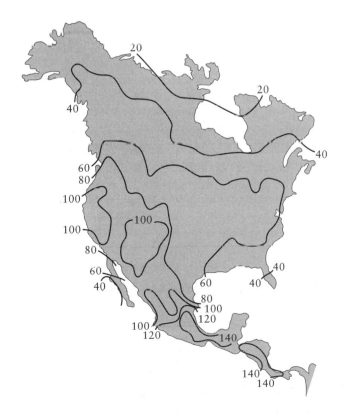

FIGURE 24.4 Species density contours for mammals in grid squares 150 miles on a side in continental North America, and for reptiles and amphibians within similar grid squares in the United States. From G. G. Simpson, *Syst. Zool.* 13:57–73 (1964); A. R. Keister, *Syst. Zool.* 20:127–137 (1971).

Species
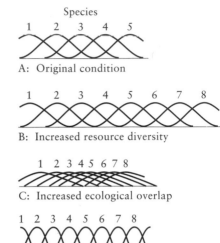

A: Original condition

B: Increased resource diversity

C: Increased ecological overlap

D: Increased specialization

FIGURE 24.5 Schematic diagram showing how resource use along a continuum can be altered to accommodate more species.

they compete with each other. Thus the niche relationships of species provide an informative measure of the structural organization of biological communities.

We may think of the community as a group of species occupying niches within a space defined by axes of resource quality and ecological condition. Within this space, adding or removing species has certain geometrical consequences. The possibilities are illustrated in Figure 24.5 by distributions of species niches along a single continuous dimension: size of prey item, perhaps, or height distribution within the intertidal zone. Furthermore, let us characterize niche relationships by measures of niche breadth and overlap, and let us assume that the availability of resources is uniform throughout niche space. Now, the addition of species to the community (which is originally in condition A in Figure 24.5) can be accommodated by one, or any combination, of three types of adjustments.

1. Without changing niche relationships, the total size of the community niche may expand in direct proportion to the number of species (condition B in the figure). Note that niche size refers to variety of resources, not their amounts.

2. Without changing niche breadth, increased diversity may be accommodated by increased niche overlap (condition C). In this case, the average productivity of each species diminishes as a consequence of increased sharing of resources.

3. Without increasing niche overlap, increased specialization may accommodate additional species in the community niche space (condition D). Here, too, average productivity decreases, because each species has access to a narrower range of resources.

Most ecologists agree that the high diversity of the tropics results at least in part from there being a greater variety of ecological roles. That is, the total community niche occupies greater volume near the equator than it does toward the poles. For example, part of the increase in the number of species of birds toward the tropics is related to an increase in fruit- and nectar-feeding species and in insectivorous species that hunt by searching for their prey while quietly sitting on perches—a type of behavior uncommon among species in temperate regions. Among mammals, the increase in number of species between temperate and tropical areas results primarily from the addition of bats to tropical communities. Nonflying mammals are no more diverse at the equator than they are in the United States and other temperate regions at a similar latitude, although their variety does decrease as one goes farther north.

One way to assess the niche diversity within a fauna is to use the morphology of species as an indication of their ecological roles—that is, to assume that differences in structure among related species reveal different ways of life. For example, the size of prey taken varies in relation to the body size of the consumer; different proportions of the appendages are related to locomotion in ways that distinguish different methods of hunting, different habitat structures, and different ways of escaping predators. Morphological analyses of communities have consistently revealed relatively constant density of species packing in morphological space. Therefore, as diversity increases in comparisons among communities, so does the total morphological volume occupied. This finding suggests that added species increase the variety of ecological roles played by members of the community.

To illustrate this principle, consider a study of bat communities in temperate and tropical localities. This study defined a morphological space having two dimensions: the ratio of ear length to forearm length and the ratio of the lengths of the third and fifth digits of the hand bones in the wing. The first dimension—a measure of ear length relative to body size—is related to the bat's sonar system and thus to the type and location of its prey. The second dimension pertains to the shape of the wing—whether it is long and thin or short and broad—and therefore serves as an index to flight characteristics of the bat and, in turn, to the types of prey it can pursue and the habitats within which it can capture prey efficiently.

When we plot each species of bat in a community on a graph whose axes are the two morphological dimensions, we can visualize the niche relationships among species (Figure 24.6). In the less diverse community in Ontario, southern Canada, the species have similar morphology; all are small insectivores. The more diverse community in Cameroun, in tropical West Africa, occupies a much larger volume of morphological space, corresponding to the greater vari-

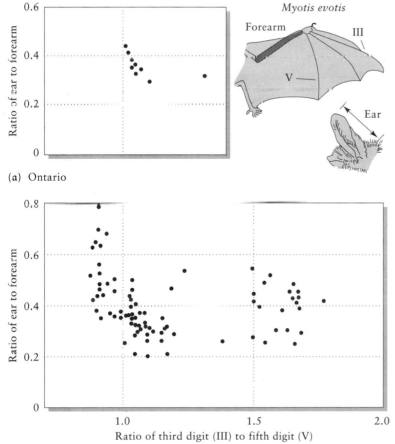

(a) Ontario

(b) Cameroun

FIGURE 24.6 Distribution in morphological space of species in the aerial-feeding bat faunas of southeastern Ontario (a) and Cameroun, West Africa (b). The horizontal axis is the ratio of the lengths of the third and fifth digits of the hand, and the vertical axis is the ratio of ear length to forearm length. After M. B. Fenton, *Can. J. Zool.* 50:287–296 (1972).

ety of ecological roles played by bats there. In addition to small insectivorous species, there are fruit-eaters, necter-eaters, fish-eaters, and large, predatory bat-eaters.

In streams and rivers, the number of species in most taxonomic groups increases from the headwaters to the mouth of the river. One presumes that as a river increases in size, it presents a greater variety of ecological opportunities and its physical conditions become more stable and therefore more reliable. Local communities reflect these changes. For example, a headwater spring in the Rio Tamesi drainage of east-central Mexico was found to support only one species of fish, a detritus-feeding platyfish (*Xiphophorus*) (Figure 24.7). Farther downstream, three species occurred: the platyfish, a detritus-feeding molly (*Poecilia*) that prefers slightly deeper water than the platyfish, and a mosquito fish (*Gambusia*) that eats mostly insect larvae and small crustacea. Species that appear in the community farther downstream include additional carnivores—among them, fish-eaters—and other fish that feed primarily on filamentous algae and vascular plants. None of the species drops out of the community downstream from any of the sampling localities. Thus diversity increases as the stream becomes larger and presents more kinds of habitats and a greater variety of food items.

Escape space

Niche dimensions are not limited to feeding and to tolerance of physical conditions. Avoidance of predation is equally important to population processes, and avenues of predator escape form dimensions of the niche along which species may diversify. We would expect predators to be most efficient when they focus their attention on portions of the niche space most densely occupied by prey species. Where many prey species use the same mechanisms to escape predation, predators having adaptations or learned behaviors that en-

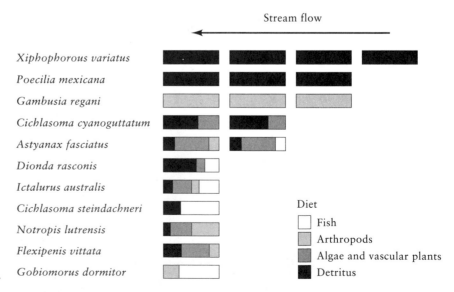

FIGURE 24.7 Food habits of fish species in four communities (vertical columns) from a headwater spring with 1 species (rightmost column) to downstream communities with up to 11 species (leftmost column). The communities sampled were in the Rio Tamesi drainage of east-central Mexico. From R. M. Darnell, *Am. Zool.* 10:9–15 (1970).

able them to exploit these prey will prosper. Thus these prey populations will suffer increased mortality. Conversely, prey having unusual adaptations for predator escape should be strongly selected. As a result, predation pressure should diversify prey with respect to escape mechanisms, and such species should therefore tend to become uniformly distributed within available escape space. The quality of a particular place within the niche space depends on predator attributes (hunting methods, body size, visual acuity, and color perception) and escape space (color and pattern of resting background for cryptic prey, availability of hiding places and other refuges, structure of the vegetation for those that rely on fleeing to escape).

One study used morphology to estimate the packing of species of moths in escape space, in this case the variety of backgrounds against which day-resting moths conceal themselves to avoid detection by diurnal visual predators. Among cryptic species, appearance has evolved to match the background against which the species rests. Hence we assume that the morphological appearance, or **aspect**, reflects the characteristics of the resting place and the searching techniques of predators to be avoided (Figure 24.8). The variety of cryptic patterns has been referred to **aspect diversity**. The appearances of moths were described by 12 characteristics, including morphology and position of the legs, coloration, and reaction to disturbance (Figure 24.9). Samples of moths were analyzed from three areas: a spruce–aspen forest in Colorado

FIGURE 24.8 Representative species of moths from Panama, photographed against the window screens to which they were attracted by ultraviolet lights. The moths show the variety of appearances and shapes exhibited by members of the community. From R. E. Ricklefs and K. O'Rourke, *Evolution* 29:313–324 (1975).

(43 species), a Sonoran desert habitat in Arizona (51 species), and a lowland rain forest habitat in Panama (203 species). The total volume of escape space used, as revealed by the variety of morphological characteristics represented in each sample, was greater in Panama than in Colorado or Arizona. Again we see that species are added to a community by expansion of the niche space used. Variation in the amount of escape space used could arise from a number of factors: variation in the diversity of escape possibilities presented by the environment, greater predation pressure, greater pressure for diversification through competition for escape space, and greater opportunity for diversification owing to a greater variety of independently evolving populations.

Although these and other examples suggest an increase in total niche space with increasing diversity, two considerations must be kept in mind. First, changes in niche breadth (specialization) and overlap may also accompany variations in diversity, although these are more difficult to measure. Second, increased community niche space may be in part a *consequence* of diversity rather than an underlying cause or permissive factor. The mere presence of species makes for more possible roles—a greater variety of interactions. For example, the larger proportion of specialized, parasitic species in collections of insects from sites with high diversity suggests enhancement of diversity through biotic interaction.

FIGURE 24.9 Diversity in appearance and behavior of moths from three localities exemplified by three characteristics related to their antipredator adaptations. Variety is greatest in the most diverse sample, that from Panama. From data in R. E. Ricklefs and K. O'Rourke, *Evolution* 29:313–324 (1975).

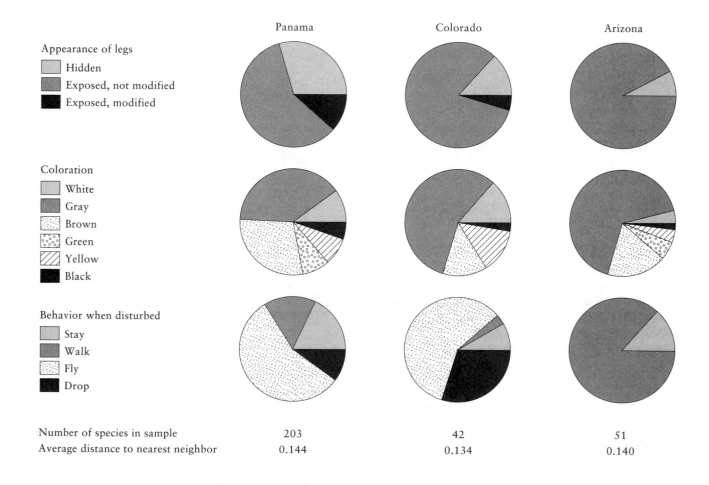

Appearance of legs
- Hidden
- Exposed, not modified
- Exposed, modified

Coloration
- White
- Gray
- Brown
- Green
- Yellow
- Black

Behavior when disturbed
- Stay
- Walk
- Fly
- Drop

	Panama	Colorado	Arizona
Number of species in sample	203	42	51
Average distance to nearest neighbor	0.144	0.134	0.140

Diversity has several components, two of which are **alpha (or local) diversity** and **gamma (or regional) diversity.** Local diversity is the number of species in small areas of more or less uniform habitat. Clearly, local diversity is sensitive to definition of habitat and to area and intensity of sampling effort. Regional diversity is the total number of species observed in all the habitats within a region. When the same species occur in all the habitats of a region, local and regional diversities are the same. When each habitat has a unique flora and fauna, regional diversity equals the average local diversity times the number of habitats in the region. Ecologists refer to the turnover of species from one habitat to the next as **beta diversity.** Therefore, gamma diversity equals alpha diversity times beta diversity. It is not practical to measure beta diversity directly because the habitat distributions of species overlap. But we can calculate the equivalent number of unique habitats recognized by species within a region from the following relationship: beta diversity equals gamma diversity divided by alpha diversity.

Where many species coexist within a region, each occurs in relatively few kinds of habitat. Changes in gamma diversity generally result from parallel changes in both alpha and beta diversity. This relationship has been most carefully noted in comparisons of islands and mainland regions, where we can examine a range of species diversity (resulting from different degrees of geographic isolation) within similar ranges of physical conditions. Islands usually have fewer species than comparable mainland areas; island species often attain greater densities than their mainland counterparts and expand into habitats that would normally fill with other species on the mainland. These phenomena, called **density compensation** and **habitat expansion,** are collectively referred to as **ecological release.** On the island of Puerto Rico, many species of birds occupy most of the habitats on the island. In Panama, which has a similar variety of tropical habitats, species occupy fewer habitats, often a single type.

Surveys of bird communities in seven tropical localities with different numbers of species — Panama, Trinidad, Jamaica, Tobago, St. Lucia, Grenada, and St. Kitts — show that where fewer species occur, each is likely to be more abundant and to live in more habitats (Table 24.2). Similar numbers of individuals of all species added together were seen in each of the seven localities, although the total numbers of species (regional diversity) differed by a factor of almost 7 between Panama and St. Kitts. In each habitat in Panama (mainland), about three times as many species (alpha diversity) were recorded, and populations of each species were about half as dense, as in the corresponding habitat on St. Kitts (small island). The beta diversity (species turnover between habitats, or equivalent number of habitats recognized) increased by a factor of almost 3 between St. Kitts and Panama.

This survey of diversity patterns suggests several general conclusions. At a global scale, our perception of biodiversity is dominated by the pronounced increase in diversity as one travels from high latitudes toward the equator. Within latitudinal belts, diversity appears to be correlated with the general topographical heterogeneity of the region and the complexity of the habitat. Islands exhibit species impoverishment. Everywhere, higher diversity is associated with greater community niche volume.

How do we explain these patterns of diversity? Some biologists have claimed that diversity increases with time and depends on age, but the strong correlation between habitat structure and diversity would seem to cast doubt on that hypothesis. Alternatively, most ecologists now believe that diversity achieves an equilibrium value in which processes that are adding and those that are subtracting species balance each other. Within a local community, species are removed by competitive exclusion and through elimination by efficient predators. The migration of species between habitats and regions, as well as the production of new species within regions, adds to the number of species in local habitats. Distinguishing among the various factors responsible for generating and maintaining diversity at the local and regional scales has proved to be exceedingly difficult.

The time hypothesis

The fossil record ought to tell us whether or not diversity has increased over time. However, the abundance and quality of fossils increases over time, paleontologists have more difficulty sampling local diversity than regional diversity, most forms cannot be distinguished as finely as species, and the record is biased with respect to habitat. Minimizing these considerations for the mo-

TABLE 24.2	*Relative abundance and habitat distribution of resident land birds in seven tropical localities within the Caribbean Basin**					
Locality	Number of species observed (regional diversity)	Average number of species per habitat (local diversity)	Habitats per species	Relative abundance per species per habitat (density)	Relative abundance per species	Relative abundance of all species
Panama	135	30.2	2.01	2.95	5.93	800
Trinidad	106	28.2	2.35	3.31	7.78	840
Jamaica	56	21.4	3.43	4.97	17.05	955
Tobago	53	21.4	3.63	4.71	17.10	906
St. Lucia	33	15.2	4.15	5.77	23.95	790
Grenada	30	15.5	4.63	5.36	24.82	745
St. Kitts	20	11.9	5.35	5.88	31.45	629

*Based on 10 counting periods in each of 9 habitats in each locality. The relative abundance of each species in each habitat is the number of counting periods in which the species was seen (maximum 10); this times the number of habitats gives the relative abundance per species; this times the number of species gives the relative abundance of all species together.

Source: G. W. Cox and R. E. Ricklefs, *Oikos* 29:60–66 (1977); J. M. Wunderle, *Wilson Bull.* 97:356–365 (1985).

ment, the data currently available suggest that during the past 65 million years, diversity has remained constant within some groups at some times and has increased in others, notably flowering plants, fish, birds, and mammals (Figure 24.10). The fossil record does not tell us unambiguously whether other groups declined at the same time, nor does it reveal whether diversity achieved local equilibrium while it increased globally.

One explanation for the diversity of the tropics is that tropical conditions appeared on the earth's surface earlier than more polar environments, allowing time for the evolution of a greater variety of plants and animals. Although this **time hypothesis** has been voiced recently, it is hardly new. It was fully stated in 1878 by the English naturalist Alfred Russel Wallace:

> *The equatorial zone, in short, exhibits to us the result of a comparatively continuous and unchecked development of organic forms; while in the temperate regions there have been a series of periodical checks and extinctions of a more or less disastrous nature, necessitating the commencement of the work of development in certain lines over and over again. In the one, evolution has had a fair chance; in the other, it has had countless difficulties thrown in its way. The equatorial regions are then, as regards their past and present life history, a more ancient world than that represented by the temperate zones, a world in which the laws which have governed the progressive development of life have operated with comparatively little check for countless ages, and have resulted in those wonderful eccentricities of structure, of function, and of instinct — that rich variety of colour, and that nicely balanced harmony of relations which delight and astonish us in the animal productions of all tropical countries.*

Because the tropical zone girdles the earth about its equator — the earth's widest point — the tropics include more area than temperate and arctic regions. The earth's climate has undergone several cycles of warming and cooling, which have been discovered by records, in sediments and fossils, of their influence on vegetation and ocean temperature. As the climate of the earth warmed, as it last did during the Oligocene epoch, perhaps 30 million years ago, the area of the tropics and subtropics expanded, reaching what is now the United States and southern Canada, and the temperate and arctic zones were squeezed into smaller areas closer to the poles. During the last 25 million years, the climate of the earth has become cooler and drier, and the tropics have contracted.

Both high and low latitudes experienced drastic fluctuations in climate during the ice ages of the last 2 million years. Temperate and arctic areas witnessed the expansion and retreat of glaciers, causing major habitat zones to be displaced geographically and, possibly, to disappear. Periods of glacial expansion were coupled with high rainfall in the tropics (pluvial periods). The Amazonian rain forest, which today covers vast regions of the Amazon River's drainage basin, was repeatedly restricted to small, isolated refuges during periods of drought. Restriction and fragmentation of the rain forest habitat

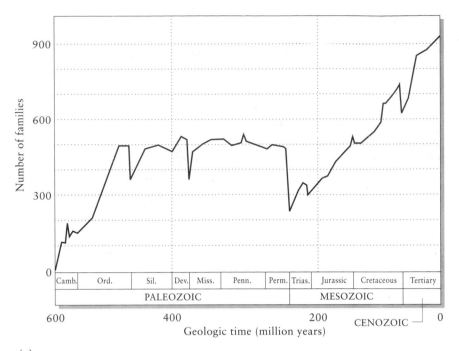

(a)

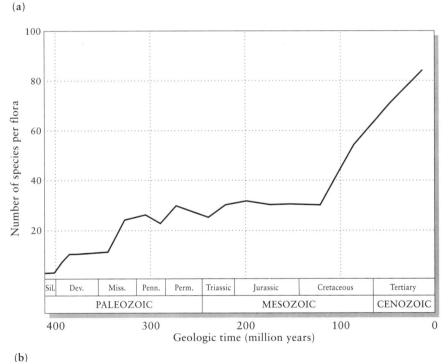

(b)

FIGURE 24.10 (a) Increase in the global total number of families of marine animals since the beginning of the Paleozoic era. Note the more or less constant diversity through the latter part of the Paleozoic era; the mass extinction at the end of the Permian period; and the continuous increase in diversity since, with the proliferation of modern marine taxa (primarily fish, mollusks, and crustaceans). From J. J. Sepkoski, Jr., *Paleobiology* 10:246–267 (1984). (b) Increase in the average number of species represented in local fossil floras of terrestrial plants. Note the rather constant diversity throughout much of the Paleozoic and Mesozoic eras, the absence of mass extinction at the end of the Permian period, and the rapid rise in diversity beginning with the appearance of flowering plants in the late Jurassic period. After A. J. Knoll, in J. Diamond and T. J. Case (eds.), *Community Ecology.* Harper & Row, New York (1986), pp. 126–141.

could have caused the extinction of many species; conversely, the isolation of populations in patches of rain forest could have facilitated the formation of new species. The alternating expansion and contraction of humid habitats accounts in part for the diversity of the present Australian avifauna. Other animals and plants were similarly affected. For example, there are about 500 species in the genus *Eucalyptus,* a type of tree or shrub native to Australia (but now widely planted around the world).

Equilibrium theories of diversity

The integration of "population thinking" into community studies during the 1950s and 1960s led ecologists to postulate that diversity might be regulated by local interactions among species. If competitive exclusion set limits on the ecological similarity of species (and thus on the strength of competition between them), then communities could become saturated with species, additions being balanced by local extinctions. Accordingly, the number of species would reach an equilibrium: new species added to the local community by regional diversification and migration of populations would be compensated for by the local exclusion of close competitors. The equilibrium point itself would be affected by physical conditions, variety of resources, predators, environmental variability, and perhaps other factors. Thus conditions in the tropics might allow greater numbers of species to coexist locally by reducing the intensity or consequences of competition.

Ecologists were attracted to this view because it placed at least part of the problem of species diversity within their conceptual domain of present-day processes taking place within small areas. The earlier idea that diversity reflected history left ecologists with little to say about the matter. The alternative, what we might call local determinism, became so attractive, however, that ecologists embraced it nearly to the point of denying that regional and historical factors influence local diversity. Instead of viewing diversity as a balance between regional production of species and local extinction, ecologists began to feel that local diversity was determined largely by local conditions. Hence their investigations concentrated almost exclusively on local interactions and niche relationships among populations.

The numbers of species on islands

During the 1960s, Robert MacArthur and E. O. Wilson developed their famous **equilibrium theory of island biogeography,** which states that the number of species on islands balances regional processes governing immigration against local processes governing extinction. On islands too small to support speciation through geographic isolation of populations, the number of species increases solely by immigration from other islands or from the mainland. Whereas we know little about rates of speciation within continents, we may reasonably assume that the greater the distance that separates islands from the mainland, the fewer the species that immigrate to those islands. Hence the advantage of the island model.

Consider an offshore island. The flora and fauna of the adjacent mainland make up the species pool of potential colonists to the island. The rate of immigration of new species to the island decreases as the number of species on the island increases; that is, as more and more of the potential mainland colonists establish themselves on the island, fewer and fewer immigrants belong to new species (Figure 24.11). When all the mainland species occur on the island, the immigration rate of new species must be zero. If species disappeared at random, the number of extinctions per unit of time would increase with the

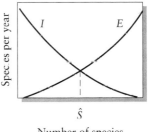

FIGURE 24.11 Equilibrium model of the number of species on islands. The equilibrium number of species $\hat{S}$ is determined by the intersection of the immigration (I) and extinction (E) curves. After R. H. MacArthur and E. O. Wilson, *Evolution* 17:373–387 (1963); MacArthur and Wilson, *The Theory of Island Biogeography*. Princeton Univ. Press, Princeton (1967).

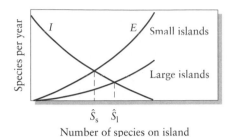

FIGURE 24.12 According to the MacArthur–Wilson equilibrium model, small islands support fewer species because of higher extinction rates. $\hat{S}_s$ = equilibrium number of species for small islands; $\hat{S}_l$ = equilibrium number of species for large islands.

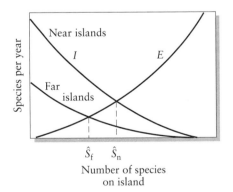

FIGURE 24.13 According to the MacArthur–Wilson equilibrium model, islands close to the mainland support more species because of higher immigration rates. $\hat{S}_f$ = equilibrium number of species on far islands; $\hat{S}_n$ = equilibrium number of species on near islands.

number of species present on the island. Where the immigration and extinction curves cross, the corresponding number of species on the island attains an equilibrium ($\hat{S}$).

Immigration and extinction rates probably do not vary in strict proportion to the number of potential colonists and the number of species established on the island. Some species undoubtedly colonize more easily than others, and they reach the island first. Thus the rate of immigration to the island initially decreases more rapidly than it would if all mainland species had equal potential for dispersal, and as a result, the relationship between immigration rate and island diversity follows a curved line. Competition between species on islands probably abets extinction, so the extinction curve rises progressively more rapidly as species diversity increases.

If the probability of extinction increased as absolute population size decreased, extinction curves would be higher for species on small islands than for those on larger islands. Therefore, small islands would support fewer species than large islands (Figure 24.12). If the rate of immigration to islands decreased with distance from mainland sources of colonists, the immigration curve would be lower for far islands than for near islands, and the equilibrium number of species for distant islands should lie to the left of the equilibrium for islands close to the mainland (Figure 24.13). These predictions have been verified for islands throughout the world.

If the number of species on an island behaved according to the equilibrium model, a change in the number of species would lead to a response tending to restore the equilibrium diversity. A natural test of this prediction began quite spectacularly in 1883 when the island of Krakatoa, located between Sumatra and Java in the East Indies, blew up after a long period of repeated volcanic eruptions. At least half the island disappeared beneath the sea, and hot pumice and ash covered its remaining area. The entire flora and fauna of the island were certainly obliterated. During the years that followed the explosion, plants and animals recolonized Krakatoa at a surprisingly high rate: within 25 years, explorers found more than 100 species of plants and 13 species of land and freshwater birds. During the next 13 years, 2 species of birds disappeared and 16 were gained, bringing the total to 27. During the next 14-year period between exploring expeditions, the number of species on the island did not change, but 5 species disappeared and 5 new ones arrived, suggesting that the number of species had reached an equilibrium, at a level that would be expected for an island the size of Krakatoa located in the tropics. Experimental studies of the colonization of mangrove islands by arthropods reinforce the pattern of colonization and attainment of an equilibrium seen on Krakatoa (Figure 24.14).

The equilibrium theory in continental communities

We can apply an equilibrium view of diversity to mainland assemblages of species as well as to those on islands. The major difference is that regional production of species augments the addition of species from outside the system by

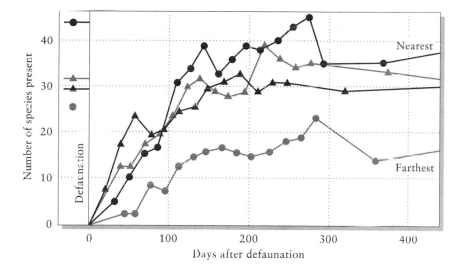

FIGURE 24.14 Recolonization curves of four small mangrove islands in the lower Florida Keys whose entire faunas, consisting almost solely of arthropods, were exterminated by methyl bromide fumigation. Estimated numbers of species present before defaunation are indicated at left. Species accumulated more slowly and achieved lower equilibrium numbers on more distant islands. From D. S. Simberloff and E. O. Wilson, *Ecology* 50:278–296 (1970).

immigration. For a large region isolated from others by barriers to dispersal (an island continent, for example), all new species must originate by speciation events within the region. The curves relating rates of species production and extinction to regional diversity might look like those drawn in Figure 24.15. The curvature of the lines would vary depending on the processes that produced them: the probability of extinction per species could increase if competitive exclusion increased with diversity, whereas it could decrease if mutualisms and alternative paths of energy flow buffered diverse communities from external perturbations; the speciation rate per species could level off if opportunities for further diversification were restricted by increasing diversity, whereas it could increase if diversity led to greater specialization and higher probability of reproductive isolation of subpopulations.

Regardless of the particular shape of the immigration, speciation, and extinction curves, many biologically reasonable models define an equilibrium level of diversity. So although such models provide a perspective, they do not explain variation in diversity: most causes can be incorporated into an equilibrium model. The time hypothesis would apply when present diversity remained far below the equilibrium. Alternatively, the number of species might saturate local communities while diversity continued to increase regionally, the growing difference being made up by increasing beta diversity. If diversity were in equilibrium, the positions of species production and extinction curves would be affected by a variety of factors, each of which could shift the equilibrium (Figure 24.16).

Competition and diversity

Intense competition promotes the exclusion of species from a community. Many ecologists have argued that less severe competition for shared resources allows more species to coexist. How may interspecific competition be reduced? Greater ecological specialization, greater resource availability, reduced

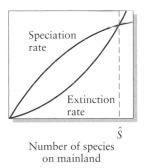

FIGURE 24.15 Equilibrium model of the number of species in a mainland region with a large area; new species are generated by the evolutionary process of speciation rather than by immigration from elsewhere. After R. H. MacArthur, *Biol. J. Linn. Soc.* 1:19–30 (1969).

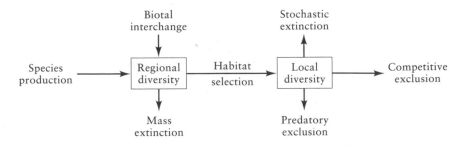

FIGURE 24.16 Diagram of the factors affecting regional and local species diversity. The numbers of species are increased at the regional level. Ecological interactions influence diversity at the local level. Each local community contains a small sample of the total regional diversity because species are habitat specialists. Thus habitat selection connects regional and local diversity.

resource demand, and intensified predation have all been cited as possibilities. In every case, however, competition between individuals of the same species, as well as between individuals of different species, influences the adjustment of populations to these factors. It is hard to imagine reduced competition as an intrinsic property of any population. Density-dependent selection tends to increase population size. Therefore, if variation in competition were to explain variation in diversity, the relative strengths of intraspecific competition and interspecific competition would have to vary in their influence on the gene pool of the population.

Perhaps differences in these relative strengths may be envisioned in terms of the geometry of niche relationships. In a niche space having few dimensions, each population has relatively few neighbors. As the number of dimensions increases, the same number of species may pack themselves differently into the niche space, increasing the number of neighbors and reducing the asymmetry of competition (Figure 24.17). This variety may prevent each individual population from adapting to compete effectively with its neighbors in communities of high niche dimension, preventing the emergence of competitive dominants. High dimensionality may occur in physically ameliorated environments, such as the wet tropics or the abyssal depths of the oceans; whereas in harsher environments, a few physical factors may dominate the character of the niche space by establishing a small number of critically important dimensions. Furthermore, because biological factors generate many niche dimensions, the dimensionality of the niche may increase as biological communities build over time, further enhancing their capacity to support many species against the evolutionary tendency of a small number of species to dominate.

(a) One dimension (asymmetrical)

(b) Two dimensions (symmetrical)

FIGURE 24.17 Schematic diagram of ecological relationships among three species, organized along (a) one dimension and (b) two dimensions. Relationships among three species positioned along one dimension are asymmetrical because the middle species faces two competitors and the outer two species interact only with one. Relationships are symmetrical on two dimensions, making it difficult for two species to exclude the third, or vice versa.

Explanations for variation in plant species diversity

Because the diversity of plants as resources rather straightforwardly influences the potential diversity of animals, the most rigorous tests of general explanations for diversity lie in their application to plant communities. The question of why there are so many different kinds of trees in the tropics has a variety of plausible answers. These may be grouped into five categories based on (1) environmental heterogeneity in space, (2) environmental variability in time, (3) environmental heterogeneity produced by disturbances to the uniform structure of the forest, (4) rates of production (by speciation or immigration) of competitively equivalent species, and (5) the favoring of rarer species by consumer pressure.

Many ecologists have argued that the diversity of trees varies in proportion to the heterogeneity of the environment. Abundant evidence suggests that tropical forest trees may be specialized to certain soil and climate conditions. But could variability in the physical environment in the tropics account for a tenfold (or more) greater diversity of plants in tropical than in temperate forests? It seems unlikely, unless plants recognize much finer differences in the environment in the tropics than they do in temperate regions.

Unpredictable temporal heterogeneity generally is not thought to promote diversity. But theoretical considerations suggest that under some circumstances, year-to-year variation in reproductive rate, such that each species predominates in some years, can lead to coexistence. The mechanism relies on a kind of frequency dependence. Suppose that a certain fraction of individual trees dies each year and that each one is replaced by an individual of the same species or by an individual of a different species in direct proportion to the production of seeds by each of the species. When a species is very rare, a bad year for seed production has little effect on the probability of seedling establishment, but a good year can bring a comparative bonanza — a little less of almost nothing is almost nothing, but a little more represents a huge relative increase. Accordingly, less common species gain a relative advantage over more common ones in years that are favorable to them. Although this mechanism may work in theory, it acts strongly only for very rare species. Any degree of competitive inferiority quickly pushes a species into the very rare category. Moreover, because every species must experience some years during which it produces more seed than all other species, relatively few species can coexist solely by this mechanism.

This model was inspired by coral reef fish, which exhibit a diversity comparable to that of tropical forest trees and exhibit little niche diversification. For this system, there has been considerable controversy over the issue of competition among species. Some observers have suggested that juvenile fish colonize patches of coral at random, with the opportunity to do so spread evenly among individuals of all species to take the place of adults that die or otherwise leave their territories in the reef. This mechanism became known as the **lottery hypothesis.** Colonization by lottery reduces competitive exclusion to chance extinction and may contribute to the coexistence of large numbers of fish in tropical reef communities. But the lottery model cannot explain the diversity of larval fish in the plankton from which coral residents come.

Other observers have related the high diversity of tropical rain forests and coral reefs to intermediate levels of disturbance to communities. We have already touched upon the **intermediate disturbance hypothesis:** disturbances caused by physical conditions, predators, or other factors open space for colonization and initiate a cycle of succession by species adapted to colonizing disturbed sites. With a moderate level of disturbance, the community becomes a mosaic of patches of habitat at different stages of regeneration; together these patches contain the full variety of species characteristic of the successional sere. For this hypothesis to account satisfactorily for differences in diversity between regions, especially the magnitude of the latitudinal gradient of tree species diversity, there must be comparable differences in level of disturbance. Rates of turnover of individual forest trees (that is, the inverse of average life span) do not differ appreciably between temperate and tropical areas (Table 24.3). Thus, although disturbance may promote diversity, it seems unlikely to account for much of the observed variation in diversity among forests or, indeed, other types of communities.

Predation and diversity

When predators reduce populations of their prey below the carrying capacity of their resources, they may reduce competition among competitors and promote coexistence. Moreover, selective predation on superior competitors may allow competitively inferior species to persist in a system. Both accidental

TABLE 24.3	Turnover of canopy trees in primary forests in tropical and temperate localities	
Locality	Turnover time (years)*	Turnover rate (percent per year)
Tropical		
Panama	62–114	0.9–1.6
Costa Rica	80–135	0.7–1.3
Venezuela	104	1.0
Gabon	60	1.7
Malaysia	32–101	1.0–3.1
Temperate		
Great Smoky Mountains	49–211	0.5–2.0
Tionesta, Pennsylvania	107	0.9
Hueston Woods, Ohio	78	1.3

*Turnover time does not include the years required to grow into the canopy, which is estimated to be 54 to 185 years for various temperate-zone species.

Source: Data from J. R. Runkle, in S. T. A. Pickett and P. S. White (eds.), *The Ecology of Natural Disturbance and Patch Dynamics,* Academic Press, Orlando, Fla. (1985), pp. 17–33; F. E. Putz and K. Milton, in E. G. Leigh, Jr., A. S. Rand, and D. M. Windsor (eds.), *The Ecology of a Tropical Forest,* Smithsonian Institution Press, Washington, D.C. (1982), pp. 95–100; N. V. L. Brokaw, in *The Ecology of Natural Disturbance and Patch Dynamics,* pp. 53–69.

and intentional experiments have revealed the immense capacity of predators to reduce populations of their prey under some conditions. Community effects have been particularly well documented in aquatic systems, where the introduction of a predatory sea star, salamander, or fish can utterly transform the community of primary consumers and producers.

From Darwin's time at least, naturalists have believed that both selective and nonselective herbivory can influence the diversity of plant species. In particular, several authors have suggested that herbivory could promote the high diversity of tropical forests. Herbivores may feed on the buds, seeds, and seedlings of abundant species so efficiently as to reduce their densities, which allows other, less common species to grow in their place. The key to this idea is that abundance per se, rather than the intrinsic quality of individuals as food items, makes a species vulnerable to consumers. Consumers locate abundant species easily, and their own populations grow to high levels.

Several lines of evidence support the **pest pressure hypothesis**. For example, attempts to establish plants in monoculture frequently have failed because of infestations of herbivores. Dense plantations of rubber trees in their native habitats of the Amazon Basin, where many species of herbivores have evolved to exploit them, have met with singular lack of success. But rubber tree plantations thrive in Malaya, where specialist herbivores are not (yet) present. Attempts to grow many other commercially valuable crops in single-species stands in the tropics have frequently met the same disastrous end that befell the rubber plantations.

The difficulties of monoculture extend beyond the tropics. Temperate-zone trees have their herbivores; few acorns escape predation by squirrels and weevils, and seedlings are attacked by herbivores and pathogens just as they are in the tropics. If pest pressure does promote greater diversity in the tropics, it must operate differently in different latitude belts. In particular, either tropical herbivores and plant pathogens must be more specialized with respect to species of host plant or their populations must be more sensitive to the density and dispersion of host populations.

The pest pressure hypothesis predicts that seedlings should establish themselves less easily close to adults of the same species than at a distance (Figure 24.18). Adult individuals may harbor populations of specialized herbivores

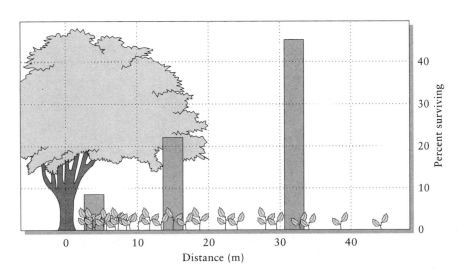

FIGURE 24.18 Schematic diagram of Janzen's pest pressure hypothesis showing that seedlings are most dense close to the parent tree but that survival of seedlings is highest at a distance, because of the detrimental effects of crowding on seedling survival. Data on survival of seedlings to 18 months of age are for the neotropical tree *Dipteryx panamensis*. From D. H. Janzen, *Amer. Nat.* 104:501–528 (1970); D. A. Clark and D. B. Clark, *Amer. Nat.* 124:769–788.

and pathogens that readily infest nearby progeny; furthermore, because most seeds germinate close to their parent, herbivores may be attracted to the abundance of seedlings there while overlooking the few that germinate at some distance. The prediction that success in germination and establishment should increase with distance has been tested in a number of studies, which have yielded varied but generally supportive results.

If seedlings establish themselves more readily at greater distance from the parent or other adults of the same species, individuals of the same species should be widely dispersed within a forest rather than randomly associated or clumped. This prediction can be tested by mapping individuals of each species and applying one of the several statistics used to test degree of dispersion. Most studies have shown that many species of trees, especially the more common ones, have significantly clumped distributions. This result weighs against a major role of pest pressure in forest community structure, but degree of clumping is relative, and herbivores and pathogens may cause greater dispersion of mature individuals than would occur otherwise.

Rates of species production

Environmental heterogeneity and consumer activities can influence local population processes in such a way as to affect species diversity. Regional processes are more difficult to contemplate, much less observe directly. Correspondingly little has been written about their role in determining local and regional diversity. In nonequilibrium systems, rates of species production and extinction, as well as the age of the region or habitat type, directly determine diversity. Where diversity attains an equilibrium, the rate of species production will influence regional diversity; it will also influence local diversity unless local communities become saturated with species at levels determined solely by local ecological conditions.

Many systematists and biogeographers have suggested that the rate of species production or the temporal history of an environment contributes to tropical diversity. But because thousands of generations may be needed for isolated populations of a species to develop barriers to reproduction and to themselves become distinct species, ecologists can neither observe the process nor experiment with it.

A few papers on theory have addressed the problem of tree species diversity in the context of species production. For example, one lottery-type model of replacement of forest trees suggested that the probability that the individual that fills a gap in the forest belongs to a particular species varies in direct proportion to the frequency of that species in the forest. Hence gaps are more likely to be filled by more abundant species than by rare ones. Without the appearance of new species, such a system will eventually become dominated by a single species, just by chance. Thus, in such a system, diversity must be maintained, and its level determined, by the rate of species production (that is, the random appearance of unique individuals).

Fragmentation of tropical forests during the periodic dry periods of the recent Ice Age (Figure 24.19) provided opportunities for allopatric speciation in

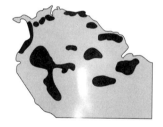

(a) Glacial

(b) Pluvial

FIGURE 24.19 Approximate distribution of lowland rain forest in South America during the height of glacial periods in the Northern Hemisphere (a) and at present (b).

the tropics at a time when harsh conditions and the restriction of habitats in temperate and arctic zones may have caused an increase in extinction rate. If differences in diversity between temperate and tropical forests express recent high rates of species production in the tropics, we would expect to find more species per genus in tropical forests than in temperate-zone counterparts. But tropical forests present their tremendous diversity as much at the family and genus levels as at the species level (Figure 24.20). In fact, the tropical forests are decidedly lacking in closely related species. Their great number of higher taxa (genera and families) belies the ancient roots of diversity there. It seems reasonable to conclude that if differences in speciation rate are responsible for differences in the diversity of forests, they have been persistent differences acting over very long periods.

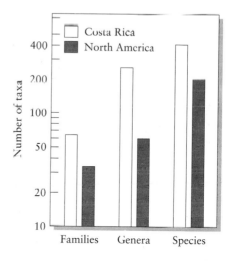

FIGURE 24.20 The diversity of trees at three taxonomic levels in a rain forest in eastern Costa Rica and in broad-leaved deciduous forests of eastern North America. From data in R. E. Ricklefs, in D. Otte and J. Endler (eds.), *Speciation and Its Consequences,* Sinauer Associates, Sunderland, Mass. (1989), pp. 599–622.

☙ SUMMARY

1. A conspicuous pattern revealed by studies of biological communities is the tendency of species diversity in tropical regions to greatly exceed that at higher latitudes.

2. Ecologists refer to the ranges of conditions and resource qualities used by each species as the niche of that species.

3. Diverse tropical communities contain a greater variety of ecological niches, as well as a greater number of species, than temperate communities.

4. In regions of high species diversity, individual species tend to be habitat specialists compared with their counterparts in regions of low diversity. Hence, although more species occupy each habitat (alpha diversity), species also distinguish differences between habitats more finely (beta diversity). The total diversity of species within a large region with many habitats is that region's gamma diversity.

5. One explanation for tropical diversity—the time hypothesis— suggests that tropical habitats are older than temperate habitats and consequently have had more time to accumulate species. According to this hypothesis, diversity is nonequilibrial, and the number of species within a region continually increases.

6. Recently, thinking about diversity has been dominated by equilibrium theories, which state that habitats become saturated by species, which reach numbers determined by local ecological conditions. Accordingly, these theories state that the greater species diversity in the tropics compared to that in higher latitudes is due to differences in resources and in interactions between species such that more species can coexist.

7. Differences in the numbers of species on islands emphasize the importance of regional processes—immigration from the mainland or other islands—to the maintenance of species diversity. On conti-

nents, immigration of species to local areas reflects, in part, the rate of production of new species, which is also a regional process.

8. Several explanations for high plant diversity in the tropics focus on the role of disturbance in creating mosaics of successional stages of different age and in creating heterogeneous conditions for seedling establishment within gaps in the forest canopy.

9. Predators may enhance diversity among their prey by reducing populations (and hence competition for resources), thereby making coexistence easier. Evidence that predators and diseases may act in a density-dependent manner also supports this hypothesis. Density-dependent and frequency-dependent predation favor the persistence of rare species and enhance diversity.

10. Several issues pertaining to community diversity have not been adequately resolved. One concerns the rate of species production, particularly whether environmental and species-specific characteristics promote speciation more strongly in tropical regions, where diversity is high.

11. If local diversity is regulated by ecological interactions within communities, diversity within a particular habitat type should be independent of the diversity of the region as a whole. The evidence only partially bears out this prediction, and processes beyond the boundaries of the community, such as species production and the exchange of species between regions, exert an important influence on biodiversity at both the local and regional levels.

❦ SUGGESTED READINGS

Case, T. J., and M. L. Cody. 1987. Testing island biogeographic theories. *American Scientist* 75:402–411.

Cody, M. L. 1975. Towards a theory of continental species diversities: bird distributions over Mediterranean habitat gradients. In M. L. Cody and J. M. Diamond (eds.), *Ecology and Evolution of Communities*. Harvard Univ. Press, Cambridge, Mass., pp. 214–257.

Connell, J. H. 1978. Diversity in tropical rain forests and coral reefs. *Science* 199:1302–1310.

Cornell, H. V., and J. H. Lawton. 1992. Species interactions, local and regional processes, and limits to the richness of ecological communities: a theoretical perspective. *Journal of Animal Ecology* 61:1–12.

Currie, D. J. 1991. Energy and large-scale patterns of animal- and plant-species richness. *American Naturalist* 137:27–49.

Fischer, A. G. 1961. Latitudinal variations in organic diversity. *American Scientist* 49:50–74.

MacArthur, R. H. 1965. Patterns of species diversity. *Biological Reviews* 40:510–533.

MacArthur, R. H. 1972. *Geographical Ecology: Patterns in the Distribution of Species*. Harper & Row, New York.

Morton, S. R., and C. D. James. 1988. The diversity and abundance of lizards in arid Australia: a new hypothesis. *American Naturalist* 123:237–256.

Ricklefs, R. E. 1987. Community diversity: relative roles of local and regional processes. *Science* 235:167–171.

Roughgarden, J. 1989. The structure and assembly of communities. In J. Roughgarden, R. M. May, and S. A. Levin (eds.), *Perspectives in Ecological Theory*. Princeton Univ. Press, Princeton, pp. 203–226.

Terborgh, J. 1992. *Diversity and the Tropical Rain Forest*. Scientific American Library, New York.

Westoby, M. 1988. Comparing Australian ecosystems to those elsewhere. *BioScience* 38:549–556.

HISTORY AND
BIOGEOGRAPHY

The earth provides an ever-changing setting for the development of biological systems. Over the millions of years of earth's history, animals and plants have witnessed changes in climate and other physical conditions, rearrangements of the geographic positions of the continents and ocean basins, the growth and wearing down of mountain ranges, the evolution of biological novelties such as new tactics of predators and disease organisms, and catastrophic impacts with extraterrestrial bodies. These changes have helped to direct the course of the evolution and diversification of organisms and have influenced the development of biological communities. The history of life reveals itself to us in the geochemical record of past environments, in fossil traces left by long-extinct taxa, and in the geographic distributions and evolutionary relationships of living species. All these lines of evidence show that life has taken tortuous and unpredictable paths and has suffered occasional setbacks.

The most obvious consequence of this history is the nonuniform distribution of animal and plant forms on the surface of the earth. Australia, for example, has many unique forms—koalas, kangaroos, and eucalyptus trees—

because of its long isolation as an island continent surrounded by ocean barriers to the dispersal of terrestrial creatures. Every other part of the earth has distinctive fauna and flora. Even the major ocean basins, interconnected as they are by continuous corridors of water, have somewhat differentiated biotas, isolated by ecological barriers of temperature and salinity.

For the ecologist, biological history raises two potential problems. The first is that the structure and functioning of organisms may be influenced as much by ancestry as by local environment. The truth of this proposition diminishes our confidence in matching the morphology, physiology, and behavior of organisms to the conditions and resources of their environments. For example, the marsupial mode of reproduction (involving, among other characteristics, an early birth and subsequent development of the young in a pouch) is uniquely a property of the marsupial line of mammalian evolution rather than a result of unique ecological properties of the continent of Australia, where marsupials are now most diverse. Biologists refer to characteristics shared by a lineage irrespective of environmental factors as **phylogenetic effects**. These effects reflect the inertia of evolution — the lack of change of some attributes in the face of change in the environment. Ecologists recognize that such effects may influence the structure and functioning of ecological systems. For example, we may ask whether, if the plants and animals of Australia were replaced by a similar number of taxa from other regions with similar climate, the new ecosystems would function in the same manner, with similar levels of biological productivity and response to environmental perturbation.

The second problem raised by biological history is that history and biogeography also affect the diversification of species. To the extent that the histories of each region of the earth have differed, we might also expect biological diversity and the development of biological communities to differ. Because of this, it may be difficult to interpret patterns of diversity solely in terms of local environmental conditions.

The simplest way to test for historical and phylogenetic effects is to compare systems that occur under a variety of similar ecological conditions in different geographic regions. If, on the one hand, the characteristics of these systems closely parallel ecological conditions and do not differ by region, we may conclude that ecology predominates over history and may safely ignore historical factors. This is the principle of convergence, which we shall discuss in greater detail later in this chapter. By **convergence**, we simply mean similarity of form and function under similar environmental conditions, despite dissimilarity in historical origin. If, on the other hand, similar habitats support different numbers of species in different regions, then ecologists must accept the impact of history on the present and incorporate historical factors into their analyses. Thus the consideration of history and biogeography is important to ecologists not only because of their intrinsic interest but also because of their potential for explaining the character of present-day ecological systems.

In this chapter, we shall first briefly examine some of the historical processes that have shaped the distribution and development of ecological systems. Then we shall examine the principle of convergence, paying particular attention to the diversity of biological communities. We shall see that history and biogeography have indeed influenced the character of local communities and have played an important role in the development of patterns of diversity.

The geological time scale

The earth formed about 4.5 billion years ago, and life arose within its first billion years. For most of the history of the earth, life forms remained primitive. The eukaryotic cell, which is the basic building block of all modern, complex organisms, is a product of the last billion years of evolution. Little record of this development exists, because very ancient life forms did not have hard skeletons or shells that form fossils. Much of the evidence of early complex life forms consists of tracks and burrows in the mud in which they lived.

All of this changed about 600 million years ago with the appearance in the fossil record of most of the modern phyla of organisms: echinoderms, arthropods, gastropods, and brachiopods rose to prominence in the oceans of that period. No one knows for sure why animals began to protect themselves with hard shells at that moment in history, but paleontologists regard the occasion as the beginning of life in its modern form. The interval of time between that critical moment and the present, occupying about one-eighth of the total history of the earth, has been divided into a series of eras, periods, and epochs (see Table 25.1). The first of these divisions is the Paleozoic era — *Paleozoic* means "old animals" — and the Cambrian is the first period within the Paleozoic era.

The divisions coincide with noticeable changes in the fauna and flora of the earth, changes easily perceived in the fossil record. Thus the end of the Cambrian period marks the disappearance of several prominent groups from the fossil record and their replacement by others not seen before. The major divisions at the end of the Paleozoic era and between the Mesozoic ("middle animals") and Cenozoic ("recent animals") eras also coincided with major extinctions of animal taxa: trilobites, among others, in the first case and dinosaurs in the second. Thus the boundaries between the various periods of paleontological time signal either major disruptions or less pervasive changes in the course of the development of life forms. In some cases, discontinuities in the rock strata at these boundaries reflect geological changes in the earth's crust. In other cases, the disruptions were caused by explosive impacts of extraterrestrial bodies on the surface of the earth.

Continental drift

The earth's surface has been very restless over its history, driven by giant convection currents in the semimolten material of the underlying mantle. The continents are islands of low-density rock floating on the denser material of the earth's interior. Currents of flow in the mantle carry the continents along like gigantic logs on the surface of water. At times in the past the continents have coalesced, and at other times they have drifted apart. This movement of the land masses on the surface of the earth is called **continental drift**. The process has two extremely important consequences for ecological systems. First, the positions of the continents and major ocean basins profoundly influence weather patterns, as we shall see later. Second, continental drift creates and breaks down barriers to dispersal alternately connecting and separating the evolving biotas of different regions of the earth.

For our purposes, we shall consider continental drift beginning with the early part of the Mesozoic era, about 200 million years ago, when all the continents were joined in a giant land mass known as **Pangaea** (Figure 25.1). By 135 million years ago, at the beginning of the Cretaceous period, the northern continents, which collectively made up **Laurasia,** had separated from the southern continents, **Gondwana.** In addition, Gondwana itself had begun to break up into three parts: West Gondwana, including Africa and South America; East Gondwana, including Antarctica and Australia; and India, which had separated from present-day Africa and was drifting toward Asia (Table 25.2 and Figure 25.2). Seventy million years later, at the end of the Cretaceous period and the Mesozoic era, South America and Africa were widely separated. The connection between Australia and South America through a temperate Ant-

TABLE 25.1	*The geological time scale*			

Era	Period	Epoch	Distinctive features	Years before present
Cenozoic	Quarternary	Recent	Modern humans	11,000
		Pleistocene	Early humans	1,700,000
	Tertiary	Pliocene	Large carnivores	5,000,000
		Miocene	First abundant grazing animals	23,000,000
		Oligocene	Large running mammals	38,000,000
		Eocene	Many modern types of mammals	54,000,000
		Paleocene	First placental mammals	65,000,000
Mesozoic	Cretaceous		First flowering plants; extinction of dinosaurs and ammonites at end of period	135,000,000
	Jurassic		First birds and mammals; dinosaurs and ammonites abundant	192,000,000
	Triassic		First dinosaurs; abundant cycads and conifers	223,000,000
Paleozoic	Permian		Extinction of many kinds of marine animals, including trilobites	280,000,000
	Carboniferous	Pennsylvanian	Great coal-forming forests; conifers; first reptiles	321,000,000
		Mississippian	Sharks and amphibians abundant; large primitive trees and ferns	345,000,000
	Devonian		First amphibians and ammonites; fishes abundant	405,000,000
	Silurian		First terrestrial plants and animals	
	Ordovician		First fishes; invertebrates dominant	495,000,000
	Cambrian		First abundant record of marine life; trilobites dominant, followed by massive extinction at end of period	570,000,000
	Precambrian		Fossils extremely rare, consisting of primitive aquatic plants	

Source: J. H. Brown and A. C. Gibson, *Biogeography.* Moseby, St. Louis (1983).

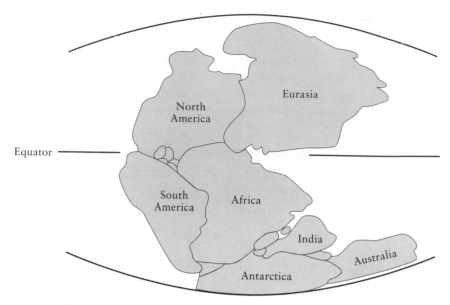

FIGURE 25.1 Positions of the continents at the beginning of the Mesozoic era, when all the land had coalesced into a single land mass, known as Pangaea. After E. C. Pielou, *Biogeography*. Wiley, New York (1979).

arctica finally dissolved about 50 million years ago. At about the same time, in the Northern Hemisphere the widening Atlantic Ocean finally separated Europe and North America, but a land bridge had already formed on the other side of the world between North America and Asia.

Many details of continental drift have yet to be resolved, particularly in such complicated areas as the Caribbean Sea, Australasia, and the Mediterranean Sea–Persian Gulf region. Nevertheless, the past history of connections between the continents endures in the distributions of animals and plants. We have only to look at the distribution of the flightless ratite birds to see the connection between the southern continents: emus and cassowaries in Australia and New Guinea, rheas in South America, ostriches in Africa, and the extinct moas of New Zealand — all descended from a common stock that inhabited Gondwana before its breakup.

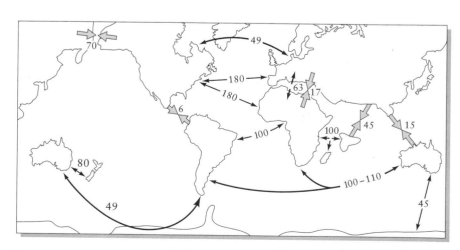

FIGURE 25.2 Estimates of the times (in millions of years before present) at which direct dispersal routes between land masses were made or broken. Hollow converging arrows show joins (the join between Australia and Asia refers to the narrowing of a gap and the appearance of stepping-stone islands). Black diverging arrows show separations. After E. C. Pielou, *Biogeography*. Wiley, New York (1979).

TABLE 25.2	*Estimated times of some of the major biogeographic events in earth history caused by continental drift*

Period or epoch	Time (million years before present)	Event
Early Triassic	200	The continental crust formed a single continent, Pangaea
Late Triassic	180	West Laurasia (North America) ↔ Africa; West Gondwana (Africa + South America) ↔ India ↔ East Gondwana (Australia + Antarctica)
Early Cretaceous	135–125	South America ↔ Africa in far south because of rotational movement
Mid-Cretaceous	110–100	South America ↔ Africa at latitude of Brazil; Africa ↔ Madagascar ↔ India; Africa, India, and Australia all drifting northward
Late Cretaceous	80	North America ↔ (Europe + Greenland); (Antarctica + Australia) ↔ (New Zealand + New Caledonia)
Very late Cretaceous	70	Contact made between northwestern North America and northeastern Siberia
Very early Paleocene	63	Africa ↔ Europe (temporarily)
Eocene	49	Dispersal route between North America and Eurasia, from being predominantly via North Atlantic, switches to Beringia because North Atlantic becomes wider and Beringia becomes warmer
Eocene	~49	Australia ↔ Antarctica
Eocene	45	India drifts into contact with Asia
Oligocene	~30	Turgai Strait (east of Ural Mountains) finally dries up
Miocene	17	Europe and Africa rejoined
Miocene	15	The narrowing gap between Australia and Southeast Asia, and the appearance of stepping-stone islands, permits plant dispersal
Pliocene	6	North America and South America joined by land bridge

Note: Double-headed arrows denote separations.
Source: E. C. Pielou, *Biogeography*. Wiley, New York (1979).

Biogeographic regions

The distributions of animals suggested to the nineteenth-century naturalist Alfred Russel Wallace, codiscoverer with Darwin of the theory of evolution by natural selection, six major biogeographic regions (Figure 25.3). We now know that these correspond to land masses isolated many millions of years ago by continental drift. Over the course of that isolation, the animals and plants of each region developed distinctive characteristics independently of evolutionary changes in other regions. The Nearctic and Palearctic regions, corresponding roughly to North America and Eurasia, maintained connections across either what is now Greenland or the Bering Strait between Alaska and Siberia throughout most of the past 100 million years. As a result, these two areas share many groups of animals and plants. The forests of Europe seem familiar to travelers from North America; few of the species are the same, but both regions have representatives of many of the same genera and families.

The continents of the Southern Hemispere, particularly Africa (the Ethiopian biogeographic region) and Australia, experienced long histories of isolation from the rest of the terrestrial world, during which time many distinctive forms of life evolved in each. Finally, the Oriental region includes the tropical regions of Southeast Asia, which were isolated from the tropical areas of Africa and South America, and contributions from the land mass of India, which drifted into contact with southern Asia about 45 million years ago. As one might expect, temperate and tropical Asia have closer affinities than temperate North America and tropical South America because of the continuous land connection between them. Indeed, the temperate forests of Asia contain a high percentage of species derived primarily from tropical forests, whereas those of temperate North America lack such species.

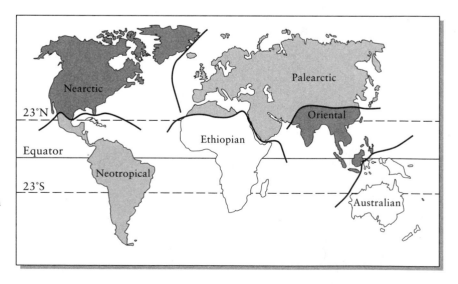

FIGURE 25.3 Major biogeographic regions of the earth. This scheme, which is widely accepted today, originated with Alfred Russel Wallace in 1876. From J. H. Brown and A. C. Gibson, *Biogeography*. Mosby, St. Louis (1983).

Climatic history

The climate patterns of the earth ultimately depend on energy received from the sun, which warms the land and seas and evaporates water. Within this framework, the distribution of heat over the surface of the earth depends largely on the circulation of the oceans, which is driven by the rotation of the earth and constrained by the positions of the continents. In periods when polar regions, which receive relatively little solar energy, are covered by oceans with circulation to tropical areas, currents distribute heat rather evenly over the surface of the earth, and temperate climates extend quite close to the poles. When the polar regions are occupied by land masses or land-locked oceans, they can become very cold indeed, as they are at present. But this has not always been the case.

When the continents coalesced into one or a few large land masses, ocean currents circulated quite freely to high latitudes, and the climate of the earth was much more equitable than it is now. Fifty million years ago, much of North America and Europe were tropical, and the Antarctic land connection between South America and Australia supported luxuriant temperate vegetation and animal life, or so the fossils in Antarctic rocks tell us. However, as Antarctica drifted over the South Pole and as the north polar ocean became trapped between North America and Eurasia, the earth's climate became more strongly differentiated into tropical (equatorial) and temperate (polar) climates. One consequence of this cooling (and drying) trend at high latitudes was the retreat to lower latitudes of plants and animals that could not tolerate freezing and, accordingly, the greater distinction between temperate and tropical biotas. During the early part of the Tertiary period, what is now temperate North America supported a mixture of tropical and temperate forms growing side by side. Today these plants and animals occupy different climate zones, the greater stratification of climate being matched by greater stratification of the biota.

Gradual changes in climate had profound effects on the geographic distributions of plants and animals. During the past 2 million years, however, the gradual cooling of the earth gave way to a series of violent oscillations in climate that had immense effects on habitats and organisms in most parts of the world. This was the Ice Age, or Pleistocene epoch — an alternation of cooling and warming that led to the advance and retreat of ice sheets over much of the Northern Hemisphere and caused cycles of wet and dry periods in the tropics. Ice came as far south as Ohio and Pennsylvania and covered much of northern Europe, driving vegetation zones southward, restricting tropical forest to isolated **refugia** of moist conditions, and generally disrupting biological communities all over the world.

One of the most striking and best documented examples of this disruption concerns the forest trees of eastern North America. Pollen grains deposited in lakes and bogs left by retreating glaciers in the northeastern United States record the coming and going of plant species. These records show plainly that the composition of plant associations in the past changed as species migrated over different routes across the landscape. After the last glaciers began to retreat about 12,000 years ago, the general pattern of reforestation began with spruce forest, which dominated the area until about 10,000 years ago, followed by associations of pine and birch, which were later replaced by more temperate species such as elm and oak.

The migration of tree species from their southern refugia since the height of the last glaciation appears on maps depicting representative species in Figure 25.4. For such species as beech and hickory, postglacial migration involved northerly range extension from southern regions across most of the eastern United States. In contrast, white pine and chestnut appear to have emerged from refugia in the Carolinas and expanded their ranges to the west as much as to the north. Thus the composition of forests during the past 10,000 years included combinations of species that do not occur anywhere in eastern North America at the present time. For many species, the environment changed too rapidly during the Ice Age, and they disappeared altogether.

Catastrophies in earth history

Although the Ice Age brought rapid change in climate and spelled the extinction of many forms of plants and animals, it pales beside the total disruption occasionally visited upon the earth by collision with asteroids and other extraterrestrial bodies. This has happened many times in the history of the earth, with consequences in direct proportion to the energy liberated by the impact. On occasion, such impacts have caused the widespread destruction of ecosystems and the extinction of many kinds of life on earth.

The most famous of these impacts occurred about 65 million years ago. Evidence now points to the vicinity of the island of Hispaniola in the Caribbean Sea as the point of impact, but evidence of the explosion and its aftermath appears in geological strata all over the world. Scientists have estimated that an asteroid 10 km in diameter traveling 25 km per second may have been responsible. Such a collision would have released enough energy to cause massive tidal waves around the world, start fires on an unprecedented scale, and throw so much dust into the air as to block the sun and cool the earth's surface for years. As a result, much of the biomass of the earth would have been destroyed immediately by the direct effects of the impact or more slowly during the weeks and months afterward, and plant production in the oceans and on the land would have slowed to a standstill. All the killing left a thin band of carbon preserved in sedimentary rocks of the time, along with thick deposits resulting from massive erosion in some areas.

One of the results of the killing was the extinction of a large fraction of the species on earth and of many larger taxa as well. These episodes are referred to as **mass extinctions.** Not all plants and animals felt the effects of the impact equally. All the dinosaurs disappeared, as did some large groups of marine organisms. Most higher taxa of plants survived, perhaps many of them as seeds in the soil, and mammals survived to fill the vacancies left by the dinosaurs.

Catastrophies of such magnitude have happened infrequently — perhaps at intervals of tens or hundreds of millions of years — yet often enough to disrupt ecosystems and change the course of community development. Each major catastrophe brings about a period of extreme environmental stress. Geologists

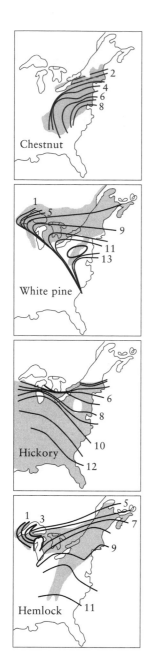

FIGURE 25.4 Migration of four species of trees in eastern North America from Pleistocene refugia to their present distributions following the retreat of the glaciers. The numerals associated with the distribution lines indicate thousands of years before the present. From M. B. Davis, *Geoscience and Man* 13:13–26 (1976).

have found evidence in the geochemistry of sediments formed after such catastrophes that thousands of years may be required for environmental conditions to return to normal; the fossil record shows that some ecosystems — tropical reefs are a case in point — may disappear for millions of years, sometimes to be rebuilt by new kinds of reef-forming organisms.

From the perspective of community development, catastrophes have several important consequences. They may eliminate species and higher taxa and thus greatly reduce the diversity of most systems. They may foster rapid evolutionary response to new types of conditions, and these changes many remain long after conditions have returned to "normal." Finally, they may create opportunities for the development of new types of biological associations. Although these effects cannot be easily identified or interpreted from present-day conditions and communities, such unique events in the past may reach down through history to influence the present.

Convergence

If long periods of isolation have led to the production of unique life forms in many regions of the earth, similar environmental conditions within each of these regions have also led to the evolution of similar solutions to these common problems. Although the plants of Mediterranean climate zones in western North America and southern Africa have different evolutionary origins reflecting more than 100 million years of isolation, they share similar life forms and similar adaptations to the winter-rainfall–summer-drought conditions. Thus the different evolutionary histories and taxonomic affinities of plants and animals of the earth's regions are in part obliterated by convergence in form and function.

Where woodpeckers are absent from a fauna, as they are from many isolated islands, other species may adapt to fill their role (Figure 25.5). Rain forests in Africa and South America are inhabited by plants and animals that have different evolutionary origins but are remarkably similar in appearance (Figure 25.6). Plants and animals of North and South American deserts resemble each other morphologically more than one would expect from their different phylogenetic origins. Similarities have also been noted in the behavior and ecology of Australian and North American lizards, despite the fact that they belong to different families and have been separated for perhaps 100 million years.

Convergence exists, and it reinforces our belief that the adaptations of organisms to their environments obey certain general rules governing structure

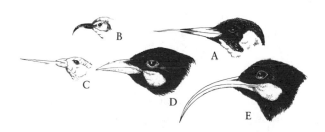

FIGURE 25.5 Unrelated birds that have become adapted to extract insects from wood. (a) The European green woodpecker excavates with its beak and probes with its long tongue. (b) The Hawaiian honeycreeper *Heterorhynchus* taps with its short lower mandible and probes with its long upper mandible. (c) The Galapagos woodpecker-finch trenches with its beak and probes with a cactus spine. The New Zealand huia (now extinct) divided foraging roles between the sexes. The male (d) excavated with its short beak, and the female (e) probed with her long beak. After D. Lack, *Darwin's Finches.* Cambridge Univ. Press, Cambridge, England (1947).

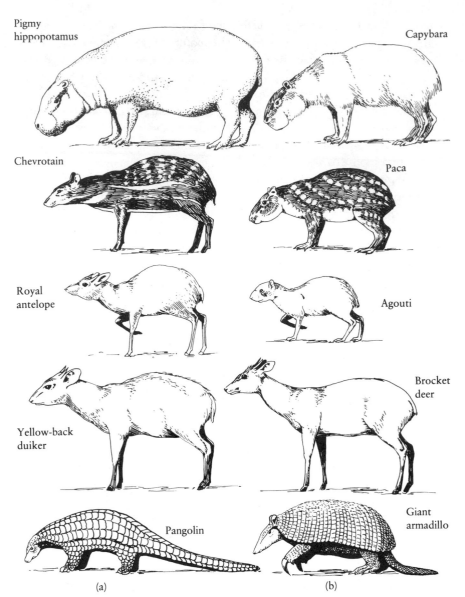

Pigmy hippopotamus

Capybara

Chevrotain

Paca

Royal antelope

Agouti

Yellow-back duiker

Brocket deer

Pangolin

Giant armadillo

(a)

(b)

FIGURE 25.6 Morphological convergence between unrelated (a) African and (b) South American rain forest mammals. Each pair is drawn to the same scale. After F. Bourlière, in B. J. Meggars, E. S. Ayensu, and W. D. Duckworth (eds.), *Tropical Forest Ecosystems in Africa and South America: A Comparative Review.* Smithsonian Institution Press, Washington, D.C. (1973), pp. 279–292.

and function. However, detailed studies often turn up remarkable differences between the plants and animals in superficially similar environments. Despite striking convergences among desert-dwelling organisms, the ancient Monte Desert of South America is the only desert region of the world that lacks bipedal, seed-eating, water-independent rodents, such as the kangaroo rats of North America and the gerbils of Asia. Among frogs and toads, however, several South American forms have carried adaptation to desert environments a step further than their North American counterparts: they construct nests of foam to protect their eggs from drying. Differences between the Australian agamid lizard *Amphibolurus inermis* and its North American iguanid analog, *Dipsosuarus dorsalis,* include diet, optimal temperature for activity, burrowing behavior, and annual cycle, even though at first glance the species are dead ringers for each other.

Coevolved relationships between species also reveal unique biogeographic quirks. One example drawn from the dispersal of seeds by ants illustrates the complexity of the problem. This interrelationship — a type of mutualism — is encouraged by the presence of edible appendages, called elaiosomes, on the seeds. As we saw in Chapter 21, ants gather these elaiosomes, with the seeds attached, and carry them into underground nests. In so doing, they effectively plant the seeds. This seed trait is uncommon in most of the world, being restricted primarily to a few species of trees in mesic environments. In Australia and South Africa, however, the trait is well represented among xerophytic shrubs, and it signals ecological and morphological features lacking in ant-dispersed plants elsewhere.

Ecologists have not resolved whether this difference between plants in Australia and those elsewhere is a consequence of the unique evolutionary history of the Australian flora. Indeed, it has been suggested that the poor soils of Australia and South Africa make it costly for plants to produce the nutritionally expensive fleshy fruits that are dispersed by birds and mammals in most parts of the world. Thus the dispersal of seeds by ants in Australia and South Africa may represent an accident of unique local *geology* rather than indicating the unique historical origins of the flora and fauna. Similar reasoning suggests that many of the distinctive attributes of the reptile fauna of Australia resulted not from its unique evolutionary history but from the absence of avian predators, whose scarcity could be attributed ultimately to the poor nutrient status of the vegetation and the paucity of insects. These issues have yet to be fully resolved. Certainly, ecologists must ensure that the comparisons they make involve habitats with closely matched physical characteristics. Otherwise, they cannot conclude unequivocally that differences in structure or function have resulted from different histories rather than different environments.

Community convergence

With regard to the principle of convergence, we may treat attributes of community organization, such as species diversity, in the same manner as the adaptations of organisms. Accordingly, we would expect independently derived communities that occupy the same habitats in different regions to have similar numbers of species. We can test this principle quite simply by comparing biodiversity in similar habitats in different parts of the world. Where ecologists have done this, they have had mixed results. In one of the first such studies, the relationship between bird species diversity and habitat complexity was found to be similar in eastern North America and Australia (Figure 25.7). This result agreed with the idea that species diversity depends on habitat type, is ultimately determined by interactions among species within the habitat, and does not display the influence of historical differences between the continents. Other studies, however, have revealed strong differences between continents, suggesting that regional and historical factors also play a role in determining diversity. We shall look at just two examples here.

The deciduous forests of eastern North America contain 253 species of trees, more than twice the number (124) found in similar habitats in Europe.

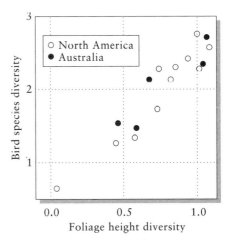

FIGURE 25.7 The relationship between bird species diversity and the structural complexity of the habitat, as indicated by the diversity of vegetation layers (foliage height diversity) in moist temperate habitats of Australia and North America. From H. F. Recher, *Amer. Nat.* 103:75–80 (1969).

TABLE 25.3	Taxonomic diversity of trees in the deciduous, broad-leaved forests of eastern North America, Europe, and eastern Asia

	NUMBER OF TREE TAXA IN MOIST TEMPERATE FORESTS IN		
Taxa	Europe	Eastern North America	Eastern Asia
Orders	16	26	37
Families	21	46	67
Genera	43	90	177
Species	124	253	729
Percent of genera predominantly tropical	5	14	32

Temperate eastern Asia, which also has a climate similar to that of eastern North America, has 729 species of trees (Table 25.3). Thus, although the climates of the three regions are similar and the forests have similar structures — they are dominated by deciduous, broad-leaved trees — species diversity varies by a factor of nearly 6 among the different regions. These figures represent the total diversity of the region, but local diversity within small areas of uniform habitat exhibits parallel differences. Thus regional diversity and local diversity appear to be closely related.

A part of the greater diversity in Asia results from a greater proportion of species (32 percent) belonging to predominantly tropical genera. A continuous corridor of forest habitat from the tropics of southeast Asia to the north facilitates the invasion of temperate ecosystems by tropical plants and animals. In the Americas, the wet tropics of Central America are separated from the moist, temperate areas of North America by a broad subtropical band of dry vegetation. In Europe, the Mediterranean Sea and arid North Africa effectively separate temperate ecosystems from tropical Africa. The fossil record suggests an ancient origin to the diversity anomaly among eastern North America, Europe, and eastern Asia. Almost twice as many genera of trees occur as fossils in eastern Asia as in Europe and in North America (Table 25.3). Europe and North America have similar numbers of genera in their fossil records, but a large proportion of these genera became extinct in Europe in association with the cooling of north temperate climates and the Ice Age. As Europe cooled, the Alps and the Mediterranean Sea posed effective barriers to southward movement, and many of the cold-intolerant varieties of plants died out. In North America, southward migration to the area bordering the Gulf of Mexico was always possible during cold periods.

A similar anomaly exists in the species diversity of mangrove forests between the Caribbean region and the Indo–West Pacific region. Mangroves are tropical forests that occur within the tidal zone along coastlines and river

FIGURE 25.8 Mangrove vegetation in an estuary on the Pacific coast of Costa Rica. Note the prop roots of *Rhizophora* trees at left and the buttressed trunks of *Pelliciera* at right. These trees are established on mud substrate within the tidal range, hence the soil is flooded periodically with salt water.

deltas (Figure 25.8). Mangrove trees tolerate high salt concentrations and anaerobic conditions in the water-saturated sediments in which they take root. Fifteen taxa of trees have independently colonized the mangrove habitat from terrestrial ancestors, and several of these have subsequently diversified there. We cannot explain the much greater diversity of the Indo – West Pacific region on the basis of habitat, because both regions have roughly equal areas of a similar variety of mangrove habitats (Figure 25.9). It appears instead that plant taxa have invaded the mangrove habitat more frequently in the Indo – West Pacific than in the Caribbean region, although the reasons for this are not clear. Possibly, terrestrial habitats fringing much of the Caribbean were arid during the latter part of the Cenozoic era. As a result, wet terrrestrial forest vegetation has

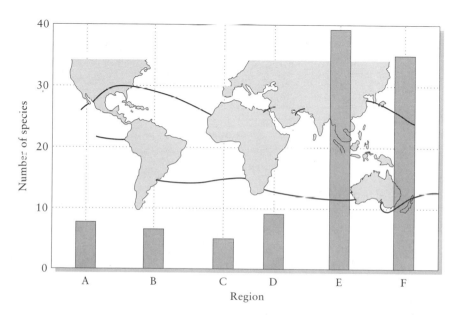

FIGURE 25.9 Limits of distribution of mangrove vegetation along the coasts of continents and islands of the world. Bars show the numbers of species of mangrove trees and shrubs in each of six regions: (A) eastern Pacific Ocean, (B) Caribbean Sea and western Atlantic Ocean, (C) eastern Atlantic Ocean, (D) western Indian Ocean, (E) eastern Indian Ocean, and (F) western Pacific Ocean. After V. J. Chapman, *Tropical Ecol.* 11:1 – 19 (1970); P. Saenger, E. J. Hegerl, and J. D. S. Davies, *Environmentalist* 3 (suppl. 3):1 – 88 (1983).

had little direct contact with the mangrove habitat and thus has provided little opportunity for terrestrial taxa to adapt gradually to mangrove conditions. This has not been a limiting factor in southeast Asia, where wet conditions have prevailed in tropical habitats throughout most of the period of evolution of modern trees.

The scales of processes regulating biodiversity

Perhaps the temperate forest and mangrove examples that we have just considered represent extremes, although similar disparities are found in marine communities between the more diverse Indo–West Pacific region and the less diverse Caribbean region, and numerous less striking examples have been reported. Nonetheless, these examples emphasize that the history and biogeographic position of a region may influence the diversity of both the entire region and the local habitat; interactions of species within local habitats make up only half of the diversity equation.

Several processes are important to the regulation of biodiversity, each with a different characteristic scale of time and space. Scale in space varies from the activity range of the individual, through the geographic and ecological dispersal of individuals within populations, to the expansion and contraction of the geographic ranges of populations. Scale in time varies according to the rates of individual and population movements, the dynamics of population interactions, and the selective replacement of genotypes within populations (evolution). Both local, contemporary processes and regional, historical processes shape community attributes. The fate of a local population depends on the balance between the tendency of intolerable physical conditions, interspecific competition, and predation to exclude the population locally and the rate of dispersal of individuals to that point from surrounding areas of population surplus. Local diversity of species depends on the balance between local rates of extinction — resulting from predators, disease, competitive exclusion, and changes in the physical environment — and regional rates of species production and immigration. Every point on earth has limited access, via dispersal, to sources of colonizing species. Local diversification depends not only on the capacity of the environment to support a variety of species but also on the access of the region to colonists, the capacity of the region to generate new forms through speciation, and its ability to sustain taxonomic diversity in the face of varying environments. Although ecology has traditionally focused on local, contemporary systems, it is now expanding its purview to embrace global and historical processes.

❦ SUMMARY

1. Life first evolved several billion years ago, but an abundant fossil record of modern life appeared about 570 million years ago, a point that marks the beginning of the Paleozoic era of geological history. The Mesozoic era, dominated on land by the reptiles, began about 220

million years ago; the age of mammals, the Cenozoic era, began 65 million years ago.

2. The positions of the continents have changed continuously throughout the evolution of life, opening and closing different pathways of dispersal between continental land masses and greatly altering the environment on earth.

3. Because animals and plants have evolved independently on different continents during prolonged periods of geographic isolation, we can distinguish biogeographic regions. The largest of these are the Neotropical, Ethiopian, and Australian, derived from the former land mass of Gondwana, and the Oriental, Palearctic, and Nearctic, derived for the most part from the Northern Hemisphere land mass of Laurasia.

4. The climate of the earth cooled considerably during the Cenozoic era, causing a contraction of tropical environments to a narrower equatorial band and an expansion of temperate and arctic environments, which are dominated by freezing temperatures for a part of the year. This cooling trend culminated in the Ice Age (alternating periods of glacial advance and retreat in the Northern Hemisphere), which caused the extinction of many species of plants and animals.

5. Infrequent global catastrophies have punctuated the general development of life when large extraterrestrial bodies hit the earth. One of the best known of these events occurred 65 million years ago, caused the extinction of the dinosaurs and other large taxa of animals, and brought the Mesozoic era to a close. Such catastrophic changes in the environment and its inhabitants have opened up new opportunities for evolution and resulted in the drastic reorganization of biological communities.

6. Despite their different histories of independent evolution, the inhabitants of similar environments on different continents often resemble one another because they adapt to similar ecological factors. This is the principle of convergence of form and function.

7. If community diversity is regulated by local interactions among species, whose outcome is determined primarily by environmental conditions, then biodiversity should also exhibit convergence between regions.

8. Several examples of nonconvergence in the diversity of temperate forests and mangrove forests demonstrate that the unique histories and biogeographic settings of each continent also influence local species diversity.

9. Biodiversity reflects a broad array of local, regional, and historical processes and events operating on a hierarchy of temporal and spatial scales. Thus, understanding patterns of species diversity requires consideration of the history of a region and the integration of

ecological study with the related disciplines of systematics, evolution, biogeography, and paleontology.

❦ SUGGESTED READINGS

Ben-Avraham, Z. 1981. The movement of continents. *American Scientist* 69:291–299.

Brooks, D. R., and D. A. McLennan. 1991. *Phylogeny, Ecology and Behavior. A Research Program in Comparative Biology.* Univ. of Chicago Press, Chicago.

Brown, J. H., and A. C. Gibson. 1983. *Biogeography.* C. V. Mosby, St. Louis.

Carlquist, S. 1981. Chance dispersal. *American Scientist* 69:509–516.

Farrell, B. D., C. Mitter, and D. J. Futuyma. 1992. Diversification at the insect–plant interface. *BioScience* 42:34–42.

Flessa, K. W. 1986. Causes and consequences of extinction. In D. M. Raup and D. Jablonski (eds.), *Patterns and Processes in the History of Life.* Springer-Verlag, Heidelberg and New York, pp. 234–257.

Marshall, L. G. 1988. Land mammals and the Great American Interchange. *American Scientist* 76:380–388.

Orians, G. H., and R. T. Paine. 1983. Convergent evolution at the community level. In D. J. Futuyma and M. Slatkin (eds.), *Coevolution.* Sinauer, Sunderland, Mass., pp. 431–458.

Otte, D., and J. Endler (eds.). 1989. *Speciation and Its Consequences.* Sinauer, Sunderland, Mass.

Ricklefs, R. E., and G. W. Cox. 1972. Taxon cycles in the West Indian avifauna. *American Naturalist* 106:195–219.

Ricklefs, R. E., and R. E. Latham. 1992. Intercontinental correlation of geographical ranges suggests stasis in ecological traits of relict genera of temperate perennial herbs. *American Naturalist* 139:1305–1321.

Stucky, R. K. 1990. Evolution of land mammal diversity in North America during the Cenozoic. *Current Mammalogy* 2:375–432.

Vermeij, G. J. 1991. When biotas meet: understanding biotic interchange. *Science* 253:1099–1104.

Ecological Applications

EXTINCTION AND CONSERVATION

Humans have an immense impact on the earth. There are so many of us (the 1992 population of 5.4 billion is increasing at a rate of almost 2 percent a year), and each individual uses so much energy and so many resources, that our activities influence virtually everything in nature. Most of the land surface of the earth and, increasingly, the oceans have come under the direct control of humankind. Virtually all areas within temperate latitudes that are suitable for agriculture have been brought under the plow or fenced. Worldwide, fully 35 percent of the land area of the earth is used for crops or permanent pastures; countless additional hectares are grazed by livestock. Tropical forests are being felled at the alarming rate of 17 million hectares (almost 2 percent of the remaining primary stands) each year. The semiarid subtropics, particularly in subsaharan Africa, have been turned into deserts by overgrazing and collecting of wood for fuel. Rivers and lakes overflow with the wastes of a consuming society. The atmosphere reeks of gases produced by chemical industries and the burning of fossil fuels.

We are fouling our nest and are rushing to exploit much of what remains to be taken. Inevitably, this deterioration of the environment will lead to a declining quality of life for all human inhabitants of the earth, as it already has for many. The animals and plants with which we share this planet, and on which we depend for all kinds of sustenance, feel the impact of human life all the more. They have been pushed aside as we have taken over land and water for our own living space and for the production of food. Their environments have been poisoned by our wastes. Entire species have succumbed to habitat destruction, hunting, and other forms of persecution.

This deterioration need not continue. Humans can live in a clean and sustaining world, but only by placing support for our own population into balance with the preservation of other species and the ecological processes that nurture us all. The science of ecology has much to say about the rational development and management of the natural world as a sustainable, self-replenishing system. What we have learned about the adaptations of organisms, the dynamics of populations, and the processes that occur in ecosystems suggests simple but urgent guidelines for living in reasonable harmony with the natural world.

First, environmental problems can never be brought under control as long as the human population continues to increase. Certainly the earth could support many more individuals than it does at present, but the quality of life would be drastically reduced in the short term, and there would be little prospect for sustainability in the long term. The world doesn't need more people; people don't need more people than exist already. For the sake of the generations of humans to come, population growth must be brought to a halt.

Studies of natural populations show that their control depends on factors that act in density-dependent fashion; these factors (which include food shortage, disease, predation, and social strife) reduce fecundity, increase mortality, or both, as the population grows. If the human population were to come under such external control, the toll in human suffering—disease, famine, warfare—would be intolerable. However, to maintain individual quality of life at a high level will require that humans exhibit a reproductive restraint that defies the entire history of evolution, during which "fitness" has been measured in terms of breeding success rather than quality of life. Only an increasing appreciation of the negative economic and environmental consequences of overpopulation will cause humanity to value the individual human experience over numbers of progeny as the two become increasingly incompatible.

Second, individual consumption of energy, resources, and food produced on higher trophic levels must be reduced. It is inconceivable that the earth could sustain the resource and energy depletion that would result if *everyone* consumed at the level now exhibited by affluent citizens of developed countries. Efficiency can be increased and superfluous consumption reduced without impairing comfort or the enjoyment of life. Insistence on a high-energy lifestyle magnifies the strain that overpopulation inflicts on the world's resources and on the quality of the environment. Each individual human can reduce her or his impact by eating lower on the food chain (reducing meat consumption), investing in energy- and resource-efficient technologies, and living closer to equilibrium with the physical world (lowering the thermostat setting in winter and raising it in summer are simple but effective).

Third, although it is inevitable that most of the world will come under human management, systems should be maintained as close to their natural state as possible to keep natural ecosystem processes intact. As a general rule, the less we alter nature, the easier it will be to sustain the world in a healthy condition. For example, many areas covered by tropical forests are unsuitable for grazing or agriculture because these activities upset natural processes of ecosystem maintenance and cause the land to deteriorate. Such areas should be left in forest as preserves, recreation areas, or sites for sustained exploitation of forest products. Deserts can be watered, and they often become tremendously productive for certain types of agriculture. But the costs of maintaining such managed systems can be extremely high as soils accumulate salts from irrigation water and as aquifers become depleted. Living with nature is always preferable to, and less costly than, going against its grain.

This chapter and the next elaborate on these themes, each addressing different aspects of applied ecology. In this chapter, we shall consider the problem of conserving species — that is, preventing their populations from dwindling to extinction. In the following chapter, we shall discuss the rational development of the natural world, which includes maintaining natural populations and ecosystem processes so that we and future generations will benefit. The problems that will arise in both efforts can be understood by applying basic principles of ecology. We must remember, however, that although solutions can be advanced from an ecological point of view, implementing them will require concerted social, political, and economic action.

Biodiversity

More than 1,400,000 species of plants and animals have been described and given Latin names (Figure 26.1). Many more, particularly in poorly explored regions of the tropics, await discovery by science. Some experts have estimated that the final species count could reach between 10 and 30 million, the vast majority of them insects. These estimates may be inflated, but it is incontestable that we share this planet with several million other kinds of plants, animals, and microbes.

Making lists of species names is one way of tabulating diversity. The concept of **biodiversity** also includes the unique attributes of all living things. Thus, although each species differs from every other in the name that science has assigned it, it also differs in the way its adaptations define its place in the ecosystem. For example, different species of plants have dissimilar tolerances of soil conditions and water stress and have disparate defenses against herbivores; they also differ in growth form and strategies for pollination and seed dispersal. Animals vary in their own obvious ways. Heterogeneity is a key to biodiversity.

The important differences between species result from genetic changes, or evolution. And evolution requires the presence of genetic variation within populations; otherwise no change can occur. Because genetic variation is crucial to the continued evolutionary response of populations to changes in the environment, **genetic diversity,** which occurs both between and within species, is another important component of biodiversity.

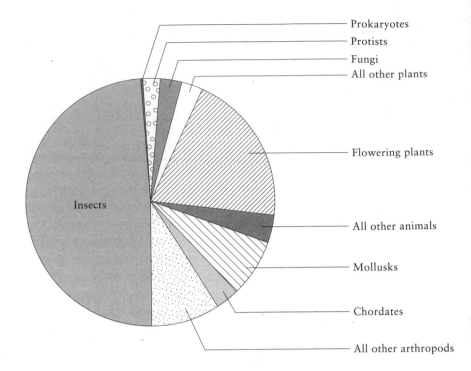

Prokaryotes
Protists
Fungi
All other plants

Flowering plants

All other animals

Mollusks

Chordates

All other arthropods

Insects

FIGURE 26.1 Proportions of 1,445,850 described species that belong to the major taxonomic groups of the five kingdoms of organisms. Data from V. Grant, *The Evolutionary Process.* Columbia Univ. Press, New York (1985).

Finally, biodiversity has a geographic component. Different regions have different numbers of species, and if diversity were a contest, tropical rain forests and coral reefs would be the clear winners. Equally important, however, is the fact that some regions boast unique species found nowhere else. Species whose distributions are limited to small areas are called **endemics,** and regions with large numbers of endemic species are said to possess a high level of **endemism.** Clearly, conservation of global biodiversity is best served by directing efforts toward areas of high endemism. Oceanic islands are well known for harboring unique forms; virtually all the birds, plants, and insects of such isolated islands as the Hawaiian and Galapagos archipelagos occur nowhere else (Figure 26.2). As a result, habitat destruction, hunting, and the introduction of alien species have a greater impact on biodiversity in these areas of high endemism than elsewhere. For example, fossil-bearing deposits have shown that more than half the avifauna of Hawaii has disappeared since human colonization of the islands. These birds occurred nowhere else. Now they are gone forever. So are the giant moas (ostrich relatives) of New Zealand. Steller's sea cow (a giant relative of the dugong and manatee) of the Bering Sea became extinct by 1768, less than 30 years after it was discovered and first hunted by Europeans.

Human activity now affects all regions of the earth (no refuge, not even in the deepest ocean abyss, exists in a pristine state), so more and more species are now vulnerable to extinction. Many will disappear before they become known to science. Some will carry to their graves valuable and irreplaceable genetic resources. Others will be missed because their presence on earth enriches our own lives. A great effort will be required to check the loss of biodiversity.

FIGURE 26.2 The Hawaiian silversword found only at high elevation on Haleakala Volcano on the island of Maui, Hawaii. Courtesy of the U.S. National Park Service.

The value of diversity

Why do we care? What concern is it of ours that a species of beetle disappears from South America? Many species already are gone. Do we really miss them? In fact, extinction occurs normally in natural systems. Why should we try to stop it? Of course, our concern does not center on natural extinction. But the rate of disappearance of certain kinds of species, particularly those subject to hunting, pollution, and destruction of habitat, is probably now at an all-time high in the history of the earth. The cause of this accelerated loss of species is directly linked to the growth and technological capacities of the human population.

The rationale for conserving biodiversity depends on the value we place on individual species. This value arises from many considerations related to our own personal interest and involvement. For many people extinction raises a *moral* issue. Some take the position that because humankind affects all of nature, it is our moral responsibility to protect nature. If morality derives from a natural law—that is, if morality is intrinsic to life itself—then we may presume that the rights of nonhuman individuals and species are as legitimate as the rights of individuals within human society. Of course, no species is guaranteed the right to perpetual existence, just as no human is guaranteed immortality. But extinction by unrestrained hunting, pollution, habitat destruction, and irresponsible spread of disease may be analogous to murder, manslaughter, genocide, and other infringements of individual human rights.

Throughout history, humans have shown even less sense of responsibility toward nature than toward each other. Whether or not species have natural rights, such rights have not been recognized in the past, nor are they likely to be generally accepted in the future. For many people, the practical problems of personal survival make it difficult to see nature in any other way than as a source of food and fuel; for some, morality is dictated more by personal greed than by concern for others—whether human or nonhuman.

In the absence of moral protection, the values of individual species must be argued from the standpoint of their *economic* and *recreational* benefits to humankind. This case depends on a number of factors. First, individual species have obvious economic importance as food resources; game species; and sources of forest and other natural products, drugs, and many organic chemicals, particularly oils and fragrances. For example, more than 100 important medicinal drugs (including codeine, colchicine, digitalin, L-dopa, morphine, quinine, strychnine, and vinblastine), which account for about one-quarter of all prescriptions filled in the United States, are extracted directly from higher plants. Some species that are of economic importance have been cultivated or domesticated and then selectively bred to enhance their desirable qualities. These are not in danger of extinction, but making room for their cultivation on a large scale has often endangered other species that are perceived as having lesser value. This recalls the classic conflict between sheep ranchers and wolves, which occasionally killed sheep and other livestock. Wolves were driven out of most of North America, often with handsome bounties on their heads. The point is that assigning economic value to species favors some over others and often does not address the issue of conserving biodiversity in a general sense.

At times, we may argue for the conservation of a habitat type by comparing the economic value of the native species that occur there and the value accrued from altering or otherwise managing the habitat. Under many circumstances, however, the short-term gain of converting forest to agriculture, for example, or of excessively exploiting a marine resource, is assumed to outweigh any long-term value of conserving the natural system for sustained income. The value of conserved species and habitats usually becomes apparent only when the long-term costs of overexploitation or habitat conversion are properly accounted for, a practice that is encouraged neither by the pragmatism of desperation nor by the notorious shortsightedness of politics.

High value may be placed on individual species because they attract **ecotourism** to an area. Many tropical countries have capitalized on this attraction by establishing parks and supporting services for tourists. In Latin America, the spectacular quetzals, macaws, and monkeys draw tourists to many areas where these species are protected. Diversity itself is often the attraction in tropical rain forests and coral reefs, with their hundreds of different species of trees, birds, or fish. In east Africa, lions, elephants, and rhinoceroses have great value because of the tourist dollars, pounds, francs, and yen they bring into countries that are badly in need of foreign currencies. Unfortunately, a few self-serving individuals prize elephants more for the value of their ivory and rhinoceroses for the value of their horn, which is made into dagger handles in some Arab countries and is believed by some Asians to have aphrodisiac properties. These considerations set up conflicts between the economic interests of the many and those of the selfish and the lawless who pander to them. Such conflicts escalate the costs of conservation, and there has been bloodshed in confrontations between poachers and government wardens.

Ecotourism has been responsible for the development and maintenance of an increasing number of parks in many parts of the world, and this impact will increase as more people become aware of the gratification that comes from experiencing nature directly, even from the comparative luxury of ecotourist hotels and camps. The capacity of ecotourism to confer enough value on species to guarantee their protection is, however, finite. People have limited money to spend, and merely increasing reserve systems will not necessarily generate more tourism. Furthermore, some areas of immense biological importance, with high diversity and endemism, either are not attractive or are inaccessible to most tourists. Deserts, semiarid regions, many islands, and most marine ecosystems fall into this category. And of course, on a simple numerical basis, most species are simply not very interesting or even perceptible to the general public. Their preservation will depend on their living in association with more treasured species or habitats.

Finally, individual species may have considerable value as *indicators of broad and far-reaching environmental change.* During the 1950s and 1960s, populations of many birds in the United States (particularly the peregrine falcon, bald eagle, osprey, and brown pelican) declined drastically to the point where several of these species had disappeared from large areas, the peregrine from the entire United States. The cause of the population declines was traced to pollution of aquatic habitats by breakdown products (residues) of DDT, a pesticide that was widely used after World War II. These pesticide residues resisted degradation and entered the aquatic food chain, where they accumulated in the fatty tissues of animals and were concentrated with each step in the food chain. The high doses consumed by top predators interfered with their physiology and reproduction, causing overly thin eggshells and the deaths of embryos. Breeding success plummeted and populations declined. The viability of the peregrine population is a sensitive indicator of the general health of the environment. Its demise sounded the alarm to environmentalists; Rachel Carson warned of a "silent spring" when no birds were left to sing.

The United States government responded by banning DDT and related pesticides, and chemical companies have since devised alternatives that have less drastic environmental effects. Bald eagles and ospreys are becoming familiar sights once again, and thanks to the helping hand of dedicated biologists, who reared birds obtained from other parts of the geographic range and released them back to the eastern United States, the peregrine has staged a spectacular comeback. This was a major victory not only for the peregrine and the cause of species conservation more generally but also for the general quality of our own environment. Unfortunately, this silver lining draws attention to a dark, ugly cloud: the manufacture and export of DDT to foreign countries is still legal and remains a source of great profit to a few. DDT will bring a bleaker future to the many who live in countries that have yet to ban the toxin.

This general background illustrates the difficulties the conservation movement faces — difficulties grounded in ignorance and greed as well as in legitimate compromises between conflicting values. Be this as it may, ecological considerations bearing on the conservation of species usually are relatively straightforward. To understand them, it will help to review what we know about extinction itself.

The types of extinction

It is useful to distinguish among three types of extinction. **Background extinction** reflects the fact that as ecosystems change, some species disappear and others take their places. This turnover of species, at a relatively low rate, appears to be a normal characteristic of natural systems. **Mass extinction** refers to the dying off of large numbers of species as a result of natural catastrophes. Volcanoes, hurricanes, and meteor impacts happen occasionally. Some occur locally, others affect the entire globe, and species that happen to be in the way disappear. **Anthropogenic extinctions** — those caused by humans — are similar to mass extinctions in the number of taxa affected and in the global dimension and catastrophic nature of such events. Anthropogenic extinction differs from mass extinction, however, in that its causes are under our control.

Most information on background and mass extinctions comes from the fossil record, which reveals the appearance and disappearance of species through geological time. Disappearance may occur in two ways. First, the species may evolve sufficiently that individuals are no longer recognized as belonging to the same taxon as their ancestors and are given a different scientific name. True extinction has not taken place, and such instances are therefore referred to as **pseudoextinction.** Second, the population may cease to exist, in which case its disappearance from the fossil record is a case of true extinction. The finer the resolution of the fossil record, the greater the probability of distinguishing between the two. Where true extinction can be demonstrated, the life spans of species in the fossil record vary according to taxon, but they generally fall within the range of 1 to 10 million years. Thus, on average, the probability that a particular species will go extinct in the period of a single year (the background extinction rate) is in the general range of 1 in a million to 1 in 10 million. If, as conservative estimates have it, on the order of 1 to 10 million species inhabit the earth, this would amount to about 1 species extinction per year at the background rate.

Mass extinctions occupy the other end of the spectrum. Natural catastrophes may cause the disappearance of a substantial proportion of species locally or globally, depending on the severity and geographic extent of the catastrophe. Such catastrophes may include prolonged drought, hurricanes of great force, and volcanic eruptions. When Krakatoa, a volcanic island in the East Indies, exploded on August 26, 1883, not an organism was left alive; any that survived the initial explosion were buried under a thick layer of volcanic debris and ash.

Some mass extinctions detectable in the fossil record are thought to have been caused by the impact with the earth of large comets or asteroids (collectively referred to as bolides), the associated tidal waves and fires, and the prolonged darkness resulting from dust and smoke in the atmosphere. Spectacular examples of mass extinctions occurred at the end of the Paleozoic era (Permian period) and the Mesozoic era (Cretaceous period). The first involved the disappearance of perhaps 95 percent of species and numerous higher taxa. The second is most famous for the extinction of the dinosaurs, but some other major groups, notably the ammonites (predaceous, nautilus-like mollusks) disappeared as well. Whatever the exact cause, these extinctions were associated with discrete, calamitous events.

And anthropogenic extinction? Are we to be regarded as a "human bolide" in our impact on the environment? Well, not yet. Many extinctions have undoubtedly gone unrecorded, and rates of extinction in many groups (particularly among large animals hunted for food and among island forms) are far above background levels. Nevertheless, if humankind turns out to be a disaster for global biodiversity, the full force of the impact will come in the future. And most important, it is preventable. Examining the causes of extinction will enable us to see why this is so.

The causes of extinction

Species disappear when deaths exceed births for a prolonged period. This much is obvious, but the statement also emphasizes that extinction may result from a variety of mechanisms that influence birth and death processes within a population. It has also been said that extinction represents the failure to adapt to changing conditions, whether because the changes occur too fast or because the population is evolutionarily unresponsive. We shall discuss four general types of factors that adversely influence population processes: (1) habitat change, (2) habitat restriction, (3) declining habitat quality, and (4) overexploitation.

Changes in climate determine physical conditions and habitat structure, which are critical to the well-being of every population. Over the long history of the earth, changes in global climate have been bought about by the drifting of the continents and associated changes in oceanic circulation. Where physical barriers to dispersal prevent the distributions of species from following the shift in climate belts, local populations may become extinct, being replaced by others that are better suited for survival in the new climate and habitat type. Local changes in climate and habitat are brought about by changing landforms, which may create rain shadows and redirect the drainages of rivers. For endemic species, these changes may bring about extinction. At present, the burning of wood and fossil fuels is increasing the carbon dioxide concentration in the atmosphere, thereby increasing the average temperature of the earth through the so-called greenhouse effect, which we shall discuss in the next chapter. This anthropogenic change in climate, which may amount to between 2 and 6°C, is likely to cause the extinction of many species, particularly plants, with narrow temperature tolerances.

Larger areas support larger populations, which are less susceptible than small populations to extinction by small-scale catastrophic events or by random variations in population size. Just by chance, every population experiences variations in births and deaths during any particular period. These cause what is known as stochastic, or random, variation in population size. The magnitude of this variation varies inversely with the number of individuals in the population. Very small populations, such as those isolated in restricted fragments of habitat, may become extinct just by chance if they suffer a series of very unlucky years. This is referred to as **stochastic extinction,** and although it is relatively unlikely except in the smallest populations, its probability increases with restriction of suitable habitat, and it is particularly troublesome for species, such as large predators, that have low population densities.

Another way in which small population size may increase the probability of extinction is by reducing the genetic variation in the population. Fewer individuals contain a smaller proportion of the total genetic variation of a larger population. Furthermore, **inbreeding** (mating among close relatives) tends to reduce genetic variation. When a population goes through a period of small population size and, as a result, exhibits reduced genetic diversity, it is said to have passed through a **bottleneck**. As a consequence, small populations with low genetic diversity may not have the capacity to respond to rapid change in the environment, which may favor some genotypes in some years and other genotypes in other years. Small glades of xeric habitat on rocky outcrops in the Ozark Mountains of Missouri, for example, support restricted populations, generally 20 to 50 individuals, of the collared lizard, a resident of southwestern deserts that colonized the Ozarks during a period of hot, dry climate 4000 to 8000 years ago. Genetic surveys have shown that the lizard populations are genetically uniform within populations but differ between populations. This is exactly the pattern expected to result from the random loss of genetic diversity within small populations.

It may be difficult to generalize problems resulting from reduced genetic diversity in populations, because there are several cases of species that have been reduced to near extinction and have lost much of their genetic variation but have recovered with spectacular growth when the population was protected. The northern elephant seal is a case in point. By 1890, hunting had reduced the population of this once-numerous species to about 20 individuals. Since then the population has increased explosively, passing 30,000 in 1970 and extending throughout much of the species's former range in California and Mexico. Several years ago, investigators could not detect any genetic differences between individuals within the species, though they used tests that reveal ample genetic variation in other species of mammals. Similarly, one of Africa's large cats, the cheetah, has no detectable genetic variation within its population. This genetic uniformity suggests that the cheetah may have gone through a population bottleneck sometime in its recent past. Nonetheless, although its reproductive success may be impaired, the cheetah population appears healthy and self-sustaining where it is not persecuted by humans.

Ultimately, of course, restriction of habitat may cause extinction by wiping out suitable places to live. Animals of the forest will disappear when all the forest has been cut.

Even when suitable habitat remains, conditions within the habitat may change and cause a population to begin a decline toward extinction. Frequently, decrease in habitat quality can be traced to the introduction of predators, competitors, or disease organisms — that is to say, biological agents of change. In many cases, the habitats affected are those of isolated islands to which organisms have been introduced from more diverse continental biotas. The Hawaiian Islands have suffered greatly from introductions of aliens, resulting in the extinction of a large proportion of the native birds and the invasion of native forest by aggressive, weedy species (Figure 26.3). Habitat quality may also be affected by various forms of pollution originating at a distance, including smog and toxic rain.

Weapons have made humans such efficient hunters that many species have literally been hunted to extinction. Within recent history, North American fatalities have included Steller's sea cow, the great auk, the passenger pigeon, and the labrador duck—all formerly abundant species, all prized for food, all vulnerable, and all slaughtered mercilessly until the last were gone. Extinction by overhunting is not, however, a recent phenomenon. Wherever humans have colonized new regions, some components of the fauna have suffered. Shortly after aboriginal man colonized Australia some 50,000 years ago, several large marsupial mammals, flightless birds, and a tortoise disappeared from the island continent. The advent of humans in the Americas 11,000 years ago was accompanied by the rapid extinction of 56 species in 27 genera of large mammals, including horses, a giant ground sloth, camels, elephants, the sabertooth tiger, a lion, and others. Madagascar, a large island off the southeastern coast of Africa, saw its first human inhabitant only 1500 years ago, yet this event brought the demise of 14 of 24 species of lemurs (mostly the large species that were suitable for food) and between 6 and 12 species of elephant birds, flightless giants found only on Madagascar. In each of these cases, a technologically superior species encountered populations unaccustomed to hunting pressures. Their lack of defenses and failure even to recognize danger spelled disaster; lack of restraint on the part of their hunters turned disaster into extinction.

Vulnerability to extinction

Why do some species seem more vulnerable to extinction than others? This question has been difficult to answer. Clearly, species that attract the attention of human exploiters are brought under great pressure. In addition, species that have evolved in the absence of hunting (particularly those on remote islands lacking most types of predators) and in the absence of disease organisms seem to fare poorly after the arrival of humans. Vulnerability is also associated with limited geographic range, restricted habitat distribution, and small local population size.

FIGURE 26.3 Exclusion of grazing within fenced areas on the island of Hawaii allows the reestablishment of native vegetation. Feral boars, originally introduced to Hawaii for hunting, have also caused the disappearance of native species by routing seedlings out of the soil. Courtesy of the U.S. Department of Agriculture, Soil Conservation Service.

But what makes one species rare and locally distributed when a close relative that exhibits similar adaptations to the environment is abundant throughout a wide geographic distribution? The difference between success and failure in natural systems may hinge on a very few percentage points of breeding success or longevity — perhaps too few for us to detect in studies of natural populations. Most species persist for a million or more years, so their populations must be fully self-sustaining and capable of recovering from the setbacks inflicted by a variable world. Whatever causes a population to embark on the slow decline to extinction may be very subtle indeed. So far, ecologists have been able to say little on this point. What they can address, however, is the problem of reversing the declining population trends of species that, in the absence of anthropogenic pressures, would be self-sustaining.

Strategies for the conservation of species

The simple solution to maintaining the population of a particular species is to guarantee the existence of a sufficient area of suitable habitat that can be kept free of alien competitors, predators, and diseases. In practice, the design of such reserves must take into account the ecological requirements of the species and the minimum size of a population that can sustain itself in the face of environmental variation. This is called the **minimum viable population,** or **MVP.** The MVP must be large enough to remain out of danger of stochastic extinction brought about by chance events. The population also must be widely enough distributed that local calamitous events such as hurricanes and fires cannot affect the entire species. At the same time, some degree of population subdivision may prevent the spread of disease from one part of the population to another.

Guaranteeing suitable habitat becomes more complex when the population has different habitat requirements during different seasons or when it undertakes large-scale seasonal migrations. In the Serengeti ecosystem of eastern Africa, fantastically large populations of grazers, such as wildebeest, zebra, and gazelle, undertake long-distance seasonal movements in search of suitable grazing as the pattern of rainfall varies seasonally within the region. It would not be possible to isolate a part of this area as a reserve, because the herds need the entire area of the Serengeti ecosystem. Thundering herds of buffalo could never be restored to the American prairies, because their migration routes are now blocked by miles of fencing and habitat converted to agriculture. Buffalo survive in a few small reserves in the American west — most notably the Greater Yellowstone ecosystem — but the natural population structure of the buffalo has been irrecoverably lost.

Long-distance migration poses similar problems for the conservation of many types of birds. Wading birds, such as sandpipers, breed on the arctic tundra, but the maintenance of their populations also depends on the conservation of the beaches and estuaries that they use during spring and fall migrations and as wintering grounds (Figure 26.4). Many of our songbirds, whose populations have been declining during recent decades, spend the winter in forests of Central and South America. Their populations have been placed in

FIGURE 26.4 Sandpipers and gulls feed on eggs laid by horseshoe crabs during May along the shores of Delaware Bay. The eggs are a major food source for migrating shorebirds. Many of the horseshoe crabs are stranded by the ebbing tide and die of exposure.

double jeopardy by forest fragmentation throughout much of their breeding range in North America and by the extensive clearing of forest in Latin America. Migration systems between Europe and Africa and between Siberia and Southeast Asia face the same problems.

Where threats of extinction come from the dwindling of suitable habitat, raising doubt about there being sufficient area to maintain minimum viable populations, the conservation strategy is relatively straightforward, even though it may be expensive and politically difficult to achieve. It is also impractical to develop a conservation strategy for every species, and the well-being of the majority will necessarily depend on conservation efforts directed toward a few of the most critically endangered or conspicuous. As habitat becomes more and more the focus of conservation efforts, however, it becomes especially important to identify the habitats that are most critical to maintaining species diversity as a whole and to determine the area of habitat required to maintain minimum viable populations of most species.

What makes an area critical for conservation? The most valuable areas are those that provide havens for the largest number of species not represented elsewhere; thus value reflects a combination of local diversity and endemism of the local flora and fauna. As a rule, endemism is highest on oceanic islands, in the tropics, and in mountainous regions. Thus such localities as Madagascar and the Hawaiian, Galapagos, and Canary Islands are extremely critical ones for conservation.

Over the area of a continent, reserves, whose number and area are necessarily limited by economic considerations, must target habitats and areas of special interest biologically. From the standpoint of conservation of biodiversity, more is to be gained by setting aside several small reserves spread out over a variety of habitats and areas of high endemism than by reserving an equal area within a single habitat type. Values other than biodiversity may dictate the setting aside of large reserves, and we shall discuss some of these factors in the next chapter. One thing is certain: the cost of setting aside larger and larger amounts of land and aquatic habitat as reserves increases out of proportion to the area itself. This is so simply because land that is least expensive in terms of economic, social, and political values is set aside first. As more land is added to a reserve system, the cost of acquiring it, in terms of purchase price and potential resources forfeited, invariably increases. It is no accident that most parks and reserves are located in remote underpopulated areas, and that establishing conservation areas becomes more difficult when it conflicts with economic interests. One example of this conflict involved the setting aside of a large tract of old-growth redwood forest in northern California as Redwoods National Park, which was vigorously opposed by the timber industry. In this case, the uniqueness of the redwood habitat and its rapid conversion by lumbering into managed tree farms greatly increased the value to society of setting aside a large area of this habitat for posterity.

Many tropical countries are in the enviable position of having large tracts of uncut forest and relatively undisturbed tropical habitats of other kinds. These have been protected fortuitously in the past by their geographic remoteness and by the small size of local human populations. It is still possible to set aside large parks and reserves in some of these countries, and several governments have moved rapidly during the past decade to preserve tracts of what remains. The problem is complicated, however, by the rapid growth of the human population, by the increased exploitation of forest products, and by the conversion of forests to agriculture. Such exploitation is justified by a legitimate need to feed people and generate export income for economic development. Thus the price of conservation is rising very rapidly in much of the world, and many of the developing countries are unable to foot the bill. Even when lands are set aside "on paper," many countries cannot afford to protect them from squatters, poachers, and self-serving politicians who may grant mining and lumbering concessions within protected lands in order to extract short-term profits. For this reason, conservation must be an international effort, and the wealth of the developed countries must be shared globally to protect the earth's biodiversity. The "haves" must simply have *less* so that the "have nots" can have *enough* without destroying the varied habitats and creatures that are our common heritage and our common trust.

The design of nature reserves

In many cases, the boundaries of biological preserves are dictated by available land area and economic considerations; basically, all that can be done is to set aside whatever remains in a relatively pristine state. In other situations, those who design the preserve may have more latitude in deciding just how to draw the boundaries of a park or other reserve. Here, several ecological principles derived from the **theory of island biogeography** help planners arrive at the best solution of the boundary problem. There are two guiding principles: the **species–area relationship** and the **edge effect.** Larger areas support more species than smaller areas because large population sizes of individual species reduce stochastic extinction, promote genetic diversity within populations, and buffer populations against disturbances. Edges should be minimized because the effects of habitat alteration extend for some distance beyond the area directly affected. For example, predators that inhabit agricultural areas, such as cowbirds (which are nest parasites of other birds), rats, and feral cats, may venture into the edges of undisturbed forest habitats. The conditions in forests adjacent to cleared land are further affected by increased wind and sunlight.

According to these considerations, when reserves are to be carved from areas of uniform habitat, such as a broad expanse of tropical rain forest, (1) larger is better than smaller, (2) one large area is better than several smaller areas that add up to the same total size, (3) corridors connecting isolated areas are desirable, and (4) circular areas are better than elongate ones with much edge. However, faced with choosing between a single large area of uniform habitat and several smaller areas each in a different habitat, planners should remember that the smaller areas will often contain a greater total number of species among them because of endemic species found in one habitat but not the others.

As always, nature reserves must be designed in accordance with the habits of their inhabitants, and requirements for special features (such as nesting sites, water holes, and salt licks) must be taken into account. In mountainous areas, many species undertake altitudinal migrations over the seasonal cycle, and so preserves set aside at different elevations must be connected by suitable corridors for travel. Roads and pipelines set in the way of migratory movements or dispersal must be bridged in some manner to allow passage.

Rescued from the brink of extinction

There have been many times when a particular species has come so close to extinction that its preservation has required exceptional intervention. Such efforts, which may cost millions of dollars, usually are directed toward species that appeal to the public. Some may question the wisdom of spending several million dollars (as happened a couple of years ago) to free three gray whales trapped in arctic ice. But the incident dramatized the empathy that many humans feel for the plight of other creatures. And many others, though perhaps less enthusiastic about spending so much to rescue individuals, are willing to devote considerable resources to the preservation of species.

In recent decades, zoological parks have become increasingly concerned with maintaining viable, genetically diverse populations of species that are endangered or even extinct in the wild. Eventually, with the development of suitable reserves, many of these populations may be reintroduced to natural settings. As the population of California condors in southern California dwindled below 30 and then below 20 individuals in the wild during the 1970s and early 1980s, management personnel had to make the difficult decision to bring the entire population into captivity. In specially constructed breeding facilities located at the Los Angeles and San Diego Zoos, the birds are protected from several mortality factors that were destroying the wild population: indiscriminate shooting, lead poisoning from slugs left in deer carcasses upon which the condors fed, and poisons and traps set out for coyote control, to which condors were attracted by baits. In addition, condors in captivity can be induced to lay up to three eggs a year, instead of the usual one, and most of the chicks are reared successfully. The objective of such a captive rearing program is to produce young birds that can be reintroduced into native habitat. The program is costly, and whether it succeeds will ultimately depend on controlling those mortality factors that threatened the population in the first place. This will require legislation, land purchase, and public education. In the particular case of the California condor, the program's success will be judged only after 30 or 40 years and the expenditure of tens of millions of dollars.

Like many other endangered species, the California condor can be saved from extinction. The experience gained through captive breeding and release programs will be useful to similar efforts in the future. The program has heightened the awareness of local residents of conservation issues and has resulted in the preservation of large tracts of condor habitat in mountainous regions of southern California. People have also come to understand that as long as care is taken, viable condor populations are compatible with other land uses, such as recreation (as long as access to nesting sites is restricted), hunting (as long as steel rather than lead bullets are used), and ranching (as long as coyote and rodent control programs are condor-safe). Concessions to condors are neither difficult nor expensive. Making them simply depends on instilling values that acknowledge natural systems as an integral part of the environment of humankind.

❦ SUMMARY

1. Humankind has an immense impact on the earth, managing or otherwise affecting most of its land surface and waters. Human activities have caused deterioration in ecological systems and the extinction of many species. The repercussions are accelerating as the human population grows toward 6 billion individuals and the per capita consumption of energy and resources increases apace.

2. The environmental crisis cannot be fully resolved until human population growth is stopped, consumption of energy and resources declines, and economic development takes ecological values into consideration.

3. Of immediate concern is the preservation of biodiversity, which encompasses the variety of living beings — plant, animal, and microbe — on earth. The concept of biodiversity recognizes genetic diversity within and between populations and acknowledges the special value of areas of endemism that are inhabited by species with restricted geographic ranges.

4. The value of individual species is rooted in generalized moral considerations, in aesthetics, in the economic and recreational benefits we derive from them, and in their role as indicators of environmental deterioration.

5. Background extinction consists of natural extinctions resulting from environmental change and from the evolutionary changes of species within communities. Mass extinctions, which appear episodically in the fossil record, reflect calamitous events in earth history, particularly the impact of meteors or other extraterrestrial bodies with the earth. Anthropogenic extinction is the disappearance of species as a result of human activities: overexploitation, the introduction of predators and disease organisms, and pollution of various kinds.

6. Habitat restriction may hasten a population's decline toward extinction by making it more vulnerable to stochastic, or random, changes in population size or by causing reduced genetic variation and thereby impairing the capacity of the population to survive environmental change.

7. Particular species can be conserved by safeguarding them from mortality factors in a large enough area to sustain the minimum viable population in the face of environmental variation.

8. Optimally designed nature preserves should include a high proportion of endemic species. For a given area of uniform habitat, reserves should be amalgamated (rather than dispersed in several small areas) to reduce stochastic extinction due to small population size, and they should be close to circular in shape to reduce edge effects.

9. In extreme cases, individual species can be rescued from the brink of extinction by massive recovery efforts that may include habitat restoration and captive breeding. Such costly programs, although they are focused on individual species, often highlight more general conservation problems and result in the conservation of habitat whose ecological value greatly exceeds that of the individual species it was preserved to save.

❦ SUGGESTED READINGS

Diamond, J., and T. J. Case. 1986. Overview: introductions, extinctions, exterminations, and invasions. In J. Diamond and T. J. Case (eds.), *Community Ecology.* Harper & Row, New York, pp. 65–79.

Glen, W. 1990. What killed the dinosaurs? *American Scientist* 78:354–370.

Hansen, A. J., T. A. Spies, F. J. Swanson, and J. L. Ohmann. 1991. Conserving biodiversity in managed forests. *BioScience* 41:382–392.

Myers, J. P., et al. 1987. Conservation strategy for migratory species. *American Scientist* 75:18–26.

Pimm, S. L. 1991. *The Balance of Nature? Ecological Issues in the Conservation of Species and Communities.* Univ. of Chicago Press, Chicago.

Rolston, H., III. 1985. Duties to endangered species. *BioScience* 35:718–726.

Simons, T., S. K. Sherrod, M. W. Collopy, and M. A. Jenkins. 1988. Restoring the bald eagle. *American Scientist* 76:252–260.

Soulé, M. E. 1985. What is conservation biology? *BioScience* 35:727–734.

Soulé, M. E. 1986. *Conservation Biology. The Science of Scarcity and Diversity.* Sinauer, Sunderland, Mass.

Terborgh, J. 1974. Preservation of natural diversity: the problem of species extinction. *BioScience* 24:715–722.

Western, D., and M. C. Pearl (eds.). 1989. *Conservation for the Twenty-First Century.* Oxford Univ. Press, Oxford.

Westman, W. E. 1990. Managing for biodiversity. *BioScience* 40:26–33.

Wilson, E. O. (ed.). 1988. *Biodiversity.* National Academy Press, Washington, D.C.

27

DEVELOPMENT AND GLOBAL ECOLOGY

The key to preserving biodiversity is to set aside large areas of the natural habitats found on earth and maintain their capacity for supporting species. Basically, this means minimizing human impacts of all kinds over representative areas of the earth's surface. As the human population grows, this goal recedes further into the distance and becomes, in the eyes of most people, less pressing than the problem of maintaining basic life-supporting systems for humans. Resolution of the conflict between natural and human values will ultimately depend on erasing the distinction between them and somehow making them more compatible. For many species, and perhaps for most of the biodiversity of the tropics, we must set aside either pristine reserves or areas carefully managed to sustain certain populations. In the end, these will be natural plant and animal parks where remnant viable populations of hundreds of thousands — perhaps millions — of species will cling to their last footholds on this planet. The setting aside of these preserves will be justified by moral and aesthetic considerations and by their considerable economic value as tourist sites, watersheds, and carbon sinks. More will be said about this later.

What of the 90 or so percent of the rest of the earth, which has been, or soon will be, converted to supporting the human population — devoted to living space, food production, forestry, mineral production, hunting, and so on? What about our own ecology? Can the earth sustain the expanding human population indefinitely at a high quality of life? To what degree are human values compatible with natural values? That is, can natural and managed ecosystems intergrade, or are preserves the only alternative to completely altered environments dominated by humans and their domesticated species?

Clearly, a sustainable biosphere will never be achieved as long as the human population continues to grow. The earth offers no new regions to colonize. Except for portions of the wet tropics, much of which cannot support dense human populations, most of the habitable area of the earth has been filled. Further population increase leads to further crowding, straining not only the structure of human society but also the life-supporting systems of the environment.

Alternatives exist, but to put any of them into effect, we must recognize certain undeniable facts.

1. We must accept that the human population of the earth will continue to increase, at least for the near future, and that most of the surface of the earth and the oceans will be devoted to supporting that population.

2. Given these premises, we must manage the planet so as to maintain natural processes in a healthy state. By paying attention to basic principles of ecology, it is possible to implement management practices that minimize interference with the ability of ecosystems to maintain themselves and respond to perturbation, while maximizing their production for human use.

3. We must realize that different ecosystems have different optimal uses and that certain exploitation and management practices are environment-friendly whereas others are not.

4. The most productive regions of the earth do not necessarily correspond to the regions of greatest population density. These imbalances can be overcome by the transport of foods, materials, and energy from one region to another, which requires a high level of international communication, cooperation, and sharing of wealth.

5. The goals of a sustainable biosphere can be met only if the costs, both short-term and long-term, of population growth and ecological mismanagement can be fully evaluated and assigned to goods and services.

Ecological processes

Throughout this book, we have discussed various processes involved in biological production and in the regulation of communities and ecosystems. These processes occur in managed as well as natural ecosystems. The key points about ecosystem function are the harnessing of energy and the continual recycling of materials. In natural systems, the primary source of energy is sunlight; recycling is accomplished by a variety of regenerative processes, some of them physical or chemical, some of them biological. In any of these processes, imbalance that leads to the accumulation or depletion of a component of the ecosystem normally sets in motion restorative processes that push the system back to a self-maintaining steady state. For example, when dead organic matter accumulates within a system, decomposing organisms increase in number and consume the excess detritus. When herbivores increase to high levels and begin to deplete their food resources, declining birth rates and increasing mortality check population growth and restore a sustainable relationship between consumer and resource.

Restorative processes may be physical, but more often they involve biological transformations. From the composition of the atmosphere to the most basic character of many habitats, plants, animals, and microbes have greatly modified the conditions of the earth's land surfaces and waters and are responsible for maintaining their qualities. When natural processes are disrupted, the environment may undergo severe change. Worse, it may lose its capacity to respond to perturbation and become permanently degraded. Thus, maintaining a sustainable biosphere requires that we conserve the ecological processes responsible for its productivity.

Threats to ecological processes

All human activities have consequences for the environment. Overexploiting a fishery is a good example. The goal is to harvest a food resource for human consumption. However, when we maximize short-term returns from the fishery—getting it while we can—fish stocks are reduced or even collapse, the fishery goes bankrupt, and our attention turns to other exploitable populations. We have seen this in the commercial whale catch, in which one species after another was hunted to near extinction, beginning with the most profitable blue whales and progressing to less and less profitable species, such as sei whales (Figure 27.1). Under the strict protection now afforded, some populations of whales are increasing. Other fisheries, such as the once immensely profitable sardine fishery of western North America and the anchovy fishery of Peru, have not recovered. It is possible that these fish stocks were pushed below the level from which they could recover and that the entire ecosystem has shifted and no longer includes the sardine.

Often the consequences of human activities are less direct and are unexpected; these may be difficult to detect or far removed in time or space. To provide a relatively straightforward example, the clearing of land for agriculture or lumber frequently leads to erosion and deposition of silt downstream from the watershed over long periods. Thus riverine habitats may be altered and reservoirs behind dams filled in. We'll look at some more examples as we consider the different kinds of threats to ecological processes.

Overexploitation

Fishing, hunting, grazing, fuel-wood gathering, lumbering, and the like are classic consumer–resource interactions. In most natural systems, these achieve steady states because as a resource becomes scarce, the efficiency of

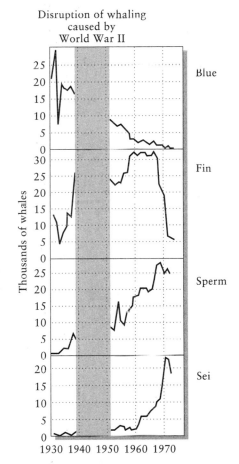

FIGURE 27.1 Commercial catches of four types of whales illustrating the shift to new, less profitable species as populations of the more profitable species are hunted to low levels. Based on R. Payne, in W. Jackson (ed.), *Man and the Environment* (2nd ed.). W. C. Brown, Dubuque, Iowa (1973), p. 143.

exploitation plummets; the consumer population then begins to decline or seek alternative resources until the consumer and the first resource are brought back into balance. The efficiency of exploitation and the ability of resources to resist exploitation are characteristics of consumers and resources that have evolved over long periods of interaction.

In economic systems, consumer–resource interactions may also come into balance, because as a resource becomes scarce and its price increases, demand for that resource drops; people either do without or find cheaper alternatives. However, because the human population's ability to exploit natural systems has been escalated out of all proportion by its ability to use tools, renewable resources may not become scarce until they are very nearly depleted and are unable to sustain even reduced exploitation. Technological skills have advanced too rapidly for nature to keep pace; humans have gained the upper hand with weapons, chain saws, cultivated plants, and domesticated animals. As a result, many of the resources that historically supported the growth of the human population, such as the vast forests and prairies of North America, have been converted to other uses. Where the fertility of the land itself has been exhausted (as in drier parts of Africa and increasingly elsewhere), this leaves the prospect of a large human population without the resource base to support itself. Where the population can no longer move to other areas or shift to new food sources, population control by starvation and associated diseases and social strife portend a grim future.

That future is already upon us in some parts of the world. In subsaharan Africa, overgrazing and fuel-wood gathering have left little vegetation to support human or any other life. In other parts of the world, vast areas of formerly productive land have been laid waste by ecologically ill-conceived practices (Figure 27.2). Although much of the tropics will sustain intensive agriculture, particularly in mountainous regions with volcanic soils, the high natural productivity of large portions of the tropics depends on the presence of the native vegetation. Old lowland soils are deeply weathered and deficient in the clay particles that retain many soil nutrients. Bedrock lies so far beneath the surface that new mineral nutrients do not enter the upper layers of soil; nutrients that do exist there are susceptible to leaching. The natural fertility of the habitat is maintained by tight recycling of nutrients between detritus and living plants. Break the cycle — by clear-cutting the forest, for example — and regenerated nutrients are lost. Over much of the Amazon Basin, forested land cleared for grazing becomes so infertile that it must be abandoned after 3 years of ranching. To be sure, forests will regenerate, but many decades or even centuries must pass before the natural fertility of the habitat is restored.

Many human populations in the tropics maintain themselves by a practice of "shifting agriculture" in which small patches of forest are cut and burned to release nutrients into the soil, planted for 2 or 3 years, and then abandoned in favor of a new patch. A particular patch recovers sufficiently to repeat this process in 50 to 100 years. Accordingly, as long as only 1 to 2 percent of the forest is cut each year, and thus perhaps 2 to 6 percent is under cultivation at any one time, the land can sustain this practice. This type of agriculture requires little input of labor, materials, or energy and takes advantage of the natural successional processes of tropical forests, but it supports only sparse human populations. When land is cultivated more intensively by tilling, fertilization, water-

FIGURE 27.2 (a) Severely overgrazed rangeland in Sandoval County, New Mexico, where the vegetation has been reduced to unproductive, largely inedible species. (b) Maintenance of large goat populations by providing water in Greenlee County, Arizona, results in overgrazing in the immediate vicinity. Courtesy of the U.S. Department of Agriculture, Soil Conservation Service.

ing, and weeding, productivity and long-term sustainability may increase greatly, but so do the inputs of labor and materials.

The situation in the Republic of Panama is typical of many tropical countries. In pre-Columbian times, shifting agriculture and locally more intense cultivation of certain crops supported a human population that probably equaled the present-day population of about 2 million. The pre-Columbian population existed in a long-term equilibrium with the environment; the present-day population most certainly does not. Why the difference?

First, the present population has grown tremendously from a relatively small size. The indigenous inhabitants were greatly reduced in number following contact with Europeans through disease, armed conflict, and disruption of society—a pattern typical of much of the earth. Thus the vast forests devoid of human impact that existed 50 to 100 years ago were a relatively recent development. In pre-Columbian times, by contrast, most of the region was intensively managed, and the human population was distributed very widely.

Second, patterns of food consumption have changed dramatically. Individuals are larger and require more food, and meat makes up a greater part of the diet. As a result, vast areas of forest have been converted to unproductive rangeland to feed cattle; this practice was originally restricted to drier parts of the country but now occurs virtually everywhere.

Finally, as in many tropical countries, land has been cleared to grow export products shipped to other parts of the world: coffee, sugar, bananas, and beef, to name a few. Most tropical countries lack abundant mineral resources and rely on agricultural exports to pay for imports of manufactured goods that have become essential to a high standard of living. The result for much of the tropics is declining natural productivity of the environment, which precludes alternative sustainable land uses.

The solution? Reduce intake, eat lower on the food chain, determine maximum sustainable yields for resource populations, consider alternative land uses for sustainable production, increase agricultural intensity on land that will bear it, improve distribution of food between areas of production and areas of need. Most of these solutions carry a price tag. Planning for sustained use cuts short-term returns, which must be made up elsewhere to maintain the wealth of a country. Increasing agricultural intensity requires disproportionately greater inputs of energy, labor, and chemical fertilizer, each contributing its own problems. Relying on crops selected for high food production often makes agriculture more vulnerable to the outbreak of disease. Segments of the human population that are forced by impoverished land to import food also must earn the money to buy it; otherwise, they will be reduced to welfare status at a cost to other segments of the population. As long as the local growth of the human population is not tied to the ability of local resources to support population growth, imbalances between human consumers and their resources will continue to proliferate.

Introduction of alien species

Both intentionally and unintentionally, humans have taken other species everywhere they have traveled. Aborigines brought dingoes (semidomesticated dogs) to Australia, Polynesians brought rats to Hawaii. Of course, the global movements of species by human agency have increased immensely since Europeans began colonizing most of the world some 500 years ago. Introductions have included edible and horticultural varieties of plants, and their pests; commercially valuable trees; domesticated species of animals for work or meat; familiar backyard animals, especially birds; disease organisms; and such commensals of human habitation and transportation as the ubiquitous cockroach and dandelion. The result has been to produce a globally distributed flora and fauna of alien species that have displaced or otherwise wreaked havoc with local biotas.

To take an extreme example, most of the island of New Zealand is inhabited by introduced plants and animals. The native forest was cut long ago and replaced by pines from North America and eucalyptus from Australia; the moas were killed off by Maori natives before Europeans arrived, and sheep now take their place; most of the birds of the countryside were transplanted from England to stave off the homesickness of the early colonists. Only at the southern tip of New Zealand do native forests of southern beech persist in the wet and remote fiordlands. Of the total New Zealand flora of 2500 species, fully 500 are naturalized introductions that account for most of the present vegetation.

Aliens can have drastic effects on native habitat. In Australia, introduced prickly pear invaded millions of hectares of native habitat, turning grassland into impenetrable thicket. In Hawaii, feral pigs rooting through forest litter have interfered with the regeneration of native forest; populations of birds in the same forests have been decimated by introduced bird pox and malaria transmitted by introduced mosquitoes. Islands, perhaps because of their low native diversity, seem particularly vulnerable to the effects of introduced species.

In many cases, alien species can bring benefits. After all, our crops and domesticated animals are alien to most of their present distributions. In other cases, even though aliens might displace native plants and animals, they do not necessarily disrupt ecosystem function. However, the effects of introduced species are often difficult to predict. In aquatic systems in particular, introduced consumers at high levels in the food chain have seriously disrupted ecosystem function and have caused basic changes in community structure. Efficient predators, such as the Nile perch introduced to Lake Victoria in East Africa and the peacock bass introduced to Gatun Lake in Panama, virtually eliminated entire trophic levels of smaller planktivorous fish. As a result, densities of zooplankton increased dramatically and algae were cleared from the water, reducing the overall productivity of the environment. In this way, a single predator — referred to as a **keystone species** in this case because of its critical place in ecosystem function — can shift the character of a habitat from one state to a qualitatively different state.

Habitat conversion

Altering the basic nature of a habitat often upsets natural processes of regeneration and control and brings about disastrous consequences. The cutting of tropical forests growing on impoverished soils breaks the tight cycling of nutrients that maintains forest productivity and greatly alters the physical structure of soil through exposure to increased leaching and sunlight. As a result, the productivity of the land decreases precipitously, and soil erosion may increase tenfold or more (Figure 27.3). By some estimates, up to 1 percent of the earth's topsoil is lost to erosion every year. In the Amazon Basin, for example, erosion rates increased from 6 to 10 metric tons (T) ha^{-1} yr^{-1} in 1960 to 18 to 190 T ha^{-1} yr^{-1} by 1985, largely as a result of deforestation and overgrazing.

FIGURE 27.3 Examples of soil erosion and gully formation on plowed farmland (a, Macon County, North Carolina) and heavily grazed pasture (b, near Bethany, Missouri). Courtesy of the U.S. Department of Agriculture, Soil Conservation Service.

Problems associated with habitat conversion are not restricted to tropical forests. The plowing of prairies destroys the dense root mats of perennial herbs that hold the soil together. A prolonged drought in the central United States during the 1920s and 1930s turned former prairies converted to agriculture into a devastated "dust bowl" of blowing soil (Figure 27.4). Mangrove forests provide natural protection for coastlines in many parts of the tropics. Where these have been cleared for fuel-wood and land reclamation, the coasts have been laid bare to rampaging hurricane-driven flood waters. The damming of rivers brings the benefits of flood control, irrigation water, and power gen-

Figure 27.4 Farmstead in Oklahoma abandoned during the height of the Dust Bowl period in 1937. Courtesy of the U.S. Department of Agriculture.

eration, but it also increases silt transport, blocks fish migrations, alters downstream water conditions, and may even change the local weather.

Irrigation

Water makes the desert bloom. Humankind has employed various irrigation schemes to increase the productivity of the land since the beginning of agriculture (Figure 27.5). Only recently, however, has irrigation been applied on immense scales to land that would otherwise be totally unsuitable for agriculture. The benefits are tremendous, but so are the costs, many of which surface only after years of profitable irrigation. The primary costs are the environmental effects of developing the dams, wells, canals, and dike work required to support irrigation, the lowering of water tables where wells are the source of irrigation water, the reduction of groundwater quality through the introduction of pesticides and fertilizers or the concentration of naturally occurring toxic elements, the salting of irrigated soils in arid zones, and the transmission of diseases by aquatic organisms. In most cases, the cost of delivering water to crops is underwritten by the population at large through taxes and other subsidies, including the burden of future environmental problems: rarely does irrigation pay its own way.

Fertilization and eutrophication

Any substance that enhances the productivity of a habitat may be considered a fertilizer. We apply fertilizers to agricultural lands to increase crop production, but a portion of these chemicals make their way into groundwater and from there to rivers, lakes, and eventually the ocean. Nitrates, phosphates, and other inorganic fertilizers have the same effect on rivers and lakes as on agricultural lands: they increase biological production. A consequence of this artificial fertilization, often called **eutrophication,** is to change the chemical and biological conditions of a body of water. Although increased production is not necessarily bad, it may cause a change in the species composition of rivers and lakes. Input of inorganic fertilizers may also turn clear, oligotrophic waters into turbid environments that are less attractive for recreation. Often, however, nutrient inputs upset seasonal cycles of nutrient use and regeneration in natural bodies of water, leading to the accumulation of organic material, high rates of bacterial decomposition, and deoxygenation of the water. Under such conditions, fish may suffocate and contribute further to the load of organic material in the water.

Direct input of organic wastes into water, such as from sewage and runoff from feedlots, poses a greater problem for water quality. Suspended or dissolved organic materials in water create what is known as **biological oxygen demand,** meaning that the decomposition of these materials by bacteria uses oxygen present in the water. When organic materials are added from outside the system, rather than being produced within the system, they may completely alter the natural balance of oxygen production by photosynthesis and oxygen consumption by respiration, because the organic inputs are unrelated to the natural productivity of the system. Under these conditions, a stream or lake may become anoxic for long periods and unsuitable for many forms of life. Before water pollution came under strict controls in the United States, large sections of our major rivers became completely anoxic, killing off local fish populations and preventing the migration of other species, such as shad and salmon, between the ocean and their headwater spawning grounds (Figure

FIGURE 27.5 Irrigation can turn desert into productive farmland. (a) Furrow irrigation on cotton in the Imperial Valley of southern California. However, the accumulation of salt that accompanies irrigation requires periodic flooding to flush salts from the soil. (b) Maricopa County, Arizona. Courtesy of the U.S. Department of Agriculture.

Figure 27.6 Fish die off in a stream near Colorado Springs, Colorado, caused by oxygen deficiency related to organic pollution. Courtesy of the U.S. Department of Agriculture, Soil Conservation Service.

27.6). The costs of such moribund rivers to fisheries and recreation, not to mention to aesthetic sensibility, were enormous. In general, natural conditions can be restored by cutting off the supply of organic nutrients either by diverting inputs to larger bodies of water that can absorb organic inputs or by improving the treatment of sewage. The costs of these solutions have been more than repaid in the long run by the benefits of enhanced water quality for fisheries, water treatment, public health, and recreation.

Toxins

Toxins are poisons that kill animals and plants by interfering with their normal physiological functions. Toxins are by no means novel. Plants have evolved chemical defenses that kill or sicken herbivores, and many animals use toxins to kill their prey. Human technology, however, has produced a stupefying array of chemical substances with adverse effects on life. Many are produced because of their toxic effects, among them a variety of insecticides, rodenticides, and herbicides. Some of these, such as the insecticide DDT (the use of which is now banned in the United States), have had damaging effects far beyond their intended victims. These undesirable side effects have occurred because of difficulties in delivering pesticides to particular targets without other species getting in the way, because of the indiscriminate action of most pesticides that are designed to attack basic physiological functions, and because many of these substances persist in the environment and may be concentrated as organisms feed on others that contain them.

Toxic substances may be divided into several classes: acids, heavy metals, organics, and radiation are the most notable. Acids are very reactive substances that produce hydrogen ions (H^+) and may be extremely toxic at high concentration (low pH). Acids affect organisms directly by interfering with physiological functions and indirectly through their influence on nutrient availability and regeneration. In particular, high acidity reduces the solubility of phosphorus in soils and waters, which tends to reduce productivity.

All environments contain natural acids, which are produced when carbon dioxide dissolves in water to form carbonic acid and when bacterial metabolism forms organic acids. Anthropogenic sources of acid are primarily of two kinds. The first occurs in coal mining areas where reduced sulfur compounds associated with coal are exposed to oxygen-rich environments. Sulfur bacteria oxidize pure sulfur and thiol (reduced) forms of sulfur to sulfates, which may then be converted to sulfuric acid in streams that drain mining areas — hence the term **acid mine drainage.** In some places, the water becomes so acid as to sterilize the aquatic environment (Figure 27.7).

A more widespread problem is **acid rain.** Coal and oil are not pure hydrocarbons; they contain sulfur and nitrogen compounds as well. After all, these fossil fuels are the remains of plants and animals that, when living, contained nitrogen and sulfur in proteins and other organic molecules. The burning of coal and oil, in addition to producing carbon dioxide and water vapor, spews nitrous oxides and sulfur dioxide into the atmosphere. When these gases dissolve in raindrops, they are converted to acid and cause acid rain. The pH of natural rain is about 6 — slightly on the acid side of neutral (pH 7) — because of naturally occurring carbonic acid. In highly industrialized areas, the pH of rain may drop to between 3 and 4, 100 to 1000 times the acidity of natural rain.

The consequences of acid rain have been acute in some regions, such as the northeastern United States, Canada, and Scandinavia, where rivers and lakes tend to be oligotrophic and thus do not contain dissolved bases to buffer acid inputs. As a result, pH may drop to as low as 4, acid enough to stunt growth or

FIGURE 27.7 Streams draining from the refuse of coal mining may be extremely acid. Courtesy of the U.S. Department of Agriculture.

even cause mortality of fish and other organisms. Acid rain may also lower the pH in soil, which increases the rate of leaching of soil nutrients and precipitates phosphorus compounds, making them unavailable for uptake by plant roots. Acid directly affects the foliage of plants as well, making leaves more susceptible to disease and frost damage. The ecological scope of the acid rain problem includes adverse effects on productivity and wildlife in both aquatic and terrestrial habitats. The solutions are primarily technological and economic: scrubbing the offending gases from the effluents of power plants and automobiles, finding alternatives to the burning of fossil fuels for energy, and reducing total demand for energy.

Heavy metals

Even in low concentrations, mercury, arsenic, lead, copper, nickel, zinc, and other **heavy metals** are toxic to most forms of life. They are introduced to the environment in a variety of ways, principally as refuse from mining and mineral smelting, as waste products of manufacturing processes, as fungicides (such as lead arsenate), and through the burning of leaded fuel. Their effects are varied but include interference with neurological function in vertebrates. Ecological studies have revealed the movement and concentration of heavy metals through the food chain and the persistence and transformation of these elements in the ecosystem. Many toxic metals, including copper and nickel particulates released to the atmosphere by smelters, eventually accumulate in soil. In the case of copper, concentrations in excess of 100 ppm adversely affect mosses, lichens, and large fungi; above 1000 ppm, earthworm abundance drops off dramatically; and most species of vascular plants cannot tolerate concentrations above 5000 ppm (0.5 percent). As fungi die out, decomposition of organic matter and mineralization of nitrogen in the soil decrease. In one study in Sweden, at copper levels of 2000 ppm, fungal populations had fallen to only 20 to 30 percent of their natural levels. The concentration of copper averages about 30 ppm in unpolluted temperate zone soils. Concentrations exceeding 1000 ppm may extend for 10 to 20 km from sites of metal smelting, with predictable effects on the diversity and productivity of local communities. These effects are mitigated to some degree by taller smokestacks, which distribute the wastes over larger areas at lower concentration, but ultimately, solving the problem will require a change in the technology of metal production to reduce toxic by-products.

High concentrations of metals in some soils have given biologists an unusual opportunity to observe evolutionary responses of plants. The refuse material from mining operations often contains concentrations of copper, lead, zinc, or arsenic in excess of 1000 ppm (and, in some cases, in excess of 10,000 ppm), which are toxic to most plants. These mine tailings are, however, often colonized by grasses and other plants that, when analyzed, reveal unusually high tolerances for toxic metals. Specific tolerances are determined by genetic factors that occur as rare mutants in populations living on normal soils. On the mine soils, these individuals have higher evolutionary fitness than intolerant forms, so their progeny proliferate and spread to form a genetically distinctive subpopulation restricted to the area of contamination. Selection has also favored reduced outcrossing and thereby restricted fertilization of metal-tolerant genotypes by pollen from intolerant plants growing on nearby uncontaminated soils! This example illustrates one facet of the immense capacity of natural systems to respond to perturbations of all sorts. Nature is extremely forgiving of human excesses — perhaps too much so for our own good.

Toxic organic compounds

Toxic organic compounds are widespread in nature as chemical defenses of plants against herbivores and as metabolic by-products of various microorganisms, such as the bacterium that causes botulism and the dinoflagellate that causes toxic red tides. Although agricultural pesticides include some natural compounds, such as nicotine and pyrethrins, most are far more deadly concoctions produced in the laboratory, to which pests have had no previous exposure and opportunity to evolve resistance. The latter include organomercurials (such as methylmercury), chlorinated hydrocarbons (DDT, lindane, chlordane, dieldrin, 2,4-D), organophosphorus compounds (parathion, malathion), carbamate insecticides, and triazine herbicides. Although these compounds do their job for agriculture and pest management, many accumulate in other parts of the ecosystem where they adversely affect plant production and wildlife populations.

Overuse and misuse of pesticides can be addressed in part by applying them properly and in the smallest effective amounts. But because insects and other pests may evolve tolerances to pesticides, just as many plants have evolved tolerances to toxic elements in soil, applications of chemicals often produce only short-term benefits, and their amounts must be increased to achieve continued results. Through ecological research, we can assess the vulnerability of natural systems to these pollutants, prescribe safe applications, and — perhaps most important — determine suitable alternatives to humankind's chemical warfare with the environment.

Biological control of pests, which relies on the principles of predator–prey interactions, often provides effective regulation of insect populations without the adverse effects of chemical pesticides. We have seen how parasitoid wasps were used to control scale insects in California citrus groves, where cyanide fumigation had been required before and was becoming ineffective because of evolved cyanide tolerance. Similarly, cactus moths effectively reduced the prickly pear cactus where it had been introduced in Australia. Some types of herbivore damage may be alleviated by inducing natural resistance in crop plants. In one case, a relatively innocuous, herbivorous mite introduced to cotton crops can induce chemical resistance that depresses subsequent infestations by more damaging species of mites. This **cross-resistance** to pest organisms is reminiscent of the immunity to dangerous diseases that sometimes develops following infection by related organisms that produce only mild symptoms.

A final type of toxic environmental pollution caused by organic compounds is that resulting from **oil spills.** Crude oil is a complex mixture of hy-

FIGURE 27.8 (a) Oil slick from a damaged tanker approaching the Caribbean coast of Panama. (b) Fringe of dead mangroves killed by the oil spill. Photographs by C. H. Hansen. Courtesy of the Smithsonian Tropical Research Institute.

drocarbons, with concentrations of nitrogen up to 1 percent and of sulfur up to 5 percent. Oil pollution occurs at the source in areas of oil production, rarely as a result of breaks in the nearly 100,000 km of oil pipeline in the world, and much more frequently in the ocean from offshore drilling and the wreckage of oil tankers (Figure 27.8).

Hydrocarbons are produced naturally in the marine system by algae at a rate of 20 to 30 million (20 to 30 × 10⁶) T yr⁻¹. These nonpetroleum hydrocarbons are very widely dispersed and do not constitute pollution, in the sense that they are natural products in low concentration. Contamination of oceans by petroleum hydrocarbons from natural seeps amounts to about 0.2 to 0.6 × 10⁶ T annually; anthropogenic sources from various types of oil spills dump between 3 and 6 × 10⁶ T into the environment each year — that is, 0.1

to 0.2 percent of global oil production. This total is small, but local effects can be devastating. Petroleum kills by coating the surfaces of organisms and, because hydrocarbons are organic solvents, by disrupting biological membranes. Over time, oil slicks disperse by evaporation of lighter fractions, emulsion of other fractions in water, and weathering and microbial breakdown of the rest. But certain types of sensitive ecosystems, such as coral reefs, may take decades to recover fully.

Radiation

Radiation comes in a broad spectrum of energy intensities, ranging from generally harmless long-wavelength radio and infrared radiation, through damaging ultraviolet radiation of shorter wavelength, to the extremely high energy levels of cosmic rays and subatomic particles released by the disintegration of atomic nuclei (radioactive decay). Natural sources create an unavoidable background level of radiation. Under some circumstances, radioactive substances, such as the radon gas present in soils in regions having granitic bedrock, can become concentrated and pose public health hazards. Such dangers are minor compared to the possibility of the extreme radiation hazards resulting from accidents at nuclear power plants, such as those that occurred at Three Mile Island and Chernobyl, from the waste products of nuclear power generation, and from nuclear war. The effects of intense radiation on life forms can be seen clearly in the results of experiments in which habitats are exposed to radiation sources.

The possibility of nuclear war is fast diminishing as the superpowers dismantle their nuclear arsenals and nations turn to economic, social, and environmental problems more pressing than national defense. Still, radioactive wastes produced by peaceful uses of the atom pose tremendous disposal problems. Depending on the waste product, radiation does not decline to harmless levels for thousands or even millions of years, far beyond the life span of waste containers, not to mention that of the institutions to which their care is entrusted. The waste disposal problem may ultimately limit the use of nuclear power, at least until new methods can be devised to deal with radioactive wastes.

Ultraviolet light is another form of radiation that is becoming a serious threat to the environment as the protective mantle of ozone in the atmosphere shows signs of depletion. We shall consider this problem in detail as we examine the atmospheric factors that threaten the entire globe.

Global threats

Because of the circulation of the atmosphere and oceans, certain types of pollution have global consequences: their effects extend far beyond the source of the pollution itself. By far the most worrisome of these changes in the environment are the destruction of the ozone layer in the upper atmosphere and the increase in carbon dioxide and other "greenhouse" gases.

Ozone

Ozone (O_3) is a molecular form of oxygen that is highly reactive as an oxidant. As a consequence, it is toxic to animal and plant life even in small concentrations. Near the earth's surface, ozone is produced by the photochemical oxidation of molecular oxygen (O_2) in the presence of nitrous oxide (NO_2). Because NO_2 is a product of gasoline combustion, ozone can reach high levels in the exhaust fumes that pollute cities, particularly in sunny parts of the world. In Los Angeles, for example, which is well known for its smog, ozone concentrations in the atmosphere at ground level can reach 0.5 ppm, which is perhaps 20 to 50 times the normal level and is damaging to health, crops, and natural vegetation.

Photochemical oxidations in the upper atmosphere also produce ozone, where it has the extremely beneficial effect of shielding the surface of the earth from ultraviolet radiation by absorbing solar radiation of short wavelength (especially in the range of 200 to 300 nm). The concentration of ozone is maintained in an approximate equilibrium by photodissociation of O_3 to O_2 and O. Unfortunately, this dissociation is accelerated by certain substances, among them chlorine atoms and simple chlorine compounds. The levels of chlorine in the upper atmosphere have been increasing because of the emission of chlorofluorocarbons (CFCs), which are used as propellants in spray cans and as coolants in air conditioning and refrigeration systems. Decreases in stratospheric ozone of 50 percent or more (so-called **ozone holes**) have been observed at high latitudes in both hemispheres. The resulting elevation of ultraviolet radiation will increase the incidence of skin cancer, because DNA also absorbs ultraviolet radiation between 280 and 320 nm in wavelength, and the energy of the absorbed radiation causes damage to DNA molecules (mutation). Of greater concern is the fact that ultraviolet light also damages the photosynthetic apparatus of plants and may cause a reduction of primary production—the base of the food chain for the entire ecosystem. This has already been observed in the oceans surrounding Antarctica. The threat is so great that most countries have banned the use of CFCs. We can hope that this action will reverse the damage that has already been done and allow atmospheric ozone to return to its natural equilibrium level.

Carbon dioxide and the greenhouse effect

Carbon dioxide (CO_2) occurs naturally in the atmosphere. Before 1850, its concentration was on the order of 280 ppm (0.028 percent). During the last 150 years, which have witnessed tremendous increases in the burning of wood, coal, oil, and gas for energy production, the level of CO_2 in the atmosphere has increased to 345 ppm (Figure 27.9). Half of this increase has occurred during the last 30 years, and the rate of increase is rising. This change in the chemistry of the atmosphere has created fears of a major warming of the earth's climate, with disastrous consequences resulting from the melting of ice caps, desertification of agricultural lands, and shifting of ecological zones faster than plants can extend their geographic ranges.

FIGURE 27.9 Global emissions of carbon dioxide from the combustion of solid (coal) and liquid (oil) fossil fuels since 1860, and the resulting increase in the concentration of carbon dioxide in the atmosphere. Accurate measurements of carbon dioxide in the atmosphere have been taken at Mauna Loa Observatory, Hawaii, which is far from regions of concentrated fossil fuel consumption, only since the late 1950s. Data from A. M. Solomon, J. R. Trabolka, D. E. Reichle, and L. D. Voorhees; R. M. Rotty and C. M. Masters, in *Atmospheric Carbon Dioxide and the Global Carbon Dioxide Cycle*. U.S. Department of Energy, Washington, D.C. (1985), pp. 1–13 and 63–80.

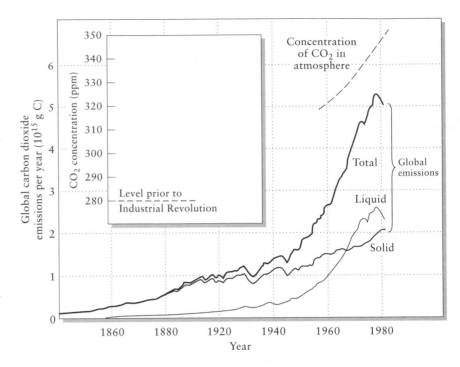

Without carbon dioxide in the atmosphere, the earth would be a very cold place. Most of the radiation absorbed from the sun at the earth's surface would be reradiated into the cold depths of space. As it is now, much of this energy, which occurs as long-wave infrared radiation peaking at a wavelength of 10,000 nm, is absorbed by carbon dioxide in the atmosphere. As the CO_2 heats up, its thermal energy is reradiated at even longer wavelengths — about half of it back toward the surface of the earth. Thus CO_2 forms an insulative blanket over the earth's surface that lets short-wavelength ultraviolet and visible light from the sun pass through but retards the loss of heat as longer wavelength infrared radiation (Figure 27.10). Glass in greenhouses works on the same principle, so the function of CO_2 in the atmosphere is known as the **greenhouse effect**.

Carbon dioxide is not the only greenhouse gas. Methane, produced in abundance by the ruminant fermentation of cattle, absorbs infrared radiation 20 times more effectively than does CO_2; the chlorofluorocarbon $CFCl_3$ is 10,000 times more effective than CO_2 and, even in its low concentration of 0.0002 ppm, is responsible for almost one-tenth of the total greenhouse gas problem.

If present levels of carbon dioxide warm the earth, more CO_2 will warm it more. At least that is the view of many climatologists. The question is, how much? And with what consequence? Current estimates of global warming in the next century suggest a rise in average global temperature of anywhere between 2 and 6°C. These predictions are based on models of the behavior of carbon dioxide and heat in the atmosphere, but scientists do not fully agree on the details of these models. Hence the differences in opinion that we hear in scientific discussions and the popular press.

The level of carbon dioxide in the atmosphere represents a balance between processes that add CO_2 and those that remove it. Before the Industrial Revolu-

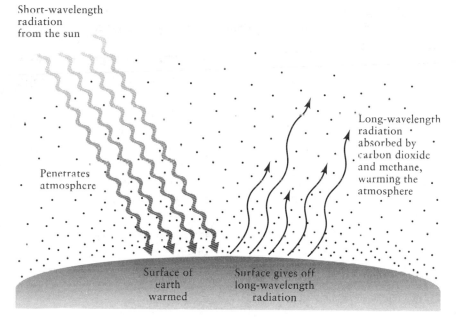

Short-wavelength radiation from the sun

Penetrates atmosphere

Long-wavelength radiation absorbed by carbon dioxide and methane, warming the atmosphere

Surface of earth warmed

Surface gives off long-wavelength radiation

FIGURE 27.10 Schematic depiction of the greenhouse effect illustrating how greenhouse gases, such as carbon dioxide and methane, allow short-wavelength light to pass but absorb longer wavelength (infrared) radiation given off by the earth.

tion, addition of CO_2 to the atmosphere by respiration of terrestrial organisms (approximately 120×10^9 T of carbon per year) was balanced by gross primary production of terrestrial vegetation, and the total amount in the atmosphere was maintained in equilibrium. During the past 150 years, increasing amounts of CO_2 have been added to the atmosphere by the burning of fossil fuels and the clearing and burning of forests. At present, forest clearing accounts for the addition of about 2×10^9 T of carbon to the atmosphere annually; the burning of fossil fuels accounts for the addition of about 5×10^9 T. Thus human activities have increased carbon flux into the atmosphere by about 6 percent —perhaps more, perhaps less, depending on whose figures we use.

The atmosphere exchanges carbon dioxide with the oceans, where excess carbon is precipitated as calcium carbonate sediments. It has been estimated that at present the oceans absorb more carbon than they release to the atmosphere by about 2.4×10^9 T annually, which is *less than half* of the anthropogenic input to the atmosphere. No matter how the arithmetic is done, it shows carbon dioxide concentrations in the atmosphere to be rising very rapidly.

And what of the consequences of increased carbon dioxide and global warming? Increased CO_2 may increase plant production, particularly in productive regions where neither water nor soil nutrients limit photosynthesis. In more arid regions, greater concentration of CO_2 may decrease the water requirements of plants, because it allows carbon dioxide to enter leaves more rapidly and stomates can then be closed to reduce water loss.

Warmer temperatures caused by the greenhouse effect will have mixed effects, some of them potentially disastrous. On the positive side, increased temperature lengthens the growing season and speeds metabolism, thereby tending to enhance production in moist environments. Balancing this benefit is increasing drought stress in arid environments, which may reduce production and accelerate the conversion of grazing and crop lands to useless wasteland (desertification) caused by overgrazing.

Many ecologists are concerned that global warming will cause climate belts to shift geographically so fast that plants will not be able to keep up. The distributions of plants expand by colonization, which is limited by the dispersal distance of seeds. As the glaciers retreated from North America and Europe at the end of the Pleistocene epoch, beginning about 18,000 years ago, the distributions of plants shifted great distances. The boundaries of distribution of trees advanced, following the receding edge of the glaciers, by as much as 100 m per year. The concern of ecologists over the present situation is that global warming may shift climate belts more rapidly than this, greatly outpacing the ability of plants to keep up. The result will be widespread disruption of natural ecological communities and, in a worst-case scenario, the extinction of a part of the flora and its associated fauna. It is impossible to gauge the accuracy of these predictions, but the potential for ecosystem deterioration, combined with the social and economic problems caused by changes in agriculture, the probable rise in sea levels, the current rate of depletion of fossil fuel resources, and the disposal of toxic materials arising from fossil fuel combustion, should prompt immediate action to reduce the burning of forests and of fossil fuels and to accelerate the search for alternative sources of energy.

Human ecology

The human population is increasing worldwide at a rate of almost 2 percent per year. Even if population growth were to stop today, staggering problems would remain. The present human population is consuming resources faster than they are being regenerated by the biosphere, all the while pouring forth so much waste that the quality of the environment in most regions of the earth is deteriorating at an alarming rate. If we are to leave a habitable world for future generations, our top priority must be to achieve a sustainable, equilibrial relationship with the rest of the biosphere. This will require putting an end to population growth, developing sustainable energy sources, providing for the regeneration of nutrients and other materials, and restoring deteriorated habitats.

In many respects, humankind has stepped beyond the bounds of the usual ecological mechanisms of restraint and regeneration. Our ability to tap nonrenewable sources of energy in the form of coal, oil, and gas deposits has *temporarily* removed conventional food – energy limitations on population growth. No longer is most of the human population supported by the land it occupies. The technological ability to reach out and continually usurp new land and resources has pushed density-dependent population feedbacks well into the future.

Our present course, however, leads in a predictable direction. It is not inviting: increasing energy, material, and food shortages; most of the population living in abject poverty, filth, and disease; the environment so badly polluted as to leave the earth gasping; social and political strife escalating as the rich and powerful try to defend their high standard of living. These are the inevitable mechanisms of population control that will eventually come into play, as they do for every species.

The future need not be like this. Where we have escaped natural restraint, we must substitute our *own* restraint. Where we produce waste products that cannot be regenerated by ecological systems, we must find ways to recycle them ourselves. Energy consumption must be scaled back, and production must be based increasingly on renewable sources, such as the sun and wind.

Above all, humankind must adopt a new attitude toward its relationship with nature. We are a part of nature, not apart from nature. To the extent that our intelligence, culture, and technology have given us the power to dominate nature, we must also use these abilities to impose self-regulation and self-restraint. This is the greatest challenge facing us. We have succeeded famously in becoming the technological species. Our survival now depends on our becoming the ecological species.

❧ SUMMARY

1. The key to the survival of the human population is the development of sustainable interactions within the biosphere. This will necessitate the control of human population growth, an increasing reliance on renewable energy sources, and the total recycling of material wastes.

2. Maintaining a sustainable biosphere requires that we conserve the ecological processes responsible for its productivity.

3. The principal local threats to natural processes are overexploitation of resources, habitat conversion, introductions of alien species, enrichment with organic wastes (eutrophication), and the production of toxic materials.

4. On a global scale, various airborne pollutants, especially the chlorofluorocarbons, have seriously reduced the concentrations of ozone (O_3) in the upper atmosphere, allowing damaging ultraviolet radiation to reach the surface of the earth at increased intensity.

5. Increased levels of carbon dioxide in the atmosphere, produced by the burning of fossil fuels, threaten to increase the average temperature of the earth by 2 to 6°C, with potentially disastrous consequences for natural ecosystems and agriculture. In addition, coastal settlements may be inundated by rising sea levels fed by melting polar ice caps.

6. Solutions to the environmental crisis will require the development of new attitudes promoting sustainability and self-restraint. The hope for such a consensus among human social, economic, and political institutions lies in making people aware of the global deterioration of the quality of human life and educating them in the basic ecological principles that must form the foundation of a self-sustaining system.

❧ SUGGESTED READINGS

Ausubel, J. H. 1991. A second look at the impacts of climate change. *American Scientist* 79:210–221.

Browder, J. O. 1992. The limits of extractivism. *BioScience* 42:174–182.

Clark, W. C. 1989. Managing planet earth. *Scientific American* 261:46–54. (See other articles in the same issue.)

Davis, G. R. 1990. Energy for planet earth. *Scientific American* 263:54–62.

Graham, R. L., M. G. Turner, and V. H. Dale. 1990. How increasing CO_2 and climate change affect forests. *BioScience* 40:575–587.

Houghton, R. A., and G. M. Woodwell. 1989. Global climate change. *Scientific American* 260:36–44.

Katzman, M. T., and W. G. Cale, Jr. 1990. Tropical forest preservation using economic incentives. *BioScience* 40:827–832.

Pimentel, D., et al. 1987. World agriculture and soil erosion. *BioScience* 37:277–283.

Pimentel, D., et al. 1991. Environmental and economic effects of reducing pesticide use. *BioScience* 41:402–409.

Repetto, R. 1990. Deforestation in the tropics. *Scientific American* 262:36–42.

Rowland, F. S. 1989. Clorofluorocarbons and the depletion of statospheric ozone. *American Scientist* 77:36–45.

Schneider, S. H. 1989. The greenhouse effect: science and policy. *Science* 243:771–782.

Vitousek, P. M., P. R. Ehrlich, A. H. Ehrlich, and P. A. Matson. 1986. Human appropriation of the products of photosynthesis. *BioScience* 36:368–373.

White, R. M. 1990. The great climate debate. *Scientific American* 263:36–43.

GLOSSARY

Acclimation. A reversible change in the morphology or physiology of an organism in response to environmental change; also called acclimatization.

Acid rain. Precipitation with high acidity (pH < 4) caused by the solution of certain atmospheric gases (sulfur dioxide and nitrous oxide) produced by combustion of fossil fuels.

Activity space. The range of environmental conditions suitable for the activity of an organism.

Adaptation. A genetically determined characteristic that enhances the ability of an individual to cope with its environment; an evolutionary process by which organisms become better suited to their environments.

Adaptive radiation. The evolution of a variety of forms from a single ancestral stock, often after colonizing an island group or entering a new adaptive zone.

Adaptive zone. A set of environmental conditions or portion of the ecological niche that requires particular adaptations for living.

Additive genetic variance (V_A). Variation in a phenotypic value within a population due to the difference in expression of alleles in the homozygous state.

Age class. Individuals in a population of a particular age.

Age-specific. Pertaining to attributes, such as survival or fecundity, that vary as a function of age.

Age structure. The distribution of individuals among age classes within a population.

Alkaloids. Nitrogen-containing compounds, such as morphine and nicotine, that are produced by plants and are toxic to many herbivores.

Allele. One of several alternative forms of a gene.

Allelopathy. Direct inhibition of one species by another using noxious or toxic chemicals.

Allochthonous. Referring to materials transported into a system, particularly minerals and organic matter transported into streams and lakes. *Compare with* Autochthonous.

Allometric constant. Slope of the relationship between the logarithm of one measurement of an organism and the logarithm of another, usually its overall size.

Allometry. Relative increase in a part of an organism or a measure of its physiology or behavior in relation to some other measure, usually its overall size.

Allopatric. Occurring in different places, usually referring to geographic separation of populations.

Alluvial. Referring to sediment deposited by running water.

Alpha diversity. The variety of organisms occurring in a particular place or habitat; often called local diversity.

Altruism. In an evolutionary sense, enhancing the fitness of an unrelated individual by acts that reduce the evolutionary fitness of the altruistic individual.

Ambient. Referring to conditions of the environment surrounding the organism.

Amino acid. One of about 30 organic acids containing the group NH_2, which are the building blocks of proteins.

Ammonification. Metabolic breakdown of proteins and amino acids with ammonia as an excreted by-product.

Anaerobic. Without oxygen.

Anion. A part of a dissociated molecule carrying a negative electrical charge.

Anisogamous. Having gametes unequal in size and behavior; usually a large, sedentary (female) and a small, motile (male) gamete. *Compare with* Isogamous.

Annual. Referring to an organism that completes its life cycle from birth or germination to death within a year.

Anoxic. Lacking oxygen; anaerobic.

Aposematism. *See* Warning coloration.

Aspect diversity. Variety of outward appearances of species that live in the same habitat and are eaten by visually hunting predators.

Assimilation. Incorporation of any material into the tissues, cells, and fluids of an organism.

Assimilation efficiency. A percentage expressing the proportion of ingested energy that is absorbed into the bloodstream.

Assimilatory. Referring to a biochemical transformation that results in the reduction of an element to an organic form, hence its gain by the biological compartment of the ecosystem.

Association. A group of species living in the same place.

Asymmetrical competition. An interaction between two species in which one exploits a particular resource more efficiently than the other; the second may persist by better avoiding predation or by subsisting on an exclusive resource.

Autecology. The study of organisms in relation to their physical environment. *Compare with* Synecology.

Autochthonous. Referring to materials produced within a system, particularly organic matter produced, and minerals cycled, within streams and lakes. *Compare with* Allochthonous.

Autotroph. An organism that assimilates energy from either sunlight (green plants) or inorganic compounds (sulfur bacteria). *Compare with* Heterotroph.

Barren. An area with sparse vegetation owing to some physical or chemical property of the soil.

Basal metabolic rate (BMR). The energy expenditure of an organism that is at rest, fasting, and in a thermally neutral environment.

Batesian mimicry. Resemblance of an edible species (mimic) to an unpalatable species (model) to deceive predators.

Benthic. Bottom dwelling in rivers, lakes, and oceans.

Beta diversity. The variety of organisms within a region arising from turnover of species among habitats.

Biennial. Requiring two years to complete the life cycle.

Biodiversity. *See* Diversity.

Biological community. *See* Community.

Biological control. Use of natural enemies, particularly parasitoid insects, bacteria, and viruses, to control pest organisms.

Biological oxygen demand (BOD). The amount of oxygen required to oxidize organic material in water samples; high values in aquatic habitats often indicate pollution by sewage and other sources of organic wastes, or the overproduction of plant material resulting from overenrichment by mineral nutrients.

Biomass. Weight of living material, usually expressed as a dry weight, in all or part of an organism, population, or community. Commonly presented as weight per unit area, a biomass density.

Biomass accumulation ratio. The ratio of weight to annual production.

Biota. Fauna and flora together.

Birth rate (*b*). The average number of offspring produced per individual per unit of time, often expressed as a function of age (*x*).

Boreal. Northern; often refers to the coniferous forest regions that stretch across Canada, northern Europe, and Asia.

Bottleneck. The condition in which a population exists for a period at very small size, during which the population may lose genetic variation by drift. *Compare with* Founder event.

Boundary layer. A layer of still or slow-moving water or air close to the surface of an object.

Breeding system. Degree of polygyny, outcrossing, and selective mating within a population; the adaptations by which organisms adjust these attributes.

Brood parasite. An organism that lays its eggs in the nest of another species or that of another individual of the same species.

C₃ photosynthesis. Photosynthetic pathway in which carbon dioxide is initially assimilated into a 3-carbon compound, phosophoglyceraldehyde (PGA), in the Calvin cycle.

C₄ photosynthesis. Photosynthetic pathway involving the initial assimilation of carbon dioxide into a 4-carbon compound, such as oxaloacetic acid (OAA) or malate.

Calcification. Deposition of calcium and other soluble salts in soils where evaporation greatly exceeds precipitation.

Calvin cycle. The basic assimilatory sequence of photosynthesis during which an atom of carbon is added to the 5-carbon ribulose bisphosphate (RuBP) molecule to produce phosphoglyceraldehyde (PGA) and then glucose.

CAM photosynthesis. Photosynthetic pathway in which the initial assimilation of carbon dioxide into a 4-carbon compound occurs at night; found in some succulent plants in arid habitats.

Carbonate ion. An anion (CO_3^{2-}) formed by the dissociation of carbonic acid or one of its salts.

Carbonic acid. A weak acid (H_2CO_3) formed when carbon dioxide dissolves in water.

Carnivore. An organism that consumes mostly flesh.

Carrying capacity (K). Number of individuals in a population that the resources of a habitat can support; the asymptote, or plateau, of the logistic and other sigmoid equations for population growth.

Caste. Individuals within a social group sharing a specialized form or behavior.

Cation. A part of a dissociated molecule carrying a positive electrical charge.

Cation exchange capacity. The ability of soil particles to absorb positively charged ions, such as hydrogen (H^+) and calcium (Ca^{2+}).

Cellulose. A long-chain molecule made up of glucose subunits found in the cell walls and fibrous structures in plants.

Chaos. Erratic change in the size of populations governed by difference equations and having high intrinsic rates of growth.

Character displacement. Divergence in the characteristics of two otherwise similar species where their ranges overlap, caused by the selective effects of competition between the species in the area of overlap.

Chemoautotroph. An organism that oxidizes inorganic compounds (often hydrogen sulfide) to obtain energy for synthesis of organic compounds; for example, sulfur bacteria.

Clay. A fine-grained component of soil, formed by the weathering of granitic rock and composed primarily of hydrous aluminum silicates.

Cleaning symbiosis. A mutualistic arrangement between two species in which one grooms the other to remove parasites.

Climax. The end point of a successional sequence, or sere; a community that has reached a steady state under a particular set of environmental conditions.

Cline. Gradual change in population characteristics or adaptations over a geographic area.

Closed-community concept. The idea, popularized by F. C. Clements, that communities are distinctive associations of highly interdependent species.

Clutch size. Number of eggs per set; usually with reference to the nests of birds.

Coadaptation. Evolution of characteristics of two or more species in response to changes in each other, often to mutual advantage. *See* Coevolution.

Coarse-grained. Referring to qualities of the environment that occur in large patches with respect to the activity patterns of an organism and, therefore, among which the organism can select. *Compare with* Fine-grained.

Codon. A sequence of three nucleotides in DNA or RNA that specifies which amino acid will be placed at a particular position in a protein.

Coevolution. Occurrence of genetically determined traits (adaptations) in two or more species selected by the mutual interactions controlled by these traits.

Coexistence. Occurrence of two or more species in the same habitat; usually applied to potentially competing species.

Cohort. A group of individuals of the same age recruited into a population at the same time.

Cohort life table. *See* Dynamic life table.

Community. An association of interacting populations, usually defined by the nature of their interaction or the place in which they live.

Compartmentalization. Subdivision of a food web into groups of strongly interacting species somewhat isolated from other such groups.

Compartment model. Representation of a system in which the various parts are portrayed as units (compartments) that receive inputs from, and provide outputs to, other such units.

Compensation point. Depth of water or level of light at which respiration and photosynthesis balance each other; the lower limit of the euphotic zone.

Competition. Use or defense of a resource by one individual that reduces the availability of that resource to other individuals, whether of the same species (intraspecific competition) or other species (interspecific competition).

Competition coefficient (a). A measure of the degree to which one consumer uses the resources of another, expressed in terms of the population consequences of the interaction.

Competitive exclusion principle. The hypothesis that two or more species cannot coexist on a single resource that is scarce relative to demand for it.

Condition. Physical or chemical attributes of the environment that, while not being consumed, influence biological processes and population growth; for example, temperature, salinity, acidity. *Compare with* Resource.

Conductance. Capacity of heat, electricity, or a substance to pass through a particular material.

Conduction. The ability of heat to pass through a substance.

Conformer. An organism that allows its internal environment to vary with external conditions.

Congeneric. Belonging to the same genus.

Connectance (C). Proportion of interspecific interactions in a community matrix not equal to zero.

Conspecific. Belonging to the same species.

Consumer. Individual or population that uses a particular resource.

Consumer chain. *See* Food chain.

Continental climate. A climate lacking the tempering effect of the ocean, usually exhibting great extremes of temperature. *Compare with* Maritime climate.

Continental drift. Movement of continents on the surface of the earth over geological time; rates of drift are on the order of centimeters per year.

Continuum. A gradient of environmental characteristics or of change in the composition of communities.

Continuum index. A scale of an environmental gradient based on changes in physical characteristics of community composition along that gradient.

Convection. Transfer of heat by the movement of a fluid (for example, air or water).

Convergence. Resemblance of organisms belonging to different taxonomic groups resulting from adaptation to similar environments.

Correlated response. Response of the phenotypic value of one trait to selection upon another trait.

Countercurrent circulation. Movement of fluids in opposite directions on either side of a separating barrier through which heat or dissolved substances may pass.

Crassulacean acid metabolism. *See* CAM photosynthesis.

Cross-resistance. Resistance or immunity to one disease organism resulting from infection by another, usually closely related, organism.

Crypsis. An aspect of the appearance of organisms whereby they avoid detection by others.

Cybernetic. Pertaining to feedback controls and communication within systems.

Cycle. Recurrent variation in a system periodically returning to its starting point.

Cyclic climax. A steady-state, cyclic sequence of communities, none of which by itself is stable.

Cyclic succession. Continual community change through a repeated sequence of stages.

Damped oscillation. Cycling with progressively smaller amplitude, as in some populations approaching their equilibria.

Death rate (d_x). The percentage of newborns dying during a specified interval. *Compare with* Mortality.

Defensive mutualism. A relationship between two species in which one defends the other against some enemy and usually receives some type of nourishment or living place in return.

Demographic. Pertaining to populations, particularly their growth rate and age structure.

Demography. Study of the age structure and growth rate of populations.

Denitrification. Biochemical reduction, primarily by microorganisms, of nitrogen from nitrate (NO_3^-) eventually to molecular nitrogen (N_2).

Density. Referring to a population, the number of individuals per unit area or volume; referring to a substance, the weight per unit volume.

Density compensation. Increase in population size in response to reduction in the number of competing populations; often observed on islands.

Density dependent. Having influence on individuals in a population that varies with the degree of crowding within the population.

Density independent. Having influence on individuals in a population that does not vary with the degree of crowding.

Deoxyribonucleic acid (DNA). A long macromolecule whose sequence of subunits (nucleotides) encodes genetic information.

Deterministic. Referring to the outcome of a process that is not subject to stochastic (random) variation.

Detoxification. Biochemical conversion of a toxic substance to harmless by-products.

Detritivore. An organism that feeds on freshly dead or partially decomposed organic matter.

Detritus. Freshly dead or partially decomposed organic matter.

Developmental response. Acquisition of one of several alternative forms by an organism depending on the environmental conditions under which it grows.

Diapause. Temporary interruption in the development of insect eggs or larvae, usually associated with a dormant period.

Diffusion. Movement of particles of gas or liquid from regions of high to low concentration by means of their own spontaneous motion.

Dimorphism. Occurrence of two forms of individuals within a population.

Dioecy. In plants, the occurrence of reproductive organs of the male and female sex on different individuals. *Compare with* Monoecy.

Diploid. Pertaining to cells or organisms having two sets of chromosomes. *See* Haploid; Meiosis.

Dispersal. Movement of organisms away from the place of birth or from centers of population density.

Dispersion. The spatial pattern of distribution of individuals within populations.

Dissimilatory. Referring to a biochemical transformation that results in the oxidation of the organic form of an element, hence its loss from the biological compartment of the ecosystem.

Dissolution. The entry of a substance into solution with water.

Distribution. The geographic extent of a population or other ecological unit.

Diversity. The number of taxa in a local area (alpha diversity) or region (gamma diversity). Also, a measure of the variety of taxa in a community that takes into account the relative abundance of each one.

Diversity index. A measure of the variety of taxa in a community that takes into account the relative abundance of each one.

DNA. *See* Deoxyribonucleic acid.

Dominance (species). The numerical superiority of a species over others within a community or association.

Dominance hierarchy. Orderly ranking of individuals in a group, based on the outcome of aggressive encounters.

Dominance variance (V_D). Variation in a phenotypic value within a population due to the unequal expression of alleles in the heterozygous state.

Dominant (gene). A gene that masks the expression of another (recessive) allele of the same gene.

Dormancy. An inactive state, including hibernation, diapause, and seed dormancy, usually assumed during an inhospitable period.

Dynamic life table. The age-specific survival and fecundity of a cohort of individuals in a population followed between birth and death of the last individual; cohort life table.

Dynamic steady state. Condition in which fluxes of energy or materials into and out of a system are balanced.

Ecocline. A geographic gradient of vegetation structure associated with one or more environmental variables.

Ecological efficiency. Percentage of energy in the biomass produced by one trophic level that is incorporated into the biomass produced by the next higher trophic level.

Ecological release. Expansion of habitat and resource use by populations in regions of low species diversity, resulting from reduced interspecific competition.

Ecology. The study of the natural environment and of the relations of organisms to each other and to their surroundings.

Ecomorphology. The study of the relationship between the ecological relations of an individual and its morphology.

Ecosystem. All the interacting parts of the physical and biological worlds.

Ecotone. A habitat created by the juxtaposition of distinctly different habitats; an edge habitat; a zone of transition between habitat types.

Ecotype. A genetically differentiated subpopulation that is restricted to a specific habitat.

Ectomycorrhyzae. Mutualistic fungal association with the roots of plants in which the fungus forms a sheath around the outside of the root.

Ectoparasite. A parasite — for example, a tick — that lives on or attached to the host's surface.

Ectothermy. Capacity to maintain body temperature by gaining heat from the environment, either by conduction or by absorbing radiation.

Edaphic. Pertaining to or influenced by the soil.

Edge effect. Change in the conditions or species composition within an otherwise uniform habitat as one approaches a boundary with a different habitat.

Egestion. Elimination of undigested food material.

Electrical potential (Eh). The relative capacity, measured in volts, of one substance to oxidize another.

Electron acceptor. A substance that readily accepts electrons and thus is capable of oxidizing another substance.

El Niño. A warm current from the tropics that intrudes each winter along the west coast of northern South America.

Eluviation. The downward movement of dissolved soil materials from the topmost (A) horizon, carried by percolating water.

Emigration. Movement of individuals out of a population. *Compare with* Immigration.

Endemic. Confined to a certain region.

Endomycorrhyzae. Mutualistic fungal association with the roots of plants in which part of the fungus resides within the root tissues.

Endoparasite. A parasite that lives within the tissues or bloodstream of its host.

Endothermy. Capacity to maintain body temperature by the metabolic generation of heat.

Energetic efficiency. The ratio of useful work or energy storage to energy intake.

Energy. Capacity for doing work.

ENSO. El Niño – Southern Oscillation; an occasional shift in winds and ocean currents, centered in the South Pacific region, with worldwide consequences for climate and biological systems.

Environment. Surroundings of an organism, including the plants, animals, and microbes with which it interacts.

Environmental grain. A concept of the spatial or temporal heterogeneity of the environment relative to the activities of an organism.

Environmental variance (V_E). Variation in a mensural trait within a population due to the influence of environmental factors.

Enzyme. Organic compound in a living cell or secreted by it that accelerates a specific biochemical transformation without itself being affected.

Epidemiology. The study of factors influencing the spread of disease through a population.

Epifaunal. Pertaining to animals living on the surface of a substrate.

Epilimnion. The warm, oxygen-rich surface layers of a lake or other body of water. *Compare with* Hypolimnion.

Equilibrium. A state of balance between opposing forces.

Equilibrium isocline. A line on a population graph designating combinations of competing populations, or predator and prey populations, for which the growth rate of one of the populations is zero.

Equilibrium theory of island biogeography. The idea that the number of species on an island exists as a balance between colonization by new immigrants and extinction of old residents.

Equitability. Uniformity of abundance in an assemblage of species. Equitability is greatest when species are equally abundant.

Escape space. Refuge from predators and parasites; often reflected in the adaptations of prey organisms to fight, flee, or escape detection.

Estuary. A semienclosed coastal water, often at the mouth of a river, having a high input of fresh water and great fluctuation in salinity.

Euphotic zone. Surface layer of water to the depth of light penetration at which photosynthesis balances respiration. *See* Compensation point.

Eusociality. The complex social organization of termites, ants, and many wasps and bees, dominated by an egg-laying queen that is tended by non-reproductive offspring.

Eutrophic. Rich in the mineral nutrients required by green plants; pertaining to an aquatic habitat with high productivity.

Eutrophication. Enrichment of water by nutrients required for plant growth; often overenrichment caused by sewage and runoff from fertilized agricultural lands and resulting in excessive bacterial growth and oxygen depletion.

Evapotranspiration. The sum of transpiration by plants and evaporation from the soil. Potential evapotranspiration is the amount of evapotranspiration that would occur, given the local temperature and humidity, if water were superabundant.

Evolution. Change in heritable traits of organisms through the replacement of genotypes within a population.

Evolutionarily stable strategy (ESS). A strategy such that, if all members of a population adopt it, no alternative strategy can invade.

Evolutionary ecology. The integrated science of evolution, genetics, adaptation, and ecology; interpretation of the structure and function of organisms, communities, and ecosystems in the context of evolutionary theory.

Excretion. Elimination from the body, by way of the kidneys, gills, and dermal glands, of excess salts, nitrogenous waste products, and other substances.

Expectation of further life (e_x). The average remaining lifetime of an individual of age x.

Experiment. Controlled manipulation of a system to determine the effect of a change in one or more factors.

Exploitation. Removal of individuals or biomass from a population by consumers.

Exploitative competition. Interaction between individuals by way of their reduction of shared resources.

Exponential growth. Continuous increase or decrease in a population in which the rate of change is proportional to the number of individuals at any given time. *See* Geometric growth.

Exponential rate of increase (r). Rate at which a population is growing at a particular time, expressed as a proportional increase per unit of time. *See* Geometric rate of increase.

External forcing function. In systems modeling, a material input from outside the system or a condition of the environment of the system that influences its structure and function.

External loading. Input of nutrients to a lake or stream from outside the system, especially sewage and runoff from agricultural lands. *Compare with* Internal loading.

Extinction. Disappearance of a species or other taxon from a region or biota.

Facilitation. Enhancement of a population of one species by the activities of another, particularly during early succession.

Facultative. Being able to adjust to a variety of conditions or circumstances; optional for the organism. *Compare with* Obligate.

Fall bloom. The rapid growth of algae in temperate lakes following the autumnal breakdown of thermal stratification and mixing of water layers.

Fall overturn. Vertical mixing of water layers in temperate lakes in autumn following breakdown of thermal stratification.

Fecundity. Rate at which an individual produces offspring.

Feedback control. *See* Internal control feedback.

Field capacity. The amount of water that soil can hold against the pull of gravity.

Filter feeder. An organism that strains tiny food particles from its aqueous environment by means of sievelike structures; for example, clams and baleen whales.

Fine-grained. Referring to qualities of the environment that occur in small patches with respect to the activity patterns of an organism, and among which the organism cannot usefully distinguish. *Compare with* Coarse-grained.

Fitness. Genetic contribution by an individual's descendants to future generations of a population.

Fixation (genetic). Increase in the proportion of a gene to unity, resulting in the elimination of all alleles.

Floristic. Referring to the species composition of plant communities.

Flux. Movement of energy or material into or out of a system.

Food chain. A representation of the passage of energy from a primary producer through a series of consumers at progressively higher trophic (feeding) levels. Thus, plant – herbivore – carnivore, and so on.

Food chain efficiency. *See* Ecological efficiency.

Food web. A representation of the various paths of energy flow through populations in the community, taking into account the fact that each population shares resources and consumers with other populations.

Founder event. Colonization of an island or patch by a small number of individuals that possess less genetic variation than the parent population. *Compare with* Bottleneck.

Frequency dependence. Referring to the condition in which the expression of a process varies with the relative proportions of phenotypes in a population.

Front. A meeting of two water masses having different characteristics.

Functional response. Change in the rate of exploitation of prey by an individual predator as a result of a change in prey density. *See also* Numerical response.

Gaia hypothesis. The idea that the biosphere has evolved to optimize the conditions for life.

Gamete. A haploid cell that fuses with another haploid cell of opposite sex during fertilization to form the zygote. In animals the male gamete is called the sperm and the female gamete is called the egg or ovum.

Gamma diversity. The inclusive diversity of all the habitat types within an area; regional diversity.

Gene. Generally, a unit of genetic inheritance. In biochemistry, gene refers to the part of the DNA molecule that encodes a single enzyme or structural protein.

Gene flow. Exchange of genetic traits between populations by movement of individuals, gametes, or spores.

Gene frequency. The proportion of a particular allele of a gene in the gene pool of a population.

Gene locus. Segment of a chromosome on which a gene resides.

Generalist. A species with broad food or habitat preferences.

Generation time. Average age at which a female gives birth to her offspring, or the average time for a population to increase by a factor equal to the net reproductive rate.

Genetic drift. Change in allele frequency due to random variations in fecundity and mortality in a population.

Genetic variance (V_G). Variation in a phenotypic value within a population due to the expression of genetic factors.

Genotype. All the genetic characteristics that determine the structure and functioning of an organism; often applied to a single gene locus to distinguish one allele, or combination of alleles, from another.

Geometric growth. Periodic increase or decrease in a population in which the increment is proportional to the number of individuals at the beginning of the period, often the breeding season. *See* Exponential growth.

Geometric rate of increase (λ). Factor by which the size of a population changes over a specified period. *See* Exponential rate of increase.

Giving-up time (GUT). *See* Optimum giving-up time.

Gondwana. A giant land mass in the Southern Hemisphere during the early Mesozoic era made up of present-day South America, Africa, India, Australia, and Antarctica.

Gonochoristic. In animals, the condition of separate male and female individuals; in plants, called dioecious.

Gradient analysis. Portrayal and interpretation of the abundances of species along gradients of physical conditions.

Grain. The scale of heterogeneity of habitats in relation to the activities of organisms.

Greenhouse effect. Warming of the earth's climate because of the increased concentration of carbon dioxide and certain other pollutants in the atmosphere.

Gross production. The total energy or nutrients assimilated by an organism, a population, or an entire community. *See also* Net production.

Gross production efficiency. The percentage of ingested food used for growth and reproduction by an organism.

Group selection. Elimination of groups of individuals with a detrimental genetic trait, caused by competition with other groups lacking the trait; often called intergroup selection, and associated with the evolution of altruism.

Guild. Species occupying similar ecological positions within the same habitat.

Habitat. Place where an animal or plant normally lives, often characterized by a dominant plant form or physical characteristic (that is, the stream habitat, the forest habitat).

Habitat compression. Restriction of habitat distribution in response to increase in number of competing species.

Habitat expansion. Increase in average breadth of habitat distribution of species in depauperate biotas, especially on islands, compared with species in more diverse biotas.

Habitat selection. Preference for certain habitats.

Handicap principle. The idea that elaborate, sexually selected displays and adornments act as handicaps that demonstrate the generally high fitness of the bearer.

Haplodiploid. A sex-determining mechanism by which females develop from fertilized eggs and males from unfertilized eggs.

Haploid. Referring to a cell or organism that contains one set of chromosomes.

Hardy–Weinberg equilibrium. The mathematical proposition that the frequencies of genes and genotypes within a population remain unchanged in the absence of selection, mutation, drift, and assortative mating.

Heat. A measure of the kinetic energy of the atoms or molecules in a substance.

Heat of melting. Amount of heat energy added to a substance to make it melt.

Heat of vaporization. Amount of energy added to a substance to make it vaporize.

Herbivore. An organism that consumes living plants or their parts.

Heritability (h^2). The proportion of variance in a phenotypic trait due to the effects of additive genetic factors.

Hermaphrodite. An organism that has the reproductive organs of both sexes.

Heterogeneity. The variety of qualities found in an environment (habitat patches) or a population (genotypic variation).

Heterotroph. An organism that uses organic materials as a source of energy and nutrients. *Compare with* Autotroph.

Heterozygous. Containing two forms (alleles) of a gene, one derived from each parent.

Hibernation. State of winter dormancy associated with lowered body temperature and metabolism.

Homeostasis. Maintenance of constant internal conditions in the face of a varying external environment.

Homeothermy. Ability to maintain constant body temperature in the face of fluctuating environmental temperature; warm-bloodedness. *Compare with* Poikilothermy.

Homologous. Having similar evolutionary origin.

Homozygous. Containing two identical alleles at a gene locus.

Horizon. A layer of soil distinguished by its physical and chemical properties.

Humus. Fine particles of organic detritus in soil.

Hydrological cycle. Movement of water throughout the ecosystem.

Hydrolysis. A biochemical process by which a molecule is split into parts by the addition of the parts of water molecules.

Hygroscopic water. Water held tightly by surface adhesion to particles in the soil, generally unavailable to plants.

Hyperdispersion. Pattern of distribution in which distances between individuals are more even than expected from random placement; overdispersion.

Hyperosmotic. Having an osmotic potential (generally, salt concentration) greater than that of the surrounding medium.

Hypertonic. Having a salt concentration greater than that of the surrounding medium.

Hypolimnion. The cold, oxygen-depleted part of a lake or other body of water that lies below the zone of rapid change in water temperature (thermocline). *Compare with* Epilimnion.

Hypo-osmotic. Having an osmotic potential (generally, salt concentration) less than that of the surrounding medium.

Hypothesis. A conjecture or explanation for a pattern or relationship embracing a mechanism for its occurrence.

Hypotonic. Having a salt concentration less than that of the surrounding medium.

Ideal free distribution. The distribution of individuals across resource patches of different intrinsic quality that equalizes the net rate of gain of each when competition is taken into account.

Identity by descent. Genes in different individuals that are direct copies of a gene in a common ancestor.

Illuviation. The accumulation of dissolved substances within a soil layer, usually the middle (B) horizon.

Immigration. Movement of individuals into a population. *Compare with* Emigration.

Inbreeding. Mating between related individuals.

Inclusive fitness. The total fitness of an individual and the fitnesses of its relatives, the latter weighted according to degree of relationship; usually applied to the consequences of social interaction between relatives.

Individual distance. The distance within which one individual does not tolerate the presence of another.

Individualistic concept. The idea espoused by H. A. Gleason that the distributions of species reflect their tolerances of physical factors and not interactions between species.

Inducible response. Any change in the state of the organism caused by an external factor; usually reserved for the response of organisms to parasitism and herbivory.

Industrial melanism. The evolution of dark coloration by cryptic organisms in response to industrial pollution, especially by soot, in their environments.

Infrared (IR) radiation. Electromagnetic radiation having a wavelength longer than about 700 nm.

Inhibition. The suppression of a colonizing population by another that is already established, especially during successional sequences.

Innate capacity for increase (r_0). The intrinsic growth rate of a population under ideal conditions without the restraining effects of competition.

Intergroup selection. *See* Group selection.

Intermediate disturbance hypothesis. The idea that species diversity is greatest in habitats with moderate amounts of physical disturbance, owing to the coexistence of early and late successional species.

Intermediate host. A host that harbors an asexual stage of the life cycle of a parasite or disease organism.

Internal control feedback. In systems modeling, the influence of one component on other components within the system.

Internal loading. Regeneration of nutrients within a system, usually referring to the sediments of a lake or river. *Compare with* External loading.

Interspecific competition. Competition between individuals of different species.

Intrasexual competition. Competition between members of the same sex, as in the case of combat between males.

Intraspecific competition. Competition between individuals of the same species.

Intrinsic rate of increase (r_m). Exponential growth rate of a population with a stable age distribution; that is, under constant conditions.

Ion. The dissociated parts of a molecule, each of which carries an electrical charge, either positive (cation) or negative (anion).

Isocline. A line on a population graph designating combinations of competing populations, or predator and prey populations, for which the growth rate of one of the populations is zero.

Isogamous. Having gametes similar in size and behavior; not differentiated into unequal (male and female) gametes. *Compare with* Anisogamous.

Iteroparity. The condition of reproducing repeatedly during the lifetime. *Compare with* Semelparity.

Key factor. An environmental factor that is particularly responsible for change in the size of a population.

Key factor analysis. A statistical treatment of population data designed to identify factors most responsible for change in population size.

Keystone species. A species, often a predator, having a dominating influence on the composition of a community, which may be revealed when the keystone species is removed.

Kinetic energy. Energy associated with motion.

Kin selection. Differential reproduction among lineages of closely related individuals based on genetic variation in social behavior.

Kranz anatomy. Arrangement of tissues in the leaves of C_4 plants in which photosynthetic cells having chloroplasts are grouped in sheaths around vascular bundles.

Laterite. A hard substance rich in oxides of iron and aluminum, frequently formed when tropical soils weather under alkaline conditions.

Laterization. Leaching of silica from soil, usually in warm, moist regions with an alkaline soil reaction.

Laurasia. A large land mass in the Northern Hemisphere during the Mesozoic era consisting of what is presently North America, Europe, and most of Asia.

Leaching. Removal of soluble compounds from leaf litter or soil by water.

Lek. A communal courtship area on which several males hold courtship territories to attract and mate with females; sometimes called an arena.

Liebig's law of the minimum. The idea that the growth of an individual or population is limited by the essential nutrient present in the lowest amount relative to requirement.

Life history. The adaptations of an organism that more or less directly influence life table values of age-specific survival and fecundity; hence, reproductive rate, age at maturity, reproductive risk, and so on.

Life table. A summary by age of the survivorship and fecundity of individuals in a population.

Life zone. A more or less distinct belt of vegetation occurring within, and characteristic of, a particular latitude or range of elevation.

Lignin. A long-chain, nitrogen-containing molecule made up of phenolic subunits, occurring in woody structures of plants and being highly resistant to digestion by herbivores.

Limestone. A rock formed chiefly by the sedimentation of shells and the precipitation and sedimentation of calcium carbonate ($CaCO_3$) in marine systems.

Limit cycle. Oscillation of predator and prey populations occurring when stabilizing and destabilizing tendencies of their interaction balance.

Limiting resource. A resource that is scarce relative to demand for it.

Limnology. The study of freshwater habitats and communities, particularly lakes, ponds, and other standing waters.

Littoral. Pertaining to the shore of the sea.

Loam. Soil that is a mixture of coarse sand particles, fine silt, clay particles, and organic matter.

Local mate competition. The situation in which males compete to mate with females at or near their place of birth, hence frequently mating with close relatives.

Logistic equation. Mathematical expression for a particular sigmoid growth curve in which the percentage rate of increase decreases in linear fashion as population size increases.

Lognormal distribution. A characterization of the number of species in logarithmically scaled abundance classes, according to which most species have moderate abundance and fewer have either extremely high or low abundance.

Lottery hypothesis. The idea that habitat patches are colonized at random from the pool of species in an area, thus maintaining the diversity of the assemblage; generally applied to coral reef habitats.

Lower critical temperature (T_c). Surrounding temperature below which warm-blooded animals must generate heat to maintain their body temperature.

Luxury consumption. Uptake of a nutrient in excess of need when the nutrient is abundant.

Maritime climate. A climate with the tempering influence of the ocean, usually exhibiting a narrow range of temperature. *Compare with* Continental climate.

Mark-recapture method. A way of estimating the size of a population by the recapture of marked individuals.

Mass extinction. Abrupt disappearance of a large fraction of a biota, thought to be caused by such environmental catastrophes as a meteor impact; significant mass extinctions occurred at the end of the Permian and Cretaceous periods.

Mating system. Pattern of matings between individuals in a population, including number of simultaneous mates, permanence of pair bond, and degree of inbreeding.

Maximum sustainable yield (MSY). The greatest rate at which individuals may be harvested from a population without reducing the size of the population; that is, at which recruitment equals or exceeds harvesting.

Mediterranean climate. A pattern of climate found in middle latitudes characterized by cool, wet winters and warm, dry summers.

Meiofauna. Animals that live within a substrate, such as soil or aquatic sediment.

Meiosis. A series of two divisions by cells destined to produce gametes, involving pairing and segregation of homologous chromosomes, and reducing chromosome number from diploid to haploid.

Melanism. Occurrence of black pigment, usually melanin.

Mesic. Referring to habitats with plentiful rainfall and well-drained soils.

Mesophyte. A plant that requires moderate amounts of moisture.

Metabolism. Biochemical transformations responsible for the building up and breaking down of tissues and the release of energy by the organism.

Metamorphosis. An abrupt change in form during development that fundamentally alters the function of the organism.

Metapopulation. A population that is divided into subpopulations between which individuals migrate from time to time. Habitat fragmentation is causing many species to assume a metapopulation structure.

Micelle. A complex soil particle resulting from the association of humus and clay particles, with negative electric charges at its surface.

Microcosm. A small, simplified system, often maintained in a laboratory, that contains the essential features of a larger natural system.

Microhabitat. The particular parts of the habitat that an individual encounters in the course of its activities.

Migration. The movement of individuals between one place and another or between subpopulations in a metapopulation.

Mimic. An organism adapted to resemble another organism or an object.

Mimicry. Resemblance of an organism to some other organism or object in the environment, evolved to deceive predators or prey into confusing the organism and that which it mimics.

Mineralization. Transformation of elements from organic to inorganic forms, often by dissimilatory oxidations.

Minimum viable population (MVP). The minimum number of individuals that can prevent a population from losing genetic variation or suffering stochastic extinction over an acceptably long period.

Mixed evolutionarily stable strategy. An evolutionary stable strategy comprising more than one phenotype within a population; generally the outcome of frequency-dependent fitnesses of the phenotypes.

Model. An organism, usually distasteful or otherwise noxious, upon which a mimic is patterned.

Model (mathematical). A quantitative representation of the relationships among the entities in a system; for example, between the species in a community.

Monoecy. In plants, the occurrence of reproductive organs of both sexes on the same individual, either in different flowers (hermaphrodite) or in the same flowers (perfect flowers). *Compare with* Dioecy.

Monogamy. A mating system in which each individual mates with only one of the opposite sex, generally involving a strong and lasting pair bond. *Compare with* Polygamy.

Morph. A specific form, shape, or structure.

Mortality (m_x). Ratio of the number of deaths to individuals at risk, often described as a function of age (x). *Compare with* Death rate.

Müllerian mimicry. Mutual resemblance of two or more conspicuously marked, distasteful species to enhance predator avoidance.

Mutation. Any change in the genotype of an organism occurring at the gene, chromosome, or genome level; usually applied to changes in genes to new allelic forms.

Mutualism. Relationship between two species that benefits both.

Mycorrhizae. Close association of fungi and tree roots in the soil that facilitates the uptake of minerals by trees.

Myxomatosis. A viral disease of some mammals, transmitted by mosquitoes, that causes a fibrous cancer of the skin in susceptible individuals.

Natural selection. Change in the frequency of genetic traits in a population through differential survival and reproduction of individuals bearing those traits.

Nectar. A sugary solution secreted by the flowers of plants to attract potential pollinators.

Negative feedback. Tendency of a system to counteract externally imposed change and return to a stable state.

Neighborhood size. The number of individuals in a population included within the dispersal distance of a single individual.

Net aboveground productivity (NAP). Accumulation of biomass in aboveground parts of plants (trunks, branches, leaves, flowers, and fruits), over a specified period; usually expressed on an annual basis (NAAP).

Net production. The total energy or nutrients accumulated by the organism by growth and reproduction; gross production minus respiration.

Net production efficiency. The percentage of assimilated food used for growth and reproduction by an organism.

Net reproductive rate (R_0). The expected number of offspring of a female during her lifetime.

Neutral equilibrium. The particular state of a system that has no forces acting upon it.

Newton's law of cooling. The principle that the rate at which a body loses heat by conduction varies in direct proportion to the difference between its temperature and that of its surroundings.

Niche. The ecological role of a species in the community; the many ranges of conditions and resource qualities within which the organism or species persists, often conceived as a multidimensional space.

Niche breadth. The variety of resources used and range of conditions tolerated by an individual, population, or species.

Niche overlap. The sharing of niche space by two or more species; similarity of resource requirement and tolerance of ecological conditions.

Niche preemption. A model in which species successively procure a proportion of the available resources, leaving less for the next.

Nitrification. Breakdown of nitrogen-containing organic compounds by microorganisms, yielding nitrates and nitrites.

Nitrogen fixation. Biological assimilation of atmospheric nitrogen to form organic nitrogen-containing compounds.

Nonrenewable resource. A resource present in fixed quantity, such as space, that can be used completely by consumers.

Normal distribution. A bell-shaped statistical distribution in which the probability density varies in proportion to $\exp(-x^2/2)$, where x is a distance (the standard deviation) from the mean.

Null model. A set of rules for generating community patterns, presupposing no interaction between species, against which observed community patterns can be compared statistically.

Numerical response. Change in the population size of a predatory species as a result of a change in the density of its prey. *See also* Functional response.

Nutrient. Any substance required by organisms for normal growth and maintenance.

Nutrient cycle. The path of an element through the ecosystem, including its assimilation by organisms and its regeneration in a reusable inorganic form.

Nutrient use efficiency (NUE). Ratio of production to uptake of a required nutrient.

Obligate. Referring to a way of life or response to particular conditions without alternatives. *Compare with* Facultative.

Oligotrophic. Poor in the mineral nutrients required by green plants; pertaining to an aquatic habitat with low productivity.

Omnivore. An organism whose diet is broad, including both plant and animal foods; specifically, an organism that feeds on more than one trophic level.

Omnivory. In the sense of food-web analysis, feeding on more than on trophic level.

Open community. A local association of species having independent and only partially overlapping ecological distributions.

Open-community concept. The idea, advocated by H. A. Gleason and R. H. Whittaker, that communities are the local expression of the independent geographic distributions of species.

Optimal foraging. A set of rules, including breadth of diet, by which organisms maximize food intake per unit of time or minimize the time needed to meet their food requirements; risk of predation may also enter the equation for optimal foraging.

Optimum giving-up time (GUT). The time that an organism should remain within a patch of resources before moving on to the next in order to maximize its rate of food intake.

Order of magnitude. A factor of 10; for example, three orders of magnitude is a factor of 1000.

Ordination. A set of mathematical methods by which communities are ordered along physical gradients or along derived axes over which distance is related to dissimilarity in species composition.

Organism concept of the community. The idea that species are functionally integrated into discrete associations in which each species has evolved to serve the well-being of the whole.

Organismic viewpoint. The idea that a community is a discrete, highly integrated association of species within which the function of each species is subservient to the whole.

Oscillation. Regular fluctuation through a fixed cycle above and below some mean value.

Osmoregulation. Regulation of the salt concentration in cells and body fluids.

Osmosis. Diffusion of substances in aqueous solution across the membrane of a cell.

Osmotic potential. The attraction of water to an aqueous solution owing to the concentration of ions and other small molecules; usually expressed as a pressure.

Outcrossing. Mating with unrelated individuals within a population.

Overdispersion. *See* Hyperdispersion.

Overlapping generations. The co-occurrence of parents and offspring in the same population as reproducing adults.

Oxic. Having oxygen.

Oxidation. Removal of one or more electrons from an atom, ion, or molecule. *Compare with* Reduction.

Oxygen dissociation curve. The relationship between the fraction of the maximum potential binding of oxygen to hemoglobin and the partial pressure of oxygen in blood; an indication of the affinity of oxygen for hemoglobin.

Oxygen tension. The partial pressure of oxygen dissolved in an aqueous solution, such as blood.

Ozone. A molecule consisting of three atoms of oxygen which, in the upper atmosphere, blocks the penetration of ultraviolet light to the earth's surface.

Ozone hole. A region of severe depletion of ozone from the upper atmosphere, usually at high latitude.

Pangaea. Supercontinent at the end of the Paleozoic era including practically all the earth's landmasses, including the future Laurasia and Gondwana.

Parasite. An organism that consumes part of the blood or tissues of its host, usually without killing the host.

Parasitoid. Any of a number of so-called parasitic insects whose larvae live within and consume their host, usually another insect.

Parental investment. An act of parental care that enhances the survival of individual offspring or increases their number.

Parent material. Unweathered rock from which soil is, in part, derived.

Parent–offspring conflict. The situation arising when the optimum level of parental investment in a particular offspring differs between the parent and that offspring. This conflict derives from the fact that offspring are genetically equivalent from the point of view of their parents but siblings carry identical copies of only half the genes of each individual.

Parity. The age-specific pattern of giving birth.

Parthenogenesis. Reproduction without fertilization by male gametes, usually involving the formation of diploid eggs whose development is initiated spontaneously.

Partial pressure. The proportional contribution of a particular gas to the total pressure of a mixture.

Pattern-climax theory. The idea that the species composition of the climax community continuously varies geographically in response to changing physical conditions of the environment.

Pelagic. Pertaining to the open sea.

Per capita. Expressed on a per-individual basis.

Perennial. Referring to an organism that lives for more than one year; lasting throughout the year.

Perfect flower. A flower having both male and female sexual organs (anthers and pistils).

Permeability. Capacity of a material to pass through something, such as a biological membrane.

Pest pressure hypothesis. The idea that individuals are vulnerable to pests and pathogens when crowded in the vicinity of their parents, permitting the coexistence of many different types of species that are attacked by different pests.

pH. A scale of acidity or alkalinity; the logarithm of the concentration of hydrogen ions.

Phenolic. Pertaining to compounds, such as lignin, based on the phenol chemical structure, a hydroxylated 6-carbon ring (C_6H_5OH).

Phenolics. Aromatic hydrocarbons produced by plants, many of which exhibit antimicrobial properties.

Phenotype. Physical expression in the organism of the interaction between the genotype and the environment; outward appearance and behavior of the organism.

Phenotypic value. The measure of a particular phenotypic trait in a particular organism.

Phenotypic variance (V_P). A statistical measure of the variation in a structure or function (phenotypic value) among individuals in a population.

Pheromones. Chemical substances used for communication between individuals.

Photic. Pertaining to surface waters to the depth of light penetration.

Photoautotroph. An organism that uses sunlight as its primary energy source for the synthesis of organic compounds.

Photoperiod. Length of the daylight period each day.

Photorespiration. Oxidation of carbohydrates to carbon dioxide and water by the enzyme responsible for CO_2 assimilation, in the presence of bright light.

Photosynthates. The organic products of photosynthesis; that is, simple carbohydrates.

Photosynthesis. Use of the energy of light to combine carbon dioxide and water into simple sugars.

Photosynthetic efficiency. Percentage of light energy assimilated by plants based either on net production (net photosynthetic efficiency) or on gross production (gross photosynthetic efficiency).

Phylogenetic effect. Resemblence of the morphology or ecology of species resulting from their common ancestry.

Phylogeny. A depiction of the evolutionary relationships among species or other taxa.

Phytoplankton. Microscopic floating aquatic plants.

Plankton. Microscopic floating aquatic plants (phytoplankton) and animals (zooplankton).

Pleiotropy. Influence of one gene on the expression of more than one trait in the phenotype.

Podsolization. Breakdown and removal of clay particles from the acidic soils of cold, moist regions.

Poikilothermy. Inability to regulate body temperature; cold-bloodedness. *Compare with* Homeothermy.

Poisson distribution. A statistical description of the random distribution of items among categories, often applied to the distribution of individuals among sampling plots.

Polyandry. A mating pattern in which a female mates with more than one male at the same time or in quick succession.

Polygamy. A mating system in which a male pairs with more than one female at the same time (polygyny) or a female pairs with more than one male (polyandry). *Compare with* Monogamy.

Polygyny. A mating pattern in which a male mates with more than one female at the same time or in quick succession.

Polygyny threshold. The difference between the intrinsic values of territories or males such that the realized values of an unmated male on a poorer territory and a mated male on a better territory are equal in the eyes of an unmated female.

Polymorphism. Occurrence of more than one distinct form of individual or genotype in a population.

Population genetics. The study of changes in the frequencies of genes and genotypes within a population.

Potential evapotranspiration (PE). The amount of transpiration by plants and evaporation from the soil that would occur, given the local temperature and humidity, if water were not limited.

Precipitation. Rainfall or snowfall. Also the change of a compound from a dissolved form to a solid form.

Predator. An animal (rarely, a plant) that kills and eats animals.

Prediction. A logical consequence of a hypothesis or outcome of a model describing some aspect of a system.

Primary consumer. A herbivore, the lowermost eater on the food chain.

Primary host. A host that harbors the sexual stage of the life cycle of a parasite or disease organism.

Primary producer. A green plant that assimilates the energy of light to synthesize organic compounds.

Primary production. Assimilation (gross primary production) or accumulation (net primary production) of energy and nutrients by green plants and other autotrophs.

Primary succession. Sequence of communities developing in a newly exposed habitat devoid of life.

Production. Accumulation of energy or biomass.

Promiscuity. Mating with many individuals within a population, generally without the formation of strong or lasting pair bonds.

Protandry. Course of development of an individual during which its sex changes from male to female.

Protogyny. The sequence of sexual change by an individual from female to male.

Proximate factor. An aspect of the environment that the organism uses as a cue for behavior; for example, day length. (Proximate factors often are not directly important to the organism's well-being.) *Compare with* Ultimate factor.

Pyramid of energy. The concept that the energy flux through a given link in the food chain decreases at progressively higher trophic levels.

Pyramid of numbers. Charles Elton's concept that the sizes of populations decrease at progressively higher trophic levels; that is, as one progresses along the food chain.

Quantitative genetics. Study of the inheritance and response to selection of continuously varying traits having polygenic inheritance.

Quantitative trait. A trait having continuous variability within a population and revealing the expression of many gene loci.

Radiation. Energy emitted in the form of electromagnetic waves.

Rain shadow. Dry area on the leeward side of a mountain range.

Random spacing. The condition in which each individual occurs without regard to the position of other individuals.

Rarefaction. A method of determining the relationship between species diversity and sample size by randomly deleting individuals from a sample.

Recessive (gene). A gene whose expression is masked by an alternative form (allele) in a diploid, heterozygous state.

Reciprocal altruism. The exchange of altruistic acts between individuals.

Recruitment. Addition of new individuals to a population by reproduction; often restricted to the addition of breeding individuals.

Redox potential (Eh). The relative capacity of an atom or compound to donate or accept electrons, expressed in volts (electrical potential). Higher values indicate more powerful oxidizers.

Reduction. Addition of one or more electrons to an atom, ion, or molecule. *Compare with* Oxidation.

Refugium. A place where a species or community may persist in the face of environmental change over the remainder of its distribution.

Regional diversity. The number of species or other taxa occurring within a large geographic area encompassing many biomes or habitats; gamma diversity.

Regulator. An organism that maintains an internal environment different from external conditions.

Regulatory response. A rapid, reversible physiological or behavioral response by an organism to change in its environment.

Relative abundance. Proportional representation of a species in a sample or a community.

Renewable resource. A resource that is continually supplied to the system so that it cannot be fully depleted by consumers.

Reproductive effort. Allocation of time or resources, or the assumption of risk, in order to increase fecundity.

Residence time. Ratio of the size of a compartment to the flux through it, expressed in units of time; thus, the average time spent by energy or a substance in the compartment.

Resource. A substance or object required by an organism for normal maintenance, growth, and reproduction. If the resource is scarce relative to demand, it is referred to as a limiting resource. Nonrenewable resources (such as space) occur in fixed amounts and can be fully used; renewable resources (such as food) are produced at a rate that may be partially determined by their use.

Respiration. Use of oxygen to break down organic compounds metabolically to release chemical energy.

Resting metabolic rate (RMR). The metabolic rate of an organism in an inactive, fasting, thermoneutral state.

Riparian. Along the bank of a river or lake.

RuBP. Ribulose bisphosphate, a 5-carbon carbohydrate to which a carbon atom is attached during the assimilatory step of the Calvin cycle in photosynthesis.

RuBP carboxylase. Enzyme in the Calvin cycle of photosynthesis responsible for the reaction of ribulose bisphosphate and carbon dioxide to form two molecules of phosphoglyceraldehyde.

Ruderal. Pertaining to or inhabiting highly disturbed sites. *See* Weed.

Rumen. An elaboration of the forepart of the stomach of certain ungulate (hoofed) mammals within which cellulose is broken down by symbiotic bacteria.

Runaway sexual selection. The situation in which females persistently choose the most extreme male phenotypes in a population, leading to continuous elaboration of secondary sexual characteristics.

Saturation point. With respect to primary production, the amount of light that causes photosynthesis to attain its maximum rate.

Search image. A behavioral selection mechanism that enables predators to increase searching efficiency for prey that are abundant and worth capturing.

Secondary host. A host that harbors an asexual stage of the life cycle of a parasite or disease organism.

Secondary plant compounds. Chemical products of plant metabolism specifically for the purpose of defense against herbivores and disease organisms.

Secondary sexual characteristics. Traits, other than the sexual organs themselves, that distinguish the sexes, including, for example, the beard of human males.

Secondary succession. Progression of communities in habitats where the climax community has been disturbed or removed.

Sedimentation. Settling of particulate material to the bottom of an ocean, lake, or other body of water.

Selection. Differential survival or reproduction of individuals in a population owing to phenotypic differences among them.

Selection differential (*S*). Difference between the mean phenotypic value of selected individuals and that of the population from which they were drawn.

Selection response (*R*). Difference between the mean phenotypic value of the offspring of selected individuals and that of the population from which the parents were drawn.

Self-incompatible. Unable to mate with oneself; in the case of a hermaphrodite, owing to structural or biochemical factors that prevent fertilization.

Selfing. Mating with oneself; applicable, of course, only to individuals (usually plants) having both male and female sexual organs.

Self-regulation. The idea that population size is regulated with respect to its resources by individual restraint from overexploitation.

Self-thinning curve. In populations of plants limited by space or other resource, the characteristic relationship between logarithm and biomass.

Semelparity. The condition of having only one reproductive episode during the lifetime. *Compare with* Iteroparity.

Senescence. Gradual deterioration of function in an organism with age, leading to increased probability of death; aging.

Sequential hermaphroditism. The condition in which an individual is first one sex and then another. *See* Protandry and Protogyny.

Sere. A series of stages of community change in a particular area leading toward a stable state.

Serial polygamy. A mating pattern in which an individual mates with several individuals of the opposite sex in succession; the broods of each mating are generally cared for, in part, simultaneously.

Serpentine. An igneous rock rich in magnesium that forms soils toxic to many plants.

Sex ratio. Ratio of the number of individuals of one sex to that of the other sex in a population.

Sexual dimorphism. The condition in which the males and females of a species have a different appearance.

Sexual selection. Selection by one sex for specific characteristics in individuals of the opposite sex, usually exercised through courtship behavior.

Shannon–Weaver index (H). A logarithmic measure of the diversity of species weighted by the relative abundance of each.

Simpson's index (D). A measure of the diversity of species weighted by the relative abundance of each.

Simultaneous hermaphroditism. The condition in which an individual has both male and female sexual organs at the same time.

Social behavior. Any direct interaction among distantly related individuals of the same species; usually does not include courtship, mating, parent–offspring, and sibling interactions.

Social dominance. Physical domination of one individual over another, initiated and sustained by aggression within a population.

Sociobiology. Study of the biological basis of social behavior.

Soil. The solid substrate of terrestrial communities resulting from the interaction of weather and biological activities with the underlying geological formation.

Soil horizon. A layer of soil formed at a characteristic depth and distinguished by its physical and chemical properties.

Soil profile. Characterization of the structure of soil vertically through its various horizons, or layers.

Soil skeleton. Physical structure of mineral soil, referring principally to sand grains and silt particles.

Solar equator. The parallel of latitude that lies directly under the sun at any given season.

Spaced distribution. The condition in which individuals avoid the proximity of others.

Specialist. An organism with restricted use of habitats or resources.

Speciation. The division of a single species into two or more reproductively incompatible daughter species.

Species. A group of actually or potentially interbreeding populations that are reproductively isolated from all other kinds of organisms.

Species diversity. *See* Diversity.

Species richness. A simple count of the number of species.

Specific heat. The amount of energy that must be added or removed to change the temperature of a substance by a specific amount. By definition, 1 calorie of energy is required to raise the temperature of 1 gram of water by 1 degree Celsius.

Spite. A social interaction in which the donor of a behavior incurs a cost in order to reduce the fitness of a recipient.

Spring bloom. An increase in phytoplankton growth during early spring in temperate lakes associated with vertical mixing of the water column.

Spring overturn. Vertical mixing of water layers in temperate lakes in spring as surface ice disappears.

Stable age distribution. Proportion of individuals in various age classes in a population that has been growing at a constant rate.

Stable equilibrium. The particular state to which a system returns if displaced by an outside force.

Standard deviation (s or σ). A measure of the variability among items in a sample, such as individuals in a population; the square root of the variance, and hence the square root of the average squared deviation from the mean.

Static life table. The age-specific survival and fecundity of individuals of different ages within a population at a given time; time-specific life table.

Steady state. Condition of a system in which opposing forces or fluxes are balanced.

Stochastic. Referring to patterns resulting from random effects.

Stratification. The establishment of distinct layers of temperature or salinity in bodies of water based on the different densities of warm and cold water or saline and fresh water.

Succession. Replacement of populations in a habitat through a regular progression to a stable state.

Superorganism. An association of individuals in which the function of each promotes the well-being of the entire system.

Superparasitism. Occurrence of more than one individual of a particular parasitoid species per host.

Survival (l_x). Proportion of newborn individuals alive at age x; also called survivorship.

Switching. A change in diet to favor items of increasing suitability or abundance.

Symbiosis. Intimate, and often obligatory, association of two species, usually involving coevolution. Symbiotic relationships can be parasitic or mutualistic.

Sympatry. Occurring in the same place, usually referring to areas of overlap in species distributions.

Synecology. The relationship of organisms and populations to biotic factors in the environment. *Compare with* Autecology.

Synergism. The interaction of two causes such that the total effect is greater than the sum of the two acting independently.

Systematics. The classification of organisms into a hierarchical set of categories (taxa) emphasizing their evolutionary interrelationships.

Systems ecology. The study of an ecological structure as a set of components linked by fluxes of energy and nutrients or by population interactions; frequently applied to ecosystems.

Systems modeling. Portrayal of a system's function by mathematical functions describing the interactions among its components.

Tannins. Polyphenolic compounds, produced by most plants, that bind proteins, thereby impairing digestion by herbivores and inhibiting microbes.

Taxonomy. The description, naming, and classification of organisms.

Temperature profile. The relationship of temperature to depth below the surface of water or the soil, or the height above the ground.

Territoriality. The situation in which individuals defend exclusive spaces, or territories.

Territory. Any area defended by one or more individuals against intrusion by others of the same or different species.

Theory of island biogeography. *See* Equilibrium theory of island biogeography.

Thermal conductance. Rate at which heat passes through a substance.

Thermal stratification. Sharp delineation of layers of water by temperature, the warmer layer generally lying over the top of the colder layer.

Thermocline. The zone of water depth within which temperature changes rapidly between the upper warm water layer (epilimnion) and lower cold water layer (hypolimnion).

Thermodynamic. Relating to heat and motion.

Thermophilic. Preferring warm or hot environments.

Three-halves power law. A generalization proposing that the relationship between the logarithms of the biomass and density of a population of plants has a slope of $-3/2$.

Time lag. Delay in the response of a population or other system to change in the environment; also called time delay.

Time-specific life table. *See* Static life table.

Tolerance. In reference to succession, the indifference of establishment of one species to the presence of others.

Torpor. Loss of the power of motion and feeling, usually accompanied by a greatly reduced rate of respiration.

Transit time. Average time that a substance or energy remains in the biological realm or any compartment of a system; ratio of biomass to productivity.

Transpiration. Evaporation of water from leaves and other parts of plants.

Transpiration efficiency. The ratio of net primary production to transpiration of water by a plant, usually expressed as grams per kilogram of water; water use efficiency.

Trophic. Pertaining to food or nutrition.

Trophic level. Position in the food chain, determined by the number of energy-transfer steps to that level.

Trophic mutualism. A symbiotic relationship between two species based on complementary methods of obtaining energy and nutrients.

Trophic structure. Organization of the community based on the feeding relationships of populations.

Ultimate factor. An aspect of the environment that is directly important to the well-being of an organism (for example, food). *Compare with* Proximate factor.

Ultraviolet (UV) radiation. Electromagnetic radiation having a wavelength shorter than about 400 nm.

Upwelling. Vertical movement of water, usually near coasts and driven by offshore winds, that brings nutrients from the depths of the ocean to surface layers.

Vapor pressure. The pressure exerted by a vapor that is in equilibrium with its liquid form.

Variance (V). A statistical measure of the dispersion of a set of values about its mean.

Veil line. The point on a lognormal distribution of species abundances below which one expects fewer than one individual per species; hence the veil line separates observed from potentially observed species.

Vertical mixing. Exchange of water between deep and surface layers.

Vesicular–arbuscular mycorrhyzae. A type of endomycorrhyzal association of a fungus with the roots of plants distinguished by the branching pattern of growth of the fungus within the root tissues.

Viscosity. The quality of a fluid that resists internal flow.

Warning coloration. Conspicuous patterns or colors adopted by noxious organisms to advertise their distastefulness or dangerousness to potential predators; aposematism.

Water potential. Force by which water is held in the soil by capillary and hygroscopic attraction.

Watershed. Drainage area of a stream or river.

Water use efficiency. *See* Transpiration efficiency.

Weathering. Physical and chemical breakdown of rock and its component minerals at the base of the soil.

Weed. A plant or animal, generally having high powers of dispersal, capable of living in highly disturbed habitats.

Wilting coefficient. The minimum water content of the soil at which plants can obtain water.

Xeric. Referring to habitats in which plant production is limited by the availability of water.

Xerophyte. A plant that tolerates dry (xeric) conditions.

Zooplankton. Tiny floating aquatic animals.

Zygote. Diploid cell formed by the union of male and female gametes during fertilization.

APPENDIX A

INTERNATIONAL

SYSTEM OF UNITS

The International System of Units (abbreviated SI after the French *Système international*) is described in detail in such publications as the National Bureau of Standards Special Publication 330, 1977 edition (U.S. Dept. of Commerce, Washington, D.C.), and M. H. Green, *Metric Conversion Handbook* (Chemical Publ. Co., New York, 1978).

The seven basic SI units are as follows.

Category of measurement	Name of SI unit	Abbreviation
Length	meter	m
Mass	kilogram	kg
Time	second	s
Electric current	ampere	A
Temperature	kelvin	K
Amount of substance	mole	mol
Luminous intensity	candela	cd

The ampere, mole, and candela are not often used in ecology and will not be discussed here. Our familiar unit of time is the second. The SI units of length and mass are metric, with simple conversions to and from the English system; for example, 1 meter equals 3.28 feet, and 1 kilogram equals 2.205 pounds. The kelvin is identical to the degree Celsius (°C) except that zero on the kelvin scale is absolute zero, the coldest temperature possible (−273°C). Most ecologists use °C instead of K, because of the convenient designations for the freezing and boiling points of water (0°C and 100°C).

In the SI system, all other measurements are derived from the basic seven. For example, the SI unit of flow is the cubic meter per second, which is derived from the base units meter and second. Some of these derived units frequently employed in ecology are as follows.

Category of measurement	Name of SI unit	Abbreviation	Expression in terms of	
			Other units	Base units
Area	square meter	m²		m²
Volume	cubic meter	m³		m³
Velocity	meter per second	m s⁻¹		m s⁻¹
Flow	cubic meter per second	m³ s⁻¹		m³ s⁻¹
Density	kilogram per cubic meter	kg m⁻³		kg m⁻³
Force	newton	N		kg m s⁻²
Pressure	pascal	Pa	N m⁻²	kg m⁻¹ s⁻²
Energy, heat	joule	J	N m	kg m² s⁻²
Power	watt	W	J s⁻¹	kg m² s⁻¹

There are also a number of other units that are not part of SI but nonetheless are widely used: the liter (L), a unit of volume equal to $1/1000$ cubic meter; the are (a), a unit of area equal to 100 square meters, and the more often used hectare (ha), equal to 100 a and 10,000 m²; and the minute, hour, and day, familiar units of time. The calorie (cal), a familiar unit of heat energy, is not derived from SI base units and should be abandoned in favor of the joule (=0.239 cal), even though ecologists continue to use the calorie and kilocalorie.

Multiples of 10 of the SI units are given special names by adding the following prefixes to the unit name.

Factor	Prefix	Abbreviation		Name
10^9	giga	G	1,000,000,000	Billion*
10^6	mega	M	1,000,000	Million
10^3	kilo	k	1000	Thousand
10^2	hecto	h	100	Hundred
10^1	deka	da	10	Ten
10^{-1}	deci	d	0.1	Tenth
10^{-2}	centi	c	0.01	Hundredth
10^{-3}	milli	m	0.001	Thousandth
10^{-6}	micro	μ	0.000 001	Millionth
10^{-9}	nano	n	0.000 000 001	Billionth

*Milliard in the United Kingdom; the British billion is a million million (10^{12}), equivalent to the U.S. trillion, commonly used only in astronomy and the federal budget.

Hence 1000 meters becomes a kilometer (km) and $^1/_{1000}$ meter is a millimeter (mm); 1000 joules is a kilojoule (kJ) and 1 million watts is a megawatt (MW).

APPENDIX B
CONVERSION FACTORS

Length

1 meter (m) = 39.4 inches (in.)
1 meter = 3.28 feet (ft)
1 kilometer (km) = 3281 feet
1 kilometer = 0.621 mile (mi)
1 micron (μ) = 10^{-6} meter
1 inch = 2.54 centimeters (cm)
1 foot = 30.5 centimeters
1 mile = 1609 meters
1 angstrom (Å) = 10^{-10} meter
1 millimicron (mμ) = 10^{-9} meter

Area

1 square centimeter (cm²) = 0.155 square inches (in.²)

1 square meter (m²) = 10.76 square feet (ft²)
1 hectare (ha) = 2.47 acres (a)
1 hectare = 10,000 square meters
1 hectare = 0.01 square kilometer (km²)
1 square kilometer = 0.386 square mile
1 square mile = 2.59 square kilometers
1 square inch = 6.45 square centimeters
1 square foot = 929 square centimeters
1 square yard (yd²) = 0.836 square meter

1 acre = 0.407 hectare

Mass

1 gram (g) = 15.43 grains (gr)
1 kilogram (kg) = 35.3 ounces (oz.)
1 kilogram = 2.205 pounds (lb)
1 metric ton (T) = 2204.6 pounds
1 ounce = 28.35 grams
1 pound = 453.6 grams
1 short ton = 907 kilograms

Time

1 year (yr) = 8760 hours (hr)
1 day (d) = 86,400 seconds (s)

566

Volume

1 cubic centimeter (cc or cm^3) = 0.061 cubic inch (in.^{3})

1 cubic inch = 16.4 cubic centimeters

1 liter (L) = 1000 cubic centimeters

1 liter = 33.8 U.S. fluid ounces (oz.)

1 liter = 1.057 U.S. quarts (qt)

1 liter = 0.264 U.S. gallon (gal)

1 U.S. gallon = 3.79 liters

1 Brit. gallon = 4.55 liters

1 cubic foot (ft^3) = 28.3 liters

1 milliliter (ml) = 1 cubic centimeter

1 U.S. fluid ounce = 29.57 milliliters

1 Brit. fluid ounce = 28.4 milliliters

1 quart = 0.946 liter

Velocity

1 meter per second (m s^{-1}) = 2.24 miles per hour (mph)

1 foot per second (ft s^{-1}) = 1.097 kilometers per hour

1 kilometer per hour = 0.278 meter per second

1 mile per hour = 0.447 meter per second

1 mile per hour = 1.467 feet per second

Energy

1 joule (J) = 0.239 calorie (cal)

1 calorie = 4.184 joules

1 kilowatt-hour (kWh) = 860 kilocalories (kcal)

1 kilowatt-hour = 3600 kilojoules

1 British thermal unit (Btu) = 252.0 calories

1 British thermal unit = 1054 joules

1 kilocalorie = 1000 calories

Power

1 kilowatt (kW) = 0.239 kilocalorie per second

1 kilowatt = 860 kilocalories per hour

1 horsepower (hp) = 746 watts

1 horsepower = 15,397 kilocalories per day

1 horsepower = 641.5 kilocalories per hour

Energy per unit area

1 calorie per square centimeter = 3.69 British thermal units per square foot

1 British thermal unit per square foot = 0.271 calorie per square centimeter

1 calorie per square centimeter = 10 kilocalories per square meter

Power per unit area

1 kilocalorie per square meter per minute = 52.56 kilocalories per hectare per year

1 footcandle (fc) = 1.30 calories per square foot per hour at 555 nm wavelength

1 footcandle = 10.76 lux (lx)

1 lux = 1.30 calories per square meter per hour at 555 nm wavelength

Metabolic energy equivalents

1 gram of carbohydrate = 4.2 kilocalories

1 gram of protein = 4.2 kilocalories

1 gram of fat = 9.5 kilocalories

Miscellaneous

1 gram per square meter = 0.1 kilogram per hectare

1 gram per square meter = 8.97 pounds per acre

1 kilogram per square meter = 4.485 short tons per acre

1 metric ton per hectare = 0.446 short ton per acre

INDEX

Abundance, lognormal
        distribution 420
    relative 419
Acacia–ant mutualism 336
Acclimation 189
Acclimatization 189
Achillea, ecotypes 313
Acid, carbonic 34
Acid mine drainage 137, 524
Acid pollution 524
Acid rain 84, 137, 524
Acidity, and phosphorus
        availability 136
Active transport 37
Activity space 183
Actual evapotranspiration 79
Adaptation 12
    to drought 55
    to heat 55
    temperature 61

Adaptive radiation 394
Additive genetic variance 300
Adiabatic cooling 74
Aerenchyma 34
Africa, conservation 516
Agave, life history 217
Age, life table 254
Age class 251, 266
Age distribution, stable 265
Age specificity 256
Age structure 265
    human population 268
Aging 218
Agression, social 231
Agriculture, shifting 516
    tropical 518
Alarm call 234
Algae, competition among 353
    coralline 133
    desiccation 93

grazing 435
photosynthesis 40, 93
succession 438
Alkaloid 334
Allele 299, 306
Allelopathy 352
Allocation of resources
        210
Allochthonous 118
Allometric constant 45
Allometry 45
Allopatry 392
Alluvial deposit 80
Alpha diversity 461
Alpine environment 74
Alpine Zone 76
Altitude and climate 74
Altruism 233
Amazon Basin, conservation
        516

Amazon, forest refuge 463
    soils of 150
Ambient temperature 180
Amino acid 297
    nonprotein 333
Ammonia 128, 133, 151
    aquatic 156
    excretion of 51
Ammonification 133
Amphibolurus 185
Anaerobic 34, 71
Anemia, sickle-cell 298
Animals, characteristics of 4
Animals, clonal 6
Anion, definition of 36
Annual aboveground net
        productivity 107
Annual reproduction 213
Anoxia, aquatic 155
Anoxic 34

Ant–acacia mutualism 336
Ant–plant relationship 396
Ant, diversity 453
Aphid, coevolution 394
Aquatic pollution 521
Aquatic production 151
Area, effect on diversity 424
Arena 227
Arizona, climate zones 76
Arrhenius, Olaf 424
Arsenic pollution 525
Artemia 37, 183
Ascaris 330
Aspect diversity 459
Assimilation efficiency 113
Assimilatory transformation 124
Association, definition of 409
Asymmetry, competitive 353, 468
cryptic 326
Atmosphere, pressure of 73
Atriplex, temperature tolerance 190
Australia, endemism 476
introduced species 364
pest management 386
rabbit population 270
Australian region 482
Autochthonous 118
Autotroph 112, 138
Autumn bloom 72
Azotobacter 135

Bacteria 138
characteristics of 7
mutualisms 336
thermophilic 30
Baker, J. R. 204
Bamboo, life history 217
Barnacle, competition 350
Barren, serpentine 75
Basal metabolic rate (BMR) 180
Bat, diversity 457
ecomorphology 457
Bates, H. W. 451
Batesian mimicry 328
Bedrock 143
Bee, pollination 338
Behavior, aggressive 351
agonistic 231
social 232
Berenbaum, M. 399
Beta diversity 461
Bicarbonate ion 34, 131
Biodiversity 451, 497
value of 499
Biogeographic region 482

Biogeography 482
island 466
Biological control 348, 367, 526
Biological oxygen demand (BOD) 156, 522
Biomass accumulation ratio 117
Biomass, forest 149
production of 106
Biomass/productivity ratio 447
Biosphere, definition of 3
sustainable 514
Bird, adaptive radiation 394
character displacement 393
clutch size 208
convergence 485
dispersal distance 254
distribution 245
diversity 452, 454
diversity on islands 461
habitat distribution 198
mating system 227
migration 201
numerical response 377
population growth 264
sexual selection 229
temperature 178
temperature regulation 181
Birth defects 218
Birth rate, stochastic 292
Biston, selection 303
Bivalve, diversity 452
Blood, salt content of 31
Blowfly, population cycle 287
BOD 156
Bog succession 432
Bottleneck, population 504
Boundary layer 42
Brine shrimp 37
Bryozoa 6, 434
Buoyancy 43
Butterfly, mimicry 328

C_3 photosynthesis 55
C_4 photosynthesis 54
Cactus moth 364
Cactus wren 187
Caecum 323
Calcium 143
requirement for 35
Calcium carbonate 36, 132
California condor, conservation of 509
Calvin cycle 55
CAM photosynthesis 54

Camouflage 304
Canavanine 333
Capacity, field 28
Capillary attraction 28
Carbohydrate, production of 106
Carbon 33
assimilation of 55
energy equivalent of 107
and energy flux 105
organic 149
oxidation of 130
Carbon cycle 129
Carbon dioxide 20, 55, 129
atmospheric pressure 33
greenhouse effect 529
solubility of 34
uptake by plants 107
Carbon-14 107
Carbonate ion 131
Carbonic acid 34, 84, 131
Carotenoid 39
Carrying capacity 273
prey 379
Carson, Rachel 501
Caste, insect 236
Cation, definition of 36
Cation exchange capacity 84
Cedar Bog Lake 118
Cellulose 115, 145, 323
Cenozoic era 478
Cercaria 330
CFCs 20, 529
Chamaephyte 97
Chaparral 94, 109
boundary 413
and fire 441
Character displacement 392
Charnov, E. L. 214
Chemical competition 352
Chemical defense 331, 399
Chemoautotroph 138
Chernobyl 528
Chinch bug 250
Chlorinated hydrocarbons 526
Chlorofluorocarbon 20, 38, 529
greenhouse effect 530
Chlorophyll 39
aquatic 155
Cichlid fish 17
Circulatory system 57
Clay 29, 83
particles 85
Cleaning symbiosis 336
Clear-cutting 148, 516
Clements, F. E. 410, 427, 435
Climate, continental 67
desert 71

and distribution 245
global variation 66
integrated description 77
maritime 67
Mediterranean 70
and plant form 98
topographic influence 73
tropical 66, 70
Climate change, and extinction 503
Climate history 483
Climax, community 435
cyclic 443
definition of 427
forest 440
transient 443
Climax vegetation 436
Climograph 78
Clone 5
Closed community concept 410
Clumped distribution 247
Clutch size 208
optimum 211
Coal burning, and acid rain 524
Coarse-grained environment 196
Codon 297
Coevolution 386, 394
competitors 391
plant defense 399
pollination 400
Coexistence, competitive 348
predator–prey 371
Cohort life table 258
Cold-blooded 178
Colonization 244
island 466
marine 433, 438
metapopulation 290
Colonizing ability 434
Color response 192
Color, warning 326
Community, boundaries of 410
climax 436
closed 410
definition of 3, 408
food web 416
open 410
trophic structure 416
Community convergence 487
Community development 427
Compartment, systems model 166
Compartment model 125
terrestrial 171
Compartmentalization, food web 419

Compensation point 39
Competition 320, 341
    asymmetric 350, 353
    character displacement 393
    chemical 352
    demonstration of 345
    and diversity 467
    exploitation 351
    genetic basis 389
    influence of predation 355
    interference 351
    interspecific 347
    intrasexual 228
    mechanisms 352
    and succession 434
    theory 346
Competitive exclusion,
    examples 349
Competitive exclusion
    principle 346
Condor, conservation of 509
Conductance, thermal 180
Conduction, thermal 42
Conductivity, thermal 27
Conflict, parent–offspring
    235
Conformer 182
Connectance, food web 419
Connectedness web 418
Connell, J. H. 350, 437
Conservation 17
    habitat 500, 508
    of species 506
Consumer 320
Consumer chain 320
Continental climate 67
Continental drift 478
Continuum concept 414
Convection 93
    thermal 42
Convergence, community 487
    evolutionary 477, 485
Cooperation, social 232
Coral reef, production of 111
Coralline alga 133
Corpora lutea 276
Countercurrent circulation
    59, 181, 183
Crassulacean acid metabolism
    (CAM) 56
Crepidula, hermaphroditism
    223
Cretaceous extinction 502
Crombie, A. C. 274
Cross-resistance 366
Crypsis 13, 304, 326, 459
Cryptophyte 98
Current, ocean 67
Curtis, J. T. 436

Cuticle, insect 192
Cyanide resistance 13
Cyanobacteria 30
Cyanogenic glycoside 334
Cycle, Calvin 55
    limit 285
    population 284, 288
    predator–prey 368, 373
    water 127
Cyclic climax 443
Cypress knees 35

Daphnia, density dependence
    275
    photoperiod response 204
    population cycle 286
Darwin, Charles 12, 228,
    263, 272, 355, 451
Darwin's finches, character
    displacement 393
Day length, response to 200
DDT 501, 523
Dean, T. A. 438
Death, programmed 216
Death rate, stochastic 292
Deciduous 203
Decomposition, detritus 145
Deer, density dependence 276
Defense, inducible 335
Defensive mutualism 336
Deforestation 20
Deletion, DNA 298
Denitrification 134, 159
Density, and plant growth 278
    population 250
    of water 27
Density compensation 461
Density dependence 274, 283
    human 496
Density-dependent factor
    274
Deoxyribonucleic acid 297
Desert 67, 99
    adaptation to 52, 55
    boundary 413
    climate of 71
    diversity 454
    homeothermy in 186
    irrigation 521
    plants 93
    production in 111
Desertification 20
Desiccation 93
Desulfomonas 136
Desulfovibrio 136
Deterministic population
    growth 290
Detritivore 320, 362
Detritus 116, 141, 143, 320

Development, temperature
    dependence of 32
Developmental flexibility 191
Developmental response 191
Diapause 203
Diatom, resource use 344
Diet breadth 197
    optimal 197
Diet quality, and teeth 322
Diffusion 57
Digestibility 115
Digestive system 323
Digitalis 333
Dimorphism, water strider 200
Dioecy 223
Dipsosaurus 186
Disc equation 375
Disease, human 330
Dispay, startle 327
Dispersal 253
    distance 254
    seed 338, 396
    in sere 447
Dispersion, population 246
Displacement, character 392
Dissimilatory transformation
    124
Distribution, ideal free 198
    population 244
Disturbance, community 432
    and diversity 469
Divergence, evolutionary 392
Diversity, alpha, beta, and
    gamma 461
    anomaly 488
    aspect 459
    biological 451
    convergence 487
    ecological 14
    effect of predation 470
    equilibrium 465
    genetic 497, 504
    indices to 422
    local and regional 461
    over geological time 464
    pest pressure 471
    regulation of 467
    role of competition 467
    species 408, 497
    and succession 448
    time hypothesis 462
    value of 499
DNA 297
Dominance, genetic 299
    social 230
Dominance genetic variance
    300
Dominant, community 420
Donor, of behavior 232

Dormancy 201
Doubling time, population 269
Dought, adaptation to 55
Drosophila, competition 390
    dispersal 253
    senescence 219
Dune succession 428, 430
Dust Bowl 520
Dynamic life table 258
Dynamics, patch 433

Ecological diversity 14
Ecological efficiency 112,
    114, 119
Ecological release 461
Ecology, definition of 1
    general principles of 11
    global 513
    human 17, 532
    landscape 77
    study of 15
    systems 163
Ecomorphology 456
Ecosystem, compartment
    model of 125
    definition of 3, 103
    energetics 117
    function, regulation of
    162
    nutrient dynamics of 142
Ecotone 410
Ecotourism 500
Ecotype 312
Ectomycorrhyzae 146
Ectoparasite 329
Ectotherm 179
Ectothermy 185
Edaphic 92
Efficiency, assimilation 112
    ecological 112, 119
    nutrient use 166
    photosynthetic 108
    transpiration 109
Egestion 113
El Niño 73
Elaiosome 396, 487
Electric fish 325
Electrical potential 130
Electrons 128
Elton, Charles 103, 368
Emigration 253
Endemic 92
Endemism 498
    and conservation 507
Endomycorrhizae 146
Endoparasite 329
Endotherm 179
Endothermy 183
Energetics, ecosystem 117

Energy, allocation of 112, 215
  flux 165
  and homeothermy 179
  kinetic 61
  and nutrient cycling 124
  in photosynthesis 106
  pyramid of 104
  rate of flux 117
Energy-flow web 418
ENSO event 73
Environment, thermal 41
Environmental variance 300
Enzyme, acclimation 189
  activity 61
  substrate affinity 62
  temperature dependence 32, 61
Epidemic, myxomatosis 386
Epidermis, insect 192
Epilimnion 71, 155
Equator, solar 66
Equilibrium, competitive 348
  evolution of 388
  Hardy–Weinberg 299
  metapopulation 290
  predator–prey 373, 379
Equilibrium diversity 465
Equilibrium isocline 373
Equilibrium population 285
Equilibrium theory of island biogeography 465
Erosion, soil 519
Escape space 458
Estuary 19, 156
  production in 109, 111
Ethiopian region 482
Euphotic zone 40, 71
Eusociality 236
Eutrophic 150
Eutrophication 155, 521
Evaporation 26, 52, 78, 127
  heat of 42
Evaporative cooling 52
Evapotranspiration 79
Evenness, species abundances 422
Evolution 12
  convergence 477
  life history 208
  plant–pathogen 398
  predator–prey 386
Evolutionary fitness 303
Excretion, nitrogen 51
Expectation of further life 256
Experimental methods 16
Exploitation competition 351
Exploitation efficiency 114
Exponential growth 263

Exponential growth rate 267, 273
External forcing function 163, 167
External loading 155
Extinction, anthropogenic 502, 505
  background 502
  catastrophic 484
  causes 503
  and diversity 463
  island 466
  local 290
  mass 484, 502
  stochastic 292, 503
  tree species 488
  vulnerability to 505

Facilitation, in succession 437
Facultative mutualism 336
Fall bloom 72
Fall overturn 71
Fecundity 208
  and density in deer 276
  evolution of 307
  heritability of 302
  life table 256
  variation in 209
Feedback, internal control 163
  negative 179
Feedlots and pollution 522
Female choice 229
Female function 223
Ferrobacillus 138
Fertilizer, and eutrophication 156
  and pollution 521
Field capacity 28
Filter feeders 322
Fine-grained environment 196
Fire, and community boundaries 412
  and seedling establishment 253
  and succession 441
Fire, effect on habitat 99
Fire climax 441
Fish, buoyancy of 44
  cichlid 17
  diversity 458, 469
  electric 325
  foraging by 198
  gills of 59
  life history 211
  structure 251
  temperature acclimation 189
  temperature regulation 183

Fishery 17
  overexploitation 515
Fitness, evolutionary 13, 303
  inclusive 233
Floristic analysis 95
Flour beetle, selection 307
Flower, perfect 223
Flowering, photoperiod response 204
Flux 167
  energy 165
Foliage height diversity 453
Food, as a resource 342
Food chain 105
  length of 119
Food chain efficiency 112
Food web 416
  stability of 417
  types of 418
Foraging, optimal 196
Forcing function, external 163
Forest, age structure 254
  disturbance and diversity 470
  gradient analysis 415
  migration 483
  nutrient cycling 172
  nutrient dynamics of 142
  nutrient regeneration 147
  production in 111, 165
  seasonal 99
  species diversity 487
  succession 440
  thorn 99
  trophic levels 120
  tropical 150, 172
Forest climax 436
Forest fragmentation 472
Forest refuge 463
Forest refugium 472
Forest succession 428
Forest–prairie edge 442
Fossil fuels 126
  greenhouse effect 530
Fossil record 502
Founder event 312
Freezing 26
Freezing tolerance 31
Freezing-point depression 31
Frequency, gene 306
Frogs, competition among 357
Front, aquatic 154
Functional response 374
Functional web 418
Fungi 144
  characteristics of 6
Furanocoumarin 399
Fynbos 396

Gallionella 138
Gamete 225
Gamma diversity 461
Gammarus 182
Gap size, and succession 433
Gastrimargus 192
Gause, G. F. 345, 346, 370
Gender 222
Gene, hemoglobin 298
  virulence 388, 398
Gene flow 314
Gene frequency 306
Gene pool 310
Genetic diversity 497, 504
Genetic drift 312
Genetic variance 300
Genetic variation 13, 312
Genetics, population 296, 306
  quantitative 300
  wing length 200
Genotype 299
  competitive 389
  frequency 299
Geographic variation, genetic 312
Geological time scale 478
Geometric growth 264
Geometric growth rate 267
Geospiza, character displacement 393
Germination, seedling 184
Gerris, wing development 199
Gill 50, 58
  countercurrent circulation in 59
Giving-up time (GUT) 197
Glacial cycles 463
Gleason, H. A. 410
Global ecology 513
Global warming 530
Glycerol 31
Glycoprotein 31
Glycoside 334
Goldfish, temperature acclimation 189
Gondwana 479
Gonochoristic 222
Gradient analysis 415
Grain, environmental 196
Granite, weathering of 83
Grasshopper, color response 191
Grassland 99
  grazing 443
  trophic levels 120
Grazing, and succession 443
Great Salt Lake 37
Greenhouse effect 529
Gross production 106

Gross production efficiency 114
Ground squirrel, behavior 234
Growth, exponential 263
  geometric 264
  indeterminate 215
  population 263
Growth rate, exponential 273
  in sere 447
Growth rings 253
Guild 408
  plant defense 335

Habitat, adaptation to 312
  and community boundaries 411
  definition of 8
  heterogeneity and diversity 469
  use of 198
Habitat change, and extinction 503
Habitat complexity and species diversity 452
Habitat conservation 500
Habitat conversion 519
Habitat expansion, on islands 461
Habitat refuge 463
Habitat selection, and dispersion 248
Habitat specialization 461
Haeckel, Ernst 1
Half-reaction 130
Haplodiploid 225
Hardy–Weinberg equilibrium 299
Hawaii, adaptive radiation 394
  habitat change 505
  introduced species 519
Hearing 324
Heat 41
  adaptation to 55
Heat of melting 27
Heat of vaporization 27
Heat stress 186
Heath, succession 444
Heavy-metal pollution 525
Hemicryptophyte 98
Hemoglobin 57
  genetics 298
Herbicides 526
Herbivore 322
  definition of 362
Herbivory 116, 367
  and succession 443
Heritability 301, 307
Hermaphrodite 223
  sequential 223

Heterogeneity, and diversity 469
  environmental 195
  temporal 199
Heterogeneous environment 195
Heterotroph 112, 138
Heterozygosity, genetic 299
Hibernation 203
Hierarchy, dominance 230
Holistic concept of community 408
Holling, C. S. 374
Homeostasis 178
Homeothermy 178, 189
Homozygosity, genetic 299
Honeycreeper, Hawaiian 394
Horizon, soil 81
Host–parasite interactions 365
Hot spring 30
Hubbard Brook Forest 84, 142, 148
Hudsonian Zone 76
Huffaker, C. B. 370
Human, age structure 268
Human ecology 532
Human parasites 330
Human population 495
  growth rate 272
Humboldt Current 67, 72
Hummingbird, temperature regulation 182
Humus 81, 84
Hunting, and extinction 505
Hurd, L. E. 438
Hydrocarbon pollution 527
Hydrogen ion 84
Hydrogen sulfide 128
Hydrogenomonas 138
Hydroid 438
Hydrologic cycle 127, 168
Hygroscopic water 54
Hymenoptera, sex determination 225
  sociality 236
Hyperdispersion 247
Hyperosmotic 37
Hyphae, fungal 7, 144
Hypo-osmotic 37
Hypolimnion 71, 155
Hypothesis, definition of 15

Ice Age 463, 483
Ice, formation of 32
  properties of 27
Ice seed 31
Ideal free distribution 198
Identity by descent 233

Iguana, desert 186
Immigration 253
  island 466
Immune response 366
Inbreeding 310, 504
  avoidance 311
Inclusive fitness 233
Index, diversity 422
Indicator species 501
Individualistic concept of community 408
Induced response 368
Inducible defense 335
Industrial melanism 303
Infrared light, perception of 324
Infrared radiation 530
Inheritance, genetic 299
Inhibition, in succession 438
Insect, aspect diversity 459
  body temperature 327
  coevolution 391, 394, 399
  competition 348
  crypsis 326
  density dependence 274
  diapause 203
  dispersal distance 254
  genetic variation 313
  parasitoid–host 387
  pollination 338
  population cycles 287, 369
  population dynamics 282
  population growth rate 270
  population variation 250
  selection 307
  social 230, 236
  ultraviolet perception 324
Insecticides 526
Insulation, thermal 42, 180
Interaction, species 320
Interference competition 351
Intermediate disturbance hypothesis 469
Internal control feedback 163, 167
Internal loading 155
Interspecific competition 347
Intertidal community 417
Intertidal zone 93
  competition in 356
  succession 434
Intertropical convergence 66, 69
Intrinsic rate of increase 267
Introduced species 244, 518
Ion, definition of 28
Iron, aquatic 155
  requirement for 35
Irrigation 521

Irruption, population 202
Island, diversity 461, 466
  endemism 498
  population extinction 293
  species-area curve 424
Isocline, equilibrium 373
Isometry 45
It[c]roparity 216

Kangaroo rat 51, 53
Keever, Catherine 440
Kelp 18
Keough, Michael 433
Kettlewell, H. B. D. 303
Key factor 283
Key-factor analysis 283
Keystone predator 418
Keystone species 519
Kidney 50
Kin selection 233
Klamath weed 367
Kozlovski, D. G. 119
Krakatoa 466, 502
Kranz anatomy 57

Lack, David 208
Lack's hypothesis 208
Lake, nitrogen cycling 169
  nutrient cycling 153
  production in 111
  seasonality of 70
  temperate 70
Lake Victoria 17
Lake Washington 156
Landscape, definition of 75
  environmental 196
Landscape ecology 77
Larrea, temperature response 191
Laterite 86, 147
Laterization 85
Latisol 86
Latitude, and species diversity 452
  variation in climate 66
Laurasia 479
Leaching 81
  detritus 144
Lead pollution 525
Leaf pubescence 55
Leaf shape 94
Learning, and predation 376
Leaves, composition of 144
Legume, chemical defense 333
  nitrogen fixation by 135
Leibig, Justus 344
Leibig's law of the minimum 344
Lek 227

Lichen 321
Life cycle, parasite 330
Life form, plant 97
Life histories 207
  heritability 302
  and succession 439, 446
Life table 256
  cohort 258
  dynamic 258
  static 258
  time-specific 258
Life zone 75
Light, intensity of 39
  ultraviolet 38, 324
Light intensity, response to 191
Lignin 115, 144, 323
Limestone 36, 129, 132
Limit cycle 285
Limiting resource 344
Limnocorral 153
Linanthus, genetic variation 314
Lincoln index 251
Lindeman, Raymond 104
Lipid reserve 287
Lipid, thermal properties 190
Lizard, competition 351
  dispersal distance 254
  life history 209
  temperature regulation 184
Loading, nutrient 155
Loam 29
Local diversity 461
Local mate competition 225
Locust, migratory 202
Logistic equation 272, 346
  time-delayed 286
Lognormal distribution 420
Lotka, Alfred J. 104, 168, 346, 372
Lottery hypothesis of diversity 469
Lower critical temperature 180
Lung 58
Luxury consumption 152
Lynx–hare cycle 368

MacArthur, Robert H. 453, 465
Magnesium, requirement for 35
Malaria 330
Male function 223
Malic acid 56
Malthus, Thomas 271
Mammal, convergence 486
  density dependence 276
  diversity 454

extinctions 505
life table 258
migration 201
population cycles 369
population growth 264
population growth rate 270
population size 381
predator–prey ratio 381
sexual dimorphism 229
temperature 178
Mangrove 51
  diversity 488
Maple, distribution of 90
Marine diversity 452
  over time 464
Marine succession 433
Maritime climate 67
Mark-recapture method 251, 303
Marsh, diversity 454
  energy flow 158
  production in 111
Mass extinction 484, 502
Mathematical model 16
Mating, assortative 311
Mating system 225
Maturity 208
  age at 211
Maximum sustainable yield 380
McIntosh, R. P. 436
Mediterranean climate 70
Melanin 192
Melanism, selection 303
Membrane, semipermeable 36
Mercury pollution 525
Meristem 5
Merriam, C. H. 75
Mesic 74
Mesozoic era 478
Metabolic rate, basal 180
  resting 180
Metabolism, and homeothermy 189
  size dependence of 45
  temperature dependence of 32
Metapopulation 289
  dynamics 291
Methane, greenhouse effect 530
Methanosomonas 138
Methylomonas 138
Micelle 84
Micrococcus 138
Microcosm 16
Microhabitat 187
Microhabitat selection 187
Micronutrient 34

Microorganisms, role in element cycling 137
Migration 201
  bird 246
  and conservation 506
  metapopulation 290
Mimicry 326
  Batesian 328
  Müllerian 328
Mine drainage, acid 524
Minimum viable population 506
Mining, and heavy-metal pollution 525
Miracidium 330
Mite, population cycle 370
  predator–prey 370
  predatory 362
Mixing, vertical 70
Model, mathematical 16
  systems 166
Mollusk, hermaphroditism 223
Monogamy 226
Morality, and conservation 499
Moreau, Reginald 208
Morphology, and ecology 456
  of predators 323
Morris, R. F. 283
Mortality, variation in 209
Mosquito, parasite vector 330
Moth, aspect diversity 459
  cactus 364
  evolution of melanism 303
Mountain sheep, life table 258
Müllerian mimicry 328
Multiple stable state 378
Mutation 13
  genetic 297
Mutation rate 298
Mutualism 320, 335
Mycelium, fungal 144
Mycorrhizae 146, 336
Myxoma virus 386
Myxomatosis 386

Natural selection 12, 303
Nature reserves, design 508
Nearctic region 482
Negative feedback 179
Neighborhood size 254
Neotropical region 482
Nest, cactus wren 187
Net production 106
Net production efficiency 114
Net reproductive rate 267
New Zealand, introduced species 519

Newton's law of cooling 180
Niche, definition of 455
  and diversity 456
Niche overlap 456
Nicholson, A. J. 287
Nicotine 333
Nile perch 17
Nitrate 128, 135
  aquatic 156
Nitrification 133
  rate of 148
Nitrite 135
Nitrobacter 134, 138
Nitrococcus 134, 138
Nitrogen, excretion of 51
  flux in forest 165
  plant growth 344
  and production 109
  requirement for 35
  soil 149
Nitrogen cycle 132, 148, 151, 155
  aquatic 169
  marsh 156
  terrestrial 171
Nitrogen fixation 135, 336
Nitrosococcus 134, 138
Nitrosomonas 134, 138
Nonrenewable resource 343
Normal distribution 420
Numerical response 376
Nutrient cycling 105, 123
Nutrient use efficiency (NUE) 166
Nutrients, aquatic 151
  and production 109
  regeneration of 144
  soil 60, 149

OAA 56
Obligate mutualism 336
Ocean current 67
Ocean, measurement of production in 108
  pollution 527
  production 111, 151
  trophic levels 120
Octave, lognormal distribution 420
Odum, Eugene P. 105
Oil burning, and acid rain 524
Oil pollution 527
Oil spill 526
Old-field succession 428, 439
Oligotrophic 150
Oligotrophic lake 156
Open community concept 410
Optimal foraging 196
Orchid pollination 338

Organic toxins 526
Organism, definition of 2
Organomercurials 526
Organophosphorus
    compounds 526
Oriental region 482
Oscillation, damped 285
    period 373
    population 285, 288
    predator–prey 368
Osmoregulation 182
Osmosis 37, 54
Osmotic potential 36, 54
Outcrossing 310
Outcrossing distance 311
Overexploitation 515
    and extinction 505
Overgrazing 516
Overturn, autumn 71
    spring 70
Ovule abortion 311
Oxaloacetic acid (OAA) 56
Oxidation 124
    chemical 33
Oxidation–reduction
    reaction 128
Oxidizer 128
Oxygen, availability of 34
    partial pressure of 58
    procurement of 57
    reduction of 130
Oxygen dissociation curve 58
Ozone 38, 529

Pack hunting 232
Paine, Robert T. 356, 417
Pairie–forest edge 442
Palatability 327
Palearctic region 482
Paleozoic era 478
Panama, climate of 70
Pangaea 479
Papilio, mimicry 328
Paramecium, competition 345
    population cycle 370
Parasite 322, 329, 362
    influence on host 284
    protozoa 330
Parasite–host interactions 365
Parasites, competition among
    348
Parasitoid, definition of 362
    evolution 387
    population cycle 369
Parent material 80
Parent–offspring conflict 235
Parental care 235
Parity 208
Partial pressure 58
Pascal 29

Patch, environmental 196
Patch dynamics 433
Patch use, optimal 197
Pathogen, plant 397
Pearl, Raymond 272
Peat 432
Peninsula effect, diversity 454
PEP 56
Peppered moth, selection 303
Perception 324
Perennial 213
Permeability, selective 37
Permian extinction 502
Peru, marine environment 72
Pest management 348, 364
Pest pressure hypothesis 471
Pesticides 501
Petroleum pollution 527
Phanerophyte 97
Phenolic 334
Phenols 144
Phenotype 299
Phenotypic value 300
Phosphate 136
    aquatic 156
Phosphoenolpyruvate (PEP)
    56
Phosphorus, aquatic 155
    cycle 135, 172
    and eutrophication 157
    flux in forest 165
    and mycorrhizae 146
    plant growth 344
    and production 109
    requirement for 35
    soil 60, 149
Photoautotroph 138
Photoperiod, response to
    200, 204
Photosynthesis 33, 40, 54, 95
    aquatic 154
    chemical balance of 106
    energetics of 106
    and leaf temperature 108
    temperature response 190
    and water loss 93
Photosynthetic efficiency 108
Photosynthetically active
    radiation (PAR) 39
Phylogenetic effect 477
Phytoplankton 151
    buoyancy of 44
    population dynamics 282
Pigment, epidermal 192
    plant 39
Pimentel, David 387, 391
Pine, fire-resistance 441
    response to light 191
Pisaster, predation by 356
Pit organ 324

Pit viper 324
Plant, acclimation 190
    allelopathy 353
    annual and perennial 213
    characteristics of 4
    competition 350
    deciduousness 203
    defense 331, 368, 399
    density dependence 277
    dioecious 223
    distribution 90, 245, 414
    diversity over time 464
    ecotype 313
    gradient analysis 415
    life history 210, 217
    life history and succession
        447
    life table 256
    outcrossing 311
    resource limitation 344
    species diversity 468
    succession 428
    temperature tolerance 190
    thermophilic 191
Plant–insect coevolution 399
Plant–water relationship 54
Plantago 184
Plasmodium 330
Pleistocene epoch 483
Plumage, insulative value 189
Pluvial period 463
Poa, life table 256
Podsolization 85
Poikilothermy 178
Poisson distribution 249
Pollen competition 311
Pollen record 483
Pollination 338, 400
Pollution, agricultural 521
    aquatic 109, 521
    and evolution 303
    industrial 303
    water 19
Polychaete 434
Polygamy 226
Polygyny 226
Polygyny threshold model 226
Polymorphism, alary 199
    mimicry 328
Population, age structure 265
    bottleneck 504
    definition of 3
    distribution 198
    estimation of 251
    fluctuations 281
    growth rate 269
    island 293
    management 276, 380
    predator–prey ratio 380
    regulation of 271

response to prey 376
    stochastic extinction 293
Population cycle 284, 368
Population density 250
    and plant growth 278
Population dynamics 281
Population equilibrium 285
Population genetics 296,
    306
Population growth, Daphnia
    275
    deterministic 290
    human 272, 495
    rate of 269
    stochastic 290
Population increase, intrinsic
    rate 267
Population projection 265
Population size, and
    extinction 503, 505
    regulation of 271
Population turnover, island
    293
Potassium, requirement for 35
    soil 84, 149
Potential evapotranspiration
    79
Potential, osmotic 36
Prairie–forest boundary 412
Precipitation 128
    global 66
    and production 109, 164
    seasonality of 69
Predation 320, 361
    and competition 355
    and diversity 458, 470
Predator 322
    keystone 418
    response to prey
        population 374
    selection by 304
Predator–prey, coexistence
    371
    cycle 368
    equilibrium 389
    evolution 387
    model 372
    stability 378
Predatory 320
Prediction, definition of 15
Pressure, atmospheric 73
    units of 29
Preston, Frank 420
Prey 320
    alternate 376
    escape abilities 325
Prey switching 376
Prickly pear cactus 364
Primary production 106
    control of 164

measurement of 107
oceanic 152
Primary succession 431
Production, global pattern 110
gross 106
measurement of 107
net 106
of oceans 151
primary 106
and species diversity 452
terrestrial 171
Promiscuity 226, 227
Protein, synthesis 297
Protozoa, parasitic 330
population cycle 370
Proximate factor 204
Pseudoextinction 502
Pseudomonas 134
Ptarmigan, temperature
regulation 189
Pubescence, leaf 55
Pyramid of energy 104, 112
Pyrethrin 333

Quality, patch 198
Quantitative genetics 300
Quantitative trait 300
Quartz 83

Rabbit, myxomatosis 386
population growth rate 270
Radiation 41
adaptive 394
mutagenesis 298
pollution 528
solar 41
Rain, acid 524
Rain forest 99
Rain shadow 67
Rainfall, seasonality of 69
Random distribution 247
Range, geographic 246
Rarefaction 423
Ratio, sex 224
Raunkaier, Christen 97, 419
Recessiveness, genetic 299
Recipient, of behavior 232
Recombination, genetic 222
Redox potential 128, 131
Reducer 128
Reduction 124
chemical 33
Redundancy, DNA 297
Reef, competition on 353
production in 111
Reef-building 132
Refugia, forest 483
Regeneration, nutrient 144,
153, 155
Regional diversity 461

Regulator 182
Relationship, genetic 233
Relative abundance 419
Renewable resource 343
Reproduction, annual 213
first 211
perennial 213
Plasmodium 330
vegetative 5
Reproductive effort 214
Reproductive rate, net 267
Reserve, design of 507
Residence time 117
Resin 368
Resource, allocation of 210,
215
definition 342
limiting 344
nonrenewable 343
renewable 343
synergistic 344
Respiration 33, 58, 129
energy of 106
Respiratory water loss 53
Respired energy 113
Response, correlated 309
developmental 191
evolutionary 307
functional 374
numerical 376
to selection 307
Rhizobium 135, 336
Ribulose bisphosphate
(RuBP) 55
Rocky shore, community 417
competition on 356
succession 434
Root biomass 151
Root mat 151
Root nodules 135
Root system 60, 95, 99
Root:shoot ratio, in sere 447
RuBP 55
Rumen 323, 336
Rust, wheat 397

Sahel 20
Salt, regulation of 182
Salt balance 50
Salt marsh 156
Sample size, and diversity 423
San Francisco Bay 19
Sand 29
Sand dune, succession 428
Sanders, Howard 423
Satiation, predator 375
Saturation point 40
Savanna 99
production in 111
Scale, ecological 8, 490

Scale insects 13
Sceloporus, life history 209
Schaffer, W. M. 214
Schistosoma 330
Schistosome 366
Schistosomiasis 330
Schoener, Thomas 352
Sea otter 18
Search image 375
Season, response to 189
Seasonal cycle, desert iguana
186
Seasonality, of climate 69
Secondary compound 331
Secondary sexual
characteristics 228
Secondary succession 431
Sediments, aquatic 141
Sedimentation 126, 151
nutrient 153
Seed dispersal 338
coevolution 396
Seedling establishment 253,
471
Seedling germination 184
Selection 306
kin 233
natural 12, 303
sexual 227
Selection differential 307
Selection response 307
Self-incompatibility 311
Self-thinning curve 278
Selfing 310
Selfishness 232
Semelparity 216
Semipermeable membrane 36
Senecio, life history 210
Senescence 218
Sere 432
Serengeti ecosystem 201
Serpentine 75
Serpentine soil 92, 412
Sewage, and eutrophication
156
and pollution 522
Sex 222
Sex ratio 224
Sexual dimorphism 228
Sexual selection 227
Shade tolerance, in sere 447
Shannon–Weaver index 422
Shark 50
Sheep, population 282
Shelford, Victor E. 409
Shreve, Forrest 413, 416
Shrubland 99
Sibling competition 235
Sickle-cell anemia 298
Sight 323

Silt 29
Simpson's index 422
Size, body 45
and endothermy 183
heritability 302
selection 309
and temperature regulation
182
Skull, structure 323
Slatyer, R. O. 437
Snail, dispersal distance 254
genetic variation 313
Snake, infrared perception 324
Social behavior, and
dispersion 248
Social group 232
Social insect 230, 236
Society 230
Sodium, requirement for 35
Soil 28, 79
availability of phosphorus
136
development 79
erosion 519
eutrophic 150
groups 86
heavy metals in 525
horizon 81
moisture 91
nutrients 60, 142, 149
oligotrophic 150
and plant distribution 91,
246
profile 81
serpentine 92
skeleton 28
tropical 85, 147
Solar constant 39
Solar equator 66
Solar radiation 41, 184
Solvent 28
Sonoran Zone 76
Sound, perception of 324
Sousa, Wayne 434
Space, activity 183
competition for 350
as a resource 342
Spaced distribution 247
Spatial variation 10
population 250
Specialization, habitat 461
Speciation, allopatric 472
Speciation rate 467
Species, number of 452, 497
Species diversity 408
Species packing 456
Species production 472
Species richness 408, 423
Species–area relationship 424
Specific heat 27

Spitefulness 232
Spodosol 85
Sponge 434
Sporozoite 330
Spring overturn 70
Stability, food web 417, 419
    predator–prey 378
Stable age distribution 265
Stable state, multiple 378
Startle display 327
State variable 167
Static life table 258
Status, social 231
Steady state 168
Stickleback, patch use by 198
Stochastic extinction 503
Stochastic population growth
    290
Stomate 55
Storage, energy 287
    resource 201
Stratification, thermal 154
Subpopulation 289
Substitution, DNA 297
Succession, causes of 436
    definition of 427
    duration of 441
    effect of wind 444
    forest 440
    old-field 439
    primary 431
    secondary 431
Sulfate 136
Sulfur, and acid mine
        drainage 524
    requirement for 35
Sulfur cycle 136
Summer plumage 189
Supercooling 31
Surface/volume ratio 46, 50
Survival, estimation of 257
    life table 256
    variation in 209
Survivorship, life table 256
Sustainable biosphere 514
Swamp vegetation 35
Swim bladder 43
Switching, prey 376
Symbiosis 321
    cleaning 336
    nitrogen fixation 135
Sympatry 392
Synergism 164, 344
Synergistic resource 344
System model 166
Systems ecology 163

Taiga 99
Tannin 331
Tansley, A. G. 103

Technology, and
    overexploitation 516
Teeth, structure 322
Temperature 30
    and activity space 186
    ambient 180
    body 53
    effect on photosynthesis
        108
    global 66
    influence on precipitation
        164
    insect body 327
    lower critical 180
    optimum 61
    regulation of 178, 189
    seasonality of 69
    and vapor pressure 43
Temperature adaptation 61
Temperature profile 70
    aquatic 156
    lake 154
Temperature regulation 327
Temperature tolerance 189
Temporal heterogeneity 199
Temporal variation 10
    population 250
Termite, sociality 236
Terpenoid 334
Territoriality, and mating
        system 227
Territory 230
Thermal conductivity 27
Thermal environment 41
Thermal stratification 154
Thermocline 71, 155
Thermodynamics 104
Thermophilic 191
Thermophilic bacteria 30
Therophyte 98
Thiobacillus 137
Thiol sulfur 136
Thornthwaite, C. W. 79
Three Mile Island 528
Three-halves power law 278
Tilman, David 342
Time, allocation of 210
    geologic 464, 478
    giving-up 197
Time delay, population 288
    population response 285
    predator–prey 369
Time hypothesis of diversity
    462
Time lag, population 288
Time-specific life table 258
Tolerance, in succession 438
Topography, and climate 73
    and distribution 245
    and diversity 454

Torpedo 325
Torpid 182
Toxins, environmental 523
Tracheae 57
Transient climax 443
Transition Zone 76
Transpiration 33, 54, 55, 78,
    95
Transpiration efficiency 109
Tree, age structure 254
    gradient analysis 415
    migration 483
    species diversity 487
    turnover 470
Tribolium, competition 345,
    390
    selection 307
Trophic level 104
Trophic mutualism 335
Trophic structure 416
    community 408
Tropics, agriculture 518
    climate of 66, 70
    nutrient regeneration 147
    soils 85
    species diversity 452
Tundra 99
    production in 111
Tunicate 434, 438
Turesson, Göte 312
Typhlodromus 364

Ultimate factor 204
Ultraviolet light 38
    perception 324
Ultraviolet radiation 529
Upwelling 70
Urea 50
    excretion of 52
Uric acid, excretion of 52
Urine, composition of 52

Vapor pressure 33
    water 43
Vaporization, heat of 27
Variance, additive 300
    calculation 301
    environmental 300
    genetic 300
Variation, genetic 13, 297,
        308, 312
    spatial 10
    temporal 10
Vector, parasite 330
Vegetation, climax 436
Vegetation map, United
        States 97
Vegetative reproduction 5
Veil line, lognormal
        distribution 421

Vertical mixing 70, 154
Vesicular–arbuscular
        mycorrhizae 146
Virulence, genetics 397
    viral 386
    wheat rust 388
Virus, myxoma 386
Viscosity, of water 43
Vitousek, Peter 165
Vole, herbivory by 367
Volterra, Vito 347, 372

Wallace, Alfred Russel 451,
        463, 482
Warm-blooded 178
Warning coloration 326
Wasp, parasitoid 363
    sex ratio 225
Water, buoyancy of 43
    cyling of 127
    density of 27
    hygroscopic 54
    and primary production 108
    properties of 26
    vapor pressure of 33, 43
Water balance 50
Water cycle 127, 168
Water loss 93
    respiratory 53
Water potential 28, 54
Water strider, wing
        development 199
Water vapor, atmospheric 128
Watershed 142
Weathering 84, 143
    nutrient input 142
Weed 434
Weight, heritability 302
    selection 309
Whale, overexploitation 515
Wheat rust 388, 397
Whittaker, Robert 415
Wildebeest, migration of 201
Wildlife management 276
Wilson, E. O. 465
Wilting coefficient 29
Wilting point 29
Wind, and succession 444
    and vertical mixing 154
Wind chill factor 42
Winter plumage 189
Wolf, population 381
Woodland 99
Wren, cactus 187

Xeric 74
Xylem element 54

Yucca, pollination 400